Probability Theory and Stochastic Processes

Probability Theory and Stochastic Processes

B. Prabhakara Rao

Professor in ECE Department,
and
Director Foreign Universities &
Alumni Relations
JNTU Kakinada.

T.S.R. Murthy

Professor of Mathematics,
Shri Vishnu Engineering College for Women,
Bhimavaram.

BSP **BS Publications**
A unit of **BSP Books Pvt., Ltd.**
4-4-309/316, Giriraj Lane, Sultan Bazar,
Hyderabad - 500 095
Phone: 040 - 23445605, 23445688

© 2014, *by Publisher*

Published by :

BSP **BS Publications**
A unit of **BSP Books Pvt. Ltd.**

4-4-309/316, Giriraj Lane, Sultan Bazar,
Hyderabad - 500 095
Phone : 040 - 23445605, 23445688
e-mail : info@bspbooks.net

ISBN : 978-93-85433-31-3 (HB)

Preface

The purpose of this book is to present a simple and lucid exposition of Probability Theory and Stochastic Processes. It is framed to meet the requirements of students preparing for professional courses. This book is useful for various B.Tech courses like Electronics and Communication Engineering, Electronics and Computer Engineering Electronics and Telematics Engineering, etc., who have Probability Theory and Stochastic Processes as a part of their curricula.

A comprehensive treatment of theory of probability, Random Variables, Probability Distributions, Stochastic Processes, Operations on One dimensional, Multi dimensional random variables, Stochastic Processes-Temporal Characteristics, Spectral Characteristics and linear System with random inputs are present in this book. This book is also useful for those preparing for competitive examinations.

The examples are given to illustrate the use of statistical methods as scientific tools in the analysis of various types of problems, which will enable the students to understand better to tackle various types of problems.

At the end of each chapter of this book a set of Quiz Questions, Exercise Questions, Review Quiz Questions and Review Exercise Questions are given to acquire proficiency and to develop confidence levels of the students.

The data in this text book has been tested extensively for B.Tech for the last four years.

Suggestions for the improvement of the book will be highly appreciated and will be incorporated.

- Authors

Acknowledgement

We are highly indebted to Prof. Tulasi Ram Das Vice Chancellor, JNTU Kakinada, Sri K.V. Vishnu Raju, Chairman, Sri Vishnu educational society who encouraged us in bringing out this work. We wish to express our sincere gratitude to Dr. Satya Prasad, Rector, JNTU Kakinada, Sri Jaya Chandran, Vice-Chairman, Sri Vishnu educational society for their cooperation and encouragement..

We would like to extend our thanks to Dr. G. Srinivasa Rao, Principal, Mr. P. Srinivasa Raju, Vice Principal, S. V. E. C. W Bhimavaram.

We express our thanks to the Faculty, Supporting Staff of their respective departments and those who are directly or indirectly helped us. We are also thankful to their respective authorities for providing the necessary facilities.

We take pleasure in expressing our thanks to the publishers of BSP Books Pvt. Ltd. for providing the necessary facilities.

We feel grateful to all those authors who directly and indirectly helped us whose contribution has made this book has a great success.

- Authors

Contents

Chapter 1

Probability

Chapter 2

Random Variable

Contents

Chapter 1

Probability

Chapter 2

Random Variable

Chapter 3

Operations on One Random Variable-Expectations

Chapter 4

Multiple Random Variables

Chapter 5

Operations on Multiple Random Variables

Chapter – 6

Random Processes

Chapter 5

Operations on Multiple Random Variables

Chapter – 6

Random Processes

Chapter – 7

Random Process-Spectral Characteristics

Chapter – 8

Linear Systems with Random Inputs 514

CHAPTER 1

Probability

1.1 Introduction

In everyday life we come across some statements like he is probably wrong, the chances of his winning the match (game) is fifty fifty, it is very likely that it will rain tonight etc. All these statements are not mathematically precise, in the sense that we cannot form any definite idea about the occurrence or non-occurrence of the events. But they give an idea of a varying degree of probability of occurrence or non-occurrence of the events.

So a probability is a quantitative measure of uncertainty a number that conveys the strength of our belief in the occurrence of an uncertain event. Since life is full of uncertainty, people have always been interested in evaluating probabilities. The Statistician I. J. Good, suggests in his "Kinds of Probability" that "the Theory of probability is much older than the human species".

While the ultimate objective of studying this subject is to facilitate calculation of probabilities in business, management, science, technology, etc., the specific objectives of this chapter are to understand the following terms.

1.2 Set Definitions

A set is a collection of objects (e.g., numbers, alphabets in English, voltages, cars, rivers, people, etc., are anything). These objects are called elements or members of the set. *A* set is generally denoted by capital letters and the elements in the set denoted by lower case letters. There are two methods of designating a set. One is *Roster (or tabular)* method, in which the elements of a set are listed within braces, e.g., the set *A* of natural numbers represented as $A = \{1, 2, 3 \ldots\}$.

The other method is *Property (Selector or Rule)* method wherein the elements of a set are described by their common characteristics. For example, if the set A be the set of all rivers in India, it would be represented as $A = \{x: x$ is a river in India$\}$ or $A = \{x / x$ is a river in India$\}$, where x is an arbitrary element of A.

A set is said to be *countable* if its elements can be put in one-to-one correspondence with the natural numbers, which are the integers 1, 2, 3, etc. If a set is not countable, it is called *uncountable.*

A set is *finite* if it contains a finite number of different elements, i.e. if the process of counting the different elements of the set comes to an end. Otherwise a set is *infinite*, i.e., if the process of counting the different elements would never come to an end.

Example: 1. $A = \{1, 2, 3, 4, 5, 6\}$ is a finite set

2. $A = \{1, 2, 3 \ldots\}$ is an infinite set

Equal Sets

Two sets A and B are said to be equal, if they contain the same elements. i.e. if and only if every element of A is an element of B.

Symbolically, $A = B$ iff $(a \in A \Rightarrow a \in B \;\wedge\; b \in B \Rightarrow b \in A)$

Example: $A = \{1, 2, 3, 4, 5, 6\}$ and $B = \{1, 2, 3, 4, 5, 6\}$ then the two sets A and B are equal.

Null Set or Empty Set or Void Set

A set is said to be *empty* if it has no elements. It is denoted by ϕ and this set is always regarded as a subset of every set. For example the set of positive integers between 1 and 2 is a null set, etc.

Singleton Set

A set contains only one element in it is called a *Singleton* set or a Unit set. For example $A = \{3\}$ is a singleton set.

Subsets

If A and B are two sets such that every element of A is also an element of B, then A is called a subset of B (or A is contained in B) and we can write $A \subset B$ If A and B are equal sets then $A \subseteq B$ or $B \supseteq A$, If $A \subseteq B$ then $a \in A \Rightarrow a \in B$

Example: The set $A = \{1, 2, 4\}$ is a subset of $B = \{1, 2, 3, 4, 5, 8\}$

A set A is said to be proper subset of B if (i) A is a subset of B, ie., every element of A is also an element of B and (ii) $A \neq B$, i.e., there is an element in B which is not in A. Therefore A is a proper subset of B is denoted by $A \subset B$. In this case B is called a superset of A.

Example: The set A = {1, 2, 4} is a proper subset of B = {1, 2, 3, 4, 5, 8}, and *B* is a superset of *A*.

The set A = {1, 2, 4} is a (Improper) subset of B = {2, 4, 1}

Universal Set

Universal set is the set which contains all the elements of all subsets under investigation in a particular context. We denote this set by *U* or *S*.

e.g., In throwing a die, we get the numbers 1, 2, 3, 4, 5, 6. Here the universal set is S = {1, 2, 3, 4, 5, 8}

Power Set

The set of all possible subsets of a set *A* is called the power set of *A* and it is denoted by the symbol *P* (*A*). If a finite set *A* has n elements, its power set contains 2^n elements.

Example: If A = {1, 2} then $P(A)$ = { ϕ, {1}, {2}, {1, 2}}, where ϕ is the null set.

1.3   Operations on Sets

Venn diagram

Operations on sets or theorem relating to sets can be well understood with the help of Venn diagram. In this diagram the Universal set *S* is denoted by a rectangular region and any subset of *S* be a region enclosed by a closed curve (or a circle) lying within the rectangular region.

Union of Sets

The union of two sets *A* and *B* denoted as $A \cup B$, as the set of all elements which belong either to *A* or to *B* or to both *A* and *B*.

Symbolically $A \cup B = \{x : x \in A \cup x \subset B\}$

In Fig. 1.1., shows set A = {1, 2, 7} and B = {1, 2, 3, 4, 5, 8} then

$$A \cup B = \{1, 2, 3, 4, 5, 7, 8\}$$

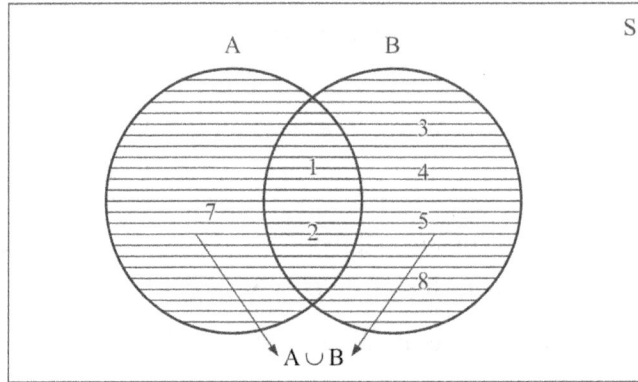

Fig. 1.1

Intersection of Sets

The intersection of two sets A and B denoted as $A \cap B$, as the set of all elements which are common to both A and B.

Symbolically $A \cap B = \{x : x \in A \cap x \in B\}$

In Fig. 1.2, shows set $A = \{1, 2, 7\}$ and $B = \{1, 2, 3, 4, 5, 7, 8\}$ then $A \cap B = \{1, 2\}$

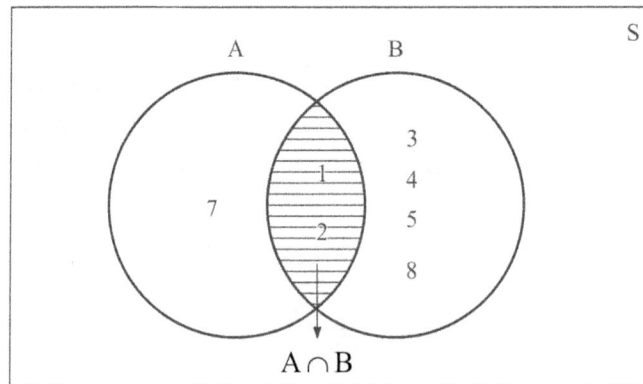

Fig. 1.2

Disjoint of Sets

If two sets A and B have no elements in common, i.e. if no element of A is in B and no element of B is in A, then A and B are said to be disjoint or mutually exclusive sets. Clearly $A \cap B = \phi$

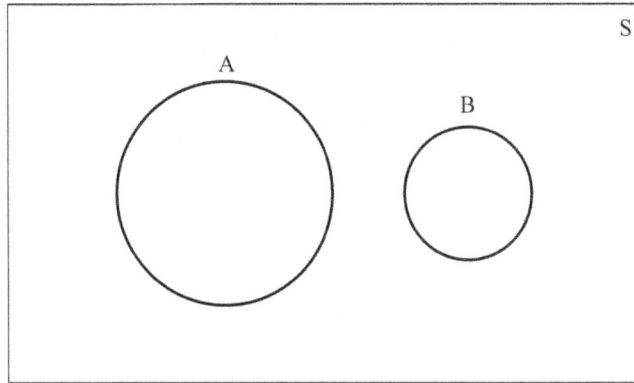

Fig. 1.3 (a)

In Fig. 1.3(b) shows set $A = \{1, 2, 6\}$ and $B = \{3, 4, 5, 7, 8\}$ are disjoint, since they have no common elements then $A \cap B = \{\ \phi\ \}$

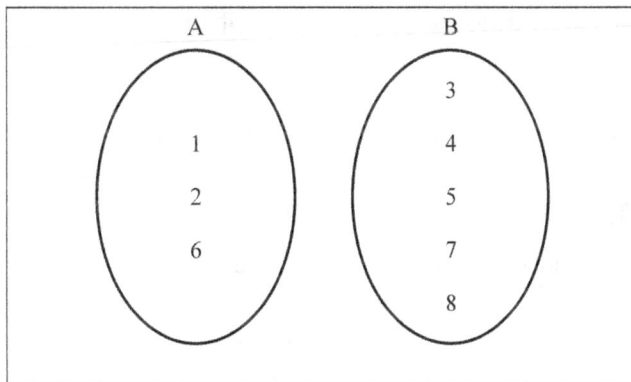

Fig. 1.3 (b)

Difference of Sets

The difference of two sets A and B is the set of elements which belong to A but which do not belong to B. We denote the difference of A and B by A-B or $A \sim B$

Symbolically, $\quad A \sim B = \{x : x \in A \cap x \notin B\}$

In Fig. 1.4, set $A = \{1, 2, 3, 4, 5, 6\}$ and $B = \{2, 3, 4, 5, 7, 8, 9\}$ then

$$A \sim B = \{1, 6\} \text{ and } B \sim A = \{7, 8, 9\}$$

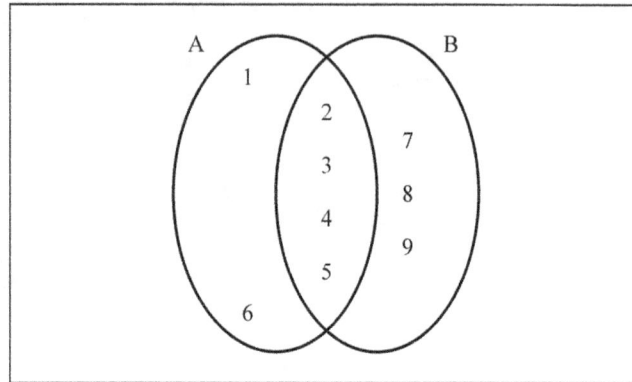

Then A ~ B = [1, 6] and B ~ A = [7, 8, 9]

Fig. 1.4

Complement of a Set (or Negation of a Set)

The complement of a set A denoted by $\overline{A}$, is the set of all elements not in A. Thus

$\overline{A} = S - A$

Since $\overline{\phi} = S$; $\overline{S} = \phi$; $A \cup \overline{A} = S$ and $A \cap \overline{A} = \phi$

1.4   Laws of Algebra of Sets

There are three main operations on sets, viz, intersection, union and complement satisfy certain laws of algebra.

1. *Commutative Law* states that for a pair of sets A and B,
    (i) $A \cap B = B \cap A$ (ii) $A \cup B = B \cup A$

2. *Distributive Law* for any three sets A, B and C we have
    (i) $A \cap (B \cup C) = (A \cap B) \cup (A \cap C)$ (ii) $A \cup (B \cap C) = (A \cup B) \cap (A \cup C)$

3. *Associative Law* for any three sets A, B and C we have
    (i) $(A \cup B) \cup C = A \cup (B \cup C) = A \cup B \cup C$
    (ii) $(A \cap B) \cap C = A \cap (B \cap C) = A \cap B \cap C$

4. *Idempotent Law* for any set A, we have
$$A \cup A = A \text{ and } A \cap A = A$$

5. *Identity Law* for any set A and ϕ is a null set, we have
$$A \cup \phi = A \text{ and } A \cap \phi = \phi$$

6. *Complement Law* for any set A and A' is complement of A, we have

   (i) $A \cup A' = S$ (ii) $A \cap A' = \phi$

   (iii) $(A')' = A$ (iv) $S' = \phi$ and   (v) $\phi' = S$

7. *De Morgan's Law* for any two sets A and B, we have

   (i) $\overline{(A \cup B)} = \overline{A} \cap \overline{B}$ and   (ii) $\overline{(A \cap B)} = \overline{A} \cup \overline{B}$

1.5   Terminology in Probability

Random Experiment: The term *experiment* refers to describe an act which can be repeated under some given conditions. The experiment whose results (output) depends on chance are called Random Experiment. For example tossing of a coin is a random experiment, and similarly throwing a die is a random experiment etc.,

Event: The output or result of a random experiment is called an event or result or outcome. For example in tossing of a coin, getting head or tail is an event and similarly in throwing a die getting 1 or 2 or 3 or 4 or 5 or 6 is event

Events are generally denoted by capital letters A, B, C, etc. The events can be splitted in to two types. One is simple event and the other is compound event. For example in tossing of a coin, getting head or tail is a simple event and similarly in throwing two dice getting a sum of 6 points is a complex or compound event. This can be splitted as (1, 5), (2, 4), (3, 3), (4, 2) and (5, 1). Each is a simple event. So Compound events can be splitted furtherly as simple events.

Mutually Exclusive Events: Two or more events are said to be mutually exclusive events if the occurrence of one event precludes (excludes or prevent) the occurrence of others. i.e., both the events cannot happen simultaneously in a single trail. For example, a person may be either alive or dead at a point of time he cannot be both alive as well as dead at the same time. Similarly, in tossing of a coin, the events head and tail are mutually exclusive and in throwing a die, all the six faces are mutually exclusive.

Equally Likely Events: Two or more events are said to be equally likely, if there is no reason to expect any one case (or any event) in preference to others. That is every outcome of the experiment has equally likely possible of occurrence are called equally likely events. For example in tossing of a coin, the events head and tail are equally likely and similarly in throwing a die, all the six faces (events) are equally likely.

Exhaustive number of Cases or Events: The total number of possible outcomes in an experiment is called exhaustive number of cases or events. For example in tossing of a coin, the exhaustive number of cases are two (i.e., Head and Tail) and similarly in throwing a die, the exhaustive number of cases are six (i.e., 1, 2, 3, 4, 5 and 6).

At Random: Means without giving any preference or priority to any case or event. For example drawing a card from a well shuffled pack of cards at random, this may be any card and similarly asking questions to students at random in a class room.

Independent and Dependent Events: Two or more events are said to be independent if the occurrence of one does not affect the occurrence of other(s). For example if a coin is tossed twice, the result of the second throw would in no way be affected by the result of the first throw.

Dependent events are those in which the occurrence or non-occurrence of one event in any trail affects the probability of other events in other trials. For example if a card is drawn from a pack of playing cards and is not replaced in the deck of cards, this will alter the probability that the second card drawn is affected.

Sample Space: The set of all possible outcomes of a random experiment is called a sample space. It is denoted by S. Each outcome in the experiment is called sample point. There are two types of sample spaces. One is finite sample space in which the number of sample points is finite and the other is infinite sample space in which the number of sample points are infinite.

For example in tossing of a coin the sample space $S = \{H, T\}$ is discrete and finite and similarly in throwing a die, the sample space $S = \{1, 2, 3, 4, 5, 6\}$ is discrete and finite sample space

For countable discrete and infinite sample points, S is the experiment 'choose randomly a positive integer' is the countably infinite set $S = \{1, 2, 3 \ldots\}$

A sample space is said to be discrete if it has only finitely many or countable infinite number of points which can be arranged in a simple sequence e_1, e_2, $\ldots$ while a sample space containing non-denumerable number of points is called a continuous sample space. All the points on a line, or a line segment, or all the points in a plane are the examples of continuous sample space.

We shall restrict ourselves only to discrete sample space

1.6   Classical Definition of Probability

Suppose that an event E can happen in m ways and non- happen in n ways all these $m + n$ ways are supposed equally likely and finite. Then the probability of happening of the event called its success, denoted by $P\ (E)$ or simply p and is defined as

$$P(E) = \frac{m}{m+n} \qquad\qquad \text{.....(1.1)}$$

and the probability of non-happening of the event called its failure, denoted by $P(\overline{E})$ or simply q and is defined as

$$P(\overline{E}) = \frac{n}{m+n} \qquad\qquad \text{.....(1.2)}$$

From eqns (1.1) and (1.2), we observe that the probability of an event can be defined as

$$P(event) = \frac{The\ number\ of favourable\ cases\ for\ the\ event}{Total\ number\ of possible\ cases}$$

It follows that $P(E) + P(\overline{E}) = \dfrac{m}{m+n} + \dfrac{n}{m+n} = \dfrac{m+n}{m+n} = 1$ or $p+q=1$. This implies that $p = 1-q$ or $q = 1-p$ Hence $0 \leq P(E) \leq 1$ and $0 \leq P(\overline{E}) \leq 1$

If $P(E) = 1$, then the event E is called certain event. For example, the death of human being or animal is certain, and then the probability of death of human being or animal is one.

If $P(E) = 0$, then the event E is called an impossible event. For example, in a women college, finding a boy student in a class room is impossible event and the probability of finding a boy student is zero and swimming in air is impossible, then the probability of swimming in air is zero.

Odds in favour of the event is the ratio of the probability of success to failure to the event in terms of success i.e., $m : n$ or $p : q$ or the ratio of number of favourable cases to the number of non-favourable cases to the event

Odds against the event is the ratio of the probability of failure to success to the event in terms of success i.e., $n : m$ or $q : p$ or the ratio number of non-favourable cases to the number of favourable cases to the event

Limitations of Classical definition of probability

The classical definition of probability is fails in the following cases. If the

(i)   various outcomes of the experiment are not equally likely or equally probable e.g., the chance that a candidate will pass in a certain examination is not 50%. Since the two possible outcomes, viz., success and failure are not equally likely.

(ii)   Exhaustive number of cases in a trail is not finite.

1.7   Probability as a Relative Frequency or Statistical (Von-Mises) Definition of Probability

Let a trail be repeated a large number of times under essentially identical conditions and let E be the event of it. The ratio of the number of times (m) the event E happens to the number of trails (n),   i.e., $\dfrac{m}{n}$ is called the relative frequency of the event E and is denoted by $R(E)$ and the probability of the event E is defined as $P(E) = \underset{n \to \infty}{Lt} \dfrac{m}{n}$

If such trail is repeated N times and if the probability of success of an event E is *P(E)* then the total number of trails favourable to E is *N.P(E)*, this product is called Mathematical Expectation.

Limitations of Statistical definition of probability

If the limit does not exist then the statistical definition of probability fails.

1.8   Axiomatic Definition of Probability

The axiomatic approach to probability closely relates to the theory of probability with the set theory. The axiomatic definition of probability includes both the Classical and Statistical definitions as particular cases and overcomes the deficiencies of each of them. The axioms thus provides a set of rules, the rules can be used to deduce theorems and the theorems can be brought together to deduce more complex theorems.

1.9   Axioms of Probability
(Probability Function of the Sample Space)

Let S be the sample space and let E be any event in S (i.e., $E \subseteq S$) we define a function P on S (for any event E the functional value $P(E)$) is a real number such that

(i)   *P(E)* ≥ 0

(ii) *P(S)* $= 1$ and

(iii) If $E_1, E_2, E_3, E_4, \ldots$ be any sequence of mutually exclusive events in the sample space S then $P(E_1 \cup E_2 \cup E_3 \cup E_4 \cup \ldots) = P(E_1) + P(E_2) + P(E_3) + \ldots$ is called a probability function on the sample space.

1.10 Usual Probability Function

In a sample space S with N sample points and E is an event of S. Let the probability of occurrence of the event E, is $P(E)$ be defined as

$$P(E) = \frac{n(E)}{n(S)}$$

Now we observe that

(i)   For every $E \subseteq S$, n$(E) \geq 0$; and $P(E) \geq 0$

(ii) P$(S) = 1 \Rightarrow$ $P(S) = \dfrac{n(S)}{n(S)} = 1$ this implies that $P(S) = 1$

(iii) If $E_1 \cap E_2 = \phi$ i.e., E_1, E_2 are mutually exclusive events in the sample space S

then $P(E_1 \cup E_2) = \dfrac{n(E_1 \cup E_2)}{n(S)} = \dfrac{n(E_1) + n(E_2)}{n(S)}$

$$= \frac{n(E_1)}{n(S)} + \frac{n(E_2)}{n(S)}$$

$$= P(E_1) + P(E_2)$$

Since the probability function obeys the axioms of probability law is called usual probability function.

We will explain the mathematical model of experiments with the following examples.

***Example* 1:** An experiment consists of observing the number of heads when tossing 4 fair coins. We develop a model for this experiment.

The sample space S consists of

$$S = \begin{cases} HHHH, HTHH, THHH, TTHH, HHHT, HTHT, THHT, TTHT \\ HHTH, HTTH, THTH, TTTH, HHTT, HTTT, THTT, TTTT \end{cases}$$

Since when tossing of four coins these are the possible outcomes. The probability of getting each event is $\frac{1}{16}$

Let X denote the number of heads in a single toss of 4 fair coins.

Given X represents the number of heads when tossing 4 fair coins. Therefore the range of X is {4, 3, 2, 1, 0}

X	4	3	2	1	0
$P(X)$	$\frac{1}{16}$	$\frac{4}{16}$	$\frac{6}{16}$	$\frac{4}{16}$	$\frac{1}{16}$

Let A be the event that $X < 2$, i.e., A = {the number of heads < 2} and B be the event that $1 < X \le 3$, i.e., B = {1 < the number of heads $\le$ 3}

Now we assign the probabilities for these events as

(i) $P(A) = P(X < 2)$:

$P(X < 2) = P(X = 0) + P(X = 1)$

$$= \frac{1}{16} + \frac{4}{16} = \frac{5}{16}$$

(ii) $P(B) = P(1 < X \le 3)$:

$P(1 < X \le 3) = P(X = 2) + P(X = 3)$

$$= \frac{6}{16} + \frac{4}{16} = \frac{5}{8}$$

Example 2: An experiment consists of observing the sum of the numbers on the faces of two dice when two dice are thrown. We develop a model for this experiment.

When two dice are thrown then the sample space S is

$$S = \begin{Bmatrix} (1,1),(1,2),(1,3),(1,4),(1,5),(1,6); & (2,1),(2,2),(2,3),(2,4),(2,5),(2,6) \\ (3,1),(3,2),(3,3),(3,4),(3,5),(3,6); & (4,1),(4,2),(4,3),(4,4),(4,5),(4,6) \\ (5,1),(5,2),(5,3),(5,4),(5,5),(5,6); & (6,1),(6,2),(6,3),(6,4),(6,5),(6,6) \end{Bmatrix}$$

There are 36 sample points in the sample space. The probability of getting each event is $\dfrac{1}{36}$

From the given data,

$X = \{2, 3, 4, 5, 6, 7, 8, 9, 10, 11, 12\}$ represents sum of the variables on the faces.

Suppose we are mainly interested in three events defined by

(i) $A = \{$Sum on the two dice is $5\}$

(ii) $B = \{ 4 \le$ Sum on the two dice $< 7\}$

(iii) $C = \{$Sum on the two $> 9\}$

Assigning the probabilities to these events,

(i) $A = \{$Sum on the two dice is $5\}$

i.e., $P(A) = P(5) = P(X = 5)$

$\qquad = P\{(1,4)\,(2,3)\,(3,2),\,(4,1)\}$

$\qquad = \dfrac{4}{36}$

Therefore $P(A) = \dfrac{4}{36}$

(ii) $B = \{ 4 \le$ Sum on the two dice $< 7\}$

i.e., $P(B) = P(4 \le X < 7)$

$\qquad = P(X = 4) + P(X = 5) + P(X = 6)$

$\qquad = P\{(1,3)\,(2,2)\,(3,1)\}$

$\qquad\quad + P\{(1,4)\,(2,3)\,(3,2),(4,1)\} + P\{(1,5)\,(2,4)\,(3,3),(4,2),\,(5,1)\}$

$\qquad = \dfrac{3}{36} + \dfrac{4}{36} + \dfrac{5}{36} = \dfrac{12}{36}$

$\qquad = \dfrac{1}{3}$

Therefore $P(B) = \dfrac{1}{9}$

(iii) $C = \{$Sum on the two dice $>10\}$

$$P(C) = P(X > 10)$$

$$= P(X = 11) + P(X = 12)$$

$$= \{(5,6)\ (6,5)\} + P\{(6,6)\}$$

$$= \frac{2}{36} + \frac{1}{36}$$

$$= \frac{3}{36}$$

$$= \frac{1}{12}$$

Therefore $P(C) = \dfrac{1}{12}$

This is the mathematical model of the experiment.

1.11   Some Theorems on Probability

1.   Probability of an impossible event is zero i.e., $P(\phi) = 0$

Proof: Since we know that $S \cup \varphi = S$

Now $P(S \cup \varphi)\quad =\quad P(S)$

$$P(S) + P(\varphi(\varphi = P(S)$$

$\because P(S) = 1$ *and S and* ϕ *are mutually exclusive events*

$$\therefore P(\phi) = 0$$

Hence probability of an impossible event is zero i.e., $P(\phi) = 0$

2.   If A is any event in a sample space S and A' is the complementary event of A then $P(A') = 1 - P(A)$ or $P(A) = 1 - P(A')$

Proof: Since we know that $S = A \cup A'$

Now $P(S) = P(A \cup A')$

$$\therefore 1 = P(A) + P(A')$$

$\because P(S) = 1$ *and A and A' are mutually exclusive events*

$$\therefore P(A') = 1 - P(A)\quad \text{or}\quad P(A) = 1 - P(A')$$

Hence, if A is any event in a sample space S and A' *is* the complementary event of A then $P(A') = 1 - P(A)$ or $P(A) = 1 - P(A')$.

3. If A is any event in the finite sample space, then prove that $P(A)$ equals to the sum of the probabilities of the individual outcomes comprising A.

Proof:

Let A be any event in the finite sample space which comprises individual outcomes $A_1, A_2, A_3, \ldots A_n$ (which are mutually exclusive events)

$$\therefore A = A_1 \cup A_2 \cup A_3 \cup \ldots \cup A_n$$

$$P(A) = P(A_1 \cup A_2 \cup A_3 \cup \ldots \cup A_n)$$

$$P(A) = P(A_1) + P(A_2) + P(A_3) + \ldots P(A_n)$$

$$\therefore P(A) = \sum_{i=1}^{n} P(A_1)$$

i.e., If A is any event in the finite sample space, then $P(A)$ equals to the sum of the probabilities of the individual outcomes comprising A.

4. Probability of any event in a sample space containing equally likely simple events is same

Proof:

Let S be sample space which comprises individual outcomes $A_1, A_2, A_3, \ldots A_n$ (which are mutually exclusive events)

$$\therefore S = A_1 \cup A_2 \cup A_3 \cup \ldots \cup A_n$$

$$P(S) = P(A_1 \cup A_2 \cup A_3 \cup \ldots \cup A_n)$$

$$P(S) = P(A_1) + P(A_2) + P(A_3) + \ldots P(A_n)$$

$$\therefore 1 = P(A_1) + P(A_2) + P(A_3) + \ldots P(A_n) \qquad \because P(S) = 1$$

Let $\quad A_1 = A_2 = A_3 = \ldots = A_n = A$

$$1 = P(A) + P(A) + P(A) + \ldots P(A)$$

$$1 = n.P(A)$$

$$\therefore P(A) = \frac{1}{n}$$

This implies that

$$P(A_1) = P(A_2) = P(A_3) = \ldots \ldots = P(A_n) = \frac{1}{n}$$

$\therefore$ Probability of any event in a sample space containing equally likely simple events is same.

1.12 Joint Probability

In some experiments, events are not mutually exclusive because of some common elements in the sample space. The probability of occurrence of these common elements in the sample space is called joint probability.

Theorem: Addition theorem on Probability (or Joint Probability):

(i) For any two events A and B, $P(A \cup B) = P(A) + P(B) - P(A \cap B)$

(ii) *For any three events A, B, and C*

$P(A \cup B \cup C) = P(A) + P(B) + P(C) - P(A \cap B) - P(B \cap C) - P(C \cap A) + P(A \cap B \cap C)$

Proof:

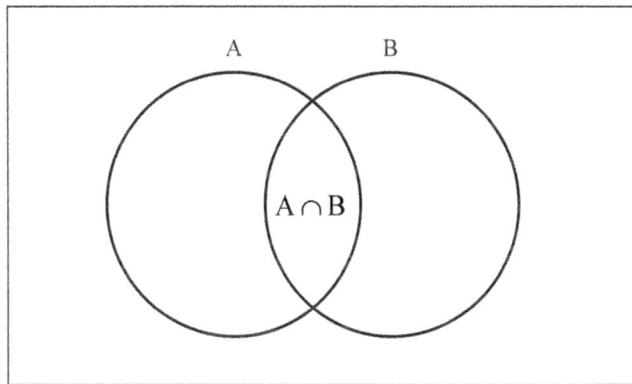

Fig. 1.5

From the Fig. 1.5,

$$(A \cup B) = (A \cap B') \cup (A \cap B) \cup (A' \cap B)$$

Here $(A \cap B')$, $(A \cap B)$ and $(A' \cap B)$ are mutually exclusive events

$$P(A \cup B) = P\{(A \cap B') \cup (A \cap B) \cup (A' \cap B)\}$$

$$= P(A \cap B') + P(A \cap B) + P(A' \cap B)$$

$$= P(A \cap B') + (A \cap B) + (A' \cap B) + P(A \cap B) - (A \cap B)$$

$$\therefore P(A \cup B) = P(A) + P(B) - P(A \cap B) \qquad \qquad ..... (1.3)$$

Since we know that from the addition theorem on probability for any two events A and B then $P(A \cup B) = P(A) + P(B) - P(A \cap B)$ we are extending for any three events A, B, and C

$$P[(A \cup B) \cup C] = P(A \cup B) + P(C) - P[(A \cup B) \cap C] \text{ from (1.3)} \qquad .....(1.4)$$

$$P[(A \cup B) \cup C] = P(A) + P(B) - P(A \cap B) + P(C) - P[(A \cap C) \cup (B \cap C)]$$

$$= P(A) + P(B) - P(A \cap B) + P(C) - [P(A \cap C) + P(B \cap C) - P(A \cap B \cap C)]$$

from (1.3)

$$= P(A) + P(B) + P(C) - P(A \cap B) - P(B \cap C) - P(C \cap A) + P(A \cap B \cap C)$$

$$\therefore P(A \cup B \cup C) = P(A) + P(B) + P(C) - P(A \cap B) - P(B \cap C) - P(C \cap A) + P(A \cap B \cap C)$$

6. If $A \subseteq B$ then prove that (i) $P(A' \cap B) = P(B) - P(A)$ and (ii) $P(A) \le P(B)$ or If A and B are two events of a sample space S such that $A \subseteq B$ then $P(B - A) = P(B) - P(A)$

Proof:

From the given data $A \subseteq B$

from the Venn diagram

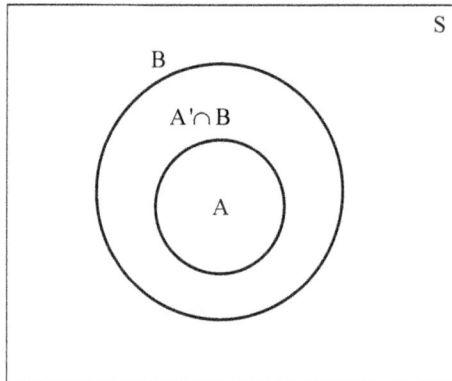

Fig. 1.6

$$P(A' \cap B) = P(B) - P(A \cap B)$$

$$= P(B) - P(A) \quad \because A \cap B = A \qquad .....(1.5)$$

Also since $P(A' \cap B) = P(B - A)$

$$\therefore \ P(B - A) = P(B) - P(A)$$

(ii)   Since,

$$P(B) - P(A) = P(A' \cap B) \ge 0$$

This implies that $P(B) \ge P(A)$ i.e., $P(A) \le P(B)$

7. If A and B are mutually exclusive events, then prove that $P(A) \leq P(B')$

Proof:

Given A and B are mutually exclusive events, then

$$P(A \cap B) = 0 \qquad \because A \cap B = \varphi$$

Now by addition theorem on probability states that for any two mutually exclusive events

$$\therefore P(A \cup B) = P(A) + P(B) \leq 1 \qquad \because 0 \leq P(A \cup B) \leq 1$$

$$P(A) + P(B) \leq 1 \Rightarrow P(A) \leq 1 - P(B) \leq P(B')$$

$$\Rightarrow P(A) \leq P(B')$$

Examples

***Example* 1:** In how many different ways can the principal of a college choose two faculty of mathematics from among six applicants and three faculty of computer science engineering from among nine applicants?

***Sol*:**

The two mathematics faculty can be chosen in ${}^6C_2 = 15$ ways

The three computer science engineering can be chosen in ${}^9C_3 = 84$ ways

Then these two things can be done in $15 \times 84 = 1,260$ ways

***Example* 2:** Three awards (research, teaching and service) will be given one year for a class of 25 graduate students in communication engineering department. If each student can receive at most one award, how many possible selections are there?

Sol:

$$25P_3 = \frac{25!}{(25-3)!} = \frac{25!}{22!} = 13,800$$

***Example* 3:** A President and a treasurer are to be chosen from a student club consisting of 60 students. How many different choices of students are possible if

   (a)   There are no restrictions

   (b)   Particular student 'A' will serve only if he is president

   (c)   Particular students B and C will serve together or not at all

   (d)   Particular students D and E will not serve together

***Sol*:**

   (a) If there is no restriction, Two students can be selected are arranged for two posts in ${}^{60}P_2 = 3540$ ways

(b) Since A will serve only if he is the president, then there are two cases.

A is selected as the president, which yields in the treasure can be selected in $494 = 49$ ways

Students are selected from the remaining 59 people which has the number of choices $^{59}P_2 = 3422$ ways

∴ The total number of choices $= 59 + 3422 = 3481$

(c) The number of selections when B and C serve together is 2.

The number of arrangements when both B and C are not chosen is $^{58}P_2 = 3306$

Therefore the total number of choices are $2 + 3306 = 3308$

(d) The number of arrangements when D serves as an officer but not E is $2 \times 58 = 116$ (where 2 is the number of positions D can take and 58 is the number of selections of the other officer from the remaining people in the club except E)

Similarly for E are 116

The number of arrangements when both D and E are not chosen is $^{58}P_2 = 3306$

Therefore the total number of choices is

$$2 \times 116 + 3306 = 3538$$

(Or)

Since D and E can serve together in only 2 ways.

Therefore the required choices are

$$3540 - 2 = 3538$$

Example 4: An electronic controlling mechanism requires six identical memory chips. In how many ways can this mechanism be assembled by placing the six chips in the six positions within the controller?

Sol: $^6P_6 = 720$

Example 5: If a test consists of 12 true – false questions, in how many different ways can a student mark the test paper with one answer to each question?

Sol:

From the given data, test consists of 12 true-false questions.

1st question can answer in 2 ways

2nd question can answer in 2 ways

Then these two questions can answer in $2 \times 2 = 2^2$ ways

∴ 12 questions can answer in $2 \times 2 \times 2 \times 2 \times \ldots \times 2 = 2^{12}$ ways

Example 6: A card is drawn from a well shuffled deck of 52 playing cards, what is the probability of drawing a

(a) black queen

(b) 3, 4, 5

(c) red card

(d) red ace or a black queen

Sol:

A pack contains 52 cards. Out of 52 cards one card is drawn in $^{52}C_1 = 52$ ways. And the total number of possible cases = 52

(a) There are 4 queens in a pack in which 2 are red and 2 are black.

Therefore the number of favourable cases for black queen are 2 and

∴ The probability of getting a black queen = 2/52

(b) There are 4 cards each of 3, 4 and 5 in a pack. Therefore the number of favourable cases are 12. and the total number of possible cases are 52.

∴ The probability of getting 3, 4 and 5 = 12/52

(c) There are 26 red cards in a pack. Therefore the number of favourable cases are 26, and the total number of possible cases are 52

∴ The probability of getting a red card = 26/52

(d) There are 4 aces in a pack out of which 2 are red and 2 are black and there are 4 queens in a pack in which 2 are red and 2 are black

∴ The numbers of favourable cases for getting a red aces or a black queen are 4

∴ The required probability = 4/52

Example 7: Four persons are chosen at random from a group containing 3 men, 2 women and 4 children. What is the chance that (i) exactly two of them will be children? (ii) at least two of them are children.

Sol:

Given that	Men	Women	Children	Total
	3	2	4	9

Total number of possible cases are $^9C_4$

(i) Two children can be selected in $^4C_2$ ways and

The remaining 2 can be selected from 3 + 2 = 5 persons in $^5C_2$ ways

Therefore the probability that exactly two of them will be children is
$$^4C_2 \times {}^5C_2/{}^9C_4 = 10/21$$

(ii)   The probability that at least two of them will be children is

$$\frac{^4C_2 \times {}^5C_2}{^9C_4} + \frac{^4C_3 \times {}^5C_1}{^9C_4} + \frac{^4C_4 \times {}^5C_0}{^9C_4}$$

$$= \frac{60}{126} + \frac{20}{126} + \frac{1}{126}$$

$$= \frac{81}{126}$$

Example 8: If 4 of 25 Electrical cables are defective and 5 of them are randomly chosen for inspection, what is the probability that only one of the defective cable will be included?

Sol:

In 25 electrical cables 4 are defective and 21 are non-defective

The number of ways of selecting 5 cables from 25 cables in $^{25}C_5$ ways

The probability that only one of the defective cable will be included is

$$= \frac{^{21}C_4 \times {}^4C_1}{^{25}C_5} = \frac{114}{253} = 0.4505$$

Example 9: If 3 of 20 circuits are connected wrongly and 4 of them are randomly chosen for inspection, what is the probability that only one of the wrong connection will be included?

Sol:

Given that out of   20   circuits 3 are wrongly connected and 17 are correctly connected. 4 circuits are selected in $^{20}C_4$ ways, and selection of 1 wrong connected circuit from 3 wrong connected   in $^3C_1$ ways and remaining 3 circuits can be taken from 17 correctly connected circuits in $^{17}C_3$ ways

The probability that only one of the wrong connected circuit will be included is

$$= \frac{^3C_1 \cdot {}^{17}C_3}{^{20}C_4} = \frac{8}{19}$$

Example 10: What is the probability that a leap year contains 53 Sundays?

Sol: A leap year consists of 366 days, of these, there are 52 complete weeks and 2 extra days. Those days may be any pair of the seven days. The days may be *(Sun, Mon)*, (Mon,

Tue), (Tue, Wed), (Wed, Thu), (Thu, Fri), (Fri, Sat), *(Sat, Sun)*. There are two favourable cases for getting one more Sunday.

So already we have 52 Sundays. For one more Sunday,

The probability of getting one more Sunday is 2/7

Hence the probability that a non-leap year contains 53 Sundays is 2/7

Example 11: A bag contains 5 black and 3 white balls. Two balls are drawn at random and the probability of drawing (i) 2 black balls (ii) 2 white balls.

Sol: Given that bag contains 8 balls viz., 5 black and 3 white balls

Two balls are drawn from 8 balls in

$$8c_2 = \frac{8.7}{1 \times 2} = 28 \text{ ways}$$

(i)   The probability of drawing 2 black balls.

2 black balls are drawn from 5 black balls in $5c_2 = 10$ ways.

Hence the probability of drawing 2 black balls

$$= \frac{No. \, of \, favourable \, cases}{Total \, no. \, of \, possible \, cases} = \frac{10}{28} = \frac{5}{14}$$

(ii)   2 white balls are drawn from 3 white balls in $3c_2 = 3$ ways

Hence the probability of drawing 2 white balls

$$= \frac{No. \, of \, favourable \, cases}{Total \, no. \, of \, possible \, cases} = \frac{3}{28}$$

Example 12: An urn contains 7 white, 6 red and 5 black balls. Two balls are drawn at random. Find the probability that they will both the black?

Sol:

Given that	Balls	*W*	*R*	*B*	Total
		7	6	5	18

Two balls are drawn at random from 18 balls in $^{18}C_2$ ways

Two black balls are drawn at random from 5 balls in $^{5}C_2$ ways

Hence the probability of getting two black balls

$$= (^{5}C_2) / (^{18}C_2)$$

$$= 10/153$$

Example **13:** Determine the probability that a non-defective bolt will be found if out of 800 bolts already examined 14 were defective.

Sol:

From the given data,

The probability of getting a defective bolt is p = $\dfrac{14}{800}$

$$= 0.0175$$

The probability of getting a non defective bolt

$$= 1 - p$$
$$= 1 - 0.0175$$
$$= 0.9825$$

Example **14:** Two digits are selected at random from the digits 1 to 9

  (i)   If the sum is odd, what is the probability that digit 2 is one of the umbers selected

 (ii)   If 2 is one of the digits selected what is the probability that the sum is odd

Sol:

From 1 to 9 digits, 5 are odd viz., (1, 3, 5, 7, 9) digits and 4 are even viz., (2, 4, 6, 8) digits

  (i)   The different possible cases for sum is odd are (1, 2), (1, 4), ... (9, 6), (9, 8) there are 20 combinations. Out of which the digit 2 occurs in 5 cases only.

        $\therefore$ The required probability = $\dfrac{5}{20} = \dfrac{1}{4}$

 (ii)   If 2 is fixed the remaining digits are 8 only, viz, {1, 3, 4, 5, 6, 7, 8, 9}

      Now the sum odd occurs with 2 from these 8 digits are 1, 3, 5, 7, 9

      The numbers of favourable cases are 5

      $\therefore$ The required probability is = $\dfrac{5}{8}$

      {Since the two digits are selected as $^{9}C_{2} = 36$ ways. Those are

$$[(1,2)], (1,3), [(1,4)], (1,5), [(1,6)], (1,7), [(1,8)], (1,9)$$
$$[(2,3)], (2,4), [(2,5)], (2,6), [(2,7)], (2,8), [(2,9)]$$
$$[(3,4)], (3,5), [(3,6)], (3,7), [(3,8)], (3,9)$$
$$[(4,5)], (4,6), [(4,7)], (4,8), [(4,9)]$$
$$[(5,6)], (5,7), [(5,8)], (5,9)$$
$$[(6,7)], (6,8), [(6,9)]$$
$$[(7,8)], (7,9)$$
$$[(8,9)]$$

Here square bracket numbers are sum odd and 2 indicates favourable cases.}

***Example* 15:** The students in a class are selected at random one after the other for an examination. Find the probability that the boys and girls are alternate, if there are

(i) 5 boys and 4 girls   and   (ii)   4 boys and 4 girls

Sol:

Given data, the selection is done at random one after the other

(i)   There are 5 boys and 4 girls

Let the boys and girls are alternate in the following manner

$$B \ G \ B \ G \ B \ G \ B \ G \ B$$

Now,

The probability that the boys and girls are alternate

$$= \frac{5! \times 4!}{9!}$$

$$= \frac{5}{9} \times \frac{4}{8} \times \frac{4}{7} \times \frac{3}{6} \times \frac{3}{5} \times \frac{2}{4} \times \frac{2}{3} \times \frac{1}{2}$$

$$= \frac{1}{126}$$

Aliter: The five boys can be arranged in $^5P_5 = 5!$ ways and 4 girls can be arranged in $^4P_4 = 4!$ ways. Then these 9 members can be arranged in 9! ways.

Therefore the required probability $= \dfrac{5! \times 4!}{9!}$

$$= \frac{1}{126}$$

(i)   There are 4 boys and 4 girls:

Let the boys and girls are alternate in the following two ways

(a) $B \ G \ B \ G \ B \ G \ B \ G$ (here choosing first boy)

(b) $G \ B \ G \ B \ G \ B \ G \ B$ (here first girl)

The probability that the boys and girls are alternate if boys first

$$= \frac{4}{8} \times \frac{4}{7} \times \frac{3}{6} \times \frac{3}{5} \times \frac{2}{4} \times \frac{2}{3} \times \frac{1}{2}$$

$$= \frac{1}{70}$$

Similarly for girls first is also $\dfrac{1}{70}$

∴ The required probability is (i.e., the probability that the boys and girls are alternate)

$$= \frac{1}{70} + \frac{1}{70} = \frac{1}{35}$$

Aliter: There are 4 boys and 4 girls are alternate. The first position can be filled in 2! ways and the 4 boys can be filled in 4! ways and 4 girls can be filled in 4! ways.

Therefore the number of favourable cases is $2! \times 4! \times 4!$

The 8 members can be seated in 8! ways.

∴ The required probability is $\dfrac{2! \times 4! \times 4!}{8!} = \dfrac{1}{35}$

Example **16:** Urn A contains 5 red and 3 white memory chips and urn B contains 2 red and 6 white memory chips. If a chip is drawn from each box what is the probability that they are both of the same colour?

Sol:

From the given data,

Urn A contains 5 red and 3 white chips = total 8 chips

Box B contains 2 red and 6 white chips = total 8 chips

Now,

$$P(\text{drawing a red chip from urn } A) = \frac{5}{8} \text{ and}$$

P(drawing a red chip from urn B) = $\dfrac{2}{8}$

Similarly,

P(getting a white chip from urn A) = $\dfrac{3}{8}$ and

P(getting a white chip from urn B) = $\dfrac{6}{8}$

∴ The probability that they are both of the same colour

$$= \frac{5}{8} \times \frac{2}{8} + \frac{3}{8} \times \frac{6}{8} = \frac{7}{16}$$

***Example* 17:** Two aeroplanes bomb a target in succession. The probability of each correctly scoring a hit is 0.3 and 0.2 respectively. The second will bomb only if the first misses the target. Find the probability that (i) the target is hit (ii) both fails to score hits.

Sol:

From the given data

$$P(A) = 0.3 \text{ and } P(B) = 0.2 \text{ this implies } P(\overline{A}) = 0.7 \text{ and } P(\overline{B}) = 0.8$$

(Here A and B be the events that represents the 1st and 2nd aeroplanes respectively)

Now

(i)   The probability that hit the target is

$= P(A$ hits the target$) + P(A$ miss the target$)$ $P(B$ hit the target$)$

$= 0.3 + 0.7 \times 0.2$

$= 0.44$

(ii)   The probability that both fails

$= P(A$ miss the target$) P(B$ miss the target$)$

$= 0.7 \times 0.8$

$= 0.56$

***Example* 18:** A box contains n tickets marked 1 through n. Two tickets are drawn without replacement. Determine the probability that the numbers on the tickets are consecutive integers.

Sol:

The number of tickets in the box is n,

Two tickets are drawn without replacement from the given tickets that can be done in $^nC_2$ ways.

Therefore the total number of possible cases $^nC_2$.

The number of favourable cases for getting the numbers on the tickets are consecutive are (n-1). {viz., (1,2), (2,3), (3,4) . . . . (n-1,n) there are (n-1) in number}

The required probability that the number on the tickets are consecutive integers

$$= \frac{n-1}{^nC_2}$$

$$= \frac{2}{n}$$

***Example* 19:** Determine the probability that for a non defective bolt will be found if out of 600 bolts already examined 12 were defective.

Sol:

Given that 12 bolts were defective out of 600 bolts

$$\therefore \text{Probability of getting a defective bolt} = \frac{12}{600} = \frac{1}{50}$$

Hence Probability of getting a non defective bolt is

$$= 1 - \frac{1}{50}$$

$$= \frac{49}{50}$$

***Example* 20:** An urn contains three boxes. Box I contains 10 diodes of which 4 are defective. Box II contains 6 diodes of which one is defective. Box III contains 8 diodes of which 3 are defective. A box is chosen and a diode is drawn. Find the probability that the diode is non defective.

Sol:

The probability of choosing any one of the boxes I, II and III from the urn $= \frac{1}{3}$

Probability of getting a defective diode from box I $= \frac{4}{10}$

$\therefore$ Probability of getting a non defective diode from box I $= \frac{6}{10}$

Probability of getting a defective diode from box II = $\dfrac{1}{6}$

∴ Probability of getting a non defective diode from box II = $\dfrac{5}{6}$

Probability of getting a defective diode from box III = $\dfrac{3}{8}$

∴ Probability of getting a non defective diode from box III = $\dfrac{5}{8}$

Hence the probability that the diode is non defective

$$= \dfrac{1}{3} \times \dfrac{6}{10} + \dfrac{1}{3} \times \dfrac{5}{6} + \dfrac{1}{3} \times \dfrac{5}{8}$$

$$= 0.678$$

***Example* 21:** Two diodes are drawn from a box containing 4 good and 6 defective diodes. Find the probability that the second diode is good if the first one is found to be defective.

Sol:

Given that there are 4 good diodes and 6 defective diodes in a box

The probability that the first diode is defective = $\dfrac{6}{10}$

and the probability that the second diode is good is $\dfrac{4}{9}$

(Since the drawing is done without replacement)

∴ The required probability that the second diode is good if the first one is found to be defective is

$$\dfrac{6}{10} \times \dfrac{4}{9} = \dfrac{4}{15}$$

***Example* 22:** An integer is chosen at random from the first 200 positive integers. What is the probability that the integer chosen is divisible by 6 or 8.

Sol:

The sample space S consists of 200 elements.

i.e., $S = \{1,2,3\ldots\ldots\ldots\ldots200\}$

∴ $n(S) = 200$

The set of numbers which are divisible by 6 is denoted by the set A

$$A = \{6, 12, 18, \ldots\ldots\ldots 198\} \quad \text{which are 33 in number}$$

$\therefore \qquad n(A) = 33 \text{ and } P(A) = \dfrac{33}{200}$

and the set of numbers which are divisible by 8 is denoted by the Set B

$$B = \{8, 16, 24 \ldots\ldots\ldots 200\} \quad \text{which are 25 in number}$$

$\therefore \qquad n(B) = 25 \text{ and } P(B) = \dfrac{25}{200}$

The set $A \cap B$ denote the numbers which are divisible by both 6 and 8

$$A \cap B = \{24, 48, 72, 96, 120, 144, 168, 192\} \quad \text{which are 8 in number}$$

$$\therefore \ n(A \cap B) = 8 \ \text{ and } \therefore \ P(A \cap B) = \frac{n(A \cap B)}{n(S)} = \frac{8}{200}$$

Now the Probability that the integer chosen is divisible by 6 or 8 is (by using addition theorem on probability $P(AUB) = P(A) + P(B) - P(A \cap B)$

$$\therefore P(AUB) = \frac{33}{200} + \frac{25}{200} - \frac{8}{200} = \frac{50}{200} = \frac{1}{4} = 0.25$$

***Example* 23:** A and B throw with a pair of dice. A wins if he throw 6 before B throws 7 and B, if he throws 7 before A throws 6. If A begins, show that his chance of winning is 30/61.

Sol: When two dice are thrown the total number of possible cases are 36 and the sample space S is

$$S = \begin{cases} (1,1), (1,2), (1,3), (1,4), (1,5), (1,6); \ \ (2,1), (2,2), (2,3), (2,4), (2,5), (2,6) \\ (3,1), (3,2), (3,3), (3,4), (3,5), (3,6); \ \ (4,1), (4,2), (4,3), (4,4), (4,5), (4,6) \\ (5,1), (5,2), (5,3), (5,4), (5,5), (5,6); \ \ (6,1), (6,2), (6,3), (6,4), (6,5), (6,6)\} \end{cases}$$

The number of possible cases for getting a sum on the two dice is 6 are (1, 5), (2, 4), (3, 3), (4, 2) and (5, 1) they are 5 in number

$\therefore$ The probability of 'A' getting the sum 6 = 5/36

The probability of 'A' not getting the sum 6 = 31/36 $\qquad \because$ 1 – 5/36 = 31/36

Similarly, the number of possible cases for getting a sum of 7 is

$$(1, 6), (2, 5), (3, 4), (4, 3), (5, 2), (6, 1) \ \text{ which are 6 in number}$$

∴ The probability of 'B' getting the sum7 is = 6/36

And the probability of B not getting the sum 7 is = 1- 6/36 = 30/36

If A starts the game:

The probability of A wins the game

= in 1st time A wins or the 3rd time A wins or 5th time A wins and so on

$$=\left\{\frac{5}{36}+\frac{31}{36}\times\frac{30}{36}\times\frac{5}{36}+\frac{31}{36}\times\frac{30}{36}\times\frac{31}{36}\times\frac{5}{36}+...\right\}$$

$$=\frac{5}{36}\left\{1+\frac{31}{36}\times\frac{30}{36}+(\frac{31}{36}\times\frac{30}{36})^2...\right\}$$

$$=\frac{5}{36}\left\{\frac{1}{1-\left(\frac{31}{36}\times\frac{30}{36}\right)}\right\}$$

$$=\frac{30}{61}$$

Note: Suppose in the question given B starts the game:

The probability of A wins the game

= in 1st time B looses the game and A wins or B looses, A looses,

B looses and A wins and so on

$$=\left\{\frac{30}{36}\times\frac{5}{36}+\left(\frac{30}{36}\times\frac{31}{36}\times\frac{30}{36}\times\frac{5}{36}\right)+...\right\}$$

$$=\frac{30}{36}\times\frac{5}{36}\left\{1+\left(\frac{31}{36}\times\frac{30}{36}\right)+\left(\frac{31}{36}\times\frac{30}{36}\right)^2+..\right\}.$$

$$=\left(\frac{30}{36}\times\frac{5}{36}\right)\left\{\frac{1}{1-\left(\frac{31}{36}\times\frac{30}{36}\right)}\right\}$$

= 0.41

***Example* 24:** Two balls are drawn in succession from a box containing 10 red 30 white, 20 blue and 15 orange balls with replacement being made after each drawing. Find the probability that

    (i) Both are white and   (ii) First is red and the second is white

Sol: From the given data box contains 10 red 30 white 20 blue and 15 orange balls. The total balls are 75.

   (i)   **Both are white**

Probability of getting first ball is white $= \dfrac{30}{75}$ and

Probability of getting second ball is white $= \dfrac{30}{75}$

Since drawing is done with replacement

Hence the probability that both the balls are white

$$= \frac{30}{75} \cdot \frac{30}{75}$$

$$= \frac{4}{25}$$

   (ii)   **First is red and the second is white**

Probability of getting first ball is red $= \dfrac{10}{75}$ and

Probability of getting second ball is white $= \dfrac{30}{75}$

Since drawing is done with replacement

Hence the probability that first is red and the second is white

$$= \frac{10}{75} \times \frac{30}{75} = \frac{4}{75}$$

***Example* 25:** Two cards are drawn from a well shuffled pack of 52 cards. Find the probability that they are both aces if the first card is (i) replaced (ii) not replaced.

Sol:

A pack contains 52 cards, the total number of aces in the pack is 4.

   (i)   Two cards are drawn one after the other and the first card is replaced before drawing the second card is then

The probability that they are both aces is

$$= \frac{4}{52} \times \frac{4}{52}$$

$$= \frac{1}{169}$$

$$= 0.00592$$

(ii)   If the first card is not replaced

The probability that they are both aces if the first card is not replaced before the second draw is

$$= \frac{4}{52} \times \frac{3}{51}$$

$$= \frac{1}{221} = 0.00452$$

Example 26: A class of 12 boys and 8 girls, three students are selected at random one after the other. Find the probability that (i) first two are boys and third is girl (ii) first and third are of the same sex and the second is of opposite sex.

Sol: There are 12 boys and 8 girls. Total number of students is 20.

(i)   The probability that first two are boys and third is girl (select three students' at random one after the other)

$$= \frac{12}{20} \times \frac{11}{19} \times \frac{8}{18}$$

$$= 0.1544$$

(ii)   The probability that first and third are of the same sex and the second is of opposite sex is as follows

B G B i.e., choose first boy, then girl and boy then the required probability

$$= \frac{12}{20} \times \frac{8}{19} \times \frac{11}{18}$$

$$= 0.1544$$

or *G B G* choose first girl, then boy and girl then the required probability

$$= \frac{8}{20} \times \frac{12}{19} \times \frac{7}{18}$$

$$= 0.0982$$

Hence the probability that first and third are of the same sex and the second is of opposite sex

$$= 0.1544 + 0.0982$$

$$= 0.2526$$

***Example* 27:** A fair die is tossed twice. Find the probability of getting a 4, 5 or 6 on the first toss and 1, 2, 3 or 4 on the second toss.

Sol: The probability of getting a 4, 5 or 6 is $\dfrac{3}{6} = \dfrac{1}{2}$

When a die is thrown the probability of getting a 1, 2, 3 or 4 is $\dfrac{4}{6} = \dfrac{2}{3}$

Hence the probability of getting a 4, 5 or 6 on the first throw and 1, 2, 3 or 4 on the second throw is

$$= \frac{1}{2} \times \frac{2}{3} = \frac{1}{3}$$

***Example* 28:** The probability that India wins a cricket test match against West Indies is known to be 2/5. If India and West Indies play three test matches, what is the probability that (i) India will lose all the three matches (ii) India will win at least one test match (iii) India will win at most one match.

Sol:

From the given data, the probability that India wins a cricket test match $= \dfrac{2}{5}$

And the probability that India lose the cricket test match $= \dfrac{3}{5}$ $\because 1 - \dfrac{2}{5} = \dfrac{3}{5}$

(i)   The probability that India will lose all the three matches is $= \dfrac{3}{5} \times \dfrac{3}{5} \times \dfrac{3}{5} = 0.6$

(ii)   The probability that India will win at least one test match (i.e., India will win one or two or all the three matches) is

$$= \frac{2}{5} \times \frac{3}{5} \times \frac{3}{5} + \frac{2}{5} \times \frac{2}{5} \times \frac{3}{5} + \frac{2}{5} \times \frac{2}{5} \times \frac{2}{5}$$

$$= \frac{18 + 12 + 6}{125}$$

$$= \frac{36}{125}$$

$$= 0.288$$

(iii)   The probability that India will win at most one match (i.e., India will win zero or one match) is

$$= \frac{3}{5} \times \frac{2}{5}$$

$$= \frac{6}{25}$$

$$= 0.24$$

***Example* 29:** The probabilities of passing in subject A, B, C and D are $3/4$, $2/3$, $4/5$ and $1/2$ respectively. To qualify in the examination a student should pass in A and two subjects among the three what is the probability of qualifying in that examination.

Sol: From the given data,

$$P(A) = \frac{3}{4} \quad P(B) = \frac{2}{3} \quad P(C) = \frac{4}{5} \quad \text{and} \quad P(D) = \frac{1}{2}$$

The probabilities of not passing in subjects B, C and D are $P(B^c) = 1/3$ $P(C^c) = 3/4$ and $P(D^c) = 1/2$. Here all the four events are independent

There are four possibilities to qualify the examination

1. Pass in Subjects A, B, C and fail in subject D
2. Pass in Subjects A, B, D and fail in subject C
3. Pass in Subjects A, C, D and fail in subject B
4. Pass in all Subjects A, B, C and D

Now the probability of qualifying in that examination is the sum of all the four cases

$$= P(A \cap B \cap C \cap D^c) + P(A \cap B \cap D \cap C^c)$$
$$+ P(A \cap C \cap D \cap B^c) + P(A \cap B \cap C \cap D)$$

$$= P(A) P(B) P(C) P(D^c) + P(A) P(B) P(D) P(C^c)$$
$$+ P(A) P(C) P(D) P(B^c) + P(A) P(B) P(C) P(D)$$

$$= \left\{ \frac{3}{4} \times \frac{2}{3} \times \frac{4}{5} \times \frac{1}{2} \right\} + \left\{ \frac{3}{4} \times \frac{2}{3} \times \frac{1}{2} \times \frac{1}{5} \right\}$$
$$+ \left\{ \frac{3}{4} \times \frac{4}{5} \times \frac{1}{2} \times \frac{1}{3} \right\} + \left\{ \frac{3}{4} \times \frac{2}{3} \times \frac{4}{5} \times \frac{1}{2} \right\}$$

$$= 0.55$$

***Example* 30:** From a pack, two cards drawn. What is the probability that either both are red or both are kings.

Sol:

The total number of possible cases for getting two cards from a well shuffled pack of 52 cards is $^{52}C_2 = 1326$

The number of favourable cases for getting 2 red cards from 26 red cards is

$$^{26}C_2 = 325$$

The number of favourable cases for getting 2 king cards from 4 red cards is $^4C_2 = 6$ (of these four kings two red cards already involved).

∴ The number of favourable cases is $^{26}C_2 + {}^4C_2 - 2$

$$= 325 + 6 - 2$$

$$= 329$$

Hence the probability of getting that either both is red or both are kings

$$= \frac{^{26}C_2 + {}^4C_2 - 2}{^{52}C_2}$$

$$= \frac{325 + 6 - 2}{1326}$$

$$= 0.248$$

Example 31: A, B, C are aiming to shoot a balloon. A will succeed 4 times out of 5 attempts. The chance of B to shoot the balloon is 3 out of 4 and that of C is 2 out of 3. If the three aim the balloon simultaneously, then find the probability that at least two of them hit the balloon.

Sol:

From the given data,

$$P(A) = \frac{4}{5}, \ P(B) = \frac{3}{4} \text{ and } P(B) = \frac{3}{4}$$

This implies $\qquad P(A^1) = \frac{1}{5}, \ P(B') = \frac{1}{4} \text{ and } P(C') = \frac{1}{3}$

The probability that at least two of them hit the balloon is

$$= \ P(A \cap B \cap C^c) + P(A \cap B^c \cap C)$$

$$+ P(A^c \cap B \cap C) + P(A \cap B \cap C)$$

$$= \ P(A) \, P(B) \, P(C^c) + P(A) \, P(B^c) \, P(C)$$

$$+ P(A^c) \, P(B) \, P(C) + P(A) \, P(B) \, P(C)$$

$$= \left\{ \frac{4}{5} \times \frac{3}{4} \times \frac{1}{3} \right\} + \left\{ \frac{4}{5} \times \frac{1}{4} \times \frac{2}{3} \right\}$$

$$+ \left\{ \frac{1}{4} \times \frac{3}{4} \times \frac{2}{3} \right\} + \left\{ \frac{4}{5} \times \frac{3}{4} \times \frac{2}{3} \right\}$$

$$= \frac{5}{6}$$

$$= 0.83$$

***Example* 32:** Out of 15 items 4 are not in good condition, and 4 are selected at random. Find the probability that (i) all are not good (ii) Two are not good

Sol: From the given data, good condition items = 11 and not in good condition items = 4
The number of possible cases for getting 4 items from 15 items is $^{15}C_4 = 1365$

(i) The probability that all are not good:

The number of favourable cases for getting 4 items from 4 not good items is $^4C_4 = 1$

Hence the required probability $= \dfrac{1}{1365} = 0.0007$

(ii) The probability that two are not good

Select the 4 items, two from good and two for not good items, then the number of favourable cases are $^{11}C_2 . {}^4C_2 = 330$

Hence the required probability $= \dfrac{330}{1365}$

$= 0.742$

1.13   Conditional Event

Let S be a sample space and E_1, E_2 are two events in S and if E_2 occurs after the occurrence of E_1, then the occurrence of the event E_2 after the occurrence of E_1 is called conditional event of E_2 given E_1. It is denoted by E_2/E_1.

Similarly we define E_1/E_2.

1.14 Conditional Probability

1. Let S be a sample space and E_1, E_2 are two events in S and if $P(E_1) \neq 0$, then the probability of occurrence of the event E_2 after the occurrence of E_1 is called conditional probability of the event E_2 given E_1 . It is denoted by $P(E_2 /E_1)$. And is defined as

2.
$$P(E_2/E_1) = \frac{P(E_1 \cap E_2)}{P(E_1)}$$

i.e.,
$$P(E_2/E_1) = \frac{n(E_1 \cap E_2)/n(S)}{n(E_1)/n(S)} = \frac{n(E_1 \cap E_2)}{n(E_1)} \quad n(E_1) \neq 0$$

similarly we can define $P(E_1/E_2) = \dfrac{P(E_1 \cap E_2)}{P(E_2)}$ and $P(E_2) \neq 0$

i.e.,
$$P(E_1/E_2) = \frac{n(E_1 \cap E_2)}{n(E_2)} \quad n(E_2) \neq 0$$

1.15　Multiplication Theorem on Probability

Let S be the sample space and let E_1, E_2 are two events in the sample space 'S' such that $P(E_1) \neq 0$ and $P(E_2) \neq 0$ then

$$P(E_1 \cap E_2) = P(E_2/E_1) \cdot P(E_1)$$
$$= P(E_1/E_2) \cdot P(E_2)$$

Proof:

Since S be the sample space and E_1, E_2 are two events in the sample space S and if $P(E_1) \neq 0$ and $P(E_2) \neq 0$ then by using the conditional probability of the event E_2 given E_1 is

$$P(E_2/E_1) = \frac{P(E_1 \cap E_2)}{P(E_1)}$$

i.e.,
$$P(E_1 \cap E_2) = P(E_2/E_1) \cdot P(E_1) \qquad\qquad .....(1.6)$$

Similarly

$$P(E_1/E_2) = \frac{P(E_1 \cap E_2)}{P(E_2)}$$

i.e.,
$$P(E_1 \cap E_2) = P(E_1/E_2) \cdot P(E_2) \qquad\qquad .....(1.7)$$

Therefore from these two equations (1.6) and (1.7) we have

$$P(E_1 \cap E_2) = P(E_2/E_1) \cdot P(E_1) = P(E_1/E_2) \cdot P(E_2)$$

This is called the multiplication theorem on probability

1.16 Statistical Independence

Independent and Dependent Events: The occurrence of one event does not affect the occurrence of other event(s) then the events are said to be independent. For example a coin is tossed twice; the result of the second toss does not affected by the result of the first toss.

The occurrence or non-occurrence of one event in any experiment affects the probability of other events in other experiment then the events are said to be dependent. For example a card is drawn from a pack of cards and is not replaced in the pack of cards, this will affect the probability that the second card drawn.

Note: The probability of drawing a king from a pack of 52 cards is 4/52, but if the card is not replaced in the pack, then the probability of drawing again a king is 3/51. So the probability is not the same. Hence the first draw affects the probability in second draw, in this case the two events are said to be dependent.

Remarks:

1. If E_1 and E_2 are two independent events, then

   (i) $P(E_1 / E_2) = P(E_1)$

   (ii) $P(E_2 / E_1) = P(E_2)$

   (iii) By multiplication theorem on probability $P(E_1 \cap E_2) = P(E_1 / E_2) . P(E_2)$, the events E_1 and E_2 are two independent then

   $$P(E_1 \cap E_2) = P(E_1) . P(E_2) \qquad \text{from equations (i) and (ii)}$$

   Similarly $E_1, E_2, E_3, E_4 \ldots$ are independent events, then

   (iv) $P(E_1 \cap E_2 \cap E_3 \cap E_4 .... \cap E_n) = P(E_1).P(E_2).P(E_3).P(E_4)....P(E_n)$

   (i)   By addition theorem on probability,

   $$P(E_1 \cup E_2) = P(E_1) + P(E_2) - P(E_1) . P(E_2)$$

   $$\because P(E_1 \cap E_2) = P(E_1) . P(E_2)$$

Theorem: If A and B are independent events then prove that (i) A' and B' (ii) A and B' are also independent.

Proof:

If A and B are independent events then

$$P(A \cap B) = P(A) \; P(B)$$

It is required to show that,

$$P(A' \cap B') = P(A') \; P(B')$$

Now

(i) $P(A' \cap B') = P[(A \cup B)'] = 1 - P(A \cup B)$

$= 1 - \left[P(A) + P(B) - P(A \cap B)\right]$

$= 1 - P(A) - P(B) + P(A)P(B)$

Since A and B are independent

$= P(A') - P(B) [1 - P(A)]$

$= P(A') - P(B)P(A')$

$\therefore P(A' \cap B') = P(A') [1 - P(B)]$

$P(A' \cap B') = P(A') P(B')$

$\therefore A'$ and B' are independent events.

(ii)   Now we shall show that $P(A \cap B') = P(A) \times P(B)$

$\because A = A \cap S$

$= A \cap (B \cup B')$

$= (A \cap B) \cup (A \cap B')$

$P(A) = P(A \cap B) + P(A \cap B')$

since $(A \cap B)$ and $(A \cap B')$ are independent

This implies,

$P(A \cap B') = P(A) - P(A \cap B)$

$= P(A) - P(A)P(B)$ since A and B are independent

$= P(A) \left[1 - P(B)\right]$

$= P(A)P(B')$

$\therefore$ $P(A \cap B') = P(A)P(B')$

Hence A' and B' are independent events

Aliter: $P(A \cap B') = P(A) - P(A \cap B)$

$= P(A) - P(A) P(B) = P(A) [1 - P(B)]$

$P(A \cap B') = P(A) P(B')$

$\therefore$ A and B' are independent

Theorem: If E_1, E_2, E_3 are mutually independent events of a sample space S, then $E_1 \cup E_2$ and E_3 are also independent events.

Proof:

Consider

$$P[(E_1 \cup E_2)/E_3] = \frac{P[(E_1 \cup E_2) \cap E_3)]}{P(E_3)}$$

$$= \frac{P[(E_1 \cap E_3) \cup P(E_2 \cap E_3)]}{P(E_3)}$$

$$= \frac{P(E_1 \cap E_3) + P(E_2 \cap E_3) - P(E_1 \cap E_2 \cap E_3)}{P(E_3)}$$

$$= \frac{P(E_1)P(E_3) + P(E_2)P(E_3) - P(E_1)P(E_2)P(E_3)}{P(E_3)}$$

$$= P(E_1). + P(E_2) - P(E_1)P(E_2)$$

$$= P(E_1 \cup E_2)$$

$\therefore \qquad P[(E_1 \cup E_2)/E_3] = P(E_1 \cup E_2)$

Hence this implies that $E_1 \cup E_2$ and E_3 are independent events.

Aliter:

$$P[(E_1 \cup E_2) \cap E_3] = P[(E_1 \cap E_3) \cup (E_2 \cap E_3)]$$

$$= P(E_1 \cap E_3) + P(E_2 \cap E_3) - P[(E_1 \cup E_3) \cap (E_2 \cap E_3)]$$

$$= P(E_1).P(E_3) + P(E_2).P(E_3)] - P[(E_1 \cap E_2 \cap E_3)]$$

$$= P(E_1).P(E_3) + P(E_2).P(E_3)] - P(E_1).P(E_2).P(E_3)]$$

$$= P(E_3)[P(E_1). + P(E_2) - P(E_1).P(E_2)]$$

$$= [P(E_1). + P(E_2) - P(E_1).P(E_2)]P(E_3)$$

$$= [P(E_1). + P(E_2) - P(E_1 \cap E_2)]P(E_3)$$

$$= [P(E_1 \cup E_2)]P(E_3)$$

This implies that $E_1 \cup E_2$ and E_3 are independent events.

Examples

***Example* 1:** If $P(A) = 1/3, P(B) = 1/4, P(A \cup B) = 1/2$ then determine

(i) $P(B/A)$ and   (ii) $P(A/B^c)$

Sol:

From the given data, $P(A) = 1/3$, $P(B) = 1/4$, $P(A \cup B) = 1/2$

   Then $P(B^c) = 3/4,$

By using addition theorem on the probability, $P(A \cup B) = P(A) + P(B) - PA \cap B)$

$$\therefore PA \cap B) = \frac{1}{3} + \frac{1}{4} - \frac{1}{2} = \frac{1}{12}$$

(i) $P(B/A) = \dfrac{P(A \cap B)}{P(A)}$

$$= \frac{1/12}{1/3} = \frac{1}{4}$$

(ii) $P(A \cap B^c) = P(A) - P(A \cap B)$

Divide on both sides by $P(B')$ then

$$\frac{P(A \cap B^c)}{P(B^c)} = \frac{P(A)}{P(B^c)} - \frac{P(A \cap B)}{P(B^c)}$$

$$\Rightarrow P(A/B^c) = \frac{1/3}{1-(1/4)} - \frac{1/12}{1-(1/4)}$$

$$= \frac{1/3}{3/4} - \frac{1/12}{3/4}$$

$$= \frac{1}{3}$$

Aliter:

(ii) $(P(A/B^c) = \dfrac{P(A \cap B^c)}{P(B^c)}$

$$= \frac{1/4}{3/4}$$

   since $P(A \cap B^c) = P(A) - P(A \cap B) = \dfrac{1}{3} - \dfrac{1}{12} = \dfrac{1}{4}$

$$= \frac{1}{3}$$

Example 2: A problem is given to three students A, B and C whose chances of solving it are 1/2, 1/3 and 1/4 respectively. What is the probability that the problem will be solved?

Sol:

Probability of Student A solving the problem is $\dfrac{1}{2}$, i.e., $P(A) = \dfrac{1}{2}$

Similarly, $P(B) = \dfrac{1}{3}$ and $P(C) = \dfrac{1}{4}$

Then $P(A^c) = 1/2$ $P(B^c) = 2/3$ and $P(C^c) = 3/4$

Probability that the problem will be solved

$$= 1 -$$

(probability that the problem will not be solved by the three students)

$$= 1 - P(A^c)\ P(B^c)\ P(C^c)$$

$$= 1 - (1/2)(2/3)(3/4)$$

$$= 1 - \dfrac{1}{4} = \dfrac{3}{4}$$

Aliter: By using addition theorem on probability

The probability that the problem will be solved (by using addition theorem on probability for three independent events) is

$$P(A \cup B \cup C) = P(A) + P(B) + P(C) - P(A).P(B) - P(B).P(C) - P(C).P(A) + P(A).P(B).P(C)$$

$$P(A \cup B \cup C) = \dfrac{1}{2} + \dfrac{1}{3} + \dfrac{1}{4} - \dfrac{1}{2} \times \dfrac{1}{3} - \dfrac{1}{3} \times \dfrac{1}{4} - \dfrac{1}{4} \times \dfrac{1}{2} + \dfrac{1}{2} \times \dfrac{1}{3} \times \dfrac{1}{4}$$

$$= \dfrac{3}{4}$$

Example 3: Three students A, B and C are in a running race. A and B have the same probability of winning and each is twice as likely to win as C. Find the probability that B or C wins.

Sol: Given that

$$P(A) = P(B) = 2\,P(C) = k \quad \text{(say)},$$

i.e., $P(A) = k$; $P(B) = k$; $P(C) = \dfrac{1}{2}k$

since we know that $P(A) + P(B) + P(C) = 1$

i.e., $\qquad k + k + \dfrac{1}{2}k = 1 \qquad\qquad \therefore k = \dfrac{2}{5}$

By using addition theorem on probability for independent events, then

$$P(B \cup C) = P(B) + P(C) - P(B)\ P(C)$$

since A, B and C are independent

$$= \frac{2}{5} + \frac{1}{5} - \frac{2}{5} \times \frac{1}{5}$$

$$= \frac{13}{25}$$

***Example* 4**: If the probability that a communication system will have high fidelity is 0.81 and the probability that it will have high fidelity and selectivity 0.18. What is the probability that a system will high fidelity will also have high selectivity?

Sol: Let the event A represents a communication system will have high fidelity then

$$P(A) = 0.81$$

Let the event $(A \cap B)$ represents high fidelity and selectivity then $P(A \cap B) = 0.18$

$\therefore$ By using conditional probability, the probability that a system will have high fidelity and also have high selectivity is

$$P(B/A) = \frac{P(A \cap B)}{P(A)}$$

$$= \frac{0.18}{0.81}$$

$$= 0.22$$

***Example* 5:** There are two boxes. In Box-I, 11 cards are there numbered 1 to 11 and Box-II, 5 cards are numbered 1 to 5. A box is chosen and a card is drawn. If the card shows an even number then another card is drawn from the same box. If card shows an odd number another card is drawn from the other box. Find the probability that (i) Both are even (ii) Both are odd (iii) What is the Probability that both cards are even that they are from box I ?

Sol: Probability of getting any box is $\dfrac{1}{2}$

Given that

In box I there 5 are even numbers and 6 are odd numbers since it contain 11 cards.

In box II there are 2 even numbers and 3 odd numbers since it contain 5 cards.

(i) Let E be the event that represents even number on both the cards

In this, a box is chosen and a card is drawn, if the first card is even then the second card is also drawn from the same box and that card is also even.

Let us assume that E_1 be the event that represents both the cards are from box-I

$$P(E_1) = \frac{1}{2} \cdot \frac{5}{11} \cdot \frac{4}{10} = \frac{1}{11} \qquad \text{Since without replacement}$$

Let us assume that E_2 be the event that represents both the cards are from box-II

$$P(E_2) = \frac{1}{2} \cdot \frac{2}{5} \cdot \frac{1}{4} = \frac{1}{20} \qquad \text{Since without replacement}$$

$\therefore \qquad P(E) = P(E_1) + P(E_2) = \frac{1}{11} + \frac{1}{20} = \frac{31}{220}$

(ii) Let E be the event that represents both the cards are odd

In this, a box is chosen and a card is drawn, if the first card is odd then the second card is also drawn from the other box and that card is also odd

Let us assume that E_1 be the event that represents 1st card is odd from box-I and the second card is odd from box-II

$$P(E_1) = \frac{1}{2} \times \frac{6}{11} \times \frac{3}{5} = \frac{9}{55} \qquad \text{Since different boxes}$$

Let us assume that E_2 be the event that represents 1st card is odd from box-II and the second card is odd from box-I

$$P(E_2) = \frac{1}{2} \times \frac{3}{5} \times \frac{6}{11} = \frac{9}{55} \qquad \text{Since different boxes}$$

$\therefore P(E) = P(E_1) + P(E_2) = 18/55$

(iii) Probability that if both cards are even then they are from box-I is

Let B_1 be the event of selecting box-I and B_2 be the event of selecting box-II

$$\therefore P(B_1) = \frac{1}{2}; \qquad P(B_2) = \frac{1}{2}$$

Let E be the event that represents an even number then it gives

The probability that both cards are even and from box-I is

$$P(E/B_1) = \frac{1}{2} \times \frac{5}{11} \times \frac{4}{10} = \frac{1}{11} \qquad \text{Since without replacement}$$

The probability that both cards are even and from box-II is

$$P(E/B_2) = \frac{1}{2} \times \frac{2}{5} \times \frac{1}{4} = \frac{1}{20} \quad \text{Since without replacement}$$

By using Baye's theorem on probability,

Probability that if both cards are even then they are from box-I is

$$P(B_1/E) = \frac{P(B_1)P(E/B_1)}{P(B_1)P(E/B_1) + P(B_2)P(E/B_2) + P(B_3)P(E/B_3)}$$

$$= \frac{\dfrac{1}{2} \times \dfrac{1}{11}}{\dfrac{1}{2} \times \dfrac{1}{11} + \dfrac{1}{2} \times \dfrac{1}{20}}$$

$$= \frac{\dfrac{1}{11}}{\dfrac{1}{11} + \dfrac{1}{20}}$$

$$= \frac{20}{31}$$

1.17 Total Probability

Let S be the sample space and let $B_1, B_2, B_3, \ldots\ldots B_n$ are 'n' mutually exclusive and exhaustive events. Let A be any event in the sample space, not necessarily joint with any $B_1, B_2, B_3, \ldots\ldots B_n$ and $P(A) > 0$, then $P(A)$ can be written   in terms of conditional probabilities as

$$P(A) \ = \sum_{i=1}^{n} P(B_i) \ P(A/B_i)$$

is called total probability of the event A.

Proof:

Since $\qquad\qquad S = B_1 \cup B_2 \cup B_3 \cup \ldots \cup B_n$

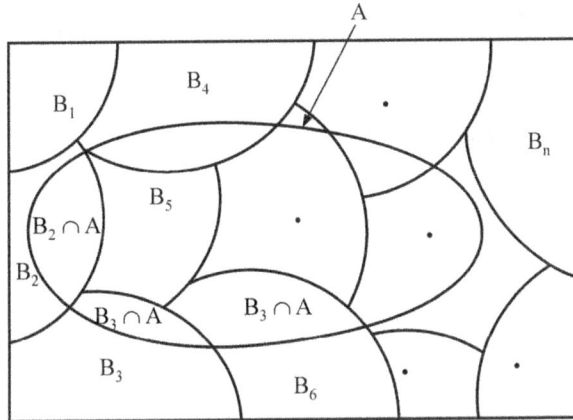

Fig. 1.16

From Fig. 1.16,

$$A = S \cap A$$

$$= A \cap (B_1 \cup B_2 \cup B_3 \cup \ldots \cup B_n)$$

i.e.,

$$P(A) = P[(A \cap B_1) \cup (A \cap B_2) \cup (A \cap B_3) \cup \ldots \cup (A \cap B_n)]$$

$$= P(A \cap B_1) + P(A \cap B_2) + P(A \cap B_3) + \ldots + P(A \cap B_n)$$

Since $(A \cap B_1), (A \cap B_2), (A \cap B_3) \ldots (A \cap B_n)$ *are mutually exclusive events*

$$= P(B_1) P(A/B_1) + P(B_2) P(A/B_2) + P(B_3) P(A/B_3) + \ldots + P(B_n) P(A/B_n)$$

∵ By using compound probability theorem, $(A \cap B) = P(B) P(A/B)$

Hence $$P(A) = \sum_{i=1}^{n} P(B_i) P(A/B_i) \qquad \ldots(1.17.1)$$

***Example* 1:** In a certain school 40% have brown hair, 25% have brown eyes and 15% have both brown hair and brown eyes. A student is selected at random from the school

(i)   If the student has brown hair, what is the probability that she has brown eyes also.

(ii)   If the student has brown eyes, what is the probability that she does not have brown hair.

Sol: Let the events *A* and *B* represents that student having brown hair and brown eyes respectively.

From the given data $P(A) = 0.4$ $P(B) = 0.25$ $P(A \cap B) = 0.15$

(i)   If the student has brown hair, the probability that she has brown eyes also is

$$P(B/A) = \frac{P(A \cap B)}{P(A)}$$

$$= \frac{0.15}{0.4}$$

$$= 0.375$$

(ii)   If the student has brown eyes, the probability that she does not have brown hair is

$$P(A'/B) = \frac{P(A' \cap B)}{P(B)}$$

$$= \frac{0.1}{0.25}$$

$$\because P(A' \cap B) = P(B) - P(A \cap B) = 0.25 - 0.15 = 0.1$$

$$= 0.4$$

***Example* 2:** A card is drawn from a well shuffled pack of cards, if the card shows up red 1 die is thrown and the result is recorded but if the card shows black two dies are thrown and their sum is recorded. What is the probability that the recorded number will be 2?

Sol: Let the events B_1 and B_2 represents red and black cards respectively when a card is drawn from a well shuffled pack. Then

$$P(B_1) = \frac{1}{2} \quad \text{and} \quad P(B_2) = \frac{1}{2}$$

Let A be the event that represents the recorded number will be 2.

If the card shows up red then one die is thrown the probability of getting 2 is

$$P(A/B_1) = \frac{1}{6}$$

If card shows up black then 2 dies are thrown the probability of getting their sum will be 2 is

$$P(A/B_2) = \frac{1}{36}$$

By the total probability theorem

$$P(A) = \sum_{i=1}^{n} P(B_i)\, P\,(A/B_i)$$

$$= P(B_1)P(A/B_1) + P(B_2)P(A/B_2)$$

$$= \frac{1}{2} \times \frac{1}{6} + \frac{1}{2} \times \frac{1}{36}$$

$$\therefore \qquad P(A) = \frac{7}{72}$$

The Probability that the recorded sum will be 2 is $\dfrac{7}{72}$.

***Example* 3:**   If A and B are two events such that $P(A) = \dfrac{1}{3}$ and $P(B) = \dfrac{3}{4}$ and

$P(AUB) = \dfrac{11}{12}$

Find $P(A/B)$ and $P(B/A)$

Sol: Since by addition theorem on probability,

$$P(A \cup B) = P(A) + P(B) - P(A \cap B)$$

$$\therefore P(A \cap B) = P(A) + P(B) - P(A \cup B)$$

$$= \frac{1}{3} + \frac{3}{4} - \frac{11}{12} = \frac{1}{6}$$

Now by conditional probability, $P(E_2/E_1) = \dfrac{P(E_1 \cap E_2)}{P(E_1)}$

Therefore $\qquad P(A/B) = \dfrac{P(A \cap B)}{P(B)} = \dfrac{1/6}{3/4} = \dfrac{2}{9}$

Similarly, $\qquad P(B/A) = \dfrac{P(A \cap B)}{P(A)} = \dfrac{1/6}{1/3} = \dfrac{1}{2}$

***Example* 4:** A bag A contains one white and one black ball and bag B contains 2 white and one black ball. A ball is transferred from the bag A to the bag B. Then a ball is drawn from bag B. Find the probability that it will be white.

Sol: The transferred ball may be white or black

Probability of drawing a white ball from bag A is $= \dfrac{1}{2}$

The box B contain 3 white balls are one red ball

The probability of drawing a white ball from bag B is $= \dfrac{3}{4}$

$\therefore$ The required probability $= \dfrac{1}{2}\dfrac{3}{4} = 3/8$

(a)   Where black ball is transferred

Probability of drawing a black ball from the bag A is $= \frac{1}{2}$

$\therefore$ Then the box B contain 2 white ball from the bag B is $= 2/4 = \frac{1}{2}$

$\therefore$ The required probability $= \dfrac{1}{2}\dfrac{1}{2} = \dfrac{1}{4}$

Since the two events are mutually exclusive

$\therefore$ Required probability $= \dfrac{3}{8} + \dfrac{1}{4} = \dfrac{3+2}{8} = 5/8$

Example 5: A box contains 100 tickets labelled 1, 2, 3, . . . 100. A ticket is drawn at random from the box. Then we have the following.

(i)   The number drawn has one digit (i.e.,) 1,2,3, . . .9, then its probability is $\dfrac{9}{100}$.

(ii) The number drawn has two digits i.e., 10, 11 ---- 99 then its probability is $\dfrac{90}{100}$.

(iii) The number drawn is greater than the number k, i.e., k + 1, - - - - 100 then its

probability is $\dfrac{100 - K}{100}$

Example 6: Two sets of candidates competing for the positions of the board of directors of a company. The probability that the first and second set will win are 0.6 and 0.4 respectively. If the first set wins the probability of introducing a new product is 0.8 and the corresponding probability in the second set wins is 0.3 what is the probability that the new product will be introduced.

Sol: Let B_1 be the event that represents 1st set wins and B_2 be the event that represents 2nd set wins

From the given data,

$$P(B_1) = 0.6 \quad \text{and} \quad P(B_2) = 0.4$$

Let the event A represents introducing a new product, from the given data

$$P\left(\frac{A}{B_1}\right) = 0.8 \quad \text{and} \quad P\left(\frac{A}{B_2}\right) = 0.3$$

Now by using total probability theorem, the probability of introducing a new product is

$$P(A) = \sum_{i=1}^{n} P(Bi)P\left(\frac{A}{B_i}\right)$$

$$= P(B_1)P(A/B_1) + P(B_2)P(A/B_2)$$

$$= 0.6\,(0.8\,) + 0.6\,(0.3)$$

$$= \frac{6}{10} \times \frac{8}{10} + \frac{6}{10} \times \frac{3}{10}$$

$$= \frac{48+18}{100} = 0.6$$

***Example* 7:** An urn A contains 5 white and 3 black balls. Another urn B contains 3 white and 5 black balls. 2 balls are takes from urn A randomly and replace in urn B. now one ball is taken from urn B what is the probability that it is a white ball?

Sol: From the given data, Urn A contains 8 balls out of which 5 are white and 3 are black.

Two balls are drawn from urn A, those two balls may be both white, both black and 1 white and 1 black.

Let B_1 be the event that represent two white balls drawn from urn A

B_2 be the event that represents two black balls drawn from urn A

B_3 be the event that represents one white and 1 black ball drawn from urn A

Now,

$$P\,(B_1) = \frac{5c_2}{8c_2} = \frac{10}{28} \quad P\,(B_2) = \frac{3c_2}{8c_2} = \frac{3}{28} \quad \text{and } P(B_3) = \frac{5c_1 \times 3c_1}{8c_2} = \frac{15}{28}$$

Now

Let A be the event that represents a white ball drawn from urn B.

$$P(A/B_1) = \frac{5c_1}{10c_1} = \frac{5}{10} \quad P(A/B_2) = \frac{3c_1}{10c_1} = \frac{3}{10} \quad \text{and } P(A/B_3) = \frac{4c_1}{10c_1} = \frac{4}{10}$$

By using total probability theorem

$$P(A) = \sum_{i=1}^{n} [P(B_i)] P\left(\frac{A}{B_i}\right)$$

For 3 events

$$P(A) = \sum_{i=1}^{3} [P(B_i)] P\left(\frac{A}{B_i}\right)$$

$$= P(B_1).P\left(\frac{A}{B_1}\right) + P(B_2).P\left(\frac{A}{B_2}\right) + P(B_3).P\left(\frac{A}{B_3}\right)$$

$$= \frac{10}{28} \times \frac{5}{10} \times \frac{3}{28} \times \frac{3}{10} + \frac{15}{28} \times \frac{4}{10}$$

$$= \frac{60+9+50}{280} = \frac{119}{280}$$

$\therefore$ The required probability is $\dfrac{119}{280}$

1.18 Baye's Theorem on Probability

Statement: Let $B_1, B_2, B_3, \ldots\ldots B_n$ are n mutually exclusive and exhaustive events with $P(B_i) \neq 0$. I = 1, 2, 3, . . . n, then for any event A which is a subset of $B_1 \cup B_2 \cup B_3 \cup \ldots\ldots \cup B_n$ such that $P(A) > 0$, then we have

$$P(B_i/A) = \frac{P(B_i)\, P(A/B_i)}{\sum_{i=1}^{n} P(B_i)\, P(A/B_i)}$$

is called Baye's theorem on probability or rule of inverse probability.

Proof:

Since $\qquad P(A \cap B_i) = P(B_i)\, P(A/B_i)$ .....(1.8)

or $\qquad P(A \cap B_i) = P(A)\, (B_i/A)$ .....(1.9)

From (1.18.1) and (1.18.2),

$$P(A)\, (B_i/A) = P(B_i)\, P(A/B_i)$$

$$P(B_i/A) = \frac{P(B_i)\, P(A/B_i)}{P(A)}$$ .....(1.10)

Therefore from total probability, $P(A) = \sum_{i=1}^{n} P(B_i) \, P(A/B_i)$ substitute in eq. (1.10), we get

$$P(B_i/A) = \frac{P(B_i) \, P(A/B_i)}{\sum_{i=1}^{n} P(B_i) \, P(A/B_i)}$$

The occurrence of a desired event A depends on the occurrence of the events $B_1, B_2, B_3, \ldots\ldots B_n$. The probabilities of A/B_i $i = 1, 2, 3, \ldots n$ are called prior probabilities $P(B_i/A)$, $i = 1, 2, 3, \ldots n$ are called posterior probabilities. Hence Baye's theorem also known as inverse probability theorem.

***Example* 1:** Box I contains 5 red balls 3 white balls, box II contains 3 red and 6 white balls. A box is chosen at random and a ball is drawn and put it into other box. A ball is drawn from the second box. Find the probability that both balls are same colour?

Sol: Probability of selecting a box from two boxes = 1/2

Given that box I contains 5 red and 3 white balls; total 8 balls in box I and box II contains 3 red and 6 white balls; total 9 balls in box II

Let B_1 be the event of selecting a red ball and B_2 be the event of selecting a white ball

$P(B_1 / Box I)$ is the probability of selecting red ball from box I = $\dfrac{1}{2} \times \dfrac{5}{8}$

$P(B_1 / Box II)$ is the probability of selecting red ball from box II = $\dfrac{1}{2} \times \dfrac{3}{9}$

Now a red ball is drawn from the second box, after it replaced in the second box is 4/10.

Similarly,

$P(B_2 / Box I)$ is the probability of selecting white ball from box I = $\dfrac{1}{2} \times \dfrac{3}{8}$

$P(B_2 / Box II)$ is the probability of selecting white ball from box II = $\dfrac{1}{2} \times \dfrac{6}{9}$

The required probability that both balls are of same color

$$= \left(\frac{1}{2} \times \frac{5}{8} \times \frac{4}{10} + \frac{1}{2} \times 1. \frac{3}{9} \times \frac{6}{9} \right) + \left(\frac{1}{2} \times \frac{3}{8} \times \frac{7}{10} + \frac{1}{2} \times \frac{6}{9} \times \frac{4}{9} \right)$$

$$= 0.5155$$

***Example* 2:** Companies B_1, B_2 and B_3 produces 30%, 45% and 25% of the cars respectively. It is known that 2%, 3% and 2% of these cars produced from B_1, B_2 and B_3 are defective

(a)   What is the probability that a car purchased is defective

(b)   If a car purchased is found to be defective, what is the probability that this car produced by the company B_1.

Sol: From the given data,

$$\therefore P(B_1) = \frac{30}{100} = 0.3; \; P(B_2) = \frac{45}{100} = 0.45;$$

$$P(B_3) = \frac{25}{100} = 0.25$$

Let A be the event that represents a defective car, then it gives

$$P(A/B_1) = 0.02; \quad P(A/B_2) = 0.03, \; P(A/B_3) = 0.02$$

Now

(a)   The probability that a car purchased is defective (by using Total probability theorem on probability) is

$$\therefore P(A) = \sum_{i=1}^{n} P(B_i) P(A/B_i)$$
$$= P(B_1)P(A/B_1) + P(B_2)P(A/B_2) + P(B_3)P(A/B_3)$$
$$= 0.3 \times 0.02 + 0.45 \times 0.03 + 0.25 \times 0.02 = 0.0245$$

(b)   If a car purchased is found to be defective, what is the probability that this car produced by the company B_1.

$$P(B_1/A) = \frac{P(B_1)P(A/B_1)}{P(B_1)P(A/B_1) + P(B_2)P(A/B_2) + P(B_3)P(A/B_3)}$$

$$= \frac{0.3 \times 0.02}{0.0245} = 0.245$$

***Example* 3:** In a certain college 25% of boys and 10% of girls are studying mathematics. The girls constitute 60% of the students. If a student is selected at random and is found to be studying mathematics. Find the probability that the student is a

(i) girl   and   (ii)   boy

Sol:

Let B_1 be the event that represents a boy student

Let B_2 be the event that represents a girl student

Given that 60% of the students are girls. It gives 40% students are boys

$$\therefore P(B_1) = \frac{40}{100} = 0.4 \; ; \quad P(B_2) = \frac{60}{100} = 0.6$$

Let A be the event that represents a student who studied mathematics given

$$P(A/B_1) = \frac{25}{100} = \frac{1}{4} \; ; \quad P(A/B_2) = \frac{10}{100} = 0.1$$

Now

(i) The probability that a student is selected at random and is found to be studying mathematics (by using Baye's theorem on probability) is a girl is

$$P(B_2/A) = \frac{P(B_2)P(A/B_2)}{P(B_1)P(A/B_1) + P(B_2)P(A/B_2)}$$

$$= \frac{0.6 \times 0.1}{0.4 \times 0.25 + 0.6 \times 0.1}$$

$$= \frac{3}{8}$$

(ii) The probability that a student is selected at random and is found to be studying mathematics is a boy is

$$P(B_1/A) = \frac{P(B_1)P(A/B_1)}{P(B_1)P(A/B_1) + P(B_2)P(A/B_2)}$$

$$= \frac{0.4 \times 0.25}{0.4 \times 0.25 + 0.6 \times 0.1}$$

$$= \frac{5}{8}$$

Aliter:

Let there are 100 students. Out of which 60 are girl students and 40 are boy students.

Mathematics studying boys are 25% of boys $= \dfrac{25}{100} \times 40 = 10$ boys

Mathematics studying girls are 10% of girls $= \dfrac{10}{100} \times 60 = 6$ girls

The required probability for

(i) girl $= \dfrac{6}{16} = \dfrac{3}{8}$ (ii) boy $= \dfrac{10}{16} = \dfrac{5}{8}$

Example 4: In a class 2% of boys and 3% of girls are having blue eyes. There are 30% girls in the class. If a student is selected at random and having blue eyes. What is the probability that the student is a

(i) girl   and   (ii) boy

Sol: Let B_1 be the event that represents a boy student

and B_2 be the event that represents a girl student

Given that 30% of the students are girls. It gives 70% students are boys

$$\therefore P(B_1) = \frac{70}{100} = 0.7 ; \quad P(B_2) = \frac{30}{100} = 0.3$$

Let A be the event that represents a student having blue eyes is given as

$$P(A/B_1) = \frac{2}{100} = 0.02 ; \quad P(A/B_2) = \frac{3}{100} = 0.03$$

Now

(i) The probability that a student is selected at random and is found to have blue eyes (by using Baye's theorem on probability) will be a girl student is

$$P(B_2/A) = \frac{P(B_2)P(A/B_2)}{P(B_1)P(A/B_1) + P(B_2)P(A/B_2)}$$

$$= \frac{0.3 \times 0.03}{0.02 \times 0.7 + 0.3 \times 0.03}$$

$$= 0.3913$$

(ii) The probability that a student is selected at random and is found to be studying mathematics is a boy is

$$P(B_1 / A) = \frac{P(B_1)P(A/B_1)}{P(B_1)P(A/B_1) + P(B_2)P(A/B_2)}$$

$$= \frac{0.02 \times 0.7}{0.02 \times 0.7 + 0.3 \times 0.03}$$

$$= 0.6087$$

Aliter: Let there are 100 students. Out of which 30 are girl students and 70 are boy students.

Blue eye boys are 2% of boys $= \dfrac{2}{100} \times 70 = 1.4$ boys

Blue eye girls are 3% of girls $= \dfrac{3}{100} \times 30 = 0.9$ girls

The required probability for a

(i)   girl having a blue eye $= \dfrac{0.9}{0.9 + 1.4} = \dfrac{0.9}{1.3} = 0.3913$

(ii)   boy having a blue eye $= \dfrac{1.4}{0.9 + 1.4} = \dfrac{1.4}{1.3} = 0.6087$

Example 5: A business man goes to hotels *X*, *Y* and *Z*; 20%, 50% and 30% of the times respectively. It is known that 5%, 4% and 8% of the rooms in *X*, *Y*, *Z* hotels have faulty plumbing. What is the probability that business man's room having faulty plumbing is assigned to hotel *Z*?

Sol: Given that,

The probabilities that the business man goes to hotels *X*, *Y* and *Z* are

$$P(X) = 0.2 \ P(Y) = 0.5 \text{ and } P(Z) = 0.3 \text{ respectively.}$$

Let *A* be the event that represents a faulty plumbing, then given that

$$P(A/X) = 0.05 \ P(A/Y) = 0.04 \text{ and } P(A/Z) = 0.08$$

Now the probability that business man's room having faulty plumbing is assigned to hotel *Z* (by using Bay's theorem on probability) is

$$P(Z/A) = \frac{P(Z)\,P(A/Z)}{P(X)\,P(A/X) + P(Y)\,P(A/Y) + P(Z)\,P(A/Z)}$$

$$= \frac{0.3 \times 0.08}{0.2 \times 0.05 + 0.5 \times 0.04 + 0.3 \times 0.08}$$

$$= \frac{0.024}{0.054}$$

$$= 0.44$$

Example 6: One factory F_1 produces 1000 articles, 20 of them being defective. Second factory F_2 produces 4000 articles, 40 of them being defective and third factory F_3 produces 5000 articles, 50 of them being defective. All these articles are put in one stockpile. One of them is chosen and is found to be defective. What is the probability that it is from factory F_1.

Sol:

Given that,

The probabilities that the three factories produce the defective articles are respectively,

$$P(F_1) = \frac{1000}{10000} = 0.1; \quad P(F_2) = \frac{4000}{10000} = 0.4 \text{ and } P(F_3) = \frac{5000}{10000} = 0.5$$

Let A be the event that represents a defective item, then given that

$$P(A/F_1) = \frac{20}{1000} = 0.02; P(A/F_2) = \frac{40}{4000} = 0.01 \text{ and } P(A/F_3) = \frac{50}{5000} = 0.01$$

respectively.

Now the probability that the defective item will be from factory F_1 (by using Baye's theorem on probability) is

$$P(F_1/A) = \frac{P(F_1)\,P(A/F_1)}{P(F_1)\,P(A/F_1) + P(F_2)\,P(A/F_2) + P(F_3)\,P(A/F_3)}$$

$$= \frac{0.1 \times 0.02}{0.1 \times 0.02 + 0.4 \times 0.01 + 0.5 \times 0.01} = \frac{0.002}{0.011} = 0.1818$$

Example 7: The contents of 3 urns are the following combinations

Urn I	1 w + 3R +2B balls
Urn II	2 w + 1R +1B balls
Urn I	3 w + 3R +3B balls

Two balls are chosen from a randomly selected urn. If the balls are one white and one red ball. What is the probability that they come from urn II.

Sol: From the given data,

Urn I (say B_1) contains   1 w + 3R +2B = 6balls

Urn II   (say B_2) contains   2 w + 1R +1B = 4 balls

Urn I   (say B_3) contains   3 w + 3R +3B = 9 balls

Two balls are chosen from a randomly selected urn

The probabilities that the urns are selected are respectively $\frac{1}{3}$.

i.e., $P(B_1) = \frac{1}{3}$; $P(B_2) = \frac{1}{3}$ and $P(B_3) = \frac{1}{3}$

Let A be the event that represents one white and one red ball, then

$P(A/B_1) = \dfrac{{}^1C_1 \times {}^3C_1}{{}^6C_2} = \dfrac{1}{5}$; $P(A/B_2) = \dfrac{{}^2C_1 \times {}^1C_1}{{}^4C_2} = \dfrac{1}{3}$ and $P(A/B_3) = \dfrac{{}^3C_1 \times {}^3C_1}{{}^9C_2} = \dfrac{1}{4}$

respectively.

Now the probability that the they came from urn II (i.e., from Urn B_2) is

$$P(B_2/A) = \frac{P(B_2)\, P(A/B_2)}{P(B_1)\, P(A/B_1) + P(B_2)\, P(A/B_2) + P(B_3)\, P(A/B_3)}$$

$$= \frac{\frac{1}{3} \times \frac{1}{3}}{\frac{1}{3} \times \frac{1}{5} + \frac{1}{3} \times \frac{1}{3} + \frac{1}{3} \times \frac{1}{4}}$$

$$= \frac{\frac{1}{3}}{\frac{1}{5} + \frac{1}{3} + \frac{1}{4}} = \frac{\frac{1}{3}}{\frac{47}{60}}$$

$$= \frac{20}{47} = 0.4255$$

Therefore the probability that they come from urn II is 0.4255

Quiz Questions

1. In a multiple choice question there are four alternative answers of which one or more than one is correct. A candidate will get marks on the questions only if he takes all the correct answers. The candidate decides to tick answers at random. If he is allowed up to three chances to answer the question, the probability that he will get marks on it is given by

   (a) $1 - \frac{3}{5}$ (b) $\frac{3}{5}$ (c) $\frac{1}{5}$ (d) None

2. The probability that a man will live 10 more years is 1/4 and the probability that his wife will live 10 more years is 1/3. Then the probability that neither will be alive in 10 years is

    (a) $\dfrac{1}{2}$ (b) $1-\dfrac{3}{5}$ (c) $\dfrac{1}{5}$ (d) None

3. An unbiased die with faces 1, 2, 3, 4, 5, and 6 is round 4 times. Out of four face values obtained the probability that the minimum face value is not less than 2 and the maximum face value is not greater than 5 is

    (a) $1-\dfrac{16}{81}$ (b) $\dfrac{16}{81}$ (c) $\dfrac{80}{81}$ (d) None

4. If $(1-3p)/2$; $(1+4p)/3$; $(1+p)/6$ are the probabilities of three mutually exclusive and exhaustive events, then the set of all values of p is

    (a) $(0, 1)$ (b) $(0,\infty)$ (c) $\left(-\dfrac{1}{4},\dfrac{1}{3}\right)$ (d) $\left(-\dfrac{1}{3},\dfrac{1}{4}\right)$

5. A pack of cards contains 4 aces, 4 kings, 4 queens and 4 jacks. Two cards are drawn at random. The probability that at least one of them is an ace is

    (a) $\dfrac{9}{20}$ (b) $1-\dfrac{3}{5}$ (c) $\dfrac{1}{5}$ (d) None

6. India plays two matches each with England and Australia. In any match the probabilities of India getting points 0, 1, 2 are 0.45, 0.05 and 0.50 respectively. Assuming that outcomes are independent, the probability of India getting at least 7 points is

    (a) 0.875   (b) 0.0875   (c) 0.0625   (d) None

7. A bag contain 10 mangoes out of which 4 are rotten, two mangoes are taken out together. If one of them is found to be good, the probability that other is also good is

    (a) $\dfrac{1}{3}$ (b) $1-\dfrac{3}{5}$ (c) $\dfrac{5}{18}$ (d) $\dfrac{4}{9}$

8. A box contains 5 brown and 4 white socks. A man pulls out two socks. The probability that they are of the same colour is

    (a) $\dfrac{1}{3}$ (b) $\dfrac{3}{5}$ (c) $\dfrac{5}{18}$ (d) $\dfrac{4}{9}$

9. You are given a box with 20 cards in it. 10 of these cards have the letter I printed on them. The other ten have the letter 7 printed on them. If you pick up 3 cards

at random and keep them in the same order, the probability of making the word IIT is

(a) $\dfrac{1}{3}$ (b) $\dfrac{4}{27}$ (c) $\dfrac{5}{38}$ (d) $\dfrac{4}{9}$

10. In order to get at least once a head with probability greater than or equal to 0.9, the number of times a coin needs to be tossed is

(a) 4 (b) 3
(c) 5 (d) None of these

11. A man alternately tosses a coin and throws a dice beginning with the coin. The probability that he gets a head in the coin before he gets a 5 or 6 in the dice is

(a) $\dfrac{3}{4}$ (b) $\dfrac{4}{27}$ (c) $\dfrac{5}{38}$ (d) $\dfrac{4}{9}$

12. The probability that a person will hit a target in a shooting practice is 0.3. If he shoots 10 times, the probability that he hit the target is

(a) 1 (b) $1-(0.7)^{10}$ (c) $(0.7)^{10}$ (d) $(0.3)^{10}$

13. The probability that at least one of the events A and B occurs is 0.7 and they occur simultaneously with probability 0.2. Then $P(A)+P(B)=$

(a) 0.8 (b) 1.4
(c) 0.6 (d) None of these

14. Out of 13 applicants for a job, there are 5 women and 8 men. It is desired to select 2 persons for the job. The probability that at least one of the selected persons will be a woman is

(a) $\dfrac{3}{4}$ (b) $\dfrac{4}{27}$ (c) $\dfrac{25}{39}$ (d) $\dfrac{4}{9}$

15. A father has 3 children with at least one boy. The probability that he has 2 boys and one girl is

(a) $\dfrac{3}{4}$ (b) $\dfrac{4}{27}$ (c) $\dfrac{5}{38}$ (d) $\dfrac{1}{3}$

16. The probability of a problem being solved by two students are 1/2, 1/3. The probability of the problem being solved is

(a) $\dfrac{4}{27}$ (b) $\dfrac{3}{4}$ (c) $\dfrac{1}{3}$ (d) $\dfrac{2}{3}$

17. One die and one coin tossed simultaneously. The probability of getting 6 on die and head on coin is

(a) $\dfrac{1}{12}$ (b) $\dfrac{3}{4}$ (c) $\dfrac{1}{3}$ (d) $\dfrac{1}{2}$

18. In throwing of two dice, the probability of getting a multiple of 4 is

(a) $\dfrac{1}{14}$ (b) $\dfrac{1}{4}$ (c) $\dfrac{1}{3}$ (d) $\dfrac{1}{2}$

19. From 4 children, 2 women and 4 men, 4 are selected. The probability that there are exactly 2 children among the selected is

(a) $\dfrac{11}{12}$ (b) $\dfrac{13}{14}$ (c) $\dfrac{10}{21}$ (d) $\dfrac{1}{21}$

20. If A and B are such events that $P(A>0$ and $P(B) \ne 1$, then $P(\overline{A}/\overline{B})$ is equal to

(a) $1-P(A/B)$ (b) $1-P(\overline{A}/B)$ (c) $\dfrac{1-P(A\cup B)}{P(\overline{B})}$ (d) $\dfrac{P(\overline{A})}{P(\overline{B})}$

21. Two dice are thrown simultaneously. The probability of obtaining a total score of 5 is

(a) $\dfrac{1}{18}$ (b) $\dfrac{1}{12}$ (c) $\dfrac{1}{2}$ (d) $\dfrac{1}{9}$

22. Three mangoes and three apples are in a box. If two fruits are chosen at random, the probability that one is a mango and the other is an apple is

(a) $\dfrac{3}{5}$ (b) $\dfrac{2}{5}$ (c) $\dfrac{1}{2}$ (d) $\dfrac{1}{9}$

23. A card is drawn at random from a pack of 100 cards numbered 1 to 100. The probability of drawing a number which is a square is

(a) $\dfrac{1}{18}$ (b) $\dfrac{1}{10}$ (c) $\dfrac{1}{2}$ (d) $\dfrac{1}{9}$

24. Seven chits are numbered 1 to 7. Four are drawn one by one with replacements. The probability that the least number on any selected chit is 5, is

(a) $(2/7)^4$ (b) $1-(2/7)^4$ (c) $(2/7)^{10}$ (d) $(3/7)^3$

25. A box contains 100 tickets numbered 1, 2, . . . 100. Two tickets are chosen at random. It is given that the maximum number on the two chosen tickets is not more than 10. The minimum number on them is 5 with probability

(a) $(13/15)^4$ (b) $(13/15)$

(c) $(3/7)^{10}$ (d) none of these

26. The probability that at least one of the events A and B occurs is 0.6. If A and B occur simultaneously with probability 0.2, then $P(\overline{A})+P(\overline{B}) =$

(a) 0.8 (b) 1.2

(c) 0.6 (d) none of these

27.   The probability of any one sample of size n being drawn out of N units is

   (a) $\dfrac{1}{Nc_n}$ (b) Nc_n (c) N^n (d) None

28.   The probability of including a specified unit/item in a sample of size n selected out of N units

   (a) $\dfrac{1}{N}$ (b) N^n (c) N_{cn} (d)   None

29.   In how many different ways can a student answer 5 questions which are having 4 options each is

   (a) 625   (b) 1024   (c) 20   (d) 400

30.   A student is to answer 5 out of 8 questions in an examination. If two out of first three, in how many ways can he choose?

   (a) 35   (b) 56   (c) 30   (d) 45

Answers

1.	(c)	2.	(a)	3.	(b)	4.	(c)
5.	(a)	6.	(b)	7.	(c)	8.	(d)
9.	(c)	10.	(a)	11.	(a)	12.	(b)
13.	(d)	14.	(c)	15.	(d)	16.	(d)
17.	(a)	18.	(b)	19.	(c)	20.	(c)
21.	(d)	22.	(a)	23.	(b)	24.	(d)
25.	(b)	26.	(b)	27.	(a)	28.	(a)
29.	(b)	30.	(c)				

Exercise Questions

1. (a)   In a single throw of three dice, find the probability of getting the same number on the three dice

   (b)   In a single throw of three dice, find the probability of getting a sum of (i) 5 and (ii) at least 5

2. (a)   In terms of probabilities $p_1 = P(A)$, $p_2 = P(B)$ and $p_3 = P(A \cap B)$, express (i) $P(A \cup B)$ (ii) $P(A/B)$ (iii) $P(\overline{A} \cap \overline{B})$ under the condition that (i) A and B are mutually exclusive (ii) A and B are mutually independent.

   (b)   Let A and B be the possible outcomes of an experiment and suppose

$$P(A) = 0.4, \quad P(A \cup B) = 0.7 \text{ and } P(B) = p$$

(i)   For what choice of p are A and B mutually exclusive

(ii)   For what choice of p are A and B independent

3.   An integer is chosen at random from two hundred digits. What is the probability that the integer is divisible by 6 or 8

4.   A man is dealt 4 spade cards from an ordinary pack of 52 cards. If he is given three more cards, find the probability p that at least one of the additional cards is also a spade.

5.   An urn contains 6 white, 4 red and 9 black balls. If 3 balls are drawn at random, find the probability that (i) two of the balls, drawn are white (ii) one is of each colour, (iii) none is red (iv) at least one is white.

6.   Assuming that any arrangement of one or more letters forms a word, how many words can be formed from (i) *RAT* and (ii) *TRAVEL* ?

7.   The probability that Jim can hit the bull on a target is 1/3. Is he certain to score a bull if he takes 3 shots? Find the probability of his scoring at least one bull in 3 shots.

8. A and B alternately cut a pack of cards and the pack is shuffled after each cut. If A start and the game is continued until one cuts diamond, what are the respective chances of A and B first cutting a diamond?

9.   A bag contains 30 balls numbered from 1 to 30. One ball is drawn at random. Find the probability that the number on the drawn ball will be a multiple of (i) 5 or 9 (ii) 5 or 6

10.   Ten unbiased coins are thrown simultaneously. Find the probability of getting at least seven heads

11.   A bag contains 6 red, 4 white and 5 black balls. A person drawn 4 balls from the bag at random. Find the probability that among the balls drawn there is at least one ball of each color.

12.   A bag contains 7 white and 3 black balls. One ball is drawn from the bag and it is replaced after noting its colour. In the second draw again one ball is drawn and its colour is noted. What is the probability that both the balls drawn are of different colours?

13.   A box contains 9 red and 5 blue flowers. One flower is drawn from the box and it is replaced after noting its colour. In the second draw again one flower is drawn and colour noted. What is the probability that both the flowers drawn are of the same colour?

14.   Five men in a group of 20 graduates. If 3 men are picked out of 20 at random what is the probability that all are graduates and what is the probability of at least one being graduate?

15.   The probability is 0.6 that a patient selected at random from a village clinic will be a male. The probability is that the patient will be male who is in for surgery is 0.2. A patient randomly selected from current resident is found to be a male, what is the probability that the patient is in the village clinic for surgery?

16.   Three electric lamps are filled in a room. Three bulbs are chosen at random from 10 bulbs having 6 good bulbs. What is 5the chance that the room is lighted?

Ans: 29/30

17.   In a group there are 3 men and 2 women. 3persons are selected at random from his group. Find the probability that one men and 2 women (or) two men & one women are selected?

Ans: 9/10

18.   Suppose the events A_1, A_2, A_3-------A_n are independent and that $P(Ai) = \dfrac{1}{i+1}$ for 1 $\leq i \leq n$. find the problem that none of events occurs.

Ans: $\dfrac{1}{n+1}\left(1-\dfrac{1}{n+1}\right)=\left(1-\dfrac{1}{n+1}\right)=\dfrac{1}{n+1}$

19.   *A* and *B* are two very weak students of statistics and their chances of solving a problem correctly are 1/8 & 1/12 respectively. If the problem of their making a common mistake is 1/ 1001. And obtain the serve answer find the chances that their answer is correct.

Ans: $\dfrac{\dfrac{1}{8}\times\dfrac{1}{12}}{\dfrac{1}{6}\times\dfrac{1}{12}+\left(1-\dfrac{1}{8}\right)\left(1-\dfrac{1}{12}\right)\times\dfrac{1}{1001}}=\dfrac{13}{14}$

20. The range of a random variable $X = \{1, 2, 3, ----\}$ and the probability are given by
$P(X = k) = \dfrac{3^{ck}}{k!}$ $(k = 1, 2, 3 ------)$ and c is a constant then $c = ?$

Ans: $\log_3(\log 2)$

21. Out of 15 items 4 are not in good condition 4 are selected at random.   Find the probability that (i) All are not good and (ii) Two are not good

22. The chance that doctor A will diagnoses a disease x correctly 60%.   The chance that the patient will die by his treatment after diagnosis is 40% and the chance of death by wrong diagnosis is 70%. A patient of doctor A, who add disease X, died. What is the chance that his disease was diagnosed correctly?

Ans: 6/13

23. N letters are to be placed in n addresses envelopes. If the letters are placed in the envelopes at random. What is the probability that at least two of the letters are not placed in the right envelopes is

Ans: $\dfrac{n!-1}{n!}$

24. If a test consists of 12 true – false question in how many different ways can a student mark the test paper with one answer to each question

Ans: 2^{12}.

Review Quiz Questions

1. A coin is tossed three times in succession, the number of sample points in sample space is

(a)   6   (b)   8   (c)   3   (d)   9

2. In a simultaneous tossing of two perfect dice the probability of obtaining 4 as the sum of the resultant faces is

(a) $\dfrac{4}{12}$ (b) $\dfrac{1}{12}$ (c) $\dfrac{3}{12}$ (d) $\dfrac{2}{12}$

3. An urn contains 9 balls, two of which are red, three blue and four black.   Three balls are drawn at random. The chance that they are of the same colour is

(a) $\dfrac{5}{84}$ (b) $\dfrac{3}{9}$ (c) $\dfrac{3}{7}$ (d) $\dfrac{7}{17}$

4. If $P(A \cap B) = \dfrac{1}{2}$, $P(\bar{A} \cap \bar{B}) = \dfrac{1}{2}$ and $2 \cdot P(A) = P(B) = p$, then the value of p is

(a) $\dfrac{1}{4}$ (b) $\dfrac{1}{2}$ (c) $\dfrac{1}{3}$ (d) $\dfrac{2}{3}$

5. If A and B are two independent events such that $P(\bar{A}) = 0.7$, $P(\bar{B}) = k$ and $P(A \cup B) = 0.8$ then k is

(a) $\dfrac{5}{7}$ (b) 1   (c) $\dfrac{2}{7}$ (d) none

6. A number is chosen at random among the first 1210 natural numbers. The probability of the number chosen being a multiple of 5 or 15 is

(a) $\dfrac{1}{5}$ (b) $\dfrac{1}{8}$ (c) $\dfrac{1}{16}$ (d) $\dfrac{1}{9}$

7. If A and B are two events, the probability that exactly one of them occurs is given by

(a) $P(A) + P(B) - 2P(A \cap B)$ (b) $P(A) + P(B) - P(A \cap B)$

(c) $P(\bar{A}) + P(\bar{B}) - 2P(\bar{A} \cap \bar{B})$ (d) $P(A \cap \bar{B}) + P(\bar{A} \cap B)$

8. The probability of drawing one white b all from a bag containing 6 red, 8 black, 10 yellow and 1 green balls is

(a) $\dfrac{1}{25}$ (b) $\dfrac{24}{25}$ (c) 1   (d) none

9. If the events S and T have equal probability and are independent with $P(S \cap T) = p \succ 0$ then $P(S)$ is

(a) $\sqrt{p}$ (b) p^2 (c) $\dfrac{p}{2}$ (d) none

10. The probability of drawing any one spade card from a pack of cards is

(a) $\dfrac{1}{52}$ (b) $\dfrac{1}{13}$ (c) $\dfrac{4}{13}$ (d) $\dfrac{1}{4}$

11. If repetitions are not allowed the number of two digit numbers, can be formed from the numbers 2,3,5,6 are ……..

(a) 10   (b) 9   (c) 12   (d) 15

12. There are 5 cards numbered 1 to 5. Two cards are drawn together. Then the probability of getting a sum is odd is ……..

(a) 2/25   (b) 4/25   (c) 6/25   (d) 8/25

13. Two persons are playing a game. Player A toss a coin. If head turns up he gets one Rupee if tail turns up he loses 1 rupee his expected gain is ....

     (a)   0   (b)   1   (c)   2   (d)   3

Review Exercise Questions

1. Find the probability of getting one red king if we select a card from a pack of 52 cards.

2. The probability of passing in subjects A, B, C, D are 3/4, 2/3, 4/5 and ½ respectively. To qualify in the examination, a student should pass in A and two subjects among the threes. What is the probability of qualifying in that examination?

3. In a class 2% of boys and 3% girls are having blue eyes. There are 30% girls in the c lass. If a student is selected and having blue eyes, what is the probability that the student is a girl?

4. The students in a class are selected at random one after the other for an examination. Find the probability that the boys and girls are alternate if there are.

     (a) 5 boys and 4 girls   (b) 4 boys and 4 girls

5. If a test consists of 12 true – false questions, in how many different ways can a student mark the test paper with one answer to each question?

6. A problem in Mechanics is given to three students A, B and C whose chances of solving it are 1/2, 1/3, 1/4 respectively. What is the probability that the problem will be solved?

7. A die is thrown 6 times if getting an even number is a success. Find the probabilities of (i) at least one success (ii) ≤ 3 success (iii) 4 of successes

8. In a family of four children find the probability that (i) all are boys (ii) Only one boy (iii) 2 boys

9. Two dice are thrown five times. Find the probability of getting a sum 7 (i) at least once   (ii) Two times   (iii) $P(1 < x < 5)$.

10. 10% of the items produced by a machine are defective. Find the probability that among 5 items there are (i). no defective items (ii). One defective item (iii). At least two defective items

Additional Examples

1. (a)   Define probability as a relative frequency and state the axioms of probability

   (b)   State and prove the Baye's theorem

   (c)   A letter is known as to have come from either *TATANAGAR* or *CALCUTTA*. On the envelope, just two consecutive letters TA are visible.   Find the probability that the letter has come from *CALCUTTA*.

Sol:

   (a)   Let a trail be repeated a large number of times under essentially identical conditions and let E be the event of it. The ratio of the number of times (m) the event E happens   to the number of trails (n),   i.e., $\dfrac{m}{n}$ is called the relative frequency of the event E and is denoted by $R(E)$ and the probability of the event E is defined as $P(E) = \underset{n \to \infty}{Lt} \dfrac{m}{n}$

Axioms of Probability

1. Let S be the sample space and let E be any event in S (i.e., $E \subseteq S$) we define a function P on S (for any event E the functional value $P(E)$) is a real number such that

   (i) $P(E) \geq 0$

   (ii) $P(S) = 1$ and

   (iii)   If $E_1, E_2, E_3, E_4, \ldots \ldots$, be any sequence of mutually exclusive events in the sample space S then $P(E_1 \cup E_2 \cup E_3 \cup E_4 \cup \ldots \ldots) = P(E_1) + P(E_2) + P(E_3) + \ldots \ldots$ is called a probability function on the sample space.

   (b)   See Baye's theorem

   (c)   In *TATANAGAR*, there are two *TA*'s and in *CULCUTTA* there are one *TA*. Therefore total number of *TA*'s is 3. These are the total number of possible cases.

   Hence the probability that the letter has come from *CALCUTTA*

   $$= \frac{Number\ of\ favourable\ cases}{Total\ number\ of\ possible\ cases} = \frac{1}{3}$$

2. (a)   Explain the following

   (i) Probability   (ii) Conditional Probability
   (iii) Independent Events and   (iv) Discrete Sample Space

(b)   State and prove the Bayes theorem

Sol:

(a)   See terminology

(b)   See Baye's theorem

3. (a)   Define the following

(i) Joint probability (ii) Conditional Probability (iii) Total probability and (iv) Independent Events

(b)   A missile can be accidently launched in two relays A and B both have failed. The probabilities of A and B failing are known to be 0.01 and 0.03 respectively. It is also known that B is more likely to fail (probability 0.06) if A has failed

(i)   What is the probability of an accidental missile launch?

(ii)   What is the probability that A will fail if B has failed?

(iii)   Are the Events A fails and B fails statistically independent?

Sol:

(a)   See terminology

(b)   From the given data,

$P(A) = 0.01$ and $P(B) = 0.03$ also given that $P(B/A) = 0.06$

Let D be the event that missile launch

(i)   The probability of an accidental missile will launch is

$$P(D) = P(A)P(B/A) + P(B)P(A/B) \dots\dots$$

Since $P(B/A) = \dfrac{P(A \cap B)}{P(A)} = 0.06$

$\Rightarrow \quad \dfrac{P(A \cap B)}{0.01} = 0.06 \Rightarrow P(A \cap B) = 0.06 \times 0.01 = 0.0006$ then

$$P(A/B) = \dfrac{P(A \cap B)}{P(B)} = \dfrac{0.0006}{0.03} = 0.02$$

$\therefore \; P(A/B) = 0.02$

$\therefore$ From equation (1)

$$P(D) = 0.01 \times 0.06 + 0.03 \times 0.02$$

$$= 0.0006 + 0.0006$$

$$= 0.0012$$

(ii) The probability that A will fail if B has failed is

i.e., $P(A/B) = 0.02$ from equation (2)

∴ The probability that A will fail if B has failed is $P(A/B) = 0.02$

(iii) The Events A fails and B fails are not statistically independent since $P(A) = 0.01$
But $P(A/B) = 0.02 \neq P(A)$

4. (a) In a bolt factory, machines A, B, C manufacture 30%, 30% and 40% of total output respectively. If the total of their outputs 4, 5 and 3 percents are defective bolts. A bolt is drawn at random from the product and is found to be defective. What are the probabilities that it was manufactured by machines A, B and C.

(b) Distinguish between Joint probability, conditional probability and total probability.

Sol:

From the given data,

$$\therefore P(A) = \frac{30}{100} = 0.3; \quad P(B) = \frac{30}{100} = 0.3 \text{ and } \quad P(C) = \frac{40}{100} = 0.4$$

Let 'D' be the event that represents a defective bolt, then

$$P(D/A) = \frac{4}{100} = 0.04; \quad P(D/B) = \frac{5}{100} = 0.05 \text{ and }$$

$$P(D/C) = \frac{3}{100} = 0.03$$

A bolt is drawn at random from the product and is found to be defective then the probabilities that it was manufactured by machines A is

By using Baye's theorem on probability

$$P(A/D) = \frac{P(A)P(D/A)}{P(A)P(D/A) + P(B)P(D/B) + P(C)P(D/C)}$$

$$= \frac{0.3 \times 0.04}{0.3 \times 0.04 + 0.3 \times 0.005 + 0.4 \times 0.03}$$

$$= \frac{0.012}{0.039} = 0.31$$

A bolt is drawn at random from the product and is found to be defective then the probabilities that it was manufactured by machines B is

By using Baye's theorem on probability

$$P(B/D) = \frac{P(B)P(D/B)}{P(A)P(D/A) + P(B)P(D/B) + P(C)P(D/C)}$$

$$= \frac{0.3 \times 0.05}{0.3 \times 0.04 + 0.3 \times 0.005 + 0.4 \times 0.03}$$

$$= \frac{0.015}{0.039}$$

$$= 0.38$$

A bolt is drawn at random from the product and is found to be defective then the probabilities that it was manufactured by machines C is

By using Baye's theorem on probability

$$P(C/D) = \frac{P(C)P(D/C)}{P(A)P(D/A) + P(B)P(D/B) + P(C)P(D/C)}$$

$$= \frac{0.4 \times 0.03}{0.3 \times 0.04 + 0.3 \times 0.005 + 0.4 \times 0.03}$$

$$= \frac{0.012}{0.039}$$

$$= 0.31$$

5. (a) What is Baye's theorem? Explain.

   (b) Determine probabilities of system error and correct system transmission of symbols for an elementary binary communication system shown in figure below, consisting of a transmitter that sends one of two possible symbols (a1 or a 0) over a channel to a receiver. The channel occasionally causes errors to occur so that a '1' show up at the receiver as a '0' and vice versa. Assume the symbols '1' and '0' are selected for a transmission as 0.6 and 0.4 respectively.

      In figure $P(B_1) = 0.6$ and $P(B_2) = 0.4$

Sol:

   (a) See Baye's theorem

   (b) From the figure,

$$P(B_1) = 0.6 \text{ and } P(B_2) = 0.4$$

$$P(A_1 / B_1) = P(A_1)$$

$$P(A_2 / B_1) = P(A_2)$$

$$P(A_1 / B_2) = P(A_1) \hspace{3cm} \text{Incomplete check}$$

$$P(A_1 / B_2) = P(A_2)$$

From there four equations A_1, A_2, B_1, B_2 are independent events

6. (a)   If A is an arbitrary event, then show that $P(\overline{A}) = 1 - P(A)$

(b)   An experiment consists of rolling a single die, two events are defined as $A = \{a$ 6 show a up\}, $B = \{a$ 2 or a 5 show up\}

(i) find $P(A)$ and $P(B)$ and (ii) Define third event C, so that $P(C) = 1 - P(A) - P(B)$

Sol:

(a)   See Probability theorem

(b)   From the given data

$P(A) = \dfrac{1}{2}$ since die having 6 number, in which 6 appears one time   and

$P(B) = \dfrac{2}{6} = \dfrac{1}{3}$ since die having 6 number, in which 2 or 5 appears one time each.   Here the number of favourable cases is 2.

(a)   Given that an experiment consists of rolling a single die, two events are defined as $A = \{a$ 6 show a up\}, $B = \{$ a 2 or a 5 show up\}

Also $P(A) = \dfrac{1}{2}$ and $P(B) = \dfrac{1}{3}$ then

$$P(C) = 1 - P(A) - P(B)$$

$$= 1 - \frac{1}{2} - \frac{1}{3}$$

$$= \frac{1}{6}$$

So we can define C in terms of the remaining events 1, 3, 4 or in any of the six events

Hence $C = \{a$ 1 show a up\} or $C = \{a$ 3 show a up\} or $C = \{a$ 4 show a up\}

Or

$C = \{a\ 1\ show\ a\ up\}$ or $C = \{a\ 2\ show\ a\ up\}$ or $C = \{a\ 3\ show\ a\ up\}$ or

$C = \{a\ 4\ show\ a\ up\}$ or $C = \{a\ 5\ show\ a\ up\}$ or $C = \{a\ 6\ show\ a\ up\}$

7. (a)   Define and explain random experiment with an example

(b)   Four cards are drawn form a well shuffled pack of playing cards. Find the probability that, (i) All are clubs (ii) Two spades and two hearts (iii) Four cards from a different suit

Sol:

(a)   See terminology

(b)   Four cards are drawn from 52 cards in $^{52}C_4$ ways

(i)   The probability that all are clubs:

In a pack there are 13 clubs; four cards are drawn from 13cards in $^{13}C_4$ ways, these are the number of favourable cases for getting four clubs

The probability that all are clubs $= \dfrac{^{13}C_4}{^{52}C_4}$

(ii)   The probability that two spades and two hearts:

In a pack there are 13 spades and 13 hearts. In this case the number of favourable cases $= {}^{13}C_2 \times {}^{13}C_2$

The probability that two spades and two hearts $= \dfrac{^{13}C_2 \times {}^{13}C_2}{^{52}C_4}$

(iii)   The probability that four cards from a different suit:

In a pack there are 4 suits. We select one card from each suit in $^{4}C_1 \times {}^{4}C_1 \times {}^{4}C_1 \times {}^{4}C_1$ ways.

The probability that four cards from a different suit $= \dfrac{^{4}C_1 \times {}^{4}C_1 \times {}^{4}C_1 \times {}^{4}C_1}{^{52}C_4}$

8. (a)   Define and explain following with example

(i) Random Experiment   (ii) Outcome   (iii) Trial and Event

(b)   A die is tossed. Find the probabilities of the event $A = \{odd\ number\ shows\ up\}$, $B = \{number\ larger\ than\ 3\ shows\ up\}$, $A \cup B$ and $A \cap B$

Sol:

(a)   See terminology

(b)   From the given data

$P(A) = \dfrac{3}{6} = \dfrac{1}{2}$ since die having 6 number, in which there are three odd number, those are 1, 3, 5. Here the number of favourable cases is 3.

and $P(B) = \dfrac{3}{6} = \dfrac{1}{2}$ since die having 6 number, in which 4, 5 and 6 are larger than the number 3.   Here the number of favourable cases is 3.

The events in $A \cup B = \{1, 3, 4, 5, 6\}$

$$\therefore P(A \cup B) = \dfrac{5}{6}$$

The events in $A \cap B = \{5\}$

$$\therefore P(A \cap B) = \dfrac{1}{6}$$

9. (a)   Discuss joint and conditional probability

   (b)   When are two events said to be mutually exclusive? Explain with an example

   (c)   Determine the probability of the card being either red or a king when one card is drawn from a regular deck of 52 cards.

Sol:

   (a)   See terminology

   (b)   See terminology

   (c)   The probability of drawing a card from a deck of 52 cards in $^{52}C_1 = 52$ways these are the total number of possible cases.

   Also there are 26 red cards and 4 kings in which there are two red king cards. Therefore $26 + 2 = 28$ favourable cases.

   Hence the probability of the card being either red or a king $= \dfrac{28}{52} = \dfrac{7}{13}$

10. (a)   Is probability relative frequency of occurrence of some event?

   Explain with an example

   (b)   Determine the probability of the card being either red or a king when one card is drawn from a regular deck of 52 cards.

Sol:

   (a)   See terminology

(b)   The probability of drawing a card from a deck of 52 cards in $^{52}C_1 = 52$ways these are the total number of possible cases.

Also there are 26 red cards and 4 kings in which there are two red king cards. Therefore 26+2 = 28 favourable cases.

Hence the probability of the card being either red or a king = $\dfrac{28}{52} = \dfrac{7}{13}$

11. (a)   What is sample space?   Explain the discrete sample space and continuous sample space with suitable example each.

(b)   In a game of dice a "shooter" can win outright if the sum of the two numbers showing up is either 7 or 11 when two dice are thrown. What is his probability of winning outright?

Sol:

(a)   (See terminology

When two dice are thrown then the sample space S is

$$S = \begin{cases} (1,1),(1,2),(1,3),(1,4),(1,5),(1,6); & (2,1),(2,2),(2,3),(2,4),(2,5),(2,6) \\ (3,1),(3,2),(3,3),(3,4),(3,5),(3,6); & (4,1),(4,2),(4,3),(4,4),(4,5),(4,6) \\ (5,1),(5,2),(5,3),(5,4),(5,5),(5,6); & (6,1),(6,2),(6,3),(6,4),(6,5),(6,6)\} \end{cases}$$

There are 36 in number.   These are the total number of possible cases.

The sum of the two dice for 7 are $(1,6),(2,5),(3,4),(4,3),(5,2),(6,1)$ these are 6 in number and for 11 are   *(5,6), (6,5)* these are 2 in number.

Therefore the total number of favourable cases for the shooter is 6+2 =8

$\therefore$ Probability of winning in the game = $\dfrac{8}{36} = \dfrac{2}{9}$

12. (a)   Explain the terms joint probability and conditional probability.

(b)   Show that conditional probability satisfies the three axioms of probability.

(c)   Two cards are drawn from a 52 cards deck (the first is not replace), (i) Given the   first card is a queen. What is the probability that the second is also a queen? (ii) Repeat part (i) for the first card a queen and second card a 7 (iii) What is the probability that both cards will be the queen?

Sol:

(a)   See terminology

(b) Let S be a sample space associated with the coin tossing experiment. And the event E_1 represents getting head and the event E_2 represents getting tail are two events in S. and $P(E_1) = \dfrac{1}{2} \neq 0$, then the probability of occurrence of the event E_2 after the occurrence of E_1 is called conditional probability of the event E_2 given E_1. It is denoted by $P(E_2 / E_1)$. And is defined as

$$P(E_2 / E_1) \; = \; \frac{P(E_1 \cap E_2)}{P(E_1)}$$

similarly we can define $P(E_1 / E_2) \; = \; \dfrac{P(E_1 \cap E_2)}{P(E_2)}$ and $P(E_2) = \dfrac{1}{2} \neq 0$

(i) $P(E_1 / E_2) \; = \; \dfrac{P(E_1 \cap E_2)}{P(E_2)} \geq 0$

(ii) E_1, E_2 are mutually exclusive events then $P(E_1 \cup E_2) = P(E_1) + P(E_2)$

(iii) $P(S) = P(E_1 \cup E_2) = P(E_1) + P(E_2) \; = \; \dfrac{1}{2} + \dfrac{1}{2} = 1$

13. (a) Distinguish between mutually exclusive events and independent events.

    (b) A letter is known to have come either from *LONDON* or *CLIFTON*. On the postmark only the two consecutive letters *ON* are legible. What is the chance that it came from London? Give step-by-step answer.

Sol:

(a) See terminology

In *LONDON* there are two *ON*'s and in *CLIFTON* there are one *ON*. Therefore total number of *ON*'s is 3.

These are the total number of possible cases.

Hence the probability that the letter has come from *LONDON*

$$= \frac{Number \; of \; favourable \; cases}{Total \; number \, of \; possible \; cases}$$

$$= \frac{1}{3}$$

14. (a) State and prove Bayes theorem of probability

(b) In a single throw of two dice, what is the probability of obtaining a sum of at least 10?

Sol:

(a) See Bayes theorem on probability

(b) When two dice are thrown then the sample space S is

$$S = \begin{Bmatrix} (1,1),(1,2),(1,3),(1,4),(1,5),(1,6); & (2,1),(2,2),(2,3),(2,4),(2,5),(2,6) \\ (3,1),(3,2),(3,3),(3,4),(3,5),(3,6); & (4,1),(4,2),(4,3),(4,4),(4,5),(4,6) \\ (5,1),(5,2),(5,3),(5,4),(5,5),(5,6); & (6,1),(6,2),(6,3),(6,4),(6,5),(6,6) \} \end{Bmatrix}$$

There are 36 in number. These are the total number of possible cases.

Probability that the sum of the two dice are at least 10

$$= 1 - \{\text{probability that the sum is greater than 10}\}$$

$$= 1 - \{\text{probability that the sum is 11 and 12}\}$$

$$= 1 - \frac{3}{36}$$

$$= \frac{11}{12}$$

Since the number of favourable cases for getting the sum on two dice 11 is 2. Those are $(5,6),(6,5)$ and 12 is 1 i.e. $(6,6)$.

Therefore the number of favourable cases is 3.

15. (a) Explain the concept of random variable

(b) What is the probability of picking an ace and a king from a deck of 52 cards?

(c) A box contains 4-point contact diodes and 6 alloy junction diodes. What is the probability that 3 diodes picked at random contain at least two point contact diodes?

Sol:

(a) See next unit random variables introduction

(b) Two cards are drawn from 52 cards in $^{52}C_2$ ways

In a pack there are 4 aces; one ace is drawn from 4 aces in $^4C_1$ ways,

And there are 4 kings, one king is drawn from 4 kings in $^4C_1$ ways. Then these two things can be done in succession in $^4C_1 \times {}^4C_1$ ways these are the number of favourable cases for getting an ace and a king

Hence the probability of picking an ace and a king from a deck of 52 cards

$$= \frac{{}^4C_1 \; {}^4C_1}{{}^{52}C_2}$$

(c) A box contains 4-point contact diodes and 6 alloy junction diodes. What is the probability that 3 diodes picked at random contain at least two point contact diodes?

Total number of diodes = 4 + 6 = 10. Three Diodes are drawn from 10 diodes in ${}^{10}C_3$ ways at least two point contact diodes means in the drawing 3 diodes (i) two are point contact diodes and (ii) all the three are point contact diodes

(i) **Two are point contact diodes:** In this case the probability that two are point contact diodes is $= \dfrac{{}^4C_2 \times {}^6C_1}{{}^{10}C_3}$

(ii) **All the three are point contact diodes:** In this case the probability that all the three are point contact diodes is

$$= \frac{{}^4C_3 \times {}^6C_0}{{}^{10}C_3}$$

Hence the required probability that 3 diodes picked at random contain at least two point contact diodes

$$= \frac{{}^4C_2 \times {}^6C_1}{{}^{10}C_3} + \frac{{}^4C_3 \times {}^6C_0}{{}^{10}C_3}$$

16. (a) A jar contains two white and three black balls. A sample of size 4 is made. What is the probability that the sample is in the order white, black, white black?

(b) A box contains 4 bad and 6 good tubes. The tubes are checked by drawing a tube at random, testing and repeating the process until all 4 bad tubes are located. What is the probability that the fourth bad tube will be located (i) on the fifth test (ii) on the tenth test?

(c) Consider the experiment of tossing four fair coins. The random variable X is associated with the number of tails showing. Compute and sketch cumulative distribution function of X

Sol:

(a) Given that jar contains 2 white and 3 black balls.

Total number of balls = 2 + 3 = 5. A sample of size 4 is drawn.

The probability that the sample is in the order white, black, white black is

$$= \frac{2}{5} \times \frac{3}{4} \times \frac{1}{3} \times \frac{2}{2}$$

$$= \frac{1}{10}$$

(b)

(c)   see in next unit random variables

17. (a)   List and explain the properties of discrete probability density function.

(b)   Three newspapers A, B and C are published in a city and a survey of readers indicates the following:

20% read A, 16% read B, 14% read C, 8% read A and B, 5% read A and C and 2% read A, B and C.

For one adult chosen at random, compute the probability that: (i) he r3ads none of the papers (ii) he reads exactly one of the papers (iii) he reads at least A and B if it is known that he reds at least one of the papers published.

18. (a)   If A and B are independent events, prove that the events $\overline{A}$ and B, A and $\overline{B}$; $\overline{A}$ and $\overline{B}$ are also independent.

(b) A_1, A_2 and A_3 are three mutually exclusive and exhaustive sets of events associated with a random experiment E_1. Events B_1, B_2 and B_3 are mutually exclusive and exhaustive sets of events associated with a random experiment E_2. The joint probabilities of occurrence of these events and some marginal probabilities are listed in the table given below.

	B_1	B_2	B_3
A_1	$\dfrac{3}{36}$	$*$	$\dfrac{5}{36}$
A_2	$\dfrac{5}{36}$	$\dfrac{4}{36}$	$\dfrac{5}{36}$
A_3	$*$	$\dfrac{6}{36}$	$*$
$P(B_j)$	$\dfrac{12}{36}$	$\dfrac{14}{36}$	$*$

(i)   Find the missing probabilities (*) in the table

(ii)   Find $P(B_3 / A_1)$ and $P(A_1 / B_3)$

(iii)   Are events A_1 and B_1 statistically independent?

Sol:

(a)   See probability theorems

(b)   See in the next unit Random variables

19. (a)   Define the following and give one example for each

(i) Sample space   (ii) Event   (iii) Mutually exclusive events
(iv) Collectively exhaustive events

(b)   In three boxes, there are Capacitors as shown in the following table:

Value in μ_f	Number in box		
↓		→	
	1	2	3
1.0	70	80	145
0.1	55	35	75
0.01	20	95	25

An experiment consists of first randomly selecting a box (assume that each box has the same probability of selection) and then randomly selecting a capacitor from the chosen box. (i) What is the probability of selecting 0.01 μ_f capacitor, given that the box 2 is chosen? (ii) If a 0.01 μ_f capacitor is chosen, what is the probability that it came from the second box?

20. Two boxes B_1 and B_2 contain 100 and 200 light bulbs respectively. The first box has 15 defective bulbs and the second 5. Suppose a box is selected at random and one bulb is picked out.

(a)   What is the probability that it is defective?

(b)   Suppose we test the bulb and it is found to be defective. What is the probability that it came from box 1?

Sol: Let B_1 be the event that represents a first box and B_2 be the event that represents a second box. Then the probability of selecting the boxes are

$$P(B_1) = \frac{1}{2} \text{ and } P(B_2) = \frac{1}{2}$$

Let A be the event that represents a defective bulb, form the given data box B_1 has 85 good and 15 defective bulbs. Similarly box B_2 has 195 good and 5 defective bulbs. Then

$$P(A|B_1) = \frac{15}{100} = 0.15. \text{ and } P(A|B_2) = \frac{5}{200} = 0.025$$

Now, by total probability theorem, the probability that the chosen bulb is defective is

$$
\begin{aligned}
P(A) &= P(B_1)P(A/B_1) + P(B_2)P(A/B_2) \\
&= \frac{1}{2}(0.15) + \frac{1}{2}(0.025) \\
&= \frac{1}{2}(0.175) \\
&= 0.0875
\end{aligned}
$$

Thus, there is nearly 9% probability that a bulb picked at random is defective.

The tested bulb is found to be defective, then the probability that it came from box 1 is (by using Baye's theorem on probability) is a girl is

$$
\begin{aligned}
P(B_1/A) &= \frac{P(B_1)P(A/B_1)}{P(B_1)P(A/B_1) + P(B_2)P(A/B_2)} \\
&= \frac{\frac{1}{2} \times 0.15}{\frac{1}{2} \times 0.15 + \frac{1}{2} \times 0.025} \\
&= \frac{0.075}{0.0875} \\
&= 0.8571
\end{aligned}
$$

CHAPTER 2

Random Variable

2.1 Introduction

We know that, when tossing of a coin, throwing a die, drawing a card from a pack 52 cards etc are called experiments. The experiments whose results depend on chance causes are called random experiments. The definition of the random variable is based on the results of a random experiment.

2.2 Definition

A random variable X can also be regarded as a real valued function defined on the sample space 'S' of a random experiment such that for each point $x \in X$ of the sample space. $p(x)$ is the probability of occurrence of the event represented by X. random variables are denoted by $X, Y, Z \ldots$.

For example when tossing a coin two times, the sample space is given by $S = \{HH, HT, TH, TT\}$

Let X represents a random variable which assumes the number of heads. So the range of the X is (2, 1, 0).

The random variable X takes the value '2', If two heads turn up.

i.e., $\qquad X = 2 \qquad$ When HH turns up

The random variable X takes the value '1', If head and tail or tail and head turns up.

i.e. $\qquad X = 1 \qquad$ When HH or TH turns up

The random variable X takes the value '0' If two tails turns up

i.e., $\qquad X = 0 \qquad$ When TT turns up

The representation of the random variable X as

Random Variable	X :	2	1	0
Outcome or Result	W :	HH	HT / TH	TT

So the random variable is a variable whose numerical value is determined by the result of the experiment.

And similarly when throwing a die once. The sample space is given by

$$S = \{1, 2, 3, 4, 5, 6\}$$

Let the random variable X represents the number multiple by 3 times, what the number turns up when a die is thrown

So the range of the sample space is $\{3, \ 6, 9, 12, 15, 18\}$.

When the random variable X takes the value '3' if 1 turns up

i.e., when throwing a die, if 1 turns up the random variable X takes the value 3.

i.e., $X = 3$ if 1 turns up

Similarly

If '2' turns up the random variable takes value 6. i.e., $X = 6$ if '2' turns up

If '3' turns up the random variable takes value 9. i.e., $X = 9$ if '3' turns up

If '4' turns up the random variable takes value 12. i.e., $X = 12$ if '4' turns up

If '5' turns up the random variable takes value 15. i.e., $X = 15$ if '5' turns up

If '6' turns up the random variable takes value 18. i.e., $X = 18$ if '6' turns up

The representation of the random variable X as

Random Variable	X :	3	6	9	12	15	18
Outcome or Result	W :	1	2	3	4	5	6

There are three types of random variables

1. Discrete random variable
2. Continuous random variable
3. Mixed random variable

Discrete Random Variable: A random variable X can assume a discrete set of values i.e., a finite set of values (as in the case of throwing a die, tossing a coin) or at most an innumerably infinite set of values (as in the case of Poisson distribution, discussed in the next chapter) is called a discrete random variable. For example the number of telephone calls arrived at an office in a finite interval of time, similarly the number of defective bolts in sample of bolts, the number of printing mistakes per page of a text book etc., are all discrete random variables.

Continuous Random Variable: A random variable X assumes all real values in an interval is called continuous random variable. In this case no enumeration of favorable and total cases is possible. For example the exact time of arrivals of telephone calls in 2.00 p.m to 4.00 p.m, similarly the height of a student arriving in to the class room, the temperature of the ragid body etc., are all continuous random variables.

Mixed Random Variable: A random variable X is said to be mixed random variable if it assumes discrete and continuous values. For example the number of telephone calls received in a particular interval of time. Here the number of telephone calls are discrete and the time of the receiving call is continuous so that it is mixed random variable.

2.3 Conditions for a Function to be a Random Variable

Random variables are not a multi valued functions. Thus every point in 'S' must correspond to only one value of the random variable. There are two conditions

1. The function X can be a random variable, the set $\{X \leq x\}$ shall be an event for any real number x. The probability of this event, denoted by $P\{X \leq x\} = \sum P(X = x)$. and

2. The probabilities of the events $P(X = \infty) = P(X = -\infty) = 0$

First we discuss Discrete Random Variable

2.4   Discrete Random Variable

A random variable X can assume a discrete set of values i.e., a finite set of values (as in the case of throwing a die, tossing a coin) or at most an innumerably infinite set of values (as in the case of Poisson distribution, discussed in the next chapter) is called a discrete random variable.

2.5   Probability Function or Probability Mass Function (p.m.f)

Let X be a discrete random variable which assumes integral set of value say, x_1, x_2, x_3, ... x_n with corresponding probabilities say, p_1, p_2, p_3, ... p_n is called a probability function of a random variable 'X'. This is denoted by p(x).

i.e., $P(x_i)$ = probability that the random variable 'X' assume the value 'x_i'

$$= P(X = x_i)$$

$$= p_i$$

The function $P(x)$ is called the frequency function or probability function of the random variable X, which satisfies the conditions, $p(x_i) \geq 0$ and $\sum p(x_i) = 1$

E.g., If the random variable X represents the number appearing on top of a die when it thrown, then

$$p(x_i) = \frac{1}{6} \ i = 1, 2, 3, 4, 5, 6.$$

In this case the probability function is a constant function.

2.6   Distribution Function (Cumulative Distribution)

If X is a random variable then $P(X \leq x)$ is called the cumulative distribution function (c.d.f) or simply distribution function and it is denoted by $F(x)$

$$\therefore \ F(x) = P(X \leq x)$$

If X is a discrete random variable then the distribution function

$$F(x) = P(X \leq x) = \sum_x P(X = x)$$

Properties of distribution function

Since $F(x) = P(X \leq x)$

1. $F(-\infty) = P(X \leq -\infty) = 0$
2. $F(+\infty) = P(X \leq +\infty) = 1$
3. $0 \leq F(x) \leq 1$
4. If $x_1 < x_2$ then $F(x_1) \leq F(x_2)$
5. $P(x_1 < X < x_2) = F(x_2) - F(x_1)$

Examples

Example 1: A random variable X has the following distribution. Determine k

$X = x$	-2	-1	0	1	2	3
$P(X = x)$	0.1	k	0.2	$2k$	0.3	k

Sol:

Since $\sum P(x) = 1$

$\Rightarrow$ $0.1 + k + 0.2 + 2\,k + 0.3 + k = 1$

$\Rightarrow$ $0.6 + 4\,k = 1$

$\therefore$ $k = 0.1$

***Example* 2:** A random variable X has the following distribution

X	1	2	3	4	8	9
$P(X)$	k	$3k$	$5k$	$7k$	$9k$	$11k$

Determine (i) k (ii) $P(X \geq 3)$

Sol:

Since $\sum P(x) = 1$

$\Rightarrow$ $k + 3\,k + 5\,k + 7\,k + 9k + 11\,k = 1$

$\Rightarrow$ $36\,k = 1$

$\therefore$ $k = \dfrac{1}{36}$

The distribution of the random variable X is

$X = x$	1	2	3	4	8	9
$P(X = x)$	$\dfrac{1}{36}$	$\dfrac{3}{36}$	$\dfrac{5}{36}$	$\dfrac{7}{36}$	$\dfrac{9}{36}$	$\dfrac{11}{36}$

(ii) $P(X \geq 3) = 1 - P(X < 3)$

$$= 1 - \{P(X = 1) + P(X = 2)\}$$

$$= 1 - \left\{\frac{1}{36} + \frac{3}{36}\right\}$$

$$= \frac{8}{9} = 0.89$$

***Example* 3:** Let the random variable X takes the values 1, 2, 3, 4, 5, 6 and its respective probabilities are $\dfrac{1}{6}, \dfrac{1}{6}, \dfrac{1}{6}, \dfrac{1}{6}, \dfrac{1}{6}, \dfrac{1}{6}$ Then the cumulative distribution function is

$X = x$:	1	2	3	4	5	6
$P(X = x)$ or $p(x)$:	$\dfrac{1}{6}$	$\dfrac{1}{6}$	$\dfrac{1}{6}$	$\dfrac{1}{6}$	$\dfrac{1}{6}$	$\dfrac{1}{6}$
$F(x) = p(X \leq x)$:	$\dfrac{1}{6}$	$\dfrac{2}{6}$	$\dfrac{3}{6}$	$\dfrac{4}{6}$	$\dfrac{5}{6}$	$\dfrac{6}{6} = 1$

Since $F(1) = p(X \leq 1) = p(X = 1) = \dfrac{1}{6}$

$F(2) = p(X \leq 2) = p(X = 1) + p(X = 2) = \dfrac{1}{6} + \dfrac{1}{6} = \dfrac{2}{6}$

$F(3) = p(X \leq 3) = p(X = 1) + p(X = 2) + p(X = 3) = \dfrac{1}{6} + \dfrac{1}{6} + \dfrac{1}{6} = \dfrac{3}{6}$ and so on

The graph of cumulative distribution function is

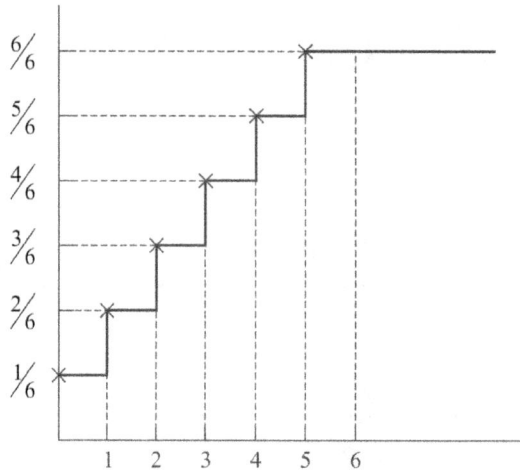

***Example* 4:** If the random variable X takes the values 1, 2, 3 and 4 such that $2.P(X = 1)$ = 3. $P(X = 2) = P(X = 3) = 5.P(X = 4)$. Find the probability distribution and cumulative distribution functions of X

Sol: Given $2.p(X = 1) = 3.p(X = 2) = p(X = 3) = 5.p(X = 4) = k$ (say)

Then $p(X = 1) = \dfrac{k}{2}$; $p(X = 2) = \dfrac{k}{3}$; $p(X = 3) = k$; $p(X = 4) = \dfrac{k}{5}$

Since we know that total probability is unity.

i.e., $\dfrac{k}{2} + \dfrac{k}{3} + k + \dfrac{k}{5} = 1$

$\Rightarrow$ $k = \dfrac{30}{61}$

Now the probability distribution of X is given by

$X = x$:	1	2	3	4
$P(X = x)$ or $P(x)$:	$\dfrac{15}{61}$	$\dfrac{10}{61}$	$\dfrac{30}{61}$	$\dfrac{6}{61}$
$F(x) = P(X \leq x)$:	$\dfrac{15}{61}$	$\dfrac{25}{61}$	$\dfrac{55}{61}$	$\dfrac{61}{61}(=1)$

Example 5: The sample space for an experiment is $S = \{0, 1, 2, 3, 4\}$. List all possible values of the following random variables

(a) $X = 2s$ (b) $X = 3s^2 + 2$ (c) $X = \sin(s\pi)$

Sol:

Given sample space $S = \{0, 1, 2, 3, 4\}$.

Now

(a) $X = 2s$

     = 0, 2, 4, 6, 8 when 0, 1, 2, 3, 4 respectively

(b) $X = 3s^2 + 2$

     = 3*0 + 2 = 2; 3*1 + 2 = 5; 3 * 4 + 2 = 14;

     3 * 9 + 2 = 29: 3 * 16 + 2 = 50

    $\therefore$ The sample space for $X = 3s^2 + 2$ is

    $X = 2, 5, 14, 29, 50$ when 0, 1, 2, 3, 4 respectively

(c) $X = \sin(s\pi) = \sin(0.\pi) = 0$; $\sin(1.\pi) = 0$; $\sin(2.\pi) = 0$; $\sin(3.\pi) = 0$ $\sin(4.\pi) = 0$

    $\therefore$ The sample space for $X = \sin(s\pi)$ is

              $X = 0, 0, 0, 0, 0$ when 0, 1, 2, 3, 4 respectively

Example 6: Given that a random variable X has the following possible values. State if X is discrete, continuous or mixed

(i) $\{-1 < x < 15\}$

(ii) $\{2, 5 < x < 7, 8, 9\}$

(iii) $\{8 \text{ for } x < 2, 6 \text{ for } x \geq 2\}$ and

(iv) $\{12, 15.2, 6.5, -7.8\}$

Sol:

(i)   Given $\{-1 < x < 15\}$ is a continuous random variable since the random variable assumes all real values in the time interval $(-1, 15)$.

(ii)   Given $\{2, 5 < x < 7, 8, 9\}$ is a mixed random variable since $X = 2, 8, 9$ are discrete and $5 < x < 7$ is continuous random variable. So that it is mixed random variable.

(iii)   Given $\{8 \text{ for } x < 2, 6 \text{ for } x \geq 2\}$ is a discrete random variable the random variable X = 8 for $x < 2$ and $X = 6$ for $x \geq 2$. Here the random variable X takes values 6 and 8 both are integers. So the random variable X is discrete.

Given $\{12, 15.2, 6.5, -7.8\}$ is a mixed random variable

***Example* 7:** A random variable X has the following probability distribution

X :	0	1	2	3	4	5	6	7	8
$p(X = x) = p(x)$:	k	$3k$	$5k$	$7k$	$9k$	$11k$	$13k$	$15k$	$17k$

   (i)   Determine the constant k

   (ii) Find $p(X < 3)$, $p(X \geq 3)$ and $p(0 < X < 5)$

   (iii) Find the smallest value of 'X' for which $p\{(X \leq x) > 0.5\}$

   (iv) Find the distribution function $F(x)$

Sol:

The probability function $p(x) = P(X = x)$ must satisfy the

$$p(x) \geq 0 \text{ and } \sum P(X = x) = 1$$

$\Rightarrow$ $k + 3k + 5k + 7k + 9k + 11k + 13k + 15k + 17k = 1$

$\Rightarrow$ $81k = 1$

$\Rightarrow$ $k = \dfrac{1}{81}$

Therefore the value of $k = \dfrac{1}{81}$

Now the probability distribution of X is

X:	0	1	2	3	4	5	6	7	8
$P(X = x)$:	$\dfrac{1}{81}$	$\dfrac{3}{81}$	$\dfrac{5}{81}$	$\dfrac{7}{81}$	$\dfrac{9}{81}$	$\dfrac{11}{81}$	$\dfrac{13}{81}$	$\dfrac{15}{81}$	$\dfrac{17}{81}$

***Example* 8:** Let X denotes the number of heads in a single toss of 4 fair coins. Determine (i) $P(X < 2)$ and (ii) $P(1 < X \leq 3)$

Sol:

Let the random variable X represents the number of heads when tossing 4 fair coins. Therefore the range of X is $\{0, 1, 2, 3, 4\}$

X	0	1	2	3	4
$P(X)$	$\dfrac{1}{16}$	$\dfrac{4}{16}$	$\dfrac{6}{16}$	$\dfrac{4}{16}$	$\dfrac{1}{16}$

Since when tossing of four coins the possible outcomes are

1. $HHHH$	5.	$HTHH$	9.	$THHH$	13.	$TTHH$		
2. $HHHT$	6.	$HTHT$	10.	$THHT$	14.	$TTHT$		
3. $HHTH$	7.	$HTTH$	11.	$THTH$	15.	$TTTH$		
4. $HHTT$	8.	$HTTT$	12.	$THTT$	16.	$TTTT$		

(i) $P(X < 2)$:

$$P(X < 2) = P(X = 0) + P(X = 1)$$

$$= \frac{1}{16} + \frac{4}{16} = \frac{5}{16}$$

(ii) $P(1 < X \leq 3)$:

$$P(1 < X \leq 3) = P(X = 2) + P(X = 3)$$

$$= \frac{6}{16} + \frac{4}{16} = \frac{10}{16}$$

***Example* 9:**

(a)   If 3 dresses drawn from a lot of 6 dress containing 2 defective dresses

Find the probability distribution of the number of defective dresses

(b)   For the discrete probability distribution

x	0	1	2	3	4	5	6	7
f	0	k	$2k$	$2k$	$3k$	k^2	$2k^2$	$7k^2 + k$

Determine (i) k (iv) Smallest value of x such that $P(X \leq x) > \dfrac{1}{2}$

Sol:

(a)   From the given data, there are 6 dresses. Out of which 2 dresses are defective. This implies that 4 dresses are non-defective. Given 3 dresses are selected.

Let the random variable X represents the number of defective dresses.

$\therefore$ The range of X is 0, 1, 2.

3 dresses are drawn from a lot of 6 dresses that can be done in $^6C_3 = 20$ ways. These are the total number of possible cases.

$$P(X=0) = \frac{^4C_3 \cdot ^2C_0}{^6C_3} = \frac{1}{5} \text{ Similarly,}$$

$$P(X=1) = \frac{^4C_2 \cdot ^2C_1}{^6C_3} = \frac{3}{5} \text{ and } P(X=2) = \frac{^4C_1 \cdot ^2C_2}{^6C_3} = \frac{1}{5}$$

The probability distribution of defective dresses is

X	0	1	2
$P(X)$	$\dfrac{1}{5}$	$\dfrac{3}{5}$	$\dfrac{1}{5}$

(b)   (i)   Since $\sum f(x) = 1$

$\Rightarrow$ $0 + k + 2k + 2k + 3k + k^2 + 2k^2 + 7k^2 + k = 1$

$\Rightarrow$ $10k^2 + 9k - 1 = 0$

$\Rightarrow$ $k = -1$ or $\dfrac{1}{10}$

$\therefore$ $k = \dfrac{1}{10}$

$\because$ probability never negative, i.e., $P(x) \geq 0$

(ii)   Smallest value of X such that $P(X \leq x) > \dfrac{1}{2}$ $(= 0.5)$ is

X	0	1	2	3	4	5	6	7
f	0	$\dfrac{1}{10} = 0.1$	$\dfrac{2}{10} =$ 0.2	$\dfrac{2}{10} = 0.2$	$\dfrac{3}{10} = 0.3$	$\left(\dfrac{1}{10}\right)^2 = 0.01$	$2\left(\dfrac{1}{10}\right)^2 = 0.02$	$7\left(\dfrac{1}{10}\right)^2 + \dfrac{1}{10}$ = 0.1
Cumulative Frequency: F	0	0.1	0.3	0.5	0.8	0.81	0.83	1

Since $F(x) = P(X \leq x)$

$F(0) = P(X \leq 0) = f(0) = 0$

$F(1) = P(X \leq 1) = f(0) + f(1) = 0.1$

$$F(2) = P(X \le 2) = f(0) + f(1) + f(2) = 0 + 0.1 + 0.2 = 0.3$$
$$F(3) = P(X \le 3) = f(0) + f(1) + f(2) + f(3) = 0.3 + 0.2 = 0.5$$
$$F(4) = P(X \le 4) = f(0) + f(1) + f(2) + f(3) + f(4) = 0.5 + 0.3$$
$$= 0.8$$

Therefore smallest value of X such that $P(X \le x) > \dfrac{1}{2}$ $(= 0.5)$ is 4

$\therefore$ $P(X \le 3) = 0.5$

Example 10: Two dice are thrown. Let X assign to each point (a, b) in S. the maximum of its number, i.e. $X(a, b) = max(a, b)$. Find the probability distribution of X is a random variable with $X(s) = \{1, 2, 3, 4, 5, 6\}$. Draw the graph of cumulative function.

Sol:

When two dice one thrown, S is the maximum of its number i.e. $X(a, b) = max(a, b)$

$$P(X = 1) = P[(1, 1)] = \frac{1}{36}$$

$$P(X = 2) = P[(2,1), (2,2), 1,2)] = \frac{3}{36}$$

$$P(X = 3) = P[(3, 1), (3, 2), (3, 3), (1, 3), (2, 3)] = \frac{5}{36}$$

$$P(X = 4) = P[(1, 4), (2, 4), (3, 4), (4, 4), (4, 1), (4, 2), (4, 3)] = \frac{7}{36}$$

$$P(X = 5) = P[(1, 5), (2, 5), (3, 5), (4, 5), (5, 5), (5, 1), (5, 2), (5, 3), (5, 4)] = \frac{9}{36}$$

$$P(X = 6) = P[(1, 6), (2, 6), (3, 6), (4, 6), (5, 6), (6, 6), (6, 1), (6, 2), (6, 3), (6, 4),$$
$$(6, 5)] = \frac{11}{36}$$

The probability distribution of the random variable X is

X:	1	2	3	4	5	6
$P(X = x)$:	$\dfrac{1}{36}$	$\dfrac{3}{36}$	$\dfrac{5}{36}$	$\dfrac{7}{36}$	$\dfrac{9}{36}$	$\dfrac{11}{36}$

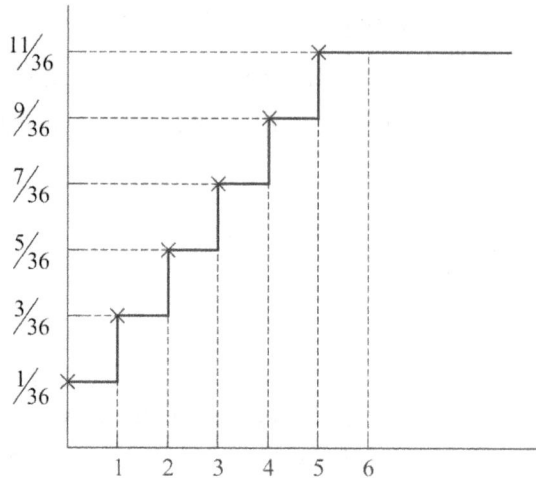

Example 11: Let X denote the minimum of two numbers that appear when a pair of dice is thrown once. Determine the discrete probability distribution

***Sol*:**

When two dice are thrown then the sample space S is

$$S = \begin{cases} (1,1),(1,2),(1,3),(1,4),(1,5),(1,6); & (2,1),(2,2),(2,3),(2,4),(2,5),(2,6) \\ (3,1),(3,2),(3,3),(3,4),(3,5),(3,6); & (4,1),(4,2),(4,3),(4,4),(4,5),(4,6) \\ (5,1),(5,2),(5,3),(5,4),(5,5),(5,6); & (6,1),(6,2),(6,3),(6,4),(6,5),(6,6)\} \end{cases}$$

From the given data, the random variable X represents the minimum of its numbers on the two dice. Then

$X = 1 = $ Min $[(1, 1), (1, 2), (2, 1), (1, 3), (3, 1), (1, 4), (4, 1), (1, 5), (5, 1),$

$\qquad\qquad (1, 6), (6, 1)]$

$\therefore \quad P(X = 1) = P(1) = \dfrac{11}{36}$

$X = 2 = $ Min $[(2, 2), (2, 3), (3, 2), (2, 4), (4, 2), (2, 5), (5, 2), (2, 6), (6, 2)]$

$\therefore \quad P(X = 2) = P(2) = \dfrac{9}{36}$

$X = 3 = $ Min $[(3, 3), (3, 4), (4, 3), (3, 5), (5, 3), (3, 6), (6, 3)]$

$\therefore \quad P(X = 3) = P(3) = \dfrac{7}{36}$

$X = 4 = $ Min $[(4, 4), (4, 5),(5, 4), (4, 6),(6, 4)]$

$\therefore \quad P(X = 4) = P(4) = \dfrac{5}{36}$

$X = 5 = $ Min $[(5, 5),(5, 6), (6, 5)]$

$$\therefore \quad P(X=5) = P(5) = \frac{3}{36}$$

$$X = 6 = \text{Min } [(6, 6)]$$

$$\therefore \quad P(X=6) = P(6) = \frac{1}{36}$$

∴ Probability distribution of the random variable X is

$X = x$	1	2	3	4	5	6
$P(X=x)$	11/36	9/36	7/36	5/36	3/36	1/36

Example 12: Two dice are thrown. Let X assign to each point (a, b) in S the maximum of its number i.e. $X(a, b) = max(a, b)$. Find the probability distribution of the random variable X with $X(S) = \{1,2,3,4,5,6\}$

Sol:

When two dice are thrown then the sample space S is

$$S = \begin{cases} (1,1),(1,2),(1,3),(1,4),(1,5),(1,6); & (2,1),(2,2),(2,3),(2,4),(2,5),(2,6) \\ (3,1),(3,2),(3,3),(3,4),(3,5),(3,6); & (4,1),(4,2),(4,3),(4,4),(4,5),(4,6) \\ (5,1),(5,2),(5,3),(5,4),(5,5),(5,6); & (6,1),(6,2),(6,3),(6,4),(6,5),(6,6)\} \end{cases}$$

From the given data,

$$X[(1,1)] = max\ (1,1) = 1;$$

$$\therefore \quad P(X=1) = P(1) = \frac{1}{36}$$

Similarly,

$$X[(2,1),(2,2),(1,2)] = 2$$

$$\therefore \quad P(X=2) = P(2) = \frac{3}{36}$$

$$X[(1,3),(3,1),(2,3),(3,2),(3,3)] = 3$$

$$\therefore \quad P(X=3) = P(3) = \frac{5}{36}$$

$$X[(1,4),(4,1),(2,4),(4,2),(3,4),(4,3),(4,4)] = 4$$

$$\therefore \quad P(X=4) = P(4) = \frac{7}{36}$$

$$X[(1,5),(5,1),(2,5),(5,2),(3,5),(5,3),(4,5),(5,4),(5,5)] = 5$$

$\therefore\ P(X=5)=P(5)=\dfrac{9}{36}$

$X[(1,6),(6,1),(2,6),(6,2),(3,6),(6,3),(4,6),(6,4),(5,6),(6,5),(6,6)]=6$

$\therefore\ P(X=6)=P(6)=\dfrac{11}{36}$

$\therefore$ Probability distribution of the random variable X is

$X=x$	1	2	3	4	5	6
$P(X=x)$	1/36	3/36	5/36	7/36	9/36	11/36

Example 13: Two dice are thrown X assign to each point. If S is the sum of the variables on the faces.   Find the probability function to the random variable X.

Sol:

When two dice are thrown then the sample space S is

$$X(S)=\begin{cases}(1,1),(1,2),(1,3),(1,4),(1,5),(1,6);\ \ (2,1),(2,2),(2,3),(2,4),(2,5),(2,6)\\(3,1),(3,2),(3,3),(3,4),(3,5),(3,6);\ \ (4,1),(4,2),(4,3),(4,4),(4,5),(4,6)\\(5,1),(5,2),(5,3),(5,4),(5,5),(5,6);\ \ (6,1),(6,2),(6,3),(6,4),(6,5),(6,6)\}\end{cases}$$

From the given data, S be the sample space represent the sum of the variables on the faces, then $S=\{2,3,4,5,6,7,8,9,10,11,12\}$

Now

$$P(2)=P(X=2)=P\,(1,1)=\dfrac{1}{36}$$

$$P\,(3)=P\,(X=3)\ =P\,\{(1,2)\,(2,1)\}\ =\dfrac{2}{36}$$

$$P\,(4)\ =P\,(X=4)=P\,\{(1,3)\,(3,1)\,(2,2)\}\ =\dfrac{3}{36}$$

$$P\,(5)\ =P\,(X=5)=P\,\{(1,4)\,(2,3)\,(3,2),(4,1)\}\ =\dfrac{4}{36}\quad\text{and so on}$$

$$P\,(12)=P\,(X=12)=P\,\{(6,6)\}=\dfrac{1}{36}$$

$\therefore$ Probability distribution of the random variable X is

$X=x$	2	3	4	5	6	7	8	9	10	11	12
$P(X=x)$	$\dfrac{1}{36}$	$\dfrac{2}{36}$	$\dfrac{3}{36}$	$\dfrac{4}{36}$	$\dfrac{5}{36}$	$\dfrac{6}{36}$	$\dfrac{5}{36}$	$\dfrac{4}{36}$	$\dfrac{3}{36}$	$\dfrac{2}{36}$	$\dfrac{1}{36}$

***Example* 14:** A random variable X takes the values -2, -1, 0 and 1 with probabilities 1/8, 1/8, ¼ and 1/2 respectively. Find and draw the probability distribution and cumulative distribution function

Sol:

The probability distribution for the above data is

X :	–2	–1	0	1
$P(X=x)$:	1/8	1/8	¼	1/2
$F(x)$:	1/8	2/8	4/8	8/8

The graph of the above distribution function is

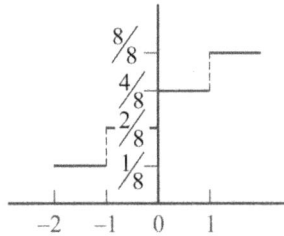

***Example* 15:** A lot containing 25 items, 5 of which are defective, 4 items are chosen at random. If X is the number of defectives found. Obtained the probability distribution of X when the items are chosen without replacement and with replacement

Sol:

Given that lot contains 25 items out of which 5 are defective

∴ Non-defective items are $25 - 5 = 20$

Hence the non-defective items are 20 and defective items are 5.

Now total items are 25 and 4 items are drawn

Given that the random variable represents the number of defective items.

The range of X is 0, 1, 2, 3, 4 (are defective items)

Without replacement:

$P(X=0)$ = Probability that no defective items is chosen

i.e., chosen four items are non-defective

$$= \frac{{}^{5}C_{0}.{}^{20}C_{4}}{{}^{25}C_{4}} = \frac{.{}^{20}C_{4}}{{}^{25}C_{4}} = \frac{1}{5}$$

$P(X=1) =$ Probability that one defective items and 3 non-defective items are chosen

$$= \frac{{}^5C_1 \cdot {}^{20}C_3}{{}^{25}C_4} = \frac{5 \cdot {}^{20}C_3}{{}^{25}C_4}$$

Similarly,

$P(X=2) =$ Probability that two defective items and 2 non-defective items are chosen

$$= \frac{{}^5C_2 \cdot {}^{20}C_2}{{}^{25}C_4} = \frac{10 \cdot {}^{20}C_2}{{}^{25}C_4}$$

$P(X=3) =$ Probability that three defective items and 1 non-defective items chosen

$$= \frac{{}^5C_3 \cdot {}^{20}C_1}{{}^{25}C_4} = \frac{10 \times 20}{{}^{25}C_4}$$

$P(X=4) =$ Probability that four defective items and zero non-defective items chosen

$$= \frac{{}^5C_4 \cdot {}^{20}C_0}{{}^{25}C_4} = \frac{5 \times 1}{{}^{25}C_4}$$

The probability distribution for the above data is

X:	0	1	2	3	4
$P(X=x)$:	$\frac{\cdot {}^{20}C_4}{{}^{25}C_4}$	$\frac{5 \cdot {}^{20}C_3}{{}^{25}C_4}$	$\frac{10 \cdot {}^{20}C_2}{{}^{25}C_4}$	$\frac{10 \times 20}{{}^{25}C_4}$	$\frac{5 \times 1}{{}^{25}C_4}$

With replacement

***Example* 16:** A shipment of 6 *TV* sets contains two defective sets. A hotel makes a random purchase of 3 of the sets. If X is the number of defective sets purchased by the hotel find the probability distribution of X.

Sol:

Given total *TV* sets = 6;

The number of defective sets = 2

Number of non de facture sets = 6 – 2 = 4

Total purchase 3 sets

Given that X represents the number of defective sets purchased

The range of X is 0, 1, 2

$P(X=0) =$ Probability that no defective set is chosen

i.e., all the chosen sets are non-defective

$$= \frac{^2C_0 \cdot {}^4C_3}{^6C_3} = \frac{1 \cdot {}^4C_3}{^6C_3} = \frac{1}{5}$$

$P(X=1)$ = Probability that one defective set and 2non-defective sets are chosen

$$= \frac{^2C_1 \cdot {}^4C_2}{^6C_3} = \frac{2 \cdot {}^4C_2}{^6C_2} = \frac{3}{5}$$

Similarly,

$P(X=2)$ = Probability that two defective sets and 1 non-defective set are chosen

$$= \frac{^2C_2 \cdot {}^4C_1}{^6C_3} = \frac{1 \cdot {}^4C_1}{^6C_3} = \frac{1}{5}$$

The probability distribution for the above data is

X	:	0	1	2
$P(X=x)$	:	$\frac{1}{5}$	$\frac{3}{5}$	$\frac{1}{5}$

***Example* 17:** A random variable X may assume 4 values with probabilities $\frac{1+3x}{4}$, $\frac{1-x}{4}$, $\frac{1+2x}{4}$, $\frac{1-4x}{4}$

Find the condition on x so that these values represent the probability unction of X

Sol:

Let the random variable X assumes the values x_1, x_2, x_3 and x_4. Given that the respective probabilities are as shown with the following probability distribution

X	:	x_1	x_2	x_3	x_4
$P(X=x)$	:	$\frac{1+3x}{4}$	$\frac{1-x}{4}$	$\frac{1+2x}{4}$	$\frac{1-4x}{4}$

I. The probability function $p(x) = P(X=x)$ must satisfy the

$$p(x) \geq 0 \text{ and } \sum P(X=x) = 1$$

$$\Rightarrow \qquad \frac{1+3x}{4} + \frac{1-x}{4} + \frac{1+2x}{4} + \frac{1-4x}{4} = \frac{4}{4} = 1$$

and $\qquad \frac{1+3x}{4} \geq 0 \qquad \Rightarrow 1+3x \geq 0$

$\Rightarrow \qquad\qquad 3x \geq -1$

$\Rightarrow \qquad\qquad x \geq -\dfrac{1}{3} \ (= -0.333)$

$\dfrac{1-x}{4} \geq 0 \qquad \Rightarrow \quad 1-x \geq 0$

$\qquad\qquad\qquad \Rightarrow \qquad x \leq 1$

$\dfrac{1+2x}{4} \geq 0 \qquad \Rightarrow \quad 1+2x \geq 0$

$\qquad\qquad\qquad \Rightarrow \qquad 2x \geq -1$

$\qquad\qquad\qquad \Rightarrow \qquad x \geq -\dfrac{1}{2} \ (= -0.5)$

$\dfrac{1-4x}{4} \geq 0 \qquad \Rightarrow \quad 1-4x \geq 0$

$\qquad\qquad\qquad \Rightarrow \qquad -4x \geq -1$

$\qquad\qquad\qquad \Rightarrow \qquad 4x \leq 1$

$\qquad\qquad\qquad \Rightarrow \qquad x \leq \dfrac{1}{4} \ (= 0.25)$

Therefore the value of x for which the probability function is defined.

$\therefore$ The range of X is $\ -\dfrac{1}{3} \leq x \leq \dfrac{1}{4}$

***Example* 18:** The probability function of an infinite discrete distribution is given by $P(X = j) = \dfrac{1}{2^j}$; $j = 1, 2 \ \ldots$. Verify that the total probability is one. Also find p(X is even); p(X is odd); $P(X \geq 3)$ and $P(X$ is divisible by 3)

Sol:

Given probability function is

$$P(X = j) = \dfrac{1}{2^j}$$

$$j = 1, 2 \ldots$$

Let the random variable X assumes the values x_1, x_2, x_3 ... Given that the respective probabilities are as shown with the following probability distribution

X	:	x_1	x_2	x_3	...	... ...
$P(X = x)$	:	$\dfrac{1}{2}$	$\dfrac{1}{2^2}$	$\dfrac{1}{2^3}$	...	... ...

I. The probability function $p(x) = P(X = x)$ must satisfy the

$$p(x) \geq 0 \text{ and } \sum P(X = x) = 1$$

$$\Rightarrow \qquad \frac{1}{2} + \frac{1}{2^2} + \frac{1}{2^3} + \ldots \ldots \ldots$$

This is a infinite geometric series.

The sum of an infinite geometric series $S_\infty = \dfrac{a}{1-r}$

Here $\qquad a = \dfrac{1}{2}$ and $r = \dfrac{1}{2}$

$$S_\infty = \frac{1/2}{1-(1/2)} = \frac{1/2}{1/2} = 1$$

The given probability distribution is 1

$$P(X \text{ is even}) = P(X = 2 \text{ or } 4 \text{ or } 6 \ldots)$$

$$= \frac{1}{2^2} + \frac{1}{2^4} + \frac{1}{2^6} + \frac{1}{2^8} + \ldots \ldots \ldots \ldots$$

This is a infinite geometric series.

The sum of an infinite geometric series $S_\infty = \dfrac{a}{1-r}$

Here $\qquad a = \dfrac{1}{2^2}$ and $r = \dfrac{1}{2^2}$

$$S_\infty = \frac{1/2^2}{1-(1/2^2)}$$

$$= \frac{1/4}{3/4}$$

$$= \frac{1}{3}$$

$\therefore \ P(X \text{ is even}) = \dfrac{1}{3}$

$P(X \text{ is odd}) = P(X = 1 \text{ or } 3 \text{ or } 5 \ldots)$

$$= \frac{1}{2} + \frac{1}{2^3} + \frac{1}{2^5} + \frac{1}{2^7} + \ldots \ldots \ldots \ldots$$

This is a infinite geometric series.

The sum of an infinite geometric series $S_\infty = \dfrac{a}{1-r}$

Here $\qquad\qquad a = \dfrac{1}{2} \quad$ and $\quad r = \dfrac{1}{2^2}$

$$S_\infty = \dfrac{1/2}{1-(1/2^2)}$$

$$= \dfrac{1/2}{3/4}$$

$$= \dfrac{2}{3}$$

$\therefore P(X \text{ is odd}) = \dfrac{2}{3}$

$P(X \geq 3) = 1 - P(X < 3) = 1 - \{P(X = 1) + P(X = 2)\}$

$$= 1 - \left\{\dfrac{1}{2} + \dfrac{1}{4}\right\}$$

$$= 1 - \left\{\dfrac{3}{4}\right\}$$

$$= \dfrac{1}{4}$$

$P(X \text{ is divisible by } 3) = P(X = 3 \text{ or } 6 \text{ or } 9 \ \ldots)$

$$= \dfrac{1}{2^3} + \dfrac{1}{2^6} + \dfrac{1}{2^9} + \ldots \quad \ldots \quad \ldots \quad \ldots$$

This is a infinite geometric series.

The sum of an infinite geometric series $S_\infty = \dfrac{a}{1-r}$

Here $\qquad\qquad a = \dfrac{1}{2^3} \quad$ and $\quad r = \dfrac{1}{2^3}$

$$S_\infty = \dfrac{1/2^3}{1-(1/2^3)}$$

$$= \dfrac{1/8}{7/8} = \dfrac{1}{7}$$

$\therefore P(X \text{ is divisible by } 3) = \dfrac{1}{7}$

***Example* 19:** For the example 2.5 in page number 4, find $P\ (1.5< X< 4.5/X> 2)$

$$P(1.5< X< 4.5/X> 2) = P(A/B)$$

$$= \frac{P(A \cap B)}{P(B)}$$

$$= \frac{P[(1.5< X< 4.5) \cap (X> 2)]}{P(X> 2)}$$

$$= \frac{P(X= 3)+ P(X= 4))}{\sum_{i=3}^{8} P(X= i)}$$

$$= \frac{\dfrac{7}{81}+ \dfrac{9}{81}}{\dfrac{7}{81}+ \dfrac{9}{81}+ \dfrac{11}{81}+ \dfrac{13}{81}+ \dfrac{15}{81}+ \dfrac{17}{81}}$$

$$= \frac{\dfrac{16}{81}}{\dfrac{72}{81}}$$

$$= \frac{16}{72}$$

$$= \frac{2}{9}$$

***Example* 20:** A random variable X has the following distribution

X	1	2	3	4	5	6	7	8
$f(x)$	k	$2k$	$3k$	$4k$	$5k$	$6k$	$7k$	$8k$

Find the value of (i) k (ii) $P(X \le 2)$ and (iii) $P(2 \le X \le 5)$

(i)   Since $\sum P(x) = 1$

$\Rightarrow$ $k+ 2\,k+ 3\,k+ 4\,k+ 5k+ 6\,k+7\,k+ 8k= 1$

$\Rightarrow$ $10k^2+ 9k\text{-}1 = 0$

$\Rightarrow$ $36\,k= 1$ $\therefore k= \dfrac{1}{36}$

(ii) $P(X \le 2)$:

$$P(X \le 2) = P(X= 1)+ P(X= 2)$$

$$= k+ 2k= 3k$$

$$= 3 \cdot \frac{1}{36}$$

$$= \frac{1}{12} = 0.08$$

(iii) $P(2 \leq X \leq 5)$:

$$P(2 \leq X \leq 5) = P(X = 2) + P(X = 3) + P(X = 4) + P(X = 5)$$

$$= 2k + 3k + 4k + 5k = 14k$$

$$= 14 \cdot \frac{1}{36} = \frac{7}{18} = 0.389$$

2.7 Continuous Random Variable

A random variable X is said to be continuous if it can take all possible values in an interval, E.g., the length of time during which a vacuum tube installed in a circuit functions is a continuous random variable, similarly, consider an experiment where a pointer on a wheel of chance is spun. Assume that the wheel is numbered from 1 to 12 is a continuous random variable

Note: A continuous random variable will have a continuous distribution function. We consider the above example a pointer on a wheel of chance is spun. Assume that the wheel is numbered from 1 to 12.

The probability of getting 1 in spinning process is

$$F(0) = P(X \leq 0) = 0;$$

$$F(1) = P(X \leq 1) = \frac{1}{12};$$

$$F(2) = P(X \leq 2) = \frac{2}{12};$$

$$F(3) = P(X \leq 3) = \frac{3}{12}$$

Similarly,

$$F(4) = \frac{4}{12}; \qquad F(5) = \frac{5}{12}; \qquad F(6) = \frac{6}{12}; \qquad F(7) = \frac{7}{12};$$

$$F(8) = \frac{8}{12}; \qquad F(9) = \frac{9}{12}; \qquad F(10) = \frac{10}{12}; \qquad F(11) = \frac{11}{12};$$

$$F(12) = \frac{12}{12}$$

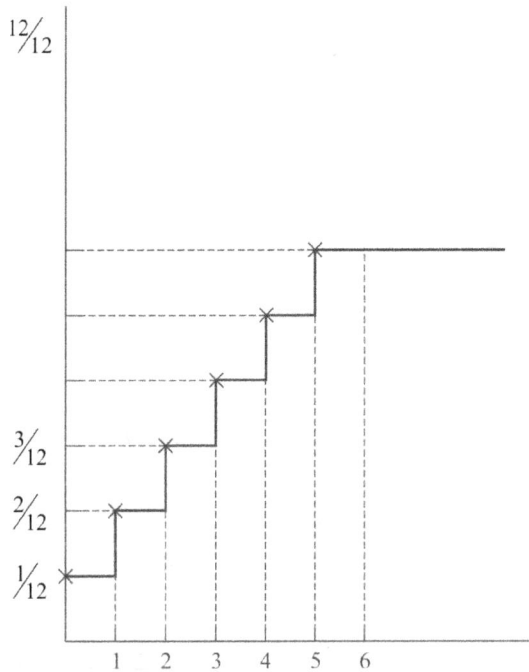

The graph of the above distribution function i.e. probability distribution function of a continuous random variable

We observed that the above graph is a ramp function.

In order to draw the pdf of this example, the derivative of the ramp function is to be known. The derivation of the ramp function is the unit step function. Therefore the pdf is drawn as

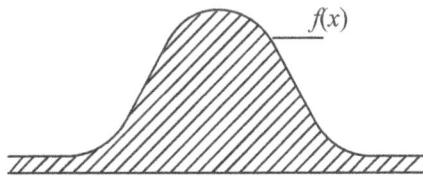

2.8   Probability Density Function

Let X is a continuous random variable such that $\left\{ x - \dfrac{1}{2dx} \le X \le x + \dfrac{1}{2dx} \right\} = f(x)dx$

then $f(x)$ is called the probability density function (pdf) of X provided $f(x)$ satisfies the following conditions

    (i) $f(x) \ge 0$ $\vee x \in R$ and   (ii) $\displaystyle\int_{-\infty}^{\infty} f(x)dx = 1$

The above two properties can be used as test to check whether the given function is a valid pdf. Both the properties must be satisfied for the validity.

The probability density function $f(x)$ is defined as the derivative of the probability distribution   function $F(x)$ of the random variable it is also called as frequency function

$$\therefore \ f(x) = \frac{d}{dx} F(x) \quad \text{this is also denoted as} \ f_X(x) = \frac{dF(x)}{dx}$$

Note: $p(a \le X \le b) = \int\limits_a^b f(x)dx$ the curve $y = f(x)$ is called the probability curve of the

random variable X.

2.9   Distribution Function (Cumulative Distribution) in Continuous Case

Let X is a random variable discrete or continuous, then $P(X \le x)$ is called the cumulative distributed function of X or distributed function of X and is denoted by $F(x)$

If X is discrete $F(x) = P(X \le x) \ = \ \sum\limits_j p_j \ $ for $\ x_j \le x$

If X is continuous $F(x) = P(X \le x) \ = \ P(-\infty \le X \le x) \ = \ \int\limits_{-\infty}^x f(x)dx$

Properties of Distribution Function

Since $F(x) = p(X \le x)$

1. $F(-\infty) = p(X \le -\infty) = 0$

2. $F(+\infty) = p(X \le +\infty) = \int\limits_{-\infty}^\infty f(x)dx \ = 1$

3. If $x_1 < x_2$ then $F(x_1) \le F(x_2)$

4. $p(x_1 < X < x_2) = \int\limits_{x_1}^{x_2} f(x)dx = F(x_2) - F(x_1)$

Examples

Example 1: A continuous random variable has the *PDF*, $f(x) = \begin{cases} k\,e^{-x} & \text{if } x \succ 0 \\ 0 & \text{elsewhere} \end{cases}$

Determine the constant k.

Sol:

Since we know that $\int\limits_{-\infty}^{\infty} f(x)dx = 1$

$\Rightarrow \qquad \int\limits_{-\infty}^{0} f(x)dx + \int\limits_{0}^{\infty} f(x)dx = 1$

$\Rightarrow \qquad 0 + \int\limits_{0}^{\infty} k\,e^{-x}\,dx = 1$

$\Rightarrow \qquad k\left[-e^{-x}\right]_{0}^{\infty} = 1$

$\Rightarrow \qquad -k\left[0-1\right] = 1$

$\Rightarrow \qquad k = 1 \Rightarrow k = 1$

Now the given $f(x)$ is

$$f(x) = \begin{cases} e^{-x} & \text{if } x > 0 \\ 0 & \text{elsewhere} \end{cases}$$

Example 2: Is $f(x) = \begin{cases} \dfrac{1}{2}(x+1) & \text{for } -1 < x < 1 \\ 0 & \text{elsewhere} \end{cases}$ represents the density of a random

variable X ?

Sol:

If $f(x)$ is a density function, then it satisfies

$$\int\limits_{-\infty}^{\infty} f(x)dx = 1$$

$\Rightarrow \qquad \int\limits_{-\infty}^{\infty} f(x)dx = \int\limits_{-\infty}^{-1} f(x)dx + \int\limits_{-1}^{1} f(x)dx + \int\limits_{1}^{\infty} f(x)dx$

$\qquad\qquad\qquad = \int\limits_{\infty}^{-1} 0.dx + \int\limits_{-1}^{1} \frac{1}{2}(x+1)\ dx + \int\limits_{1}^{\infty} 0.dx$

$$= \frac{1}{2}\left[\frac{x^2}{2} + x\right]_{-1}^{1}$$

$$= \frac{1}{2}\left[\left(\frac{1^2}{2} + 1\right) - \left(\frac{(-1)^2}{2} - 1\right)\right]$$

$$= \frac{1}{2}\left[\left(\frac{3}{2}\right) - \left(-\frac{1}{2}\right)\right]$$

$$= \frac{1}{2}\cdot\frac{4}{2}$$

$$= 1$$

Hence $f(x) = \begin{cases} \dfrac{1}{2}(x+1) & for \quad -1 < x < 1 \\ 0 & elsewhere \end{cases}$ is a density function.

***Example* 3:** If $f(x)$ is a probability density function defined as

$$f(x) = \begin{cases} kx^3 & in \quad 0 \le x \le 3 \\ 0 & else \ where \end{cases}$$

Find the value of k and find the probability between $x = \dfrac{1}{2}$ and $x = \dfrac{3}{2}$

Sol:

Given p.d.f is $f(x) = \begin{cases} kx^3 & in \quad 0 \le x \le 3 \\ 0 & else \ where \end{cases}$

If $f(x)$ is a density function, then it satisfies $\displaystyle\int_{-\infty}^{\infty} f(x)dx = 1$

$$\Rightarrow \qquad \int_{-\infty}^{0} f(x)dx + \int_{0}^{3} f(x)dx + \int_{3}^{\infty} f(x)dx = 1$$

$$\Rightarrow \qquad k.\int_{0}^{3} x^3 \, dx = 1$$

$$\Rightarrow \qquad k\left[\frac{x^4}{4}\right]_{0}^{3} = 1$$

$$\Rightarrow \qquad k\left[\left(\frac{3^4}{4}-0\right)\right]=1$$

$$\Rightarrow \qquad \frac{81}{4}k=1$$

$$\therefore \qquad k=\frac{4}{81}$$

Now

$$f(x) \;=\; \begin{cases} \dfrac{4}{81}x^3 & in\;\;0\le x\le3 \\[2mm] 0 & else\;where \end{cases}$$

$$P\left(\frac{1}{2}\le X\le\frac{3}{2}\right):$$

$$P\left(\frac{1}{2}\le X\le\frac{3}{2}\right) \;=\; \int_{\frac{1}{2}}^{\frac{3}{2}} f(x)dx$$

$$=\frac{4}{81}\int_{\frac{1}{2}}^{\frac{3}{2}} x^3 dx$$

$$=\frac{4}{81}\left[\frac{x^4}{4}\right]_{\frac{1}{2}}^{\frac{3}{2}} \;=\; \frac{1}{81}\left[x^4\right]_{\frac{1}{2}}^{\frac{3}{2}}$$

$$=\frac{1}{81}\left[\left(\frac{3}{2}\right)^4-\left(\frac{1}{2}\right)^4\right]$$

$$=\frac{1}{81}\left[\frac{80}{16}\right]=\frac{5}{81}=0.0617$$

***Example* 4:** If $f(x)=k\,e^{-|x|}$ is p.d.f in $-\infty\le x\le\infty$ find the values of k and the probability between 0 and 4

Sol: Since we know that $\int\limits_{-\infty}^{+\infty} f(x)dx = 1$

$$\Rightarrow \qquad \int\limits_{-\infty}^{+\infty} f(x)dx = 1$$

$$\Rightarrow \qquad \int\limits_{-\infty}^{0} k\,e^{-|x|}dx + \int\limits_{0}^{+\infty} k\,e^{-|x|}dx = 1$$

$$\Rightarrow \qquad \int\limits_{-\infty}^{0} k\,e^{x}dx + \int\limits_{0}^{+\infty} k\,e^{-x}dx = 1$$

$$\because \qquad |x| = \begin{cases} -x & \text{if } x \text{ lies } -\infty \text{ to } 0 \\ x & \text{if } x \text{ lies } 0 \text{ to } \infty \end{cases}$$

$$\therefore \qquad k\left[e^{x}\right]_{-\infty}^{0} + k\left[-e^{-x}\right]_{0}^{\infty} = 1 \Rightarrow k[1-0] - k[0-1] = 1$$

$$2k = 1 \Rightarrow k = \frac{1}{2}$$

$$\therefore \qquad f(x) = \frac{1}{2}e^{-|X|}$$

Now

$$P(0 \le X \le 4) = \int\limits_{0}^{4} f(x)\,dx$$

$$= \int\limits_{-0}^{4} \frac{1}{2}e^{-|X|}\,dx$$

$$= \frac{1}{2}\int\limits_{0}^{4} e^{-X}\,dx$$

$\because |x| = x$ because x lies 0 to 4

$$= \frac{1}{2}\left[\frac{e^{-X}}{-1}\right]_{0}^{4} = -\frac{1}{2}\{e^{-4} - e^{0}\}$$

$$= \frac{1}{2}\{1 - e^{-4}\}$$

$$\approx 0.491$$

***Example* 5:** A continuous random variable X has the distribution function

$$F(x) = \begin{cases} 0 & if \ x \le 1 \\ k(x-1)^4 & if \ 1 \le x \le 3 \\ 1 & if \ x > 3 \end{cases}$$

Determine (i) k (ii) the probability density function of X

Sol:

Since
$$f(x) = \frac{d}{dx}F(x),$$

$$\therefore f(x) = \begin{cases} 0 & if \ x \le 1 \\ 4k(x-1)^3 & if \ 1 \le x \le 3 \\ 0 & if \ x > 3 \end{cases}$$

i.e.,
$$f(x) = \begin{cases} 4k(x-1)^3 & if \ 1 \le x \le 3 \\ 0 & else \ where \end{cases}$$

Since $f(x)$ is a density function, $\int\limits_{-\infty}^{\infty} f(x)dx = 1$ and the range of X is 1 to 3

$$\int\limits_{-\infty}^{1} f(x).dx + \int\limits_{1}^{3} f(x) \ dx = 1$$

i.e.,
$$\int\limits_{-\infty}^{1} 0.dx + \int\limits_{1}^{3} 4k(x-1)^3 \ dx = 1$$

$$\Rightarrow \qquad 4k \int\limits_{1}^{3}(x-1)^3 \ dx = 1$$

$$\Rightarrow \qquad 4k\left[\frac{(x-1)^4}{4}\right]_{1}^{3} = 1 \Rightarrow k\left[(x-1)^4\right]_{1}^{3} = 1$$

$$\Rightarrow \qquad k\left[(3-1)^4 - (1-1)^4\right] = 1 \Rightarrow k = \frac{1}{16}$$

The probability density function $f(x)$ is

$$f(x) = \begin{cases} \frac{1}{4}(x-1)^3 & if \ 1 \le x \le 3 \\ 0 & else \ where \end{cases}$$

Example 6:   Probability density function of a random variable X is

$$f(x) = \begin{cases} \dfrac{1}{2}\sin x & \text{if } 0 \le x \le \pi \\ 0 & \text{elsewhere} \end{cases}$$

Find the probability between 0 and $\dfrac{1}{2}$

Sol:

Given that $f(x) = \begin{cases} \dfrac{1}{2}\sin x & \text{if } 0 \le x \le \pi \\ 0 & \text{elsewhere} \end{cases}$

The probability distribution of the random variable X between 0 and $\dfrac{\pi}{2}$ is

$$\int\limits_0^{\pi/2} f(x)\, dx \ = \ \frac{1}{2}\int\limits_0^{\pi/2} \sin x\, dx$$

$$= \ \frac{1}{2}\Big[-\cos x\Big]_0^{\pi/2}$$

$$= \ -\frac{1}{2}\Big[\cos\frac{\pi}{2} - \cos 0\Big]$$

$$= \ -\frac{1}{2}\big[0 - 1\big]$$

Hence $\displaystyle\int\limits_0^{\pi/2} f(x)\, dx \ = \ \frac{1}{2}$

Example 7: For the continuous probability function $f(x) = kx^2 e^{-x}$ is a density function, when $x \ge 0$.

Find the value of k

Sol:

Since $f(x)$ is a density function, $\displaystyle\int\limits_{-\infty}^{\infty} f(x)\, dx = 1$

$$\int\limits_{-\infty}^{\infty} f(x)\, dx \ = \ \int\limits_{-\infty}^{0} f(x)\, dx + \int\limits_{0}^{\infty} f(x)\, dx = 1$$

$$\Rightarrow \qquad \int_{0}^{\infty} kx^2 e^{-x} dx = 1 \Rightarrow k \int_{0}^{\infty} x^2 e^{-x} dx = 1$$

$$\Rightarrow \qquad k \left[x^2 \left(\frac{e^{-x}}{-1} \right) - 2x \left(\frac{e^{-x}}{1} \right) + 2 \left(\frac{e^{-x}}{-1} \right) \right]_{0}^{\infty} = 1$$

$$\Rightarrow \qquad k \left[(x^2.0 - 2x.0 + 2.0) - (0 - 0 - 2.1) \right] = 1$$

$$\Rightarrow \qquad 2k = 1 \quad \Rightarrow \quad k = \frac{1}{2}$$

$$\therefore \qquad f(x) = \frac{1}{2} x^2 e^{-x}$$

***Example* 8:** If X is a continuous random variable with distribution

$$f(x) = \begin{cases} \dfrac{1}{6}x + k & if\ 0 \le x \le 3 \\ 0 & elsewhere \end{cases}$$

Determine the value of k and $P(1 \le X \le 2)$

Sol:

(i) Since $\qquad \displaystyle\int_{-\infty}^{\infty} f(x)dx = 1$

$$\Rightarrow \qquad \int_{-\infty}^{0} f(x)dx + \int_{0}^{3} f(x)dx + \int_{3}^{\infty} f(x)dx = 1$$

$$\Rightarrow \qquad 0 + \int_{0}^{3} (\frac{1}{6}x + k)dx + 0 = 1$$

$$\Rightarrow \qquad \frac{1}{6}\left[\frac{x^2}{2} \right]_{0}^{3} + k [x]_{0}^{3} = 1 \quad \Rightarrow \quad \frac{1}{6}\left[\frac{9}{2} \right] + k(3) = 1$$

$$\Rightarrow \qquad \frac{3}{4} + 3k = 1 \qquad \Rightarrow \quad 3k = \frac{1}{4}$$

$$\therefore \qquad k = \frac{1}{12}$$

Now

$$f(x) = \begin{cases} \dfrac{1}{6}x + \dfrac{1}{12} & if\ 0 \le x \le 3 \\ 0 & elsewhere \end{cases}$$

(ii) $$P(1 \leq X \leq 2) = \int_{1}^{2} f(x) dx$$

$$= \int_{1}^{2} (\frac{1}{6}x + \frac{1}{12}) \, dx$$

$$= \frac{1}{6} \int_{1}^{2} x \, dx + \frac{1}{12} \int_{1}^{2} dx$$

$$= \frac{1}{6} \left[\frac{x^2}{2} \right]_{1}^{2} + \frac{1}{12} [x]_{1}^{2}$$

$$= \frac{1}{6} \left[\frac{4}{2} - \frac{1}{2} \right] + \frac{1}{12} [2 - 1]$$

$$= \left[\frac{1}{4} \right] + \left[\frac{1}{12} \right] = \frac{4}{12}$$

$$= \frac{4}{12} = 0.25$$

∴ $$P(1 \leq X \leq 2) = 0.25$$

Example 9: A continuous random variable has the PDF,

$$f(x) = \begin{cases} 2e^{-2x} & \text{if } x \succ 0 \\ 0 & \text{elsewhere} \end{cases}$$

Find the probabilities that it will take on a value   (i) between 1 and 3   and (ii) greater than 0.5

Sol: $$P(1 < X < 3) = \int_{1}^{3} f(x) dx$$

$$= \int_{1}^{3} 2 e^{=2x} \, dx$$

$$= 2 \int_{1}^{3} e^{=2x} \, dx$$

$$= 2 \left[\frac{e^{-2x}}{-2} \right]_{1}^{3} = -\{e^{-6} - e^{-2}\}$$

$$= e^{-2} - e^{-6} = 0.1353 - 0.002478 = 0.1328$$

$\therefore$ $\qquad$ $P(1 \le X \le 2) = 0.1328$

$$P(X > 0.5) = \int\limits_{0.5}^{\infty} f(x)dx = 1 - \int\limits_{-\infty}^{0.5} f(x)dx$$

$$= 1 - \left[\int\limits_{-\infty}^{0} f(x)dx + \int\limits_{0}^{0.5} f(x)dx \right]$$

$$= 1 - \left[\int\limits_{0}^{0.5} f(x)dx \right] \qquad\qquad \because \int\limits_{-\infty}^{0} f(x)dx = 0$$

$$= 1 - \int\limits_{0}^{0.5} 2\,e^{=2x}\,dx$$

$$= 1 - 2 \int\limits_{0}^{0.5} e^{=2x}\,dx$$

$$= 1 - 2 \left[\frac{e^{-2x}}{-2} \right]_{0}^{0.5}$$

$$= 1 + \{e^{-1} - e^{0}\}$$

$$= 1 + \{e^{-1} - 1\}$$

$$= e^{-1} = 0.3678$$

$\therefore$ $\qquad$ **$P\ (X > 0.5) = 0.3678$**

Example 10: A continuous random variable X is defined as

$$f(x) = \begin{cases} \dfrac{1}{16}(3+x)^2 & \text{if } -3 \le x \le -1 \\[2mm] \dfrac{1}{16}(6-2x^2) & \text{if } -1 \le x \le 1 \\[2mm] \dfrac{1}{16}(3-x)^2 & \text{if } 1 \le x \le 3 \\[2mm] 0 & \text{else where} \end{cases}$$

Verify that $f(x)$ is a density function of the random variable X.

Sol:

Since *f(x)* is a density function, $\int_{-\infty}^{\infty} f(x)dx = 1$

$$\int_{-\infty}^{\infty} f(x)dx$$

$$= \int_{-\infty}^{-3} 0.dx + \int_{-3}^{-1} \frac{1}{16}(3+x)^2 dx + \int_{-1}^{1} \frac{1}{16}(6-2x^2)\ dx + \int_{1}^{3} \frac{1}{16}(3-x)^2\ dx + \int_{3}^{\infty} 0.dx$$

$$= \int_{-\infty}^{-3} f(x)dx + \int_{-3}^{-1} f(x)dx + \int_{-1}^{1} f(x)dx + \int_{1}^{3} f(x)dx + \int_{3}^{\infty} f(x)dx$$

$$= \frac{1}{16} \int_{-3}^{-1} (3+x)^2 dx + \frac{1}{16} \int_{-1}^{1} (6-2x^2)\ dx + \frac{1}{16} \int_{1}^{3} (3-x)^2\ dx$$

$$= \frac{1}{16} \left\{ \left[\frac{(3+x)^3}{3} \right]_{-3}^{-1} + \left[6x - \frac{2x^3}{3} \right]_{-1}^{1} - \left[\frac{(3-x)^3}{3} \right]_{1}^{3} \right\}$$

$$= \frac{1}{16} \left\{ \left[\frac{8}{3} - 0 \right] + \left[(6-\frac{2}{3}) - (-6+\frac{2}{3}) \right] - \left[0 - \frac{8}{3} \right] \right\}$$

$$= 1$$

∴ $\int_{-\infty}^{\infty} f(x)dx = 1,$ Hence *f(x)* is a density function.

***Example* 11:** Let *F(x)* be the distribution function of a random variable *X* given by

$$F(x) = \begin{cases} cx^3 & when\ \ 0 \le x \le 3 \\ 1 & when\ \ x > 3 \\ 0 & when\ x > 3 \end{cases}$$

If *P(X = 3) = 0*, determine (i) c (ii) mean and (iii) *P(X > 1)*

Sol:

Since $f(x) = \dfrac{d}{dx} F(x)$, then

$$f(x) = \begin{cases} 3\,cx^2 & when\ \ 0 \le x \le 3 \\ 0 & when\ \ x > 3 \\ 0 & when\ x > 3 \end{cases}$$

This can be written as $f(x) = \begin{cases} 3cx^2 & when \quad 0 \le x \le 3 \\ 0 & other\ wise \end{cases}$

Since $f(x)$ is a density function, $\int\limits_{-\infty}^{\infty} f(x)dx = 1$ and the range of X is 0 to 3

$\Rightarrow$ $\qquad \int\limits_{-\infty}^{0} f(x).dx + \int\limits_{0}^{3} f(x)\ dx = 1$

$\Rightarrow$ $\qquad \int\limits_{-\infty}^{0} 0.dx + \int\limits_{0}^{3} 3\,cx^2\ dx = 1$

$\Rightarrow$ $\qquad 3c \int\limits_{0}^{3} x^2\ dx = 1$

$\Rightarrow$ $\qquad 3c \left[\dfrac{x^3}{3} \right]_0^3 = 1 \qquad \Rightarrow \qquad c\left[27 - 0 \right] = 1$

$\Rightarrow$ $\qquad 27\,c = 1 \Rightarrow c = \dfrac{1}{27}$

Therefore the probability density function is

$$f(x) = \begin{cases} 3.\dfrac{1}{27}x^2 & when \quad 0 \le x \le 3 \\ 0 & other\ wise \end{cases}$$

i.e., $\qquad f(x) = \begin{cases} \dfrac{1}{9}x^2 & when \quad 0 \le x \le 3 \\ 0 & other\ wise \end{cases}$

The range of x is 0 to 3

$$P(X > 1) = \int\limits_{1}^{\infty} f(x)dx$$

$$= \int\limits_{1}^{3} f(x)dx + \int\limits_{3}^{\infty} f(x)dx = \int\limits_{1}^{3} f(x)dx + \int\limits_{3}^{\infty} 0.dx$$

$$= \dfrac{1}{9} \int\limits_{1}^{3} x^2\ dx$$

$$= \frac{1}{9}\left[\frac{x^3}{3}\right]_1^3$$

$$= \frac{1}{9}\left[\frac{1}{3}(27-1)\right]$$

$$= \frac{26}{27} = 0.9629$$

$\therefore$ $P(X > 1) = 0.9629$

***Example* 12:** A continuous random variable X has the probability density function

$$f(x) = \begin{cases} k\,x\,e^{-\lambda x} & \text{when } x \geq 0, \ \lambda > 0 \\ 0 & \text{other wise} \end{cases}$$

Determine the value of k

***Sol*:**

Since $f(x)$ is a density function, $\displaystyle\int_{-\infty}^{\infty} f(x)dx = 1$

$$\int_{-\infty}^{\infty} f(x)dx = \int_{-\infty}^{0} f(x)dx + \int_{0}^{\infty} f(x)dx = 1$$

$\Rightarrow$ $\displaystyle\int_{0}^{\infty} k\,x\,e^{-\lambda x}dx = 1 \Rightarrow k\int_{0}^{\infty} x\,e^{-\lambda x}dx = 1$

$\Rightarrow$ $\displaystyle k\left[x\left(\frac{e^{-\lambda x}}{-\lambda}\right) - \left(\frac{e^{-\lambda x}}{\lambda^2}\right)\right]_0^{\infty} = 1$

$\Rightarrow$ $\displaystyle k\left[(x.0 \ -0)-(0-\frac{1}{\lambda^2})\right] = 1$

$\Rightarrow$ $\displaystyle \frac{1}{\lambda^2}k = 1 \Rightarrow k = \lambda^2$

$\therefore$ $f(x) = \begin{cases} \lambda^2 x\,e^{-\lambda x} & \text{when } x \geq 0, \ \lambda > 0 \\ 0 & \text{other wise} \end{cases}$

***Example* 13:** A random variable gives measurements X between 0 and 1 with a probability function

$$f(x) = \begin{cases} 12x^3 - 21x^2 + 10\,x & \text{when } 0 \leq x \leq 1 \\ 0 & \text{else where} \end{cases}$$

Find (i) $P\left(X \le \dfrac{1}{2}\right)$ (ii) $P\left(X > \dfrac{1}{2}\right)$ and (iii) a number k such that $P(X \le k) = \dfrac{1}{2}$

Sol:

Given
$$f(x) = \begin{cases} 12x^3 - 21x^2 + 10\,x & when \quad 0 \le x \le 1 \\ 0 & else\ where \end{cases}$$

(i) $P\left(X \le \dfrac{1}{2}\right)$

$$P\left(X \le \dfrac{1}{2}\right) = \int_{-\infty}^{\frac{1}{2}} f(x)dx$$

$$= \int_{-\infty}^{0} f(x)dx + \int_{0}^{\frac{1}{2}} f(x)dx = 0 + \int_{0}^{\frac{1}{2}} f(x)dx$$

$$= \int_{0}^{\frac{1}{2}} (12x^3 - 21x^2 + 10\,x)\,dx$$

$$= \left[12\dfrac{x^4}{4} - 21\dfrac{x^3}{3} + 10\dfrac{x^2}{2}\right]_{1}^{3} = \left[3x^4 - 7x^3 + 5x^2\right]_{0}^{\frac{1}{2}}$$

$$= \left[3\left(\dfrac{1}{2}-0\right)^4 - 7\left(\dfrac{1}{2}-0\right)^3 + 5\left(\dfrac{1}{2}-0\right)^2\right]$$

$$= \left[3\left(\dfrac{1}{2}\right)^4 - 7\left(\dfrac{1}{2}\right)^3 + 5\left(\dfrac{1}{2}\right)^2\right]$$

$$= 0.5625$$

(ii) $P\left(X > \dfrac{1}{2}\right) = \int_{\frac{1}{2}}^{\infty} f(x)dx = \int_{\frac{1}{2}}^{1} f(x)dx + \int_{1}^{\infty} f(x)dx$

$$= \int_{-\infty}^{0} f(x)dx + \int_{0}^{\frac{1}{2}} f(x)dx = \int_{\frac{1}{2}}^{1} f(x)dx + 0$$

$$= \int_{\frac{1}{2}}^{1} (12x^3 - 21x^2 + 10\,x)\,dx$$

$$= \left[12\frac{x^4}{4} - 21\frac{x^3}{3} + 10\frac{x^2}{2} \right]_{\frac{1}{2}}^{1} = \left[3x^4 - 7x^3 + 5x^2 \right]_{\frac{1}{2}}^{1}$$

$$= \left[3\left(1 - \left(\frac{1}{2}\right)^4\right) - 7\left(1 - \left(\frac{1}{2}\right)^3\right) + 5\left(1 - \left(\frac{1}{2}\right)^2\right) \right]$$

$$= \left[3\left(\frac{15}{16}\right) - 7\left(\frac{7}{8}\right) + 5\left(\frac{3}{4}\right) \right]$$

$$= 0.4375$$

(iii) $P(X \le k) = \dfrac{1}{2}$:

$$P(X \le k) = \int_{-\infty}^{K} f(x)dx = \frac{1}{2}$$

$$\Rightarrow \int_{-\infty}^{0} f(x)dx + \int_{0}^{K} f(x)dx = \frac{1}{2}$$

$$\Rightarrow \int_{0}^{K} f(x)dx = \frac{1}{2} \qquad\qquad \because \int_{-\infty}^{0} f(x)dx = 0$$

$$\Rightarrow \int_{0}^{K} (12x^3 - 21x^2 + 10\,x)\,dx = \frac{1}{2}$$

$$\Rightarrow \left[12\frac{x^4}{4} - 21\frac{x^3}{3} + 10\frac{x^2}{2} \right]_{0}^{K} = \frac{1}{2}$$

$$\Rightarrow \left[3x^4 - 7x^3 + 5x^2 \right]_{0}^{K} = \frac{1}{2}$$

$$\Rightarrow \left[3K^4 - 7K^3 + 5K^2 \right] = \frac{1}{2}$$

$$\Rightarrow 6K^4 - 14K^3 + 10K^2 = 1$$

$$\therefore \quad k = 0.452$$

Example 14: A continuous random variable X has the probability density function

$$f(x) = \begin{cases} kx & for \ \ 0 \le x < 2 \\ 2k & for \ \ 2 \le x < 4 \\ k(6-x) & for \ \ 4 \le x < 6 \\ 0 & elsewhere \end{cases}$$

Find k of the density function.

Sol: If $f(x)$ is a density function, then it satisfies $\displaystyle\int_{-\infty}^{\infty} f(x)dx = 1$

$\Rightarrow \displaystyle\int_{-\infty}^{\infty} f(x)dx = \int_{-\infty}^{0} f(x)dx + \int_{0}^{2} f(x)dx + \int_{2}^{4} f(x)dx + \int_{4}^{6} f(x)dx + \int_{6}^{\infty} f(x)dx = 1$

$\Rightarrow \displaystyle\int_{-\infty}^{0} f(x)dx + \int_{0}^{2} f(x)dx + \int_{2}^{4} f(x)dx + \int_{4}^{6} f(x)dx + \int_{6}^{\infty} f(x)dx = 1$

$\Rightarrow \displaystyle\int_{-\infty}^{0} 0.dx + \int_{0}^{2} kx \, dx + \int_{2}^{4} 2k \, dx + \int_{4}^{6} k(6-x)dx + \int_{6}^{\infty} 0.dx = 1$

$\Rightarrow \displaystyle k.\int_{0}^{2} x \, dx + 2k \int_{2}^{4} dx + k \int_{4}^{6} (6-x)dx = 1$

$\Rightarrow k\left[\dfrac{x^2}{2}\right]_0^2 + 2k[x]_2^4 + k\left[6(x)_4^6 - \left(\dfrac{x^2}{2}\right)_4^6\right] = 1$

$\Rightarrow k\left[\dfrac{4}{2}\right] + 2k[2] + k\left[6(2) - \left(\dfrac{36}{2} - \dfrac{16}{2}\right)\right] = 1$

$\Rightarrow 2k + 4k + k(\ 12\text{-}10) = 1$

$\Rightarrow 8k = 1 \quad \Rightarrow k = \dfrac{1}{8}$

Now the probability density function becomes

$$f(x) = \begin{cases} \dfrac{1}{8}x & for \ \ 0 \le x < 2 \\ 2.\dfrac{1}{8} = \dfrac{1}{4} & for \ \ 2 \le x < 4 \\ \dfrac{1}{8}(6-x) & for \ \ 4 \le x < 6 \\ 0 & elsewhere \end{cases}$$

***Example* 15:** If the probability density function of a random variable is given by

$$f(x) = \begin{cases} k(1-x^2) & \text{if } 0 \le X \le 1 \\ 0 & \text{elsewhere} \end{cases}$$

Find the value of k and the probabilities that a random variable X will take on a value (i) between 0.1 and 0.2   (ii) greater than 0.5

Sol:

Since we know that $\displaystyle\int_{-\infty}^{\infty} f(x)dx = 1$

$$\Rightarrow \quad \int_{-\infty}^{0} f(x)dx + \int_{0}^{1} f(x)dx + \int_{1}^{\infty} f(x)dx = 1$$

$$\Rightarrow \quad 0 + \int_{0}^{1} k(1-x^2)dx + 0 = 1$$

$$\Rightarrow \quad k\left[x - \frac{x^3}{3} \right]_{0}^{1} = 1$$

$$\Rightarrow \quad k\left[1 - \frac{1^3}{3} \right] = 1$$

$$\Rightarrow \quad \frac{2}{3}k = 1 \quad \Rightarrow \quad k = \frac{3}{2}$$

(i) $P(0.1 < X < 0.2) = \displaystyle\int_{0.1}^{0.2} f(x)dx$

$$= \int_{0.1}^{0.2} \frac{3}{2}(1-x^2)dx$$

$$= \frac{3}{2} \int_{0.1}^{0.2} (1-x^2)dx$$

$$= \frac{3}{2} \left[x - \frac{x^3}{3} \right]_{0.1}^{0.2}$$

$$= \frac{3}{2} \left[\left(0.2 - \frac{(0.2)^3}{3} \right) - \left(0.1 - \frac{(0.1)^3}{3} \right) \right]$$

$$= \frac{3}{2} (0.0977) = 0.149$$

(ii) $$P(x > 0.5) = \int\limits_{0.5}^{\infty} f(x)dx$$

$$= \frac{3}{2} \int\limits_{0.5}^{1} (1 - x^2)dx$$

$$= \frac{3}{2} \left[x - \frac{x^3}{3} \right]_{0.5}^{1}$$

$$= \frac{3}{2} \left[\left(1 - \frac{(1)^3}{3} \right) - \left(0.5 - \frac{(0.5)^3}{3} \right) \right]$$

$$= \frac{3}{2} (0.208) = 0.3125$$

***Example* 16:** Let $f(x) = 3x^2$ when $0 \le x \le 1$ be the probability density function of a continuous random variable X. Determine a and b such that   (i) $P(X \le a) = P(X > a)$ and (ii) $P(X > b) = 0.05$

Sol:

Given $f(x) = 3x^2$ defined in $0 \le x \le 1$

(i) $P(X \le a) = P(X > a)$

$$P(X \le a) = \int\limits_{0}^{a} 3x^2 dx = \int\limits_{a}^{1} 3x^2 dx$$

$\Rightarrow$ $3.\frac{1}{3}\left(x^3\right)_0^a = 3.\frac{1}{3}\left(x^3\right)_a^1$

$\Rightarrow$ $a^3 = 1 - a^3$

$\Rightarrow$ $a^3 = \frac{1}{2}$

$\therefore$ $a = \left(\frac{1}{2}\right)^3$

(ii) $P(X > b) = 0.05$:

    $P(X > b) = 0.05$

$$\Rightarrow \qquad \int\limits_{b}^{1} 3x^2 dx = 0.05$$

$$\Rightarrow \qquad 3.\frac{1}{3}\left(x^3\right)_{b}^{1} = 0.05$$

$$\Rightarrow \qquad 1 - b^3 = 0.05$$

$$\Rightarrow \qquad b^3 = 0.95 \quad \text{i.e., } b = \left(\frac{19}{20}\right)^{1/3}$$

$$\therefore \qquad b = 0.983$$

2.10 Some Probability Distribution

So far, we have discussed about the random variables and their characteristics. In this section we present some important special distributions which are useful in the physical situations. The total probability is distributed according to some definite probability rule to the variables which can be expressed mathematically. This study will also enable us to fit a mathematical model or a function of the form y = p (x) to the variables. The probability distributions can be classified as discrete distributions and continuous distributions.

First we study discrete probability distributions

2.11 Binomial Distribution

A random variable X which takes two values 0 and 1 with probabilities p and q respectively i.e., p(X = 1) = p and p(X = 0) = q where $p = 1 - q$ is called a Bernoulli variate and is said to have a Bernoulli distribution.

The probability distribution of Binomial distribution is

$$P(X = x) \;=\; {}^{n}C_{x}\, p^{x}\, q^{n-x} \qquad x = 0, 1, 2, \ldots\ldots.\text{n.}$$

where n and p are called the parameters of the Binomial distribution and ${}^{n}C_{x}$ is called the binomial coefficient.

Examples

***Example* 1:** A fair coin tossed 10 times. Find the probability of occurrence of 5 or 6 heads.

Sol:

Given coin is fair coin so $p = \dfrac{1}{2}$ (probability of getting head) and $q = \dfrac{1}{2}$ (probability of getting tail),

Now

The probability distribution of Binomial distributions is

$$P(X = x) \ = \ {}^nC_x \, p^x \, q^{n-x} \qquad x = 0, 1, 2, \ldots \ldots .n.$$

Now the random variable X represents the number of heads. The required probability is $P(5 \le X \le 6)$.

$$\therefore P(5 \le X \le 6) = P(X = 5) + P(X = 6)$$

$$= {}^{10}C_5 \left(\tfrac{1}{2}\right)^5 \left(\tfrac{1}{2}\right)^{10-5} + {}^{10}C_6 \left(\tfrac{1}{2}\right)^6 \left(\tfrac{1}{2}\right)^{10-6}$$

$$= {}^{10}C_5 \left(\tfrac{1}{2}\right)^{10} + {}^{10}C_6 \left(\tfrac{1}{2}\right)^{10}$$

$$= \left(\tfrac{1}{2}\right)^{10} \left[{}^{10}C_5 + {}^{10}C_6 \right]$$

$$= 0.451$$

Example 2: A binary source generates digits 1 and 0 randomly with probability 0.6 and 0.4 respectively.

(i) What is the probability that two 1's and three 0's will occur in a five digit sequence?

(ii) What is the probability that at least three 1's will occur in a five digit sequence?

Sol:

Let the random variable X represents the number of 1's obtained in a five digit sequence.

p is the probability that the binary source generates digit 1

q is the probability that the binary source generates digit 0

$\therefore$ from the given data $p = 0.6$ and $q = 0.4$, n = 5

(i) The probability that two 1's (and three zeros) occur in a five digit sequence is

$$P(X = 2) = {}^5C_2 (0.6)^2 (0.4)^{5-2} = 10 \times (0.6)^2 (0.4)^3 = 0.23$$

Since the probability distribution of binomial distribution functions

$$P(X = x) = {}^nC_x \, p^x \, q^{n-x} \quad x = 0, 1, 2, \ldots \ldots .n.$$

(ii) The probability that at least three one's will occur in a five digit sequence is

$$P(X \ge 3) = 1 - P(X < 3) = 1 - \{ P(X \le 2) \}$$

$$= 1 - \{ P(X = 0) + P(X = 1) + P(X = 2) \}$$

$$= 1 - \{ {}^5C_0 (0.6)^0 (0.4)^{5-0} + {}^5C_1 (0.6)^1 (0.4)^{5-1} + {}^5C_2 (0.6)^2 (0.4)^{5-2} \}$$

$$- 1 - 0.37$$

$$= 0.683.$$

***Example* 3:** If the probability is 0.20 that any one person will dislike the taste of a new tooth-paste, what is the probability that 5 of 18 randomly selected persons will dislike it?

Sol:

From the given data $p = 0.20$ is the probability that any one person dislike a new tooth-paste

$$\therefore q = 0.80 \quad n = 18$$

Let the random variable X represents that number of persons dislike the tooth-paste.

$$\therefore \qquad P(X = 5) = {}^{18}C_5 (0.20)^5 (0.80)^{18-5}$$

$$= {}^{18}C_5 (0.20)^5 (0.80)^{13}$$

$$= 0.1507$$

***Example* 4:** In a precision bomb attack there is a 50% chance that any one bomb will strike the target. Two direct hits are required to destroy the target completely. How many bombs must be dropped to give a 99% chance or better of completely destroying the target?

Sol:

From the given data $p = 50\% = \dfrac{50}{100} = \dfrac{1}{2}$ is the probability that any one bomb will strike the target. It gives $q = \dfrac{1}{2}$.

Also given two direct hits one required to destroy the target completely.

Let n be the number of bombs should be dropped to ensure 99% chance or more to completely destroying the target.

Let the random variable X represents that the number of bombs strike the target and the probability distributions Binomial distributions.

$$P(X = x) = {}^{n}C_x \, p^x \, q^{n-x} \qquad x = 0, 1, 2, \ldots \ldots .n.$$

P[out of n bombs at least two bombs strike the target directly] ≥ 0.99

i.e., $\qquad P[X \geq 2] \geq 0.99$

Consider $\qquad P(X \geq 2) = 1 - P(X < 2)$

$$= 1 - \{P(X = 0) + P(X = 1)\}$$

$$= 1 - \left\{ {}^nC_0 \left(\frac{1}{2}\right)^0 \left(\frac{1}{2}\right)^{n-0} + {}^nC_1 \left(\frac{1}{2}\right)^1 \left(\frac{1}{2}\right)^{n-1} \right\}$$

$$= 1 - \left\{ \left(\frac{1}{2}\right)^n \left[{}^nC_0 + {}^nC_1 \right] \right\}$$

$$= 1 - \frac{1}{2^n} \left({}^{n+1}C_1 \right)$$

$\therefore \qquad {}^nC_r + {}^nC_{r+1} = {}^{n+1}C_{r+1}$

$$= 1 - \frac{1}{2^n}(n+1)$$

$$= 1 - \frac{n+1}{2^n}$$

$\therefore$ from (1)

$$1 - \frac{n+1}{2^n} \geq 0.99$$

$$0.01 \geq \frac{n+1}{2^n} \Rightarrow \frac{n+1}{2^n} \leq 0.01$$

$$n + 1 \leq (0.01)2^n$$

$$2^n \geq \frac{n+1}{0.01} (ie)2^n \geq 100 + 100n$$

$\therefore \qquad 2^n \geq 100 + 100n$

By trail and error, this inequality satisfied by n = 11.

Hence the minimum number of bombs needed to destroy the target completely is 11.

2.12 Poisson Distribution

Poisson distribution is a limiting case of Binomial distribution under the following assumptions.

(i) n, the number of trails is indefinitely large i.e., $n \to \infty$.

(ii) p, the probability of success for each trail is indefinitely small i.e., $p \to 0$.

(iii) $np = \lambda(say)$ is finite i.e., $p = \frac{\lambda}{n}$; $q = 1 - p$ and λ is a positive real quantity.

Note: Poisson distribution is a good approximation of the binomial distribution when 'n' is greater than or equal to 20 and p is less than or equal to 0.05.

Definition: A random variable X is said to follow a Poisson distribution if it assumes only non-negative values and its probability distribution is given by

$$p(X = x) = \frac{e^{-\lambda}.\lambda^x}{x!} \qquad x = 0,1,2\ldots$$

If the above conditions holds good, we can substitute the mean of the binomial distribution np in place of the mean of the Poisson distribution 'λ'.

Poisson distribution describes

(i)   The number of printing mistakes per page of a large text book.

(ii) The number of defective bulbs produced by a reputed factory.

(iii) The number of telephone calls received from in a given time internal etc.

Examples

Example 1: If a random variable X has a Poisson distribution such that $p(1) = p(2)$. Find $p(4)$.

Sol:

The probability distribution of Poisson distribution is

$$P(X = x) = \frac{e^{-\lambda}.\lambda^x}{x!} \qquad x = 0,1,2 \ldots$$

From the given data $P(X = 1) = P(X = 2)$

$$\Rightarrow \qquad \frac{e^{-\lambda}\lambda^1}{1!} = \frac{e^{-\lambda}\lambda^2}{2!}$$

$$\Rightarrow \qquad \frac{\lambda}{1} = \frac{\lambda^2}{2} \Rightarrow \lambda = 2$$

Now $\qquad\qquad p(4) = P(x = 4)$

$$= \frac{e^{-\lambda}.\lambda^4}{4!}$$

$$= \frac{e^{-2}.2^4}{4!}$$

$$= \frac{16 \times 0.1353}{1 \times 2 \times 3 \times 4} = 0.0902$$

$\therefore$ p(4) = 0.0902

Example 2: Assume automobile arrivals at a gasoline station are Poisson and occur at an average rate of 50/hour. The station has only one gasoline pump. If all cars are assumed to require one minute obtain fuel. What is the probability that a waiting line will occur at the pump?

Sol:

The arrivals of cars on an average 50 cars / hour

$$\therefore \lambda = \frac{50}{60} \text{ cars per minute i.e., } \lambda = \frac{5}{6} \text{ cars/min.}$$

A waiting line will occur if two or more cars will arrive in any one minute internal.

The required probability is $P(X \geq 2)$.

For a Poisson distribution probability distribution is

$$P(X = x) = \frac{e^{-\lambda} . \lambda^x}{x!} \qquad x = 0, 1, 2 \dots \infty.$$

$$\therefore \qquad P(X \geq 2) = 1 - P(X < 2)$$

$$= 1 - \{P(X = 0) + P(X = 1)\}.$$

$$= 1 - \left\{ 1 - \frac{e^{-5/6} . (5/6)^0}{\angle 0} + \frac{e^{-5/6} . (5/6)^1}{\angle 1} \right\}$$

$$= 1 - \{0.4346 + 0.3622\}$$

$$= 1 - 0.7968 = 0.2032$$

∴ The probability that a waiting line will occur at the pump is 0.2032.

Example 3: A random variable X is known to be Poisson with $\lambda = 4$. What is the probability that $0 \leq x \leq 5$.

Sol:

For a Poisson probability distribution

$$P(X = x) = \frac{e^{-\lambda} . \lambda^x}{\angle x} \qquad x = 0, 1, 2 \dots \infty.$$

Given $x = 4$

$$P(0 \leq X \leq 5) = P(X = 0) + P(X = 1) + \dots + P(X = 5)$$

$$= \frac{e^{-4} . 4^0}{\angle 0} + \frac{e^{-4} . 4^1}{\angle 1} + \dots + \frac{e^{-4} 4^5}{\angle 5}$$

$$= e^{-4}\left[1+4+\frac{4^2}{1.2}+\frac{4^3}{1.2.3}+\frac{4^4}{1.2.3.4}+\frac{4^5}{1.2.3.4.5}\right]$$

$$= e^{-4}\left[\frac{150+240+320+320+256}{30}\right]$$

$$= e^{-4}[42.87]$$

$$= 0.7851$$

***Example* 4:** 10% of the bolts produced by a certain machine turn out to be defective. Find the probability that in a sample of 10 bolts selected at random, exactly two of them will be defective using (i) Binomial distribution (ii) Poisson distribution and comment on the result.

Sol:

From the given data, Probability of selecting a defective bolt is $p = 10\% = 0.01$ then $q = 0.9$.

Also given that n = 10.

(i) By using binomial distribution

$$P(X=x) = {}^nC_x\ p^x\ q^{n-x} \qquad\qquad x = 0,1,2\cdots n$$

Now $$P(X=2) = {}^{10}C_2.(0.1)^2(0.9)^{10-2}$$

$$= \frac{10\times 9}{1\times 2}\times(0.1)^2\times(0.9)^8$$

$$= 45\times(0.1)^2\times(0.9)^8$$

$$= 0.1937$$

(ii) By using Poisson distribution, the probability distribution function is

$$P(X=x)=\frac{e^{-\lambda}.\lambda^x}{\angle x} \qquad x = 0, 1, 2 \ldots.$$

The mean $\lambda = n\,p = 10\times 0.1 = 1$

$\therefore \qquad$ $$P(X=2) =\frac{e^{-1}.1^2}{\angle 2}=\frac{e^{-1}}{2}$$

$$=\frac{0.1353}{2}$$

$$= 0.1839$$

Note: Clearly there is a significant difference between the two probability distributions of Binomial and Poisson. Even though Poisson distribution is an approximation to Binomial distribution and it is applicable for large n and less probability of success.

Example 5: The probability that an individual suffers a bad reaction due to certain injection is 0.001. Determine the probability that out of 2000 individuals (i) exactly 3 (ii) more than 2 individuals will suffer a bad reaction.

Sol:

From the given data $p = 0.001$ and $n = 2000$

The mean $\lambda = $ n p $= 0.001 \times 2000 = 2$

The probability distribution of Poisson distribution is

$$p(X = x) = \frac{e^{-\lambda}.\lambda^x}{\angle x} \qquad x = 0, 1, 2 \ldots \infty$$

(i) Probability that exactly 3 will suffer a bad reaction is

$$P(X = 3) = \frac{e^{-2}.2^3}{\angle 3} = \frac{8}{6} e^{-2}$$

$$= \frac{4}{3} \times 0.1353$$

$$= 0.1804$$

(ii) Probability that more than 2 individuals will suffer a bad reaction is

$$P(X \geq 2) = 1 - P(X \leq 2) \qquad \qquad \ldots (1)$$

$$P(X \leq 2) = P(X = 0) + P(X = 1) + P(X = 2)$$

$$= \frac{e^{-2}.2^0}{\angle 0} + \frac{e^{-2}.2^1}{\angle 1} + \frac{e^{-2}.2^2}{\angle 2}$$

$$= e^{-2}[1 + 2 + 2]$$

$$= 5 \times 0.1353$$

$$= 0.6665$$

$\therefore$ From (1) $\qquad P(X > 2) = 1 - 0.6665$

$$= 0.3335$$

Example 6: The probability of getting no misprint in a page of a book in e^{-4}. Determine the probability that a page of a book contains more than or equal to 2 misprints.

Sol:

From the given data, $P(X = 0) = e^{-4}$

i.e.,
$$\frac{e^{-\lambda} \lambda^0}{\angle 0} = e^{-4}$$

This implies $\lambda = 4$

Now we shall find the probability that a page of a book contains more than 2 misprints is

$$P(X \geq 2) = 1 - P(X < 2) \qquad \qquad \text{..... (1)}$$

Now
$$P(X < 2) = P(X = 0) + P(X = 1)$$

$$= \frac{e^{-4} . 4^0}{\angle 0} + \frac{e^{-4} . 4^1}{\angle 1}$$

$$= e^{-4}(1 + 4)$$

$$= 5e^{-4}$$

$$= 0.183 \times 5 = 0.0916$$

$\therefore \qquad P(X \geq 2) = 1 - 0.0916$

$$= 0.9084$$

Example 7: Assuming that one in 80 births in case of twins. Calculate the probability of 2 or more births of twins on a day when 30 births occur using

(i) Binomial and   (ii) Poisson approximate.

Sol:

From the given data, the probability of getting twin births in 80 births is $\dfrac{1}{80}$

i.e., $p = 0.0125$ then $q = 0.9875$ and also given $n = 30$.

Let the random variable X represents the number of twin births takes place.

(i) By using Binomial distribution,

$$P(X = x) = {}^nC_x \, p^x \, q^{n-x} \qquad \qquad x = 0, 1, 2 \cdots n$$

$$P(x \geq 2) = 1 - P(x < 2)$$
$$= 1 - \{P(x=0) + P(x=1)\}$$
$$= 1 - \{^{30}C_0.(0.0125)^0 (0.9875)^{30} + ^{30}C_1.(0.0125)^1 (0.9875)^{29}\}$$
$$= 1 - \{(0.9875)^{29} [0.9875 + 30.(0.0125)]\}$$
$$= 1 - (0.9875)^{29} (1.3625)$$
$$= 1 - 0.6943 \times 1.3625$$
$$= 1 - 0.9459$$
$$= 0.054$$

(ii) The probability distribution of Poisson distribution is

$$P(X=x) = \frac{e^{-\lambda} \lambda^x}{\angle x} \qquad x = 0, 1, 2 \ldots$$

$$\lambda = np = 30 \times 0.0125 = 0.375$$

$$P(X \geq 2) = 1 - P(X < 2)$$
$$= 1 - \{P(X=0) + P(X=1)\}$$
$$= 1 - \left\{ \frac{e^{-0.375}(0.375)^0}{\angle 0} + \frac{e^{-0.375}(0.375)^1}{\angle 1} \right\}$$
$$= 1 - e^{0.375} [1 + 0.375]$$
$$= 1 - 0.6873 \times 1.375$$
$$= 1 - 0.945 = 0.055$$

2.13 Poisson Process

***Example* 8:** Passengers arrive at an airport checkout counter at an average rate of 1.5 per minute. Find the probabilities that

(i) Almost 4 will arrive at a given time and

(ii) At least 3 will arrive during an internal of 2 minutes.

Sol:

From the given date, $\lambda = 1.5$ *and* $t = 1$ $\therefore \lambda t = 1.5$

The probability distribution of Poisson distribution is

$$P(X = x) = \frac{e^{-\lambda t}.(\lambda t)^x}{\angle x} \qquad\qquad x = 0, 1, 2 \dots$$

(i) The probability that almost 4 will arrive at a given time

$$P(X \leq 4) = P(X = 0) + P(X = 1) + P(X = 2) + P(X = 3) + P(X = 4)$$

$$= \frac{e^{-1.5}(1.5)^0}{0!} + \frac{e^{-1.5}(1.5)^1}{1!} + \frac{e^{-1.5}(1.5)^2}{2!} + \frac{e^{-1.5}(1.5)^3}{3!} + \frac{e^{-1.5}(1.5)^4}{4!}$$

$$= e^{-1.5}\left[1 + 1.5 + \frac{(1.5)^2}{2} + \frac{(1.5)^3}{6} + \frac{(1.5)^4}{24}\right]$$

$$= e^{-1.5}\left[2.5 + 1.125 + 0.5625 + 0.2109\right]$$

$$= 0.223 \times 4.3984$$

$$= 0.9814$$

(ii) The probability that at least 3 will arrive during an internal of 2 minutes is

Here t = 2 $\therefore \lambda t = 1.5 \times 2 = 3$

$$P(X \geq 3) = 1 - P(X \leq 2)$$

$$= 1 - \{P(X = 0) + P(X = 1) + P(X = 2)\}$$

$$= 1 - \left[\frac{e^{-3}.3^0}{0!} + \frac{e^{-3}.3^1}{1!} + \frac{e^{-3}.3^2}{2!}\right]$$

$$= 1 - e^{-3}\left(1 + 3 + \frac{9}{2}\right)$$

$$= 1 - e^{-3} \times 8.5$$

$$= 1 - 0.0498 \times 8.5$$

$$= 0.5768$$

Example 9: If a bank receives on the average 6 bad cheques per day. What are the probabilities that it will receive

(i) Four bad cheques on any given day.

(ii) 10 bad cheques over any two consecutive days?

Sol:

From the given data, $\lambda = 6$, $t = 1$

For Poisson distribution, the probability function is

$$P(X = x) = \frac{e^{-\lambda t}.(\lambda t)^x}{x!} \qquad x = 0, 1, \ldots$$

(i)   The probability that it will receive four bad cheques on any given day

$$P(X = 4) = \frac{e^{-6}.6^4}{4!}$$

$$= 0.134$$

(ii)   The probability that it will receive 10 bad cheques over any two consecutive days is

Here for two consecutive days $\therefore t = 2$

$$\therefore \lambda t = 6 \times 2 = 1.2$$

$$P(X = 10) = \frac{e^{-12}.12^{10}}{10!}$$

$$= 0.1048$$

Example 10: A telephone switch board operator expects to come across on average 6 ghost calls per day. Calculate the probability of

(i)   4 calls being ghost calls on any day

(ii)   10 calls being ghost calls per any two consecutive days?

Sol:

From the given data $\lambda = 6$ i.e., average number of ghost calls on any day.

The probability distribution of Poisson distribution is

$$P(X = x) = \frac{e^{-\lambda t}(\lambda t)^x}{\angle x} \qquad x = 0, 1, 2 \ldots .$$

(i)   The probability of 4 calls being ghost calls on any day

Here $t = 1\ day$ then $\lambda t = 6$

$$P(X = 4) = \frac{e^{-6}.6^4}{4!}$$

$$= \frac{3.2125}{24}$$

$$= 0.1339$$

(ii)   The probability of 10 calls being ghost calls on any two consecutive days

Here $t = 2$ then $\lambda t = 6 \times 2 = 12$

$$\therefore P(X = 10) = \frac{e^{-12} . 12^{10}}{\angle 10}$$

$$= \frac{3,80433.43}{1.2.3.4.5.6.7.8.9.10}$$

$$= \frac{3,80,433.43}{36,28,800}$$

$$= 0.1048$$

***Example* 11:** A manufacture of cotton pins knows that 5% of his product is defective. If he sells cotton pins in boxes of 100 and guarantees that not more than 10 pins will be defective. What is the probability that a box will fail to meet the guaranteed quality?

Sol:

From the given data, the probability of defective cotton pins

$$p = 5\% = 0.05 \text{ and given } n = 100.$$

$\therefore$ The average number of defective pins $\lambda = np = 100 \times 0.05 = 5$

$$\therefore \lambda = 5$$

The probability distribution of Poisson distribution is

$$P(X = x) = \frac{e^{-\lambda} . \lambda^x}{\angle x}.$$

Let the random variable X represents the number of defective pins

Now the required probability is $P(X > 10)$

$$\therefore P(X > 10) = 1 - P(X \leq 10)$$

$$= 1 - \{P(X = 0) + P(X = 1) + \cdots + P(X = 10)\}$$

$$= 1 - \left\{ \frac{e^{-5} . 5^0}{\angle 0} + \frac{e^{-5} . 5^1}{\angle 1} + - - + \frac{e^{-5} . 5^{10}}{\angle 10} \right\}$$

$$= 1 - e^{-5} \left[1 + 5 + \cdots + \frac{5^{10}}{10!} \right]$$

$$= 0.01$$

Example 12: A car hire form has two cars which it hires out day by day. The number of demands for a car on each day is distributed as Poisson variant with mean 1.5. Calculate the proportion of days on which

  (i) Neither car is used and

  (ii) Some demand is refused.

Sol:

The average number of demands of car on each day is $\lambda = 1.5$

  Let the random variable X represents the number of demands for a car on any day.

  The probability distribution of Poisson distribution is

$$P(X = x) = \frac{e^{-\lambda} . \lambda^x}{\angle x} \qquad x = 0, 1, 2 \ldots.$$

  (i) The proportion of days on which probability that neither car is used is

$$P(X = 0) = \frac{e^{-1.5} . (1.5)^0}{\angle 0}$$

$$= e^{-1.5} = 0.2231.$$

  (ii) The Proportion of days on which some demand is refused (means the demand is more than 2 cars) is

$$P(X > 2) = 1 - P(X \le 2)$$

$$= 1 - \{P(X = 0) + P(X = 1) + P(X = 2)\}$$

$$= 1 - \left\{ \frac{e^{-1.5} . (1.5)^0}{0!} + \frac{e^{-1.5} (1.5)^1}{1!} + \frac{e^{-1.5} (1.5)^2}{2!} \right\}$$

$$= 1 - e^{-1.5} (1 + 1.5 + 1.125)$$

$$= 1 - 0.8087$$

$$= 0.1916$$

Example 13: Baby are born in a populated state at the rate of one birth every 12 minutes. The time between births follows an exponential distribution. Find the

  (i) average number of births per day

  (ii) The probability that the number of births will occur in any one day

  (iii) The probability of issuing 50 birth certificates that issued during the first 2 hours of the 3 hours period.

Sol:

From the given data, $\lambda t = 1$ *and* $t = 12$; 5 *births in 1 hour*

The birth rate per day is computed as $24 \times 5 = 120$ births per day

$$\therefore \lambda = 120 \text{ births / day.}$$

(i)   The average number of births per year in the State is

$$\lambda t = \lambda(365) = 120 \times 365 = 43{,}800 \ \ \textit{births / year}$$

(ii)   The probability that number of births in any one day is calculated from the Poisson distribution

$$P(X = x) = \frac{e^{-\lambda t} \cdot (\lambda t)^x}{\angle x} \qquad\qquad x = 0,1,2 \ \ldots .$$

$$\therefore P(x = 0) = \frac{e^{-120}(120)^0}{\angle 0} = 0 \quad \text{(approximately)}$$

(iii) P[issuing 50 birth certificates in 3 hrs. / 40 certificates were issued during the first 2 hours] is equivalent to

P[10 births i.e., $50 - 40 = 10$ in 1 hours i.e., $3 - 2$ hours]

We shall calculate p [10 births in 1 hour].

Given $\lambda = 5$ births / hour

$$P[X = 10] = \frac{e^{-5}(5)^{10}}{\angle 10} = 0.01813.$$

***Example* 14:** Suppose that a printer circuit board (PCB) manufacturing company has been averaging 35 errors per month and that the company has 2 PCB errors in a day. Arbitrarily choosing a time unit of one day, the average number of errors per day is 35 / $30 = 1.1667$. The PCB manufacturing company would like to know

(i)   The probability of no error in a day and

(ii)   Probability of 2 PCB errors in a day.

Sol:

From the given data,

The average no. of errors per day $\lambda = 1.1667$ for the Poisson distribution.

$$P(X = x) = \frac{e^{-\lambda} \cdot \lambda^x}{\angle x} \qquad\qquad x = 0, 1, 2 \ldots .$$

(i)   The probability of no error in a day is

$$P(X = 0) = \frac{e^{-1.1667}.(1.1667)^0}{\angle 0}$$

$$= e^{-1.1667}$$

$$= 0.31139$$

(ii) Probability of 2 PCB errors in a day is

$$P(X = 2) = \frac{e^{-1.1667}.(1.1667)^2}{2!} = 0.2119$$

2.14 Uniform Distribution (Rectangular Distribution)

A random variable X is said to have a continuous uniform distribution over an interval (a, b), if its probability density function is a constant k (say) over the range of the random variable X is given by

$$f(x) = \begin{cases} k & if \quad a < x < b \\ 0 & other\ wise \end{cases}$$

Here $$k = \frac{1}{b-a}$$

Since the total probability is always unity then we have

$$\int_a^b f(x)dx = 1 \;\Rightarrow\; \int_a^b k.dx = 1 \;\Rightarrow\; k(x)_a^b = 1$$

$$\Rightarrow \qquad k(b\text{-}a) = 1 \quad \Rightarrow \quad k = \frac{1}{b-a}$$

$$\therefore \qquad f(x) = \begin{cases} \dfrac{1}{b-a} & if \quad a < x < b \\ 0 & other\ wise \end{cases}$$

Here a and b are the parameters of the uniform distribution on the internal (a, b)

This can also represented as $X \sim u(a,b)$. i.e., X follows uniform distribution in (a, b). This can be soon as

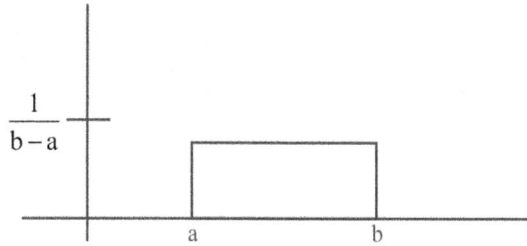

The distribution function of the uniform distribution is given by

$$F(x) = p(X \le x) = \int_{-\infty}^{x} f(x)dx$$

$$p(x < a) = \int_{-\infty}^{a} f(x)dx = 0$$

$$p(a \le X \le b) = \int_{a}^{b} f(x)dx = \int_{a}^{b} \frac{1}{b-a}dx$$

$$F(x) = \begin{cases} 0 & if\ x < a \\ \dfrac{x\text{-}a}{b\text{-}a} & a < x < b \\ 1 & if\ x \ge b \end{cases}$$

The probability distribution function of uniform distribution graph is

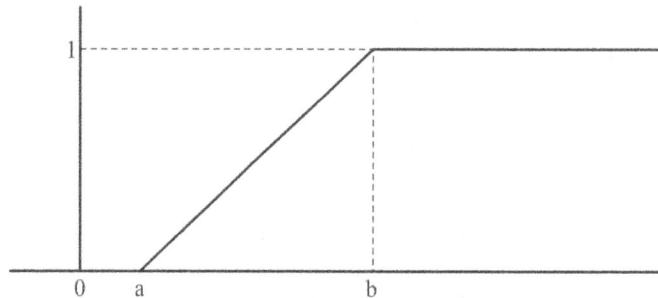

Example 1: Buses arrive at a specified stop at 14 minutes interval starting at 7.00 am that is they arrive at 7.00, 7.15, 7.30, 7.45 and so on. If a passenger arrives at the stop at a random time that is uniformly distributed between 7.00 and 7.30 am. Find the probability that a passenger waits

(i)   Less than 5 minutes for a bus and

(ii) At least 12 minutes for a bus.

Sol:

Given that passengers arrives at the bus stop at a random time is uniformly distributed between 7.00 and 7.30.

Let the random variable X represents the uniform distribution on

$$f(x) = \frac{1}{b-a} \qquad 0 < X < 30$$

$$= \frac{1}{30-0}$$

$$= \frac{1}{30} \qquad 0 < x < 30$$

(i) The passenger will have to wait less than 5 minutes, if he arrives at the stop between 7.10 and 7.15 or 7.25 and 7.30.

$\therefore$ Required probability $= P(10 < X < 15) + P(25 < X < 30)$

$$= \int_{10}^{15} f(x)dx + \int_{25}^{30} f(x)dx$$

$$= \int_{10}^{15} \frac{1}{30} dx + \int_{25}^{30} \frac{1}{30} dx$$

$$= \frac{1}{30} \left[(x)_{10}^{15} + (x)_{25}^{30} \right]$$

$$= \frac{1}{30} \left[(15-10) + (30-25) \right]$$

$$= \frac{1}{30} (5+5)$$

$$= \frac{1}{3}$$

(ii) The passenger will have to wait less than 12 minutes, if he arrives at the stop between 7.00 and 7.03 or 7.15 and 7.18

$\therefore$ The reamed probability $= P(0 < X < 3) + P(15 < X < 18)$

$$= \int_{0}^{3} \frac{1}{30} dx + \int_{15}^{18} \frac{1}{30} dx$$

$$= \frac{1}{30} [3+3]$$

$$= \frac{1}{5}$$

2.15 Geometric Distribution

Let us assume that, the probability of success in an experiment be p and probability of failure be $q = 1 - p$. Supose we have a series of independent trails, if the first success occurred in k^{th} trail, it has proceeded by $k - a$ failure trails. Means that probability of success on the kth trail is given by

$$P(X = k) = \begin{cases} q^{K-1} p & if\, k = 1, 2 \ldots \\ 0 & otherwise \end{cases}$$

If $k = 1$, the first trail be success i.e., $P(X=1) = q^{1-1}.p = p$

If $k = 2$ the first trail be failure and second trail be success

i.e., $P(X = 2) = q^{2-1}.p = q.p$

The geometric distribution is a special case of negative binomial distribution. Also the geometric distribution is a negative binomial distribution where the number of success k is equal to 1.

Definition: A random variable is said to have a geometric distribution its probability distribution is given by

$$P(X = k) = \begin{cases} q^{K-1} p & k = 1,2,3... \\ 0 & otherwise \end{cases}$$

Note: The probability distribution function of the geometric distribution is

k	:	1	2	3	4	----
$P(X = k)$	:	p	qp	q^2p	q^3p	----

The terms of the probability distribution follows geometric series, that's why it is called the geometric probability distribution.

Examples

***Example* 1:** Gopal is a basket ball player. He is a 70% free throw shooter. That his probability of making a free throw is 0.70. What is the probability that Gopal makes his first free throw on his fifth shot?

Sol:

From the given data, the probability of getting on free throw is

$$p = 0.70 \text{ then } q = 0.3$$

The probability distribution of geometric distribution is

$$P(X = k) = q^{k-1} p \qquad k = 1,2 \ldots 5$$

Now the probability that Gopal makes his first free throw on his fifth shot is

$$P(X=1) = (0.3)^{5\text{-}1}\,(0.7)$$
$$= (0.3)^4\,(0.7)^1$$
$$= 0.00567$$

Example 2: Suppose that a trainee soldier shoots a target in an independent fashion. The probability that the target is hit on any shot is 0.8.

    (i)   What is the probability that the target would be hit on 6th attempt?

    (ii)   What is the probability that the it takes him less then 5 shots?

    (iii) What is the probability that it takes him an even number of shots?

Sol:

From the given data, $p = 0.8$ then $q = 0.2$

    The probability distribution of geometric distribution is

$$P(X=k) = q^{k\text{-}1}\,p \qquad k = 1,2 \ldots$$

    (i)   The probability that the target would be hit on 6th attempt

$$P(X=6) = (0.2)^{6\text{-}1}\,(0.8)$$
$$= (0.2)^5\,(0.8)$$
$$= 0.000256$$

    (ii)   The probability that it takes him less then 5 shots

$$P(X<5) = P(X=1) + P(X=2) + P(X=3) + P(X=4)$$
$$= (0.2)^5\,(0.8) + (0.2)^{5\text{-}2}\,(0.8) + (0.2)^{5\text{-}3}\,(0.8) + (0.2)^{5\text{-}4}\,(0.8)$$
$$= (0.8)\,[(0.2)^4 + (0.2)^2 + (0.2)^1]$$
$$= (0.8)\,(0.2)\,[(0.2)^3 + (0.2)^2 + (0.2) + 1]$$
$$= 0.1997 \qquad\qquad\qquad \textbf{\textit{Check Ans:}}\ 0.9984$$

    (iii) The probability the that it takes him an even number of shots is

$$\sum_{k=1}^{\infty} q^{2k-1}\,p = \sum_{k=1}^{\infty} (0.2)^{2k-1}(0.8)$$

$$= (0.2)(0.8)\sum_{k=2}^{\infty} (0.2)^{2k-2}$$

$$= 0.15.\sum_{k=2}^{\infty} \left[(0.2)^2\right]^{k-1}$$

$$= 0.16\left[(0.2)^1 + \left\{(0.2)^2\right\} + \left\{(0.2)^2\right\}^{|3} + \cdots\right]$$

$$= 0.16 \times 0.2\left[1 + (0.2)^2 + \left[(0.2)^2\right]^{|2} + \cdots\right]$$

$$= 0.032\left[\frac{1}{1-(02)^2}\right]$$

$$= 0.0333$$

2.16 Gamma Distribution

Gamma distribution is a two parameter family of continuous probability distribution. It is a probability model for waiting times. For instance, Gamma distribution were fitted to rainfall amounts from different storms, and differences in amounts from seeded and unseeded storms were reflected in differences in estimated parameters α and β. It also plays an important role in reliability problems.   Gamma and exponential distributions allows the gamma to be involved similar types of problem.

Definition: A continuous random variable X is said to have a gamma distribution assuming only non negative values, then its probability density function is given by

$$f(x) = \frac{\beta}{\Gamma(\alpha)}(\beta x)^{\alpha-1}e^{-\beta x} \quad 0 < x < \infty \ , \ \ \alpha > 0, \ \beta > 0$$

$$= 0 \text{ other wise}$$

where α and β are parameters of the distribution, where Gamma function is given by

$$\Gamma(\alpha) = \int\limits_0^\infty x^{\alpha-1}e^{-x}dx \qquad \alpha > 0$$

If $\beta = 1$ then the probability density function becomes

$$f(x) = \frac{x^{\alpha-1}}{\Gamma(\alpha)}e^{-x} \qquad 0 < x < \infty, \ \ \alpha > 0$$

$$= 0 \qquad \text{other wise}$$

α is called the shape parameter and β is called rate parameter. Graph of Gamma function for different values of α and β are shown in figure.

Note : If $\alpha = 1$ then the Gamma distribution becomes exponent distribution.

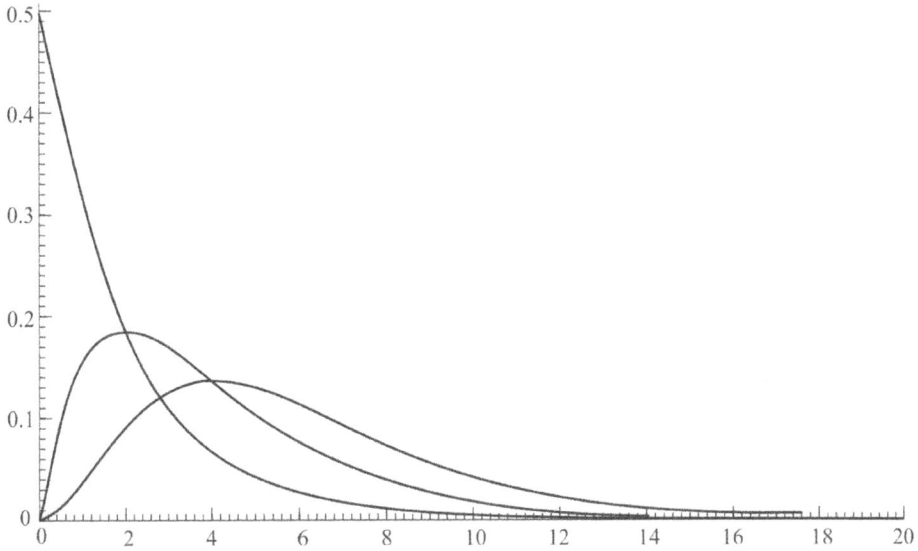

Probability density function of gamma distribution

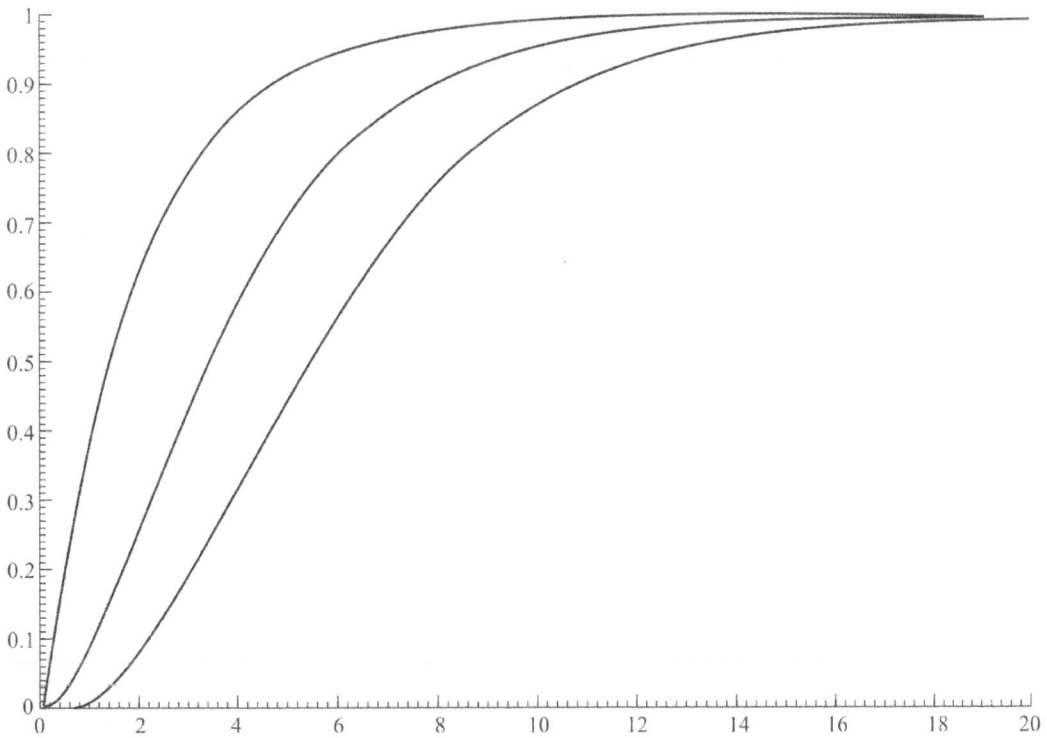

Cumulative distribution function of Gamma distribution

Gamma PDF (gamma = 0.5)

Gamma PDF (gamma = 1)

Gamma PDF (gamma = 2)

Gamma PDF (gamma = 5)

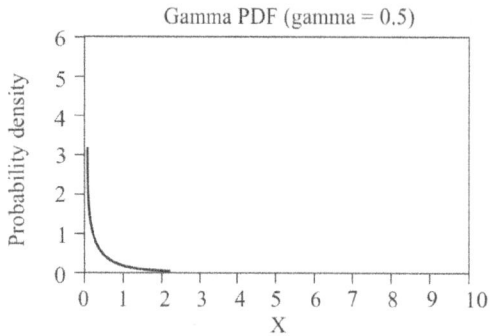

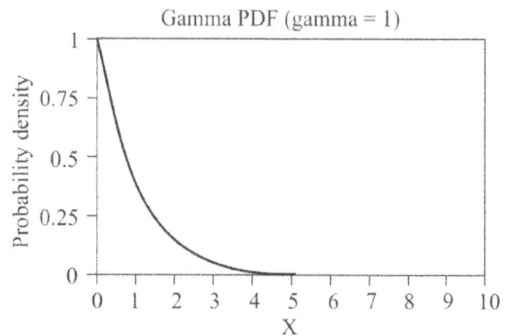

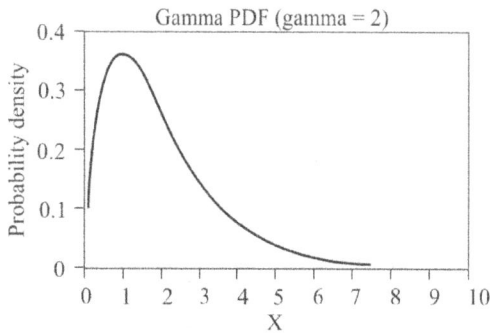

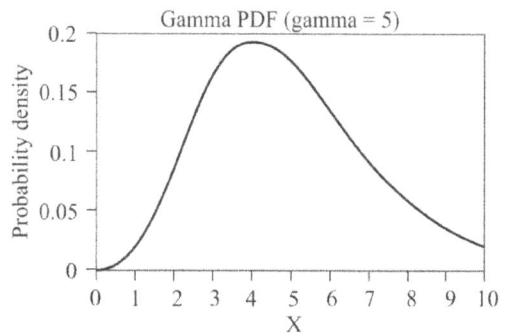

Probability distribution function of Gamma distribution

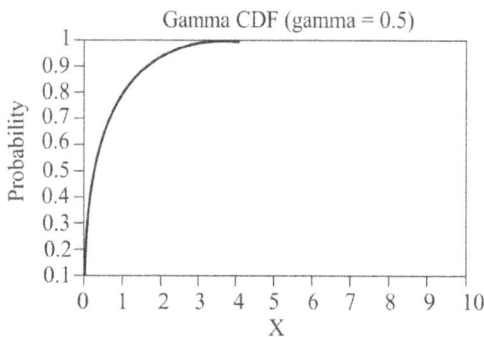

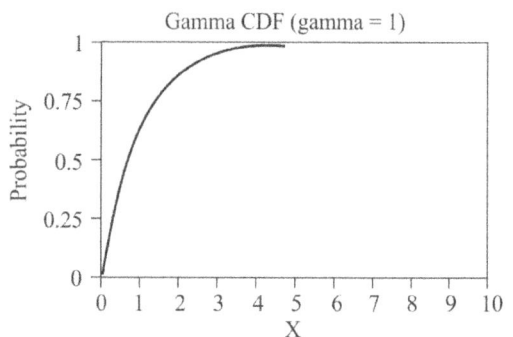

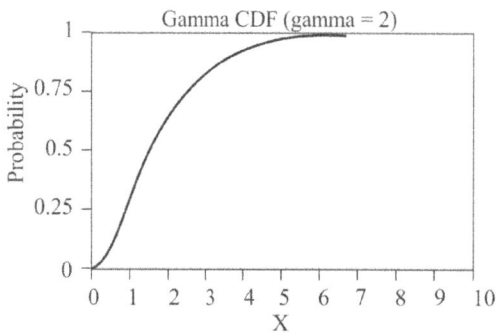

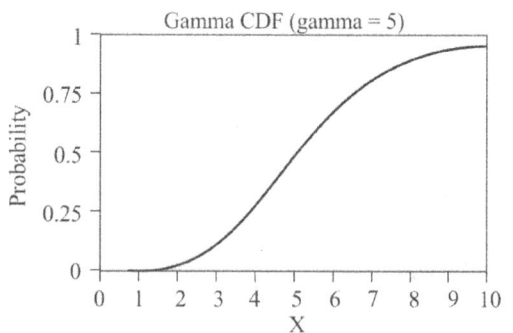

Gamma CDF (gamma = 0.5)

Gamma CDF (gamma = 1)

Gamma CDF (gamma = 2)

Gamma CDF (gamma = 5)

Cumulative distribution function of Gamma distribution

Examples

Example **1:** In a biomedical study with rats a dose–response investigation is used to determine the effect of the dose of a toxicant on their survival time. The toxicant is one that is frequently discharged in to the atmosphere from jet fuel. For a certain dose of the toxicant the study determines that the survival time in weeks has a gamma distribution with $\alpha = 5$ and $\beta = 10$. What is the probability that a rat survives no longer than 60 weeks?

Sol:

The probability density function of Gamma distribution is

$$f(x) = \frac{\beta}{\Gamma(\alpha)} \ (\beta x)^{\alpha-1} \ e^{-\beta x}$$

Given $\alpha = 5$ and $\beta = 10$

Then

$$f(x) = \frac{10}{\Gamma(5)} \ (10x)^{5-1} \ e^{-10x}$$

$$= \frac{10}{4!}(10x)^4 \ e^{-10 \ x}$$

$$= \frac{5}{12}(10x)^4 e^{-10x}$$

Let the random variable X represents the survival time in weeks run the remind probability is

$$P(X \le 60) = \int_0^{60} \frac{5}{12}(10x)^4 \ e^{-10x} \ dx.$$

$$= \frac{5}{12} \int_0^{60}(10x)^4 e^{-10x} dx$$

Put $10x = t$ then $10 \ dx = dt$

When $x = 0$ then $t = 0$ and   When $x = 60$ then $t = 600$

$$= \frac{5}{12} \int_0^{600} t^4 \ e^{-t} \ dt$$

$$= \frac{1}{24} \int_0^{600} t^4 e^{-t} dt \qquad .....(1)$$

Consider

$$\int_0^{600} t^4 \, e^{-t} \, dt = \left[t^4 \frac{e^{-t}}{-1} - 4t^3 \, e^{-t} + 12 \, t^2 \frac{e^{-t}}{-1} - 24t \, e^{-t} + 24 \frac{e^{-t}}{-1} \right]_0^{600}$$

$$= \left[-t^4 \, e^{-t} - 4t^3 \, e^{-t} \, 12 \, t^2 \, e^{-t} - 24 \, t \, e^{-t} - 24 \, e^{-t} \right]_0^{600}$$

$$= \left\{ -e^{-t} \left[t^4 + 4 + 3 + 12 \, t^2 + 24 \, t + 24 \right] \right\}_0^{600}$$

$$= 0 - \text{some thing}$$

Note: For general values of $\alpha = 5$ and $\beta = 10$ the integral is evaluated by using tables of in complete Gamma integrals. The in complete Gamma functions of the form $\int_0^\infty \frac{e^{-x} \, x^{4-1}}{\Gamma(n)} dx$, which have been tabulated for different values of α and n.

***Example* 2:** The daily consumption of milk in a city, in excess of 20,000 liters, is approximately distributed as a Gamma variate with the parameters $\alpha = 2$ and $\beta = \dfrac{1}{10,000}$. The city has a daily stock of 30,000 liters. What is the probability that the stock is insufficient on a particular day?

Sol:

From the given data, $\alpha = 2$ and $\beta = \dfrac{1}{10,000}$.

The probability density function of Gamma distribution is

$$f(x) = \frac{\beta}{\Gamma(\alpha)} \, (\beta x)^{\alpha-1} \, e^{-\beta x}$$

Let the random variable X represents the daily milk consumption in a city and the random variable $Y = X - 20,000$ has a gamma distribution with p d f.

$$g(y) = \frac{1}{(10,000)^2 \Gamma(2)} \, y^{2-1} \, e^{-y/10,000}$$

$$= \frac{y \, e^{-y/10,000}}{(10,000)^2} \qquad 0 < y < \infty.$$

Since the daily milk stock of the city is 30,000 liters.

The stock is insufficient on a particular day is given by

$$P(X > 30,000) = P(Y > 10,000)$$

$$= \int_{10,000}^{\infty} g(y) \, dy$$

$$= \int\limits_{10,000}^{\infty} \frac{y\,e^{-y/10,000}}{(10,000)^2}\,dy$$

Put $t = \dfrac{y}{10,000}$ then $dt = \dfrac{dy}{10,000}$

When $y = 10,000$ then $t = 1$ and when $y = \infty$ then $t = \infty$

$$= \int\limits_{1}^{\infty} t\,e^{-t}\,dt$$

$$= \left\{ t.\left(\frac{e^{-t}}{-1}\right)\right\}_{1}^{\infty} - \left(\frac{e^{-t}}{1}\right)_{1}^{\infty}$$

$$= \left(-t\,e^{-t}\right)_{1}^{\infty} - \left(e^{-t}\right)_{1}^{\infty}$$

$$= \left(0 - e^{-1}\right) - \left(0 - e^{-1}\right)$$

$$= e^{-1} + e^{-1}$$

$$= 2\,e^{-1}$$

$$= 2\,(0.36788)$$

$$= 0.7358$$

∴ The probability that the stock of milk is un sufficient on a particular

Day $= 0.7358$

2.17 Gaussian/Normal Distribution

The most popular continuous probability distribution is the *normal distribution.* It is an indispensable tool for the analysis and the interpretation of the basic data obtained by observation and experience. This is often called *Gaussian distribution.*

Definition: A continuous random variable X is said to have normal distribution with parameters μ and σ, its probability density function is defined as

$$f(x) = \frac{1}{\sigma\sqrt{2\pi}}\,e^{-\frac{1}{2}\left(\frac{x-\mu}{\sigma}\right)^2} \qquad -\infty \leq X \leq \infty, \quad \sigma > 0$$

The graph of the curve is

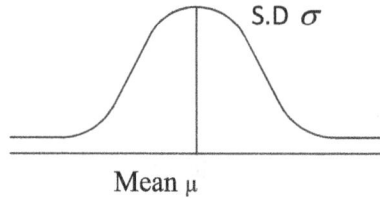

Mean μ

Normal distribution is a bell shaped curve which is symmetrical about the mean μ. The shape of the curve depends on the parameter σ standard deviation. As the standard deviation becomes smaller, the normal distribution becomes sleeper. It is is larger, the normal distribution has a high tendency to flatten out or become broader.

If we take $Z = \dfrac{x - \mu}{\sigma}$ and $Z = x$ only when $\mu = 0$ and $\sigma = 1$ then it becomes

$$f(Z) = \frac{1}{\sqrt{2\pi}} e^{-\frac{1}{2}Z^2} \qquad -\infty \le Z \le \infty$$

Is called a standard normal distribution with mean zero and standard deviation 1.

Its graph is

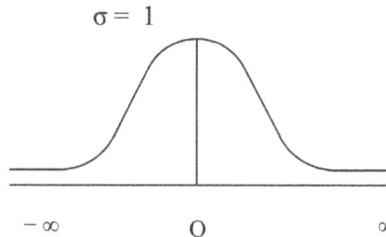

Applications of Gaussian distribution

1. Life of items subjected to wear and tear like bulbs, batteries, currency notes, typres, etc.

2. Length and diameter of certain manufactured products like pipers, screws, discs, etc

3. Breaking streangth of cables, matalic wires, bursting/tensile strength of paper and plastic bags, etc.

4. Heights and weights of children/students/people in a town/locality etc

5. So many examples are given in example problems, etc

Characteristics of the Gaussian distribution:

1. The normal probability curve with mean μ and standard deviation σ is given by

$$f(x) = \frac{1}{\sigma\sqrt{2\pi}} e^{-\frac{1}{2}\left(\frac{x-\mu}{\sigma}\right)^2} \qquad -\infty \leq X \leq \infty$$

2. The curve is bell - shaped and symmetrical about the line $x = \mu$

3. Mean, median and mode of the normal distribution coincide and the normal distribution has uni-modal

4. $f(x)$ decreases rapidly as x increases

5. X-axis is an asymptote to the curve

6. The maximum probability occurs at the point $x = \mu$ and is $\dfrac{1}{\sigma^2\sqrt{2\pi}}$

7. Mean deviation about mean $= \dfrac{4}{5}\sigma$

8. Since $f(x)$ being the probability, can never be negative, so that no portion of the curve lies below the x-axis

9. The linear combination of independent normal variates is also a normal variate.

10. The points of inflexion of the curve are at $x = \mu \pm \sigma$

11.   (i)   Area of the normal curve between $\mu - \sigma$ and $\mu + \sigma$ is 0.6826

        i.e., $P(\mu - \sigma < X < \mu + \sigma) = 0.6826$

   (ii)   Area of the normal curve between $\mu - 2\sigma$ and $\mu + 2\sigma$ is 0.9544

        i.e., $P(\mu - 2\sigma < X < \mu + 2\sigma) = 0.9544$

   (iii)   Area of the normal curve between $\mu - 3\sigma$ and $\mu + 3\sigma$ is 0.9973

        i.e., $P(\mu - 3\sigma < X < \mu + 3\sigma) = 0.9973$

Normal distribution as a limiting form of Binomial Distribution

Normal distribution is another limiting form of the binomial distribution under the following conditions:

   (i)   The number of trails n, is indefinitely large, i.e., $n \to \infty$

   (ii)   Neither p nor q is very small

For large n, the calculation of binomial probabilities is very difficult. In such a case we can use normal curve and the required probability is computed.

Examples

***Example* 1:** If X is a normal variate, find the area A

(i) to the left of $Z = -1.78$ (ii) to the right of $Z = -1.45$ (iii) corresponding to $-0.8 \le Z \le 1.53$ and (iv) to the left of $Z = -2.52$ and the right of $Z = 1.83$

Sol:

    (i)   The area to the left of $Z = -1.78$ is

    i.e., $P(Z \le -1.78) = F(-1.78)$

                                 $= 0.0375$

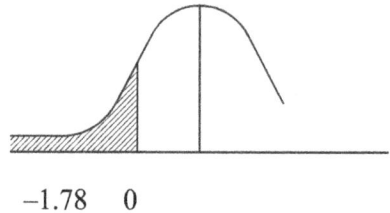

                                         $-1.78 \quad 0$

(ii)   The area to the right of $Z = -1.45$ is

        i.e., $P(Z > -1.45) = 1 - P(Z \le -1.45)$

                            $= 1 - F(-1.45)$

                            $= 1 - 0.0735$

                            $= 0.9265$

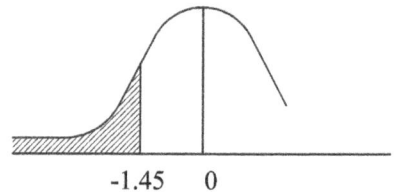

                                        $-1.45 \quad 0$

(iii) The area corresponding to $-0.8 \le Z \le 1.53$ is:

        i.e., $P(-0.8 \le Z \le 1.53) = F(1.53) - F(-0.8)$

                                          $= 0.9397 - 0.2119$

                                          $= 0.7251$

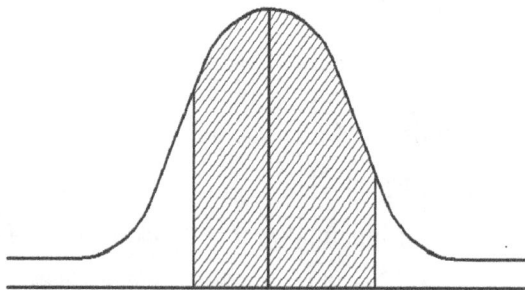

(iv)   The area to the left of $Z = -2.52$ and to the area to the right of $Z = 1.83$ is:

i.e., $P(Z \leq -2.52) = F(-2.52)$

$$= 0.0059$$

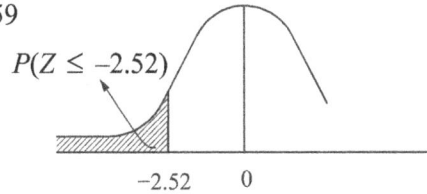

$P(Z \leq -2.52)$

$-2.52 \quad 0$

i.e., $P(Z > 1.83) = 1 - P(Z \leq 1.83)$

$$= 1 - F(1.83)$$

$$= 1 - 0.9664$$

$$= 0.0336$$

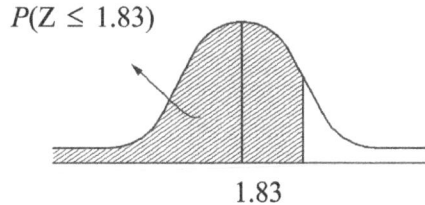

$P(Z \leq 1.83)$

1.83

***Example* 2:** The weights of 300 students are normally distributed with men 68 kgs and standard deviation 3 kgs. How many students have masses (i) greater than 72 kgs (ii) less than or equal to 64 kgs (iii) Between 65 and 71 kgs inclusive.

Sol:

Given $N = 300, \ \mu = 68 \ $ and $\sigma = 3$

Let the random variable X represents the weights of students

The standard normal variate $Z = \dfrac{x - \mu}{\sigma}$

(i) $P(X > 72)$:

When $x = 72$, then $Z = \dfrac{72 - 68}{3} = 1.33$

$P(X > 72) = P(Z > 1.33) = 1 - P(Z \leq 1.33) = 1 - F(1.33)$

$$= 1 - 0.9082 = 0.0918$$

$\therefore$ The number of students have masses greater than 72 kg are

$$= N. \ P(X > 72)$$

$$= 300 \times 0.0918 = 27.54$$

$$\approx 28$$

(ii)   P $(X \leq 64)$:

When x = 64 , then $Z = \dfrac{64 - 68}{3} = -1.33$

P $(X \leq 64) = $ P $(Z \leq -1.33) = $ F(-1.33)

$$= 0.9082$$

∴ The number of students have masses greater than 72 kg are

$$= \text{N. P } (X \leq 64)$$

$$= 300 \times 0.9082$$

$$= 27.246$$

$$\approx 27$$

(i)   P $(65 \leq X \leq 71)$:

When x = 65 , then $Z = \dfrac{65 - 68}{3} = -1$

When x = 71 , then $Z = \dfrac{71 - 68}{3} = 1$

P $(65 \leq X \leq 71)$ = P $(-1 < Z < 1) = $ F$(1) - $ F(-1)

$$= 0.8413 - 0.1587$$

$$= 0.6826$$

∴ The number of students have masses between 65 and 71 kgs inclusive are

$$= N. \, P(65 \leq X \leq 71)$$

$$= 300 \times 0.6826 = 204.78$$

$$\approx 205$$

***Example* 3:** Given that the mean heights of students in a class is 158 cms. With standard deviation of 20 cms. Find how many students heights lie between 150 cms and 170 cms. If they are 100 students in the class.

Sol:

Given $N = 100, \quad \mu = 158 \quad$ and $\sigma = 20$

Let the random variable X represents the weights of students

The standard normal variate $Z = \dfrac{x-\mu}{\sigma}$

$$P(150 < X < 170):$$

When $x = 150$, then $Z = \dfrac{150-158}{20} = -0.4 \ (\text{say } z_1)$

When $x = 170$, then $Z = \dfrac{170-158}{20} = 0.6 \ (\text{say } z_1)$

$$P(150 < X < 170) = P(-0.4 < Z < 0.6) = F(0.6) - F(-0.4)$$

$$= 0.7257 - 0.3446$$

$$= 0.3811$$

$\therefore$ The number of students height lies between 150 and 170 cms. are

$$= N.P(150 < X < 170)$$

$$= 100 \times 0.3811 = 38.11$$

$$\approx 38$$

Example 4: The income of a group of 10,000 persons was found to be normally distributed with mean Rs. 750 p.m. and standard deviation Rs. 50. Show that of this group about 95% had income exceeding Rs. 668 and only 5% had income exceeding Rs. 832. What was the lowest income among the richest 100?

Sol:

Given $N = 10000$; $\mu = 750$ and $\sigma = 50$

The standard normal variate $Z = \dfrac{x-\mu}{\sigma}$

(i) The number of persons had income exceeding 668 is

i.e., $P(X > 668):$

When $x = 668$ then $Z = \dfrac{668-750}{50} = -1.64$

$$P(X > 668) = P(Z > -1.64) = 1 - P(Z \le -1.64) = 1 - F(-1.64)$$

$$= 1 - 0.0505 = 0.9495 \approx 0.95$$

$\therefore$ The number of persons had income exceeding 668 is

$$= N.P(X > 668)$$

$$= 10000 \times 0.95 = 9500$$

So out of 10,000 persons 9,500 had income exceeding 668. Hence the percentage is 95%.

∴ The income of a group of 10,000 persons was about 95% had income exceeding Rs. 668

(ii)   The number of persons had income exceeding 832 is:

i.e., $P(X > 832)$ and

When $x = 832$ then $Z = \dfrac{832 - 750}{50} = 1.64$

$$P(X > 832) = P(Z > 1.64) = 1 - P(Z \leq 1.64) = 1 - F(1.64)$$
$$= 1 - 0.9495 = 0.0505 \approx 0.05$$

∴ The number of persons had income exceeding 668 is

$$= N.P(X > 832)$$
$$= 10000 \times 0.05 = 500$$

So out of 10,000 persons 500 had income exceeding 832. Hence the percentage is 5%.

∴ The income of a group of 10,000 persons was about 5% had income exceeding Rs. 832

(iii)   The lowest income among the richest one is:

Let ' x_1 ' be the lowest income among the richest 100 persons.

$$P(X > x_1) = \frac{100}{10000} = 0.01$$

When $x = x_1$ then $Z = \dfrac{x_1 - 750}{50} = z_1 \text{ (say)}$

$$P(X > x_1) = P(Z > z_1) = 1 - P(Z \leq z_1) = 0.01$$

⇒ $P(Z \leq z_1) = 0.99 \Rightarrow z_1 = 2.33$

∴ $\dfrac{x_1 - 750}{50} = 2.33 \Rightarrow x_1 - 750 = 116.5 \Rightarrow x_1 = 866.5$

$$\approx 867$$

Hence the lowest income of the richest 100 persons is Rs. 867/-

Example 5: In a sample of 100 cases, the mean of a certain test is 14 and standard deviation is 2.5. Assuming the distribution to be normal. Find (i) how many students score between 12 and 15? (ii) how many score above 18? (iii) how many score below 18?

Sol:

Given $N = 1000$, $\mu = 14$ and $\sigma = 2.50$

The standard normal variate $Z = \dfrac{x - \mu}{\sigma}$

(i)   The probability that the students scores between 12 and 15:

i.e., $P(12 < X < 15):$

When $x = 12$, then $Z = \dfrac{12 - 14}{2.5} = -0.8$

When $x = 15$, then $Z = \dfrac{15 - 14}{2.5} = 0.4$

$\therefore$ $P(12 < X < 15) = P(-0.8 < Z < 0.4) = F(0.4) - F(-0.8)$

$= 0.6554 - 0.2119$

$= 0.4435$

$\therefore$ The number of students score between 12 and 15 is

$= N.\ P(12 < X < 15)$

$= 1000 \times 0.4435 = 443.5$

≈ 443 or 444

(ii) The probability that the students score more than 18:

i.e., $P(X > 18):$

When $x = 18$, then $Z = \dfrac{18 - 14}{2.5} = 1.6$

$P(X > 18) = P(Z > 1.6) = 1 - P(Z \leq 1.6) = 1 - F(1.6)$

$= 1 - 0.9452$

$= 0.0548$

$\therefore$ The number of students score more than 18

$= N.\ P(X > 18)$

$= 1000 \times 0.0548 = 54.8 \approx 55$

(iii) The probability that the students score below 18:

i.e., $P(X < 18):$

When $x = 18$, then $Z = \dfrac{18 - 14}{2.5} = 1.6$

$P(X < 18) \quad = P(Z < 1.6) = F(1.6)$

$= 0.9452$

∴ The number of students score more than 18

$$= N.\ P(X < 18)$$

$$= 1000 \times 0.9452$$

$$\approx 945$$

2.18 Exponential Distribution

A continuous random variable X is said to follow exponential distribution with parameter $\theta > 0$ if its probability density function is given by

$$f(x) = \begin{cases} \theta\, e^{\theta x} & x \geq 0 \\ 0 & Otherwise \end{cases}$$

The probability density finds its applications in Queueing theory, random processes, reliability etc. In queueing theory, the service pattern of the units/persons considered to be exponential. In reliability the life time of an equipment, successive arrivals in telephone booths, etc.

The situations arise to apply exponential distribuion as the amount of time until some specific event occurs. For example the amount of time startingf from a car comes for service at a service station, a patient comes at an emergency reception, etc., are all follows exponential distribution.

2.19 Rayleigh Density function

The most important properties of the guassian distribution involve two or more independent normal variables.

Supose X and Y are two independent gaussian random variables with zero mean and same variance σ^2, then

$$f_X(x) = \frac{1}{\sigma\sqrt{2\pi}}\, e^{-\frac{1}{2}\left(\frac{x}{\sigma}\right)^2}$$

and

$$f_Y(y) = \frac{1}{\sigma\sqrt{2\pi}}\, e^{-\frac{1}{2}\left(\frac{y}{\sigma}\right)^2}$$

Since the random variables X and Y are independent, then

$$f_X(x) \times f_Y(y) = f(x.y)$$

$$\Rightarrow \qquad f(x.y) = \frac{1}{\sigma\sqrt{2\pi}}e^{-\frac{1}{2}\left(\frac{x}{\sigma}\right)^2} \times \frac{1}{\sigma\sqrt{2\pi}}e^{-\frac{1}{2}\left(\frac{y}{\sigma}\right)^2}$$

$$= \frac{1}{\sigma^2 2\pi}e^{-\frac{1}{2\sigma^2}\left(x^2+y^2\right)} \qquad\qquad .....(1)$$

Put $x = r\cos\theta$ and $y = r\sin\theta$ $0 \le r < \infty;$ $0 \le \theta < 2\pi$

$\therefore \qquad\qquad r = \sqrt{x^2 + y^2}$ and $dxdy = rdrd\theta$

The corresponding joint probability density function is $P(r,\theta)$

Now $\qquad\qquad f(x.y)\ dxdy = P(r,\theta)\ drd\theta$

$\therefore \qquad\qquad P(r,\theta)\ drd\theta = \frac{r}{\sigma^2 2\pi}e^{-\frac{1}{2\sigma^2}r^2}$

$$P(r) = \int_0^{2\pi} P(r,\theta)d\theta$$

$$= \int_0^{2\pi} \frac{r}{\sigma^2 2\pi}e^{-\frac{1}{2\sigma^2}r^2}\ d\theta$$

$$= \frac{r}{\sigma^2 2\pi}e^{-\frac{1}{2\sigma^2}r^2}\int_0^{2\pi} d\theta$$

$$= \frac{r}{\sigma^2 2\pi}e^{-\frac{1}{2\sigma^2}r^2}\left[\theta\right]_0^{2\pi}$$

$$= \frac{r}{\sigma^2}e^{-\frac{1}{2\sigma^2}r^2} \qquad\qquad 0 \le r < \infty$$

This is called Rayleigh distribution. Therefore the sum of two normal variables with zero mean and same variance is a Rayleigh variable.

Hence the Rayleigh density function is

$$f(r) = \begin{cases} \dfrac{r}{\sigma^2}e^{-\frac{1}{2\sigma^2}r^2} & for\ r \ge 0 \\ 0 & Otherwise \end{cases}$$

The other form of Rayleigh density function is

$$f(r) = \begin{cases} \dfrac{2}{b}(x-a)e^{-\frac{1}{b}(x-a)^2} & \text{for } x \geq a \\ 0 & \text{Otherwise} \end{cases}$$

Its graph is

The distribution function of Rayleigh function is

$$F(x) = \begin{cases} 1 - e^{-\frac{x^2}{\sigma^2}} & \text{for } x \geq 0 \\ 0 & \text{Otherwise} \end{cases}$$

The other form of the distribution function of Rayleigh function is

$$F(x) = \begin{cases} 1 - e^{-\frac{(x-a)^2}{b}} & \text{for } x \geq a \\ 0 & \text{Otherwise} \end{cases}$$

Its graph is

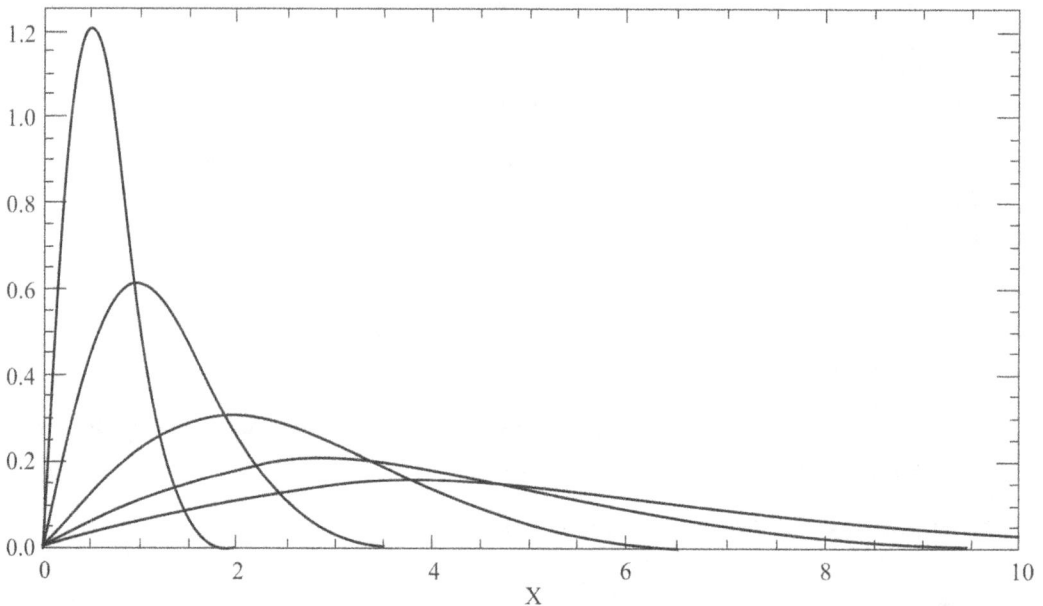

Probability distribution function of Rayleigh distribution

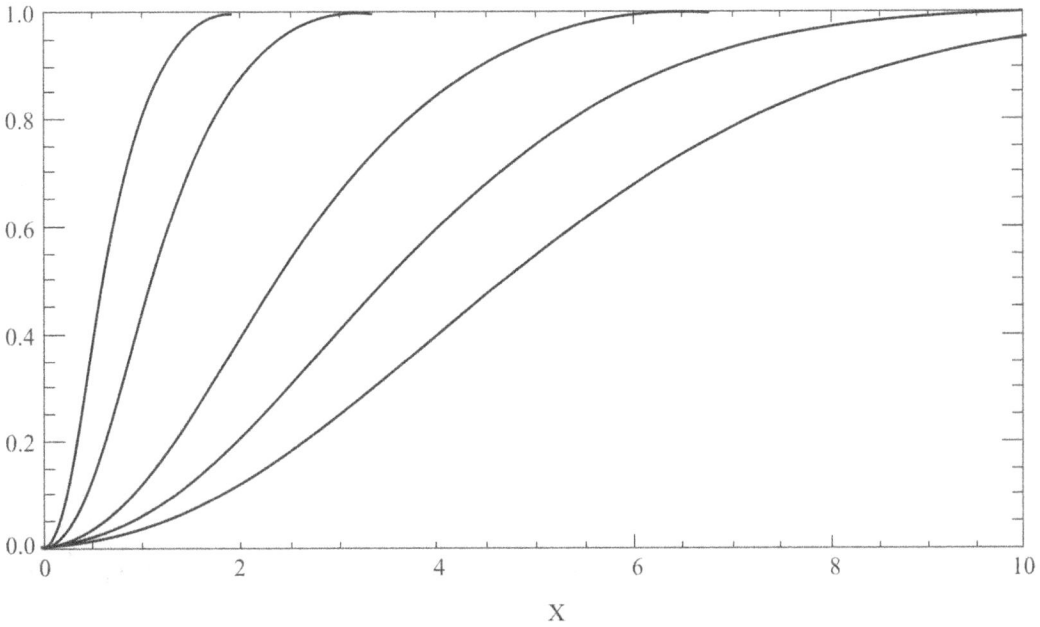

Cumulative distribution function of Rayleigh distribution

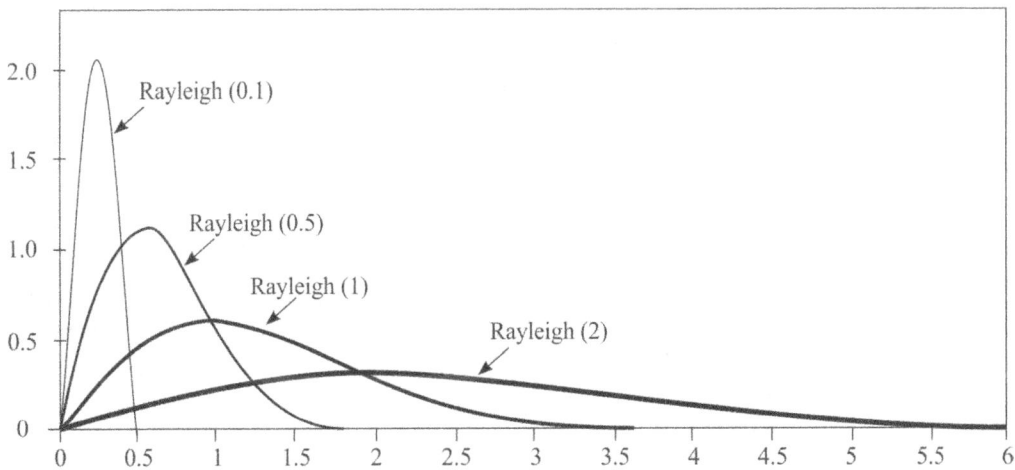

The Rayleigh density function describes the envelope of one type of noise when passed through a band pass filter. It is also is important in analysis of errors in various measurement systems.

2.20 Conditional Distribution

Since by the definition of conditional probability of the event A given B is

$$P[A/B] = \frac{P(A \cap B)}{P(B)} = \frac{P(A,B)}{P(B)}$$

This concept is also applied to the conditional distribution function.

Suppose that A and B are two events of the same experiment E.

Let A be an event $(X \leq x)$ of the random variable X, for given B. Then the conditional distribution of X given B is

$$P[X \leq x/B] = \frac{P[(X \leq x) \cap B]}{P[B]}$$

This is also denoted by $F_X(x/B)$

Here $P[(X \leq x) \cap B]$ denotes the joint probability of the event $[(X \leq x), B]$ i.e., $P[(X \leq x), B]$

Properties of Conditional distribution

Conditional distribution also satisfies the properties same as that of ordinary distribution function.

1. $F_X(-\infty/B) = 0$ same as the property of CDF

   Since $F_X(-\infty/B) = P(X \leq -\infty/B) = 0$ because $P(X \leq -\infty) = 0$

   Hence $F_X(-\infty/B) = 0$

2. $F_X(+\infty/B) = 1$

   Since $F_X(+\infty/B) = P(X \leq +\infty/B) = 1$ because $P(X \leq +\infty) = 1$

   Hence $F_X(+\infty/B) = 1$

3. $0 \leq F_X(x/B) \leq 1$

   Same as the property of CDF, $0 \leq F(x) \leq 1$

4. If $x_1 < x_2$ then $F_X(x_1/B) \leq F_X(x_2/B)$

   Same as the property of CDF, If $x_1 < x_2$ then $F(x_1) \leq F(x_2)$

5. If $x_1 < x_2$, $p(x_1 < X < x_2/B) = F_X(x_2/B) - F_X(x_1/B)$

   Same as the property of CDF, If $x_1 < x_2$ then $p(x_1 < X < x_2) = F(x_2) - F(x_1)$

2.21 Conditioning Event

Since by definition of conditional distribution,

$$F_X(x/B) = P[X \leq x/B] = \frac{P[(X \leq x) \cap B]}{P[B]}$$

The event B can be defined in terms of X or in terms of some other random variable.

(a)   If the event B is defined in terms of X, let $B = (X \leq b)$ where $-\infty < b < \infty$

$$\therefore F_X(x/X \leq b) = P[X \leq x/X \leq b] = \frac{P[(X \leq x) \cap X \leq b]}{P[X \leq b]}$$

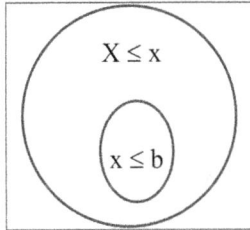

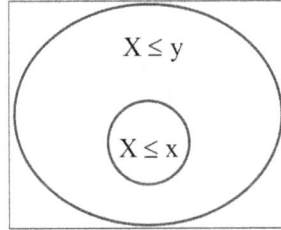

Fig. 1   **Fig.2**

There are two cases arise, whether $b \leq x$ or $x < b$

Case 1. If $b \leq x$ then $P[(X \leq x) \cap X \leq b] = P[X \leq b]$ then,

$$F_X(x/X \leq b) = P[X \leq x/X \leq b] = \frac{P[X \leq b]}{P[X \leq b]} = 1$$

Case 2. If $x < b$ then $P[(X \leq x) \cap X \leq b] = P[X \leq x] = \frac{F_X(x)}{F_X(b)}$

$$\therefore F_X(x/X \leq b) = \begin{cases} \dfrac{F_X(x)}{F_X(b)} & if\, x < b \\ 1 & if\, x \geq b \end{cases}$$

(b)   If the event B is defined in terms of another variable Y, i.e., If the random variable X is conditioned by a second random variable Y, where $Y \leq y$, is called *point conditioning*.

$$\therefore F_{X/Y}(x/y) = P[X \leq x/Y \leq y] = \frac{P[(X \leq x) \cap (Y \leq y)]}{P[Y \leq y]}$$

Provided $P[Y \leq y] \neq 0$

This is also simply represented as $F_{X/Y}(x/y) = \dfrac{F_{X,Y}(x,y)}{F_Y(y)}$

If the random variable Y is defined in such a way that its value lies between two constants y_a and y_b.

i.e., between the interval y_a and y_b. ($y_a \leq Y \leq y_b$), this is called *interval conditioning*.

2.22 Conditional Density Function

This is also similar definition of ordinary density function.

$$f_X(x/y) = \frac{d}{dx}\left[F_X(x/Y)\right]$$

In the above case of $F_X(x/X \leq y) = \begin{cases} \dfrac{F_X(x)}{F_Y(y)} & \text{for } x < y \\[2mm] 1 & \text{for } x \geq y \end{cases}$

i.e., The corresponding conditional density function is given by

$$f_X(x/X \leq y) = \begin{cases} \dfrac{F_X(x)}{\displaystyle\int_{-\infty}^{y} f_X(x)dx} & \text{for } x < y \\[3mm] 0 & \text{for } x \geq y \end{cases}$$

Conditional Density Properties

Conditional density also satisfies the properties same as that of ordinary density function.

1. $f_X(x/B)) \geq 0$ same as the property, $f(x) \geq 0$

2. $\displaystyle\int_{-\infty}^{\infty} f_X(x/B)dx = 1$ same as the property, $\displaystyle\int_{-\infty}^{\infty} f(x)dx = 1$

3. $f_X(x/B) = \displaystyle\int_{-\infty}^{x} F_X(x/B)dx$

4. $P(x_1 < X < x_2/B) = \displaystyle\int_{x_1}^{x_2} f_X(x/B)dx$

Example

Example 6: If $P(x) = \begin{cases} 0.2x & for\ x=1,2,3. \\ 0 & Otherwise \end{cases}$ find (i) $P(X=1\ or\ 2)$ and

(ii) $P(\frac{3}{2}<X<\frac{5}{2}/X>1)$

Sol:

Given $P(x) = \begin{cases} 0.2x & for\ x=1,2,3. \\ 0 & Otherwise \end{cases}$

(i) $P(X=1\ or\ 2):$

$P(X=1\ or\ 2) = P(X=1)+P(X=2)$

$= 0.2\times1+0.2\times2$

$= 0.6$

(ii) $P(\frac{3}{2}<X<\frac{5}{2}/X>1):$

$$P(\frac{3}{2}<X<\frac{5}{2}/X>1) = \frac{P\left[(\frac{3}{2}<X<\frac{5}{2})\cap(X>1)\right]}{P[X>1]}$$

Since $P\left[(\frac{3}{2}<X<\frac{5}{2})\cap(X>1)\right] = P[(1.5<X<2.5)\cap(X>1)]$ is same as is

same as $P[X=2]$

$P[X=2] = 0.2\times2=0.4$

And $P[X>1]= 1-P[X\leq1] = 1-P[X=1] = 1-0.2 = 0.8$

Since the random variable X is discrete.

$$P(\frac{3}{2}<X<\frac{5}{2}/X>1) = P(\frac{3}{2}<X<\frac{5}{2}/X>1) = \frac{0.4}{0.8} = 0.5$$

Quiz Questions

1. The mean of the probability distribution of the number of heads obtained in two flips of a balanced coin is

(a)   1   (b)   1/2   (c)   4/5   (d)   8/25

2. If the probability distribution of the random variable X is

$X = x$	0	1	2	3	4	5
$P(X = x)$	1/25	2/25	3/25	1/5	1/5	9/25

$$P(1 < x < 4) =$$

(a)   1/25   (b)   2/25   (c)   4/25   (d)   8/25

3. The variance of the probability distribution of problem (2) is

(a)   25   (b)   27   (c)   21   (d)   20

4. If a coin so weighed that probability of getting head is 2/3, X is the random variance assignment to the no. of the head in tossing a coin. Mean is

(a)   1/3   (b)   2/3   (c)   ½   (d)   ¾

5. Probability of getting 2 heads in tossing 5 coins is

(a)   1/256   (b)   3/256   (c)   5/256   (d)   7/256

6. If probability of getting 1 or 6 is a success. The probability of getting at least once 1 or 6 if we throw a dice four timer is

(a)   61/81   (b)   63/81   (c)   65/81   (d)   59/81

7. A sample of 3 items is selected from a box having 6 items of which 3 are detective variance of the distribution of defective items is

(a)   1/20   (b)   3/20   (c)   5/20   (d)   9/20

8. If X is the random variable assigned to number of heads in tossing two coins than the variance is

(a)   0   (b)   1   (c)   ½   (d)   2

9. If the distribution is

X	1	2	3	5	7
$f(x)$	0.2	0.3	0.2	0.25	0.05

Then $E(X) =$ (a) 1   (b)   2   (c)   3   (d)   5

10.   A fair die is thrown. Let X denote the twice the number appearing then the standard deviation at X

(a)   2.5   (b)   2.9   (c)   3.4   (d)   4

11.   If the probability density function of a random variance X is

$$f(x) = \begin{cases} 2x(x-1) & 1 \le x \le 4 \\ 0 & elesewhere \end{cases}$$

then the mean μ is

(a)   28   (b)   56   (c)   32   (d)   64

12.   The probability density function at a random variance X is $\frac{1}{2}$ sin x in $0 \le x \le \pi$ = 0 else, where then the mode of x is

(a)   0   (b) π (c) $\pi/4$ (d) $\pi/2$

13.   The probability density function at a random variance X is $\frac{1}{2}$ sin x in $0 \le x \le \pi$ = 0 else, where then the median is

(a)   0   (b) $\pi/4$ (c) $\pi/2$ (d) Π

14.   If the probability density function at a random variance X is $f(x) = 2x(x-1)$ in $1 \le x \le 4 = 0$ else , where then the variance σ^2 is

(a)   251   (b)   262   (c)   281.6   (d)   246.4

15.   If X is a continuous random variable and $Y = aX + b$ then the expected valued of $Y =$

(a) $E(X)$ (b) $aE(X) + E(b)$

(c) $aE(X) + b$ (d) $E(a+ (X))$

16.   If X is a continuous random variable and $Y = aX + b$ then $Var(Y) =$

(a) $A^2 V(X)$ (b) $aV(X) + b$

(c) $aV(X) + b$ (d) $a^2 V(X)$

17.   If the probability density of a continuous random variable X is $f(x) = 2 e^{-2x}$, $x > 0$. Then $P(x \ge 1/2)$ in $0 < x \le \infty$ is

(a)   0.133   (b)   0.24   (c)   0.368   (d)   0.425

18.   Two dice are thrown let X assign to each point (a,b) in S the maximum of its numbers $X(a. b) = \max(a, b)$ then $P(X = 6) =$

(a)   5/36   (b)   7/36   (c) $\frac{1}{4}$ (d)   11/36

19.   The mean of uniform probability distribution $f(x) = 1/n$ for $x = 1, 2, 3 \ldots \ldots n$ is

(a)   N   (b) n^2 (c) $n - 1/2$ (d) $n + 1/2$

20.   If a continuous random variable has the probability density X is 3/8 then the variance is

(a)   9/320   (b)   11/320   (c)   21/320   (d)   19/320

21. The mean of 20, 22, 25, 28, 30, 55 is

(a)   31   (b)   25   (c)   30   (d)   none

22. The mean of the random variable X is

X	8	10	15	20	8
Frequency	5	8	8	4	5

(a)   30   (b)   12   (c)   25   (d)   none

23. If σ is the SD of $x_1, x_2 \ldots x_n$ then the SD of $ax_1, ax_2 \ldots ax_n$ is

(a) $|a|\sigma$ (b) $a^2\sigma$ (c) $a^2\sigma^2$ (d)   none

24. The expectation of the sum of the points tossing a pair of fair dice is

(a)   1   (b)   7   (c)   0   (d)   none

25. The variance of the sum obtained in tossing a pair of fair dice is . . .

(a) $\dfrac{3}{40}$ (b) $\dfrac{6}{35}$ (c) $\dfrac{35}{6}$ (d)   none

26. The expectation of a discrete random variable X whose probabilities is given

by $f(x) = (1/2)^x$; $x = 1, 2, 3$ - - - -

(a)   2   (b)   1   (c)   0   (d)   none

27. A continuous random variable X has the probability density given by

$$f(x) = \begin{cases} 2e^{-2x} & x > 0 \\ 0 & x \le 0 \end{cases}$$

then $E(X)$ is

(a) $\dfrac{3}{40}$ (b) $\dfrac{9}{4}$ (c) $\dfrac{1}{2}$ (d)   none

28. σ and μ respectively denotes Standard deviation and mean of a random variable

X and $Z = \dfrac{x - \mu}{\sigma}$ then $E(Z)$ is

(a) $\dfrac{1}{4}$ (b)   1   (c)   0   (d)   none

29. σ and μ respectively denotes Standard deviation and mean of a random variable X and $Z = \dfrac{x-\mu}{\sigma}$ then $V(Z)$ is

(a) $\dfrac{1}{4}$ (b)   1   (c)   0   (d)   none

30.   The density of a continuous random variable X is $f(x) = \begin{cases} \dfrac{4x(9-x^2)}{81} & 0 \le x \le 3 \\ 0 & otherwise \end{cases}$

then the mode is

(a) $\sqrt{3}$ (b)   1.62   (c) $2\sqrt{3}$ (d)   none

(*Hint:* Mode occur only when $f(x)$ is maximum. The relative maximum of $f(x)$ occurs when f'(x)=0 i.e., $f'(x) = \dfrac{36x - 4x^2}{81} = 0 \Rightarrow x = \sqrt{3}$)

31.   The density of a continuous random variable X is $f(x) = \begin{cases} \dfrac{4x(9-x^2)}{81} & 0 \le x \le 3 \\ 0 & otherwise \end{cases}$

then the median approximately is

(a) $\sqrt{3}$ (b)   1.62   (c) $2\sqrt{3}$ (d)   none

(*Hint:* The median is that value a for which $P(X \le a) = \frac{1}{2}$. Now for $0 < a < 3$.

$$P(X \le a) = \frac{4}{81} \int_0^a x(9-x^2)dx = \frac{1}{2} \Rightarrow 2a^4 - 36a^2 + 81 = 0$$

The required median must lie between 0 and 3 is given by $a^2 = 9 - \dfrac{9}{\sqrt{2}}$ from

which a = 1.62 approximately.)

32.   In a binomial distribution the mean and variance are 5 and 4 respectively. Then the number of trails is

(a)   20   (b)   25   (c)   30   (d)   35

33. In a family there are 6 children, any particular child being a boy is $\frac{1}{2}$. The probability that there are 3 boys and 3 girls is

    (a) $\frac{5}{16}$ (b) $\frac{1}{2}$ (c) $\sqrt{3}$ (d)   None

34. In a binomial distribution $n = 20$, $p = 0.25$ then its mean is

    (a)   5   (b)   15   (c)   3   (d)   35

Answers

1. (a)	2. (d)	3. (b)	4. (b)	5. (c)
6. (c)	7. (d)	8. (c)	9. (c)	10. (c)
11. (d)	12. (c)	13. (c)	14. (c)	15. (c)
16. (d)	17. (c)	18. (d)	19. (d)	20. (d)
21. (c)	22. (b)	23. (a)	24. (b)	25. (c)
26. -	27. (c)	28. (c)	29. (b)	30. (a)
31. (b)	32. (b)	33. (a)	34. (a)	

Review Quiz Questions

1. If the probability distribution of the random variable X is

$X = x$	0	1	2	3	4	5	6
$P(X=x)$	1/25	1/25	3/25	1/5	1/5	9/25	1/25

    $$P(1 < x < 3) =$$

    (a)   1/25   (b)   2/25   (c)   4/25   (d)   none

2. The variance of the probability distribution of problem (1) is

    (a)   25   (b)   27   (c)   21   (d)   none

3. If a coin so weighed that probability of getting head is 1/3, X is the random variable assignment to the number of the head in tossing a coin. Mean is

    (a)   1/3   (b)   2/3   (c)   ½   (d)   none

4. Probability of getting 2 heads in tossing 5 coins is

    (a)   1/256   (b)   3/256   (c)   5/256   (d)   none

5. If probability of getting 2 or 4 is a success. The probability of getting at least once 2 or 4 if we throw a dice four timer is

  (a)   61/81   (b)   63/81   (c)   65/81   (d)   none

6. A sample of 3 items is selected from a box having 5 items of which 2 are detective mean of the distribution of defective items is

  (a)   1/20   (b)   3/20   (c)   5/20   (d)   none

7. If the distribution is

X	1	2	3	5	7	9	11
$f(x)$	0.1	0.3	0.2	0.15	0.05	0.01	0.1

  Then $E(X) =$

  (a)   1   (b)   2   (c)   3   (d)   none

8. A fair die is thrown. Let X denote the twice the number appearing then the standard deviation at X

  (a)   2.5   (b)   2.9   (c)   3.4   (d)   4

9. If the probability density function of a random variance X is

$$f(x) = \begin{cases} x^2(x-1) & 1 \le x \le 3 \\ 0 & elsewhere \end{cases}$$

  then the mean μ is

  (a)   28   (b)   56   (c)   32   (d)   none

10. The probability density function at a random variance X is ½ cosx in $0 \le x \le \pi = 0$ else, where then the mode of x is

  (a)   0   (b) π (c) $\pi / 4$ (d)   none

11. The probability density function at a random variance X is ½ cosx in $0 \le x \le 2\pi = 0$ else, where then the median is

  (a)   0   (b) $\pi / 4$ (c) $\pi / 2$ (d)   none

12. If the probability density function at a random variance X is $f(x) = x(x-1)$ in $1 \le x \le 2 = 0$ else , where then the variance σ^2 is

  (a)   251   (b)   262   (c)   281.6   (d)   none

13. If X is a continuous random variable and $Y = 2aX + b$ then the expected valued of $Y =$

  (a) $E(X)$ (b) $a E(X) + E(b)$

  (c) $a E(X) + b$ (d)   none

14. If X is a continuous random variable and $Y = X + b$ then $Var\ (Y)$ is

   (a) $A^2\ V(X)$ (b) $a\ V(X) + b$ (c) $a\ V(X) + b$ (d)   None

15. The mean of uniform probability distribution $f(x) = 1/n$ for $x = 1, 2, 3 \ldots \ldots n$ is

   (a)   N   (b) n^2 (c) $n - 1/2$ (d) $n + 1/2$

16. If a continuous random variable has the probability density X is $3/8$ then the variance is

   (a)   9/320   (b)   11/320   (c)   21/320   (d)   19/320

17. If σ is the SD of $x_1, x_2 \ldots x_n$ then the SD of $bx_1, bx_2 \ldots .bx_n$ is

   (a) $|b|\sigma$ (b) $b^2\ \sigma$ (c) $b^2\ \sigma^2$ (d)   none

18. The variance of the sum obtained in tossing a pair of fair dice is . . .

   (a) $\dfrac{3}{40}$ (b) $\dfrac{6}{35}$ (c) $\dfrac{35}{6}$ (d)   none

19. The expectation of a discrete random variable X whose probabilities is given

   by $f(x) = (1/2)^x$; $x = 1, 2, 3 ----$

   (a)   2   (b)   1   (c)   0   (d)   none

20. A continuous random variable X has the probability density given by

$$f(x) = \begin{cases} e^{-x} & x > 0 \\ 0 & x \le 0 \end{cases}$$

   then $E(X)$ is

   (a) $\dfrac{3}{40}$ (b) $\dfrac{9}{4}$ (c) $\dfrac{1}{2}$ (d)   none

21. σ and μ respectively denotes Standard deviation and mean of a random variable

   X and $Z = \dfrac{x - \mu}{\sigma}$ then $V(Z)$ is

   (a) $\dfrac{1}{4}$ (b)   1   (c)   0   (d)   none

22. The density of a continuous random variable X is

$$f(x) = \begin{cases} \dfrac{2x(3 - x^2)}{9} & for\ 0 \le x \le 2 \\ 0 & otherwise \end{cases}$$ then the mode is

   (a) $\sqrt{3}$ (b)   1.62   (c) $2\sqrt{3}$ (d)   none

23. In a binomial distribution the mean and variance are 5 and 4 respectively. Then the number of trails is

   (a) 20   (b) 25   (c) 30   (d) 35

24. In a family there are 5 children, any particular child being a boy is $\dfrac{1}{2}$. The probability that there are 3 boys and 2 girls is

   (a) $\dfrac{5}{16}$ (b) $\dfrac{1}{2}$ (c) $\sqrt{3}$ (d) None

25. In a binomial distribution $n = 20$, $p = 0.25$ then its mean is

   (a) 5   (b) 15   (c) 3   (d) 35

Exercise Questions

1. If a random variable has the probability density
$$f(x) = \begin{cases} 2e^{-2x} & \text{for } x > 0 \\ 0 & \text{for } x \le 0 \end{cases}$$

   Find the probability that it will take on a value
   - (i) Between   1 and 3
   - (ii) Greater than .5

2. A random variable x has the following probability functions

Values of $X = x$	0	1	2	3	4	5	6
$P(x)$	k	$3k$	$5k$	$7k$	$9k$	$11k$	$13k$

   - (i) Find k (ii)   Evaluate $P(X < 4)$
   - (iii) $P(X \ge 5)$ (iv) $P(3 < X \le 6)$
   - (v) What is the smallest value of x for which $P(X \le x) > 1/2$?

3. If $f(x) = k\,e^{-|x|}$ is probability density function in $-\infty \le x \le \infty$. Find the value at k find the variance of the random variable and also find probability between 0 and 4

4. A continuous random variable X has a probability density function $f(x) = 3x^2$, $0 \le x \le 1$ find a and b
   - (i) $P\{x \le a\} = P\{x > a\}$
   - (ii) $P\{X > b\} =$

5. A continuous random variables X has the distribution function

$$F(x) = \begin{cases} 0 & \text{if } x \le 1 \\ k(x-1)^4 & \text{if } 1 < x \le 3 \\ 1 & \text{if } x > 1 \end{cases}$$

　　Find

　　(i)　the value of k and　　(ii)　The probability density function of X

6. The length of the time (in minutes) that a certain lady speaks on the telephone is found to be random phenomenon with a probability function specified by the function.

$$F(x) = \begin{cases} Ae^{-x/5} & \text{for } x \ge 0 \\ 0 & \text{otherwise} \end{cases}$$

7. Find the value of A that makes $F(x)$ is a probability distribution function

　　(b) What is the probability that the no. of minutes that she will talk over the phone is
　　(i)　More than 10 minutes
　　(ii)　Less than 10 minutes
　　(iii)　Between 5 and 10 minutes

8. Two dice are thrown X assign to each point if s the sum of the variables on the faces. Find mean and variance of the random variable.

9. A fair coin is tossed until a head or five tails occurs. Find the expected number E of tosses of the coin.

10. For the probability discrete distribution

X	0	1	2	3	4	5	6	7
F	0	k	$2k$	$2k$	$3k$	k^2	$2k^2$	$7k^2+k$

　　Determine (i) k (ii) Mean (iii) Variance　(iv)　smallest　value　of　x　such　that $P(X \le x) > 1/2$

11. If X is a continuous random variable and $Y = aX+b$. Prove that　　$E(Y) = a\,E(X) + b$ and $V(Y) = a^2\,V(X)$

12. The probability density function is

$$y = \begin{cases} k(3x^2 - 1) & -1 \le x \le 2 \\ 0 & \text{otherwise} \end{cases}$$

　　Find the value of k and find the $p(-1 \le x \le 0)$

13. A continuous random variable X has the pdf, $f(x) = ke^{-\gamma x}$ if $x \ge 0,\ \gamma \ge 0$. Determine the constant k, find mean and variance.

Answers

1.   (i)   0.133;   (ii)   0.368
2. $k = 1/49$: 16/49: 24/49:33/49; $x = 4$.
3.   0.49   4. $a = (1/2)^{1/3}$: $b = (19/20)^{1/3}$

5.   (i) $k = 1/16$ (ii) $f(x) = \begin{cases} \dfrac{(x-1)^3}{4} & 1 \le x \le 3 \\ 0 & otherwise \end{cases}$

6.   (a) $A = 1/5$ (b) (i) $1/e^2$ (ii) $(1 - e^{-1})$ (iii) $(e - 1/e^2)$
7.   Mean $= 7$, Variance $= 5.8$ 8. $E(x) = 1.9$.
9.   (i) $k = 1/10$ (ii)   3.66   (iii)   37.7   (iv)   4
11. $k = 1/6$; $P(-1 \le x \le 0) = 1/3$
12.   mean $= 2/\gamma$; Variance $= 2/\gamma2$

Review Exercise Questions

1. If a random variable has the probability density

$$f(x) = \begin{cases} e^{-x} & \text{for x > 0} \\ 0 & \text{for x} \le 0 \end{cases}$$

   Find the probability that it will take on a value
       (i)   Between   2 and 4   (ii)   Greater than 0.6
2. A random variable x has the following probability functions

Values of $X = x$	0	1	2	3	4	5	6	7
$P(x)$	k	$3k$	$5k$	$7k$	$9k$	$11k$	$13k$	$15k$

       (i)   Find k
       (ii)   Evaluate $P(x < 3)$
       (iii) $P(X \ge 4)$
       (iv) $P(2 < X \le 7)$
       (v)   What is the smallest value of x for which $P(X \le x) > 1/4$?
3. If $f(x) = A\, e^{-2|x|}$ is probability density function in $-\infty \le x \le \infty$. Find the value at A find the variance of the random variable and also find probability between 0 and 2

A continuous random variable 'X' has a probability density function $f(x) = 5x^2$, $0 \le x \le 2$ find a and b

(a) $P\{X \le a\} = P\{X > a\}$ (b) $P\{X > b\} = P\{X \le b\}$

4. A continuous random variables X has the distribution function

$$F(x) = \begin{cases} 1 & \text{if } x \le 1 \\ k(x-1)3 & \text{if } 1 < x \le 2 \\ 0 & \text{if } x > 1 \end{cases}$$

Find (i) The value of k and   (ii) The probability density function of X

5. The length of the time (in minutes) that a certain lady speaks on the telephone is found to be random phenomenon with a probability function specified by the function.

$$F(x) = \begin{cases} ke^{-x/2} & \text{for } x \ge 0 \\ 0 & \text{otherwise} \end{cases}$$

(a) Find the value of k that makes $F(x)$ is a probability distribution function

(b) What is the probability that the no. of minutes that she will talk over the phone is
     (i)   More than 10 minutes
     (ii)   Less than 10 minutes
     (iii)   Between 5 and 10 minutes

6. Two dice are thrown X assign to each point if s the sum of the variables on the faces. Find mean and variance of the random variable.

7. A fair coin is tossed until a head or four heads occurs. Find the expected number E of tosses of the coin.

8. For the probability discrete distribution

X	0	1	2	3	4	5	6
F	0	k	2k	2k	3k	k^2	$2k^2$

Determine (i) k (ii) Mean (iii)Variance (iv) smallest value of x such that

$$P(X \le x) > 1/5$$

9. The probability density function is

$$y = \begin{cases} A(3x^3 - 1) & -1 \le x \le 3 \\ 0 & \text{otherwise} \end{cases}$$

Find the value of k and find the $p(-1 \le x \le 0)$

10. A continuous random variable X has the pdf, $f(x) = Ae^{-\gamma x}$ if $x \ge 0$, $\gamma \ge 0$. Determine the constant k, find mean and variance.

CHAPTER 3

Operations on One Random Variable-Expectations

3.1   Introduction

It is more convenient to describe a random variable by parameters which are representatives of the distribution of the random variable and it consists of too many details.   In this, the random variable is used as a statistical tool for describing the characteristics of some real, physical random models.   In this section, we deal some important statistical tools.   Expectation (i.e., for average physical situations it will be helpful for the actual study of the distribution) moments (for the characterization of the distribution function) variance etc are performed on random variables.

3.2   Expected Value of a Random Variable

Let X be a random variable then the expected value of a random variable X is denoted by E (X) and is defined as

(i)   If X is a discrete random variable and p_i is the corresponding probability then

$$\mathrm{E}\ (X) = \sum_{i=1}^{n} x_i\ p_i$$

$$= \sum x_i\ P(X = x_i)$$

$$= \sum x_i\ P(x_i)$$

(ii)   If X is a continuous random variable and $f(x)$ is the corresponding pdf then

$$E(X) = \int_{-\infty}^{\infty} xf(x)dx$$

Provided $E(X)$ is finite.

Note: The expectation of a random variable is also known as mathematical expectation, or average or simple average or Arithmetic mean.

Properties

1.   If 'a' is any constant, then $E(aX) = a\,E(X)$

    Since $E(X) = \sum\limits_{i=1}^{n} x_i\,p(x = x_i) = a.\sum x_i\,p(x = x_i)$

                     $= a.\,E(X)$

2.   If a is any constant, then $E(a) = 0$

    Since $E(X) = \sum\limits_{i=1}^{n} x_i\,p(X = x_i)$

               $\therefore E(a) = \sum\limits_{i=1}^{n} a\,p(X = x_i) = a\sum\limits_{i=1}^{n} p(X = x_i)$

                               $= a.1$

                               $= a$

3.   If $\bar{x}$ is the mean of the random variable X then $E\,(X - \bar{x}) = 0$.

    Since $E\,(X - \bar{x}) = \sum\left(x_i - \bar{x}\right)pi$

                    $= \sum p_i x_i - \bar{x}\sum pi$

                    $= E(X) - \bar{x}$

                    $= E(X) - E(X)$

                    $= 0$

4.   If $f(x)$ is the probability density function of a random variable X is symmetrical about a point a i.e., $f(a - x) = f(a + x)$ then $E(X) = a$.

5.   If $Y = aX + b$ then $E(X) = a\,E(X) + b$

6.   The mathematical expectation of the sum of the random variables $X_1, X_2 \,.... X_n$ is equal to the sum of their expectations, provided all the expectations exist.

    i.e., $E(X_1 + X_2 + X_3 + ... + X_4) = E(X_1) + E(X_2) + ..._+ E(X_n)$

7. The mathematical expectation of the product of a number of independent random variable is equal to the product of their individual expectations.

i.e., $X_1, X_2 .... X_n$ are independent random variables then

$E(X_1. X_2.X_3 .... X_n) = E(X_1). E(X_2),.... E(X_n)$ If all the expectations exist.

Examples

***Example* 1:** Calculate the mean of the following distribution

$X = x$:	0.3	0.2	0.1	0	1	2	3
$P(X = x)$:	0.05	0.10	0.30	0	0.30	0.15	0.1

Sol: From the given data, the Probability distribution of the random variable X is

$X = x$:	0.3	0.2	0.1	0	1	2	3
$P(X = x)$:	0.05	0.10	0.30	0	0.30	0.15	0.1

Mean of the random variable X is

$$E(X) = \sum_{i=1}^{n} x_i p_i(x_i)$$

$$= 0.3 \times 0.05 + 0.2 \times 0.10 + 0.1 \times 0.30 + 0 \times 0 +$$
$$1 \times 0.30 + 2 \times 0.15 + 3 \times 0.10$$

$$= 0.965$$

***Example* 2:** Find the expectation of the points on a die.

Sol: Let the random variable X assume the values x_i $i = 1,2.... 6$ are 1, 2, 3, 4, 5, 6 each of which occurs with the same probability $\dfrac{1}{6}$ then the probability distribution of the random variable X is

X :	1	2	3	4	5	6
$P(X = x)$:	$\dfrac{1}{6}$	$\dfrac{1}{6}$	$\dfrac{1}{6}$	$\dfrac{1}{6}$	$\dfrac{1}{6}$	$\dfrac{1}{6}$

Than the mathematical expectation of X is $E(X) = \sum x_i p_i$

$$= 1. \frac{1}{6} + 2. \frac{1}{6} + 3. \frac{1}{6} + 4. \frac{1}{6} + 5. \frac{1}{6} + 6. \frac{1}{6}$$

$$= \frac{1}{6} (1 + 2 + 3 + 4 + 5 + 6)$$

$$= \frac{1}{6} \times \frac{6.7}{1.2}$$

$$= \frac{7}{2}$$

$$= 3.5$$

***Example* 3:** A discrete random variable X has possible values $x_i = i^2$, $i = 1, 2, 3, 4, 5$. Which occur with probabilities 0.4, 0.25, 0.15, 0.1 and 0.1 respectively. Find the mean value of X.

Sol: Given probability distribution is $x_i^2 = i^2$

$$i = 1, 2, 3, 4, 5$$

X :	1	4	9	16	25
$P(X = x_i)$:	0.4	0.25	0.15	0.1	0.1

Then the mathematical expectation of X is $E(X) = \sum x_i p_i$

$$= 1.(0.4) + 4(0.25) + 9(0.15) + 16(0.1) + 25(0.1)$$
$$= 0.4 + 1 + 1.35 + 1.6 + 2.5$$
$$= 6.85$$

***Example* 4:** A number is chosen at random from the set 1, 2, 3,.... 100 and another number is chosen at random from the set 1, 2, ... 50. What is the expected value of the product?

Sol:

Let $x \in X$ be the number chosen at random from the set 1, 2, 3.... 100 each of which occurs with the same probability $\frac{1}{100}$ then the probability distribution of the random variable X is

X :	1	2	3	--- --- ---	100
$P(X=x_i)$:	$\frac{1}{100}$	$\frac{1}{100}$	$\frac{1}{100}$	--- ---- ---	$\frac{1}{100}$

Then $E(X) = \sum x_i \, p_i$

$$= 1.\frac{1}{100} + 2.\frac{1}{100} + 3.\frac{1}{100} + \text{-----} + 100.\frac{1}{100}$$

$$= \frac{1}{100} [1 + 2 + - - - - + 100]$$

$$= \frac{1}{100} \cdot \frac{100 \times 101}{2}$$

$$= \frac{101}{2}$$

Similarly y be the number chosen at random from the set 1, 2, 3 - - - - 50 each of which occur with the same probability $\frac{1}{50}$ then the probability distribution of the random variable Y is

Y	:	1	2	3	---	---	---	50
$P(Y = y)$	:	$\frac{1}{50}$	$\frac{1}{50}$	$\frac{1}{50}$	---	---	---	$\frac{1}{50}$

Then

$$E(Y) = \sum x_i \, p_i$$

$$= 1. \frac{1}{50} + 2. \frac{1}{50} + 3. \frac{1}{50} + - - - - 50. \frac{1}{50}$$

$$= \frac{1}{50} [1 + 2 + - - - - + 50]$$

$$= \frac{1}{50} \cdot \frac{50 \times 51}{2}$$

$$= \frac{51}{2}$$

Since X and Y are two independent random variables. Then the expected value of the product is

$$E(XY) = E(X) \ E(Y)$$

$$= \frac{101}{2} \times \frac{51}{2}$$

$$= \frac{5151}{4}$$

***Example* 5:** The natural numbers are the positive values of a random variable X i.e., $X_n = n, n = 1, 2 \ldots \ldots$. These numbers occurs with probabilities $p(x_n) = \left(\dfrac{1}{2}\right)^n$. Find the expected value of X.

Sol: From the given probability distribution $P(x_n) = \left(\dfrac{1}{2}\right)^n$ of the random variable $X_n = n$, $n = 1, 2 \ldots$ is

X	:	1	2	3	- - - - - - - - -
$P(X = x_n)$	:	$\dfrac{1}{2}$	$\left(\dfrac{1}{2}\right)^2$	$\left(\dfrac{1}{2}\right)^3$	- - - - - - - - -

Then $E(X) = \sum x_i \, p_i$

$$= 1.\frac{1}{2} + 2.\left(\frac{1}{2}\right)^2 + 3.\left(\frac{1}{2}\right)^3 + - - - \quad - - - \quad - - -$$

$$= \left(1 \times \frac{1}{2}\right) + \left(2 \times \frac{1}{4}\right) + \left(3 \times \frac{1}{8}\right) + --- \quad --- \quad ---$$

Since $\left(1 \times \dfrac{1}{2}\right) + \left(2 \times \dfrac{1}{4}\right) + \left(3 \times \dfrac{1}{8}\right) + ---$ ---   --- is Arithmetico Geometric progression.

The ratio of G.P is $r = \dfrac{1}{2}$

Let $S = \left(1 \times \dfrac{1}{2}\right) + \left(2 \times \dfrac{1}{4}\right) + \left(3 \times \dfrac{1}{8}\right) + ---$ ---   ---   .....(3.1)

$$\frac{1}{2} S = \left(1 \times \frac{1}{4}\right) + \left(2 \times \frac{1}{8}\right) + \left(3 \times \frac{1}{16}\right) + --- \quad --- \quad --- \quad\quad .....(3.2)$$

Subtract (2) from (1)

$$S - \frac{1}{2}S = \frac{1}{2} + \frac{1}{4} + \frac{1}{8} + --- \quad --- \quad --- \quad \Rightarrow$$

$$\Rightarrow \quad \frac{1}{2}S = \frac{1}{2} + \frac{1}{4} + \frac{1}{8} + --- \quad --- \quad ---$$

$$= \frac{1}{2}\left[1 + \frac{1}{2} + \frac{1}{4} \quad --- \quad --- \quad --- \right]$$

$$\therefore \qquad S = 1 + \frac{1}{2} + \frac{1}{4} + \text{---} \quad \text{---} \quad \text{---}$$

This is in Geometric progression. Here $a = 1$, $r = \dfrac{1}{2}$

The sum of the infinite series

$$S_\infty = \frac{a}{a-r}$$

$$= \frac{1}{1-\dfrac{1}{2}}$$

$$= \frac{1}{(1/2)}$$

$$= 2$$

$$\therefore \qquad E(X) = 2$$

***Example* 6:** Two unbiased dice are thrown. Find the expected value of the sum of the numbers of points on them.

Sol: Let the random variable X represents the sum of the numbers of points on two dice when thrown.

So from the previous knowledge the probability distribution of the random variable X and with its probability is

X :	2	3	4	5	6	7	8	9	10	11	12
$P(X=x)$:	$\dfrac{1}{36}$	$\dfrac{2}{36}$	$\dfrac{3}{36}$	$\dfrac{4}{36}$	$\dfrac{5}{36}$	$\dfrac{6}{36}$	$\dfrac{5}{36}$	$\dfrac{4}{36}$	$\dfrac{3}{36}$	$\dfrac{2}{36}$	$\dfrac{1}{36}$

$$\therefore \qquad E(X) = \sum x_i \, p(X = x_i)$$

$$= 2 \times \frac{1}{36} + 3 \times \frac{2}{36} + \text{----} + 12 \times \frac{1}{36}$$

$$= 7$$

$\therefore$ The expected value of the sum of the numbers of points on them is 7

***Example* 7:** A coin is tossed until a head appears. What is the expectation of number of tosses?

Sol: Let the random variable X represents the number of tosses required to get the head. The probability distribution is

$$X \quad : \quad 1 \quad 2 \quad 3 \quad 4 \quad \text{---} \quad \text{---} \quad \text{---}$$

$$\text{Out come } X \quad : \quad T \quad HT \quad HHT \quad HHHT \quad \text{---} \quad \text{---} \quad \text{---}$$

$$P(X=x) \ : \ \frac{1}{2} \quad \frac{1}{2} \times \frac{1}{2} \quad \left(\frac{1}{2}\right)^3 \quad \left(\frac{1}{2}\right)^4 \quad \text{---} \quad \text{---} \quad \text{---}$$

$$\therefore \quad E\,(X) = \sum x_i \, P(X = x_i)$$

$$= \left(1 \times \frac{1}{2}\right) + \left(2 \times \frac{1}{4}\right) + \left(3 \times \frac{1}{8}\right) + \text{---} \quad \text{---} \quad \text{---}$$

This is an infinite Arithmetico Geometry series. The ratio of $r = \dfrac{1}{2}$

$$\text{Let} \quad S = \left(1 \times \frac{1}{2}\right) + \left(2 \times \frac{1}{4}\right) + \left(3 \times \frac{1}{8}\right) + \text{---} \quad \text{---} \quad \text{---} \qquad \text{.....(3.3)}$$

$$\frac{1}{2}S = \left(1 \times \frac{1}{4}\right) + \left(2 \times \frac{1}{8}\right) + \left(3 \times \frac{1}{16}\right) + \text{---} \quad \text{---} \quad \text{---} \qquad \text{.....(3.4)}$$

Subtract (3.4) from (3.3)

$$S - \frac{1}{2}S = \frac{1}{2} + \frac{1}{4} + \frac{1}{8} + \text{---} \quad \text{---} \quad \text{---} \quad \Rightarrow$$

$$\Rightarrow \quad \frac{1}{2}S = \frac{1}{2} + \frac{1}{4} + \frac{1}{8} + \text{---} \quad \text{---} \quad \text{---}$$

$$= \frac{1}{2}\left[1 + \frac{1}{2} + \frac{1}{4} \quad \text{---} \quad \text{---} \quad \text{---}\right]$$

$$\therefore \quad S = 1 + \frac{1}{2} + \frac{1}{4} + \text{---} \quad \text{---} \quad \text{---}$$

This is in Geometric progression. Here $a = 1, r = \dfrac{1}{2}$

The sum of the infinite series

$$S_\infty = \frac{a}{a-r}$$

$$= \frac{1}{1 - \dfrac{1}{2}}$$

$$= \frac{1}{\frac{1}{2}}$$

$$= 2$$

$$\therefore \qquad E\ (X) = 2$$

Hence the expected number of tosses is 2

***Example* 8:** Find the mean of the geometric distribution

Sol: Since $\qquad E(X) = \sum x\ P(X = x)$

For the geometric distribution

$$E(X = k) \quad = \sum_{k=1}^{\infty} k\ P(X = k)$$

$$= \sum_{k=1}^{\infty} kq^{k-1}p \ = \ p\sum_{k=1}^{q} k\ q^{k-1}$$

$$= \ p.\sum_{k=1}^{\infty} \frac{d}{dq}\left(q^{k}\right) = \ p.\frac{d}{dq}\left[\sum_{k=1}^{\infty}q^{k}\right]$$

$$= \ p.\frac{d}{dq}\left[q + q^{2} + q^{3} + ----\right]$$

$$= \ p.\frac{d}{dq}\left[q\left(1 + q + q^{2} + ----\right)\right]$$

$$= \frac{d}{dq}\left\{q.\frac{1}{1-q}\right\} = \ p.\frac{d}{dq}\left\{\frac{q}{1-q}\right\}$$

$$= p\left\{\frac{(1-q)1 - q(-1)}{(1-q)^{2}}\right\}$$

$$= p\left[\frac{1}{(1-q)^{2}}\right] = \text{p.} \ \frac{1}{p^{2}}$$

$$= \frac{1}{p}$$

$\therefore$ The mean of the geometric distribution is $\dfrac{1}{p}$.

***Example* 9:** Find the mean of the random variable whose probability density function is given by $f(x) = \dfrac{3}{5}10^{-5}$ (100-x) $0 \le x \le 100$.

Sol: Since X is a continuous random variable, then the mean

$$E(X) = \int_{-\infty}^{\infty} x f(x) dx$$

$$= \int_{0}^{100} x \frac{3}{5} 10^{-5} (100 - x)$$

$$= \frac{3}{5} 10^{-5} \int_{0}^{100} x (100 - x) \, dx.$$

$$= \frac{3 \times 10^{-5}}{5} \left[100 \int_{0}^{100} x dx - \int_{0}^{100} x dx \right]$$

$$= \frac{3 \times 10^{-5}}{5} \left[100 \left(\frac{x^2}{2} \right)_{0}^{100} - \left(\frac{x^3}{3} \right)_{0}^{100} \right]$$

$$= \frac{3 \times 10^{-5}}{5} \left[50 \times (100)^2 - \frac{1}{3} (100)^3 \right]$$

$$= \frac{3 \times 10^{-5}}{5} \left[5 \times 10 \times 10^2 - \frac{1}{3} \times 10^3 \right]$$

$$= \frac{3}{5} \times 10^{-5} \times 10^3 - \frac{3}{5} \times \frac{1}{3} \times 10^{-5} \times 10^3$$

$$= 3 \times 10^{-2} = \frac{14}{7} \times 100.50$$

$$= \frac{7}{250}$$

***Example* 10:** A wire of length a units is divided into two parts. If the first part of length X then find $E(X)$

Sol: Since the position of the point of division are equally likely, so X is uniformly distributed in $(0,a)$.

The pdf of uniform distribution

$$f(x) = \frac{1}{b - a}$$

$$= \frac{1}{a-0}$$

$$= \frac{1}{a}$$

$$\therefore \qquad f(x) = \frac{1}{a}$$

Now

$$E(X) = \int_{-\infty}^{\infty} xf(x)dx$$

$$= \int_{0}^{a} x\frac{1}{a}dx$$

$$= \frac{1}{a}\left(\frac{x^2}{2}\right)_{0}^{a}$$

$$= \frac{1}{a}\cdot\frac{a^2}{2}$$

$$= \frac{a}{2}$$

$$\therefore \qquad E(X) = \frac{a}{2}$$

***Example* 11:** Let X be a continuous random variable with probability density function (pdf)

$$f(x) = \frac{8}{x^3}, \ x > 2 \text{ find } E(W) \text{ where } w = \frac{x}{3}$$

Sol: Give $f(x) = \frac{8}{x^3}$ for $x > 2$

Now

$$E(X) = \int_{-\infty}^{\infty} xf(x)dx$$

$$= \int_{2}^{\infty} x\frac{8}{x^3}dx$$

$$-8.\int_{2}^{\infty} x^2\frac{8}{x^3}dx$$

$$= 8 \left(\frac{x^{-1}}{-1} \right)_{2}^{\infty}$$

$$= -8 . \left(\frac{1}{x} \right)_{2}^{\infty}$$

$$= -8 \left[0 - \frac{1}{2} \right] = 4$$

Now

$$E(w) \quad = E \left(\frac{x}{3} \right)$$

$$= \frac{1}{3} E(X)$$

$$= \frac{1}{3} . 4 = \frac{4}{3}$$

$$\therefore \qquad E(w) = \frac{4}{3}$$

***Example* 12:** A continuous random variable x has p.d.f $f(x) = kx(2-x) \quad 0 \le x \le 2$. Find the mean of the distribution.

Sol: Given f(x) $= kx(2-x) = k(2x-x^2) \qquad 0 \le x \le 2$

First to find the value of k since we know that

$$\int_{\infty}^{\infty} f(x) dx = 1$$

$$\Rightarrow \qquad k \int_{0}^{2} (2x - x^2) dx = 1$$

$$\Rightarrow \qquad k \left[2 \left(\frac{x^2}{2} \right)_{0}^{2} - \frac{1}{3} \left(x^3 \right)_{0}^{2} \right] = 1$$

$$\Rightarrow \qquad k \left[4 - \frac{8}{3} \right] = 1$$

$$\Rightarrow \qquad k \left[\frac{12 - 8}{3} \right] = 1$$

$$\Rightarrow \qquad k \left[\frac{4}{3} \right] = 1$$

$$\Rightarrow \qquad k = \frac{3}{4}$$

The mean of the random variable x is

$$E(X = \int_{-\infty}^{\infty} xf(x)dx$$

$$= \int_{0}^{2} x. \; \frac{3}{4}\left(2x-x^2\right)dx$$

$$= k\frac{3}{4}\int_{0}^{2}\left(2x^2 - x^3\right)dx$$

$$= k\frac{3}{4}\left[\frac{16}{3} - 4\right]$$

$$= \frac{3}{4}\left[\frac{2}{3}(8) - \frac{1}{4}.16\right]$$

$$= \frac{3}{4}\left[\frac{16}{3} - 4\right]$$

$$= \frac{3}{4}\left[\frac{16-12}{3}\right]$$

$$= \frac{3}{4}\;\frac{4}{3}$$

$$= 1$$

Example 13: Find the mean of the Normal Distribution

Sol: The probability density function of normal distribution is

$$f(x) = \frac{1}{\sigma\sqrt{2\pi}}e^{-\frac{1}{2}\left(\frac{x-\mu}{\sigma}\right)^2} \qquad\qquad -\infty \le X \le \infty$$

$$\text{Mean} = E(X) = \int_{\infty}^{\infty} xf(x)\,dx \;=\; \int_{-\infty}^{\infty} x\,\frac{1}{\sigma\sqrt{2\pi}}e^{-\frac{1}{2}\left(\frac{x-\mu}{\sigma}\right)^2}\,dx$$

Put $\dfrac{x-\mu}{\sigma} = z$ then $dx = \sigma\,dz$ and $x = \mu+\sigma z$ also limits unaltered

$$= \frac{1}{\sigma\sqrt{2\pi}} \int_{-\infty}^{\infty} (\mu + \sigma z)\, e^{-\frac{1}{2}z^2}\, \sigma\, dz$$

$$= \frac{\mu}{\sqrt{2\pi}} \int_{-\infty}^{\infty} e^{-\frac{1}{2}z^2}\, dz + \frac{\sigma}{\sqrt{2\pi}} \int_{-\infty}^{\infty} z\, e^{-\frac{1}{2}z^2}\, dz$$

$$= \frac{\mu}{\sqrt{2\pi}} 2 \int_{0}^{\infty} e^{-\frac{1}{2}z^2}\, dz + 0$$

$\because z\, e^{-\frac{1}{2}z^2}$ is an odd and $e^{-\frac{1}{2}z^2}$ is even function

$$= \frac{2\mu}{\sqrt{2\pi}} \frac{\sqrt{\pi}}{\sqrt{2}} \qquad\qquad \because \int_{0}^{\infty} e^{-\frac{1}{2}z^2}\, dz = \sqrt{\frac{\pi}{2}}$$

$$= \mu \qquad\qquad\qquad\qquad\qquad .....(3.5)$$

$\therefore$ The mean of the Normal Distribution is 'μ'

***Example* 14:** Find the Mean of the Gamma Distribution

Sol: Since
$$E(X) = \int_{-\infty}^{\infty} x\, f(x)\, dx$$

$$= \int_{0}^{\infty} x. \frac{\beta}{\Gamma(\alpha)} (\beta x)^{\alpha-1} e^{-\beta x} dx.$$

$$= \frac{\beta}{\Gamma(\alpha)} \int_{0}^{\infty} x (\beta x)^{\alpha-1} e^{-\beta x} dx$$

Put $t = \beta x$ then $dt = \beta\, dx$. The limits are unattended and $x = \dfrac{t}{\beta}$ and $dx = \dfrac{1}{\beta} dt$

$$= \frac{\beta}{\Gamma(\alpha)} \int_{0}^{\infty} \left(\frac{t}{\beta}\right). t^{\alpha-1}.e^{-t} \frac{1}{\beta}\, dt$$

$$= \frac{1}{\beta\, \Gamma(\alpha)} \int_{0}^{\infty} e^{-t}. t^{\alpha}\, dt$$

$$= \frac{1}{\beta\, \Gamma(\alpha)} \int_{0}^{\infty} e^{-t} t^{(\alpha+1)-1} dt$$

$$= \frac{1}{\beta\, \Gamma(\alpha)}\, \Gamma(\alpha + 1)$$

$$= \frac{\alpha \, \Gamma(\alpha)}{\beta \, \Gamma(\alpha)} \qquad\qquad \because \; \Gamma(\alpha+1) = \alpha \, \Gamma(\alpha)$$

$$= \frac{\alpha}{\beta}$$

$\therefore$ Mean of the Gamma distribution is $\dfrac{\alpha}{\beta}$.

3.3 Function of a Random Variable

Let X be a random variable expressed as a function of another random variable Y, then

$X = g(Y)$. Here g is a function defined on the range of Y with values in the range of X such a function has a specific rule, i.e., for every value of y of Y there exists X has a value $g(y)$ uniquely.

For e.g., let X represents the number of heads when tossing of two coins at a time, then the probability distribution of heads, X is

$$X \; : \quad 0 \qquad 1 \qquad 2$$
$$P(X{=}x) : \quad \frac{1}{4} \qquad \frac{1}{2} \qquad \frac{1}{4}$$

The event Y of both coins becomes known, the number of heads X is completely determined by $X = g(Y)$, where g is the function defined by

Outcome y : $\qquad\qquad\qquad TT \quad HT \quad TH \quad HH$

Number of Heads $X = g(Y)$: $\quad 0 \qquad 1 \qquad 1 \qquad 2$

This relationship is shown as

Possible values of Y	Possible values of $X = g(Y)$
TT	0
HT	1
TH	1
HH	2

Here the probability of each possible value of x of X is the sum of the probabilities of those y for which $g(y) = x$. For $x = 0$ and 2 there is a unique y giving $g(y) = x$. So

$P(X = x) = \dfrac{1}{4}$ for these x. But there are two results of y given $g(y) = 1$, so

$P(X = 1) = \dfrac{1}{4} + \dfrac{1}{4} = \dfrac{1}{2}$. This is the example for function of a random variable.

3.4 Moments

If X is a random variable then $E(X^r)$ is called r^{th} order moments of X about the origin and is denoted by μ_r^1.

i.e., $\mu_r^1 = E(X^r)$

And $E\left[(X-\mu)^r\right]$ is called r^{th} order central moment of X about mean and is denoted by μ_r

$E\{|X|^r\}$ and $E\{|X-\mu|^r\}$ (=(denoted by) β_n) are called absolute moments of X and

$E\{(X-a)^n\}$ and $E\{|X-a|^n\}$ are called generalized moments of X.

Note:

Since $E(X) = \begin{cases} \sum xp(X=x) & \text{if } X \text{ is dicrete random variable} \\ \int\limits_{-\infty}^{\infty} xf(x)dx & \text{if } X \text{ is continuous random variable} \end{cases}$

$E(X^r) = \begin{cases} \sum x^r P(X=x) & \text{if } X \text{ is discrete random variable} \\ \int\limits_{-\infty}^{\infty} x^r f(x)dx & \text{if } X \text{ is contiuous random variable} \end{cases}$

There are two types of moments
   (i)   Moments about origin and
   (ii)   Moments about mean

3.5 Moments about the Origin and the Mean

For each positive integer r ($r = 0, 1, 2 \ldots$), the rth moment of the random variable X about the origin is denoted by $\mu_r^/$ and is defined as

$$\mu_r^/ = E(x^r) = \begin{cases} \sum x^r P(X=x) & \text{if } X \text{ is discrete random variable} \\ \int\limits_{-\infty}^{\infty} x^r f(x)dx & \text{if } X \text{ is continuous random variable} \end{cases}$$

3.6 Moments about the Mean

For each positive integer r ($r = 0, 1, 2, ...$), the rth moment of the random variable X about the mean is demoted by μ_r and is defined as

$$\mu_r = E\left[(x-\mu)^2\right] = \begin{cases} \sum (x-\mu)^r P(X=x) & \text{if } X \text{ is discret andom variabele} \\ \int\limits_{-\infty}^{\infty} (x-\mu)^r f(x)\, dx & \text{if } X \text{ is continuous random variable} \end{cases}$$

The moments determined about the mean value is called "Central moments"

Note: Relation between moments about origin and central moments

$$\mu_r = \mu_r^1 - {}^rC_1\, \mu_{r-1}^1\, \mu_1^1 + {}^rC_2\, \mu_{r-2}^1 \left(\mu_1^1\right)^2 - .... + (-r)^r \left(\mu_1^1\right)^r.$$

$$\mu_1 = \mu_1^1 - \mu_1^1 = 0$$

$$\mu_2 = \mu_2^1 - \left(\mu_1^1\right)^2 = \text{var}iance$$

$$\mu_3 = \mu_3^1 - 3\mu_2^1\, \mu_1^1 + 2(\mu_1^1)^3$$

$$\mu_4 = \mu_4^1 - 4\mu_3^1\, \mu_1^1 + 6\, \mu_2^1 \left(\mu_1^1\right)^2 - 3(\mu_1^1)^4$$

3.7 Variance of a Random Variable

Variance is an important characteristic of a random variable X, and it is used to measure dispassion (or variation) of the random variable X.

The variance or dispassion of a random variable X is demoted by σ_x^2 and is defined as

$$\sigma_x^2 = Var\,(X) = E(X^2) - [E(X)]^2 \qquad\qquad .....(3.6)$$

Here $E(X^2)$ is called the second moment and is denoted by $\mu_2^{/}$ of the distribution. The first moment is $E(X)$ and is denoted by $\mu_1^{/}$ or $\bar{x}$ and the variance is called the second central moment μ_2. We shall discuss the moment's detail.

Equation (1) is obtained from the definition

$$\sigma_x^2 = E\left[(X-\mu)^2\right]$$

$$= E\left[x^2 - 2\mu x + \mu^2\right]$$

$$= E(X^2) - 2\,\mu\,E(X) + \mu^2$$

$$= E(X^2) - 2 \mu \cdot \mu + \mu^2 \qquad \qquad \because E(X) = \mu$$

$$= E(X^2) - 2 \mu^2 + \mu^2$$

$$= E(X^2) - \mu^2$$

$$\sigma_x^2 = E(X^2) - [E(X)]^2 \qquad \qquad \qquad .....(3.7)$$

$$Var\ (X) = \sigma_x^2 = E\Big[X - \{E(X)\}\Big]^2$$

$$= E\ \{(X - \mu)\}^2$$

where μ is the mean random variable X.

1. If X is a discrete random variable then the variance is given by

$$\sigma_x^2 = \sum_{i=1}^{n} (x_i - \mu)^2\ p(X = x_i) \qquad \qquad .....(3.8)$$

2. If X is a continuous random variable, then the variance is given by

$$\sigma_x^2 = \int_{-\infty}^{\infty} (x_i - \mu)^2 f(x)\ dx \qquad \qquad .....(3.9)$$

Note: The variance σ_x^2 of the random variable X is further simplified as

Properties:

1. If a is a constant, and X is a random variable then *Var (a X)* $= a^2$ *Var (X)*

   Proof:

   Since $\qquad\qquad Var(aX) = \Big\{E(ax)^2\Big\} - \Big\{E(ax)\Big\}^2 \qquad$ by equation (3.9)

   $$= a^2 E\Big(x^2\Big) - \{aE(x)\}^2$$

   $$= a^2 E\Big(x^2\Big) - a^2 \{E(x)\}^2$$

   $$= a^2 \Big\{E\Big(x^2\Big) - \{E(x)\}^2\Big\}$$

   $$= a^2\ Var\ (X)$$

   $\therefore \qquad\qquad V\ (ax) = a^2\ \sigma_x^2$

2. If a is a constant then $V(a) = 0$

   Proof:

   Since $\qquad\qquad V(X) = E\Big(x^2\Big) - \{E(x)\}^2$

Now $V(a) = E(a^2) - \{E(a)\}^2$

$\quad\quad\quad = a^2 - \{a\}^2$

$\quad\quad\quad = 0$

3. If a is a constant then $Var\ (X + a) = Var\ (X)$ i.e., $\sigma_x^2\ (X + a) = \sigma_x^2$

   Proof:

$$V(X + a) = E\left[(X+a)^2\right] - \left[E\{X+a\}\right]^2$$

$$= E\left[X^2 + 2aX + a^2\right] - \left[E[X] + E(a)\right]^2$$

$$= E(X^2) + 2a\ E(X) + a^2 - \left\{[E(X)]^2 + 2E(X).a + a^2\right\}$$

$$= E(X^2) + 2a\ E(x) + a^2 - [E(X)]^2 - 2aE(X) - a^2$$

$$= E(X^2) - [E(X)]^2$$

$$= Var\ (X)$$

$\therefore\quad\quad \sigma_{x+a}^2 = \sigma_x^2$

4. If X and Y are two random variables then

$$\sigma_{x+y}^2 = \sigma_x^2 + \sigma_y^2 + 2\{E(XY) - E(X)E(Y)\}$$

$$V(X + Y) = E\left[(X+Y)^2\right] - \left[E\{X+Y\}\right]^2$$

$$= E\left[X^2 + 2XY + Y^2\right] - [E(X) + E(Y)]^2$$

$$= E(X^2) + 2\ E(XY) + E(Y^2) - \left\{[E(X)]^2 + [E(Y)]^2 + 2E(X)E(Y)\right\}$$

$$= E(X^2) - \{E(X)\}^2 + E(Y^2) - \{E(Y)\}^2 + 2\ \{E(XY) - E(X)E(Y)\}$$

$$= \sigma_x^2 + \sigma_y^2 + 2\{E(XY) - E(X)E(Y)\} \quad\quad\quad .....(3.10)$$

Similarly,

$$V(X-Y) = \sigma_x^2 + \sigma_y^2 - 2\{E(XY) - E(X)E(Y)\} \qquad \qquad \dots(3.11)$$

Note: X and Y are independent random variables then $E(XY) = E(X).E(Y)$

And from the equations (3.10) and (3.11) then $V(X \pm Y) = \sigma_x^2 + \sigma_y^2$

Standard Deviation: Standard deviation is the positive square root of the variance of the random variable X and is denoted by σ_x.

$$\therefore \sigma_x = +\sqrt{V(X)}$$

3.8 Skewness

A frequency distribution is said to be symmetrical when the values of the variable equidistant from their mean, have equal frequencies. If a frequency distribution is not symmetrical, it is said to be asymmetrical or skewed or any deviation from symmetry is called skewness. Skewness may be the positive or negative. A distribution is said to be positively skewed if the frequency curve has a longer tail towards the higher values of x.

i.e. The Skewness of the density function is also called as "third central moment" of the random variable and is given by

$$\mu_3 = E\left[(x-\mu)^3\right]$$ The normalized 3rd central moment is known as the skewness of

the density function or *Coefficient of skewness*.

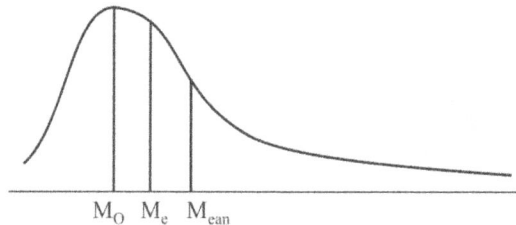

Fig. 3.1

In this figure, the distribution is positively skewed *mean>median>mode*

A distribution is said to be negatively skewed if the frequency curve has a longer tail towards the lower values of X. In this negatively skewed distribution mean < median > mode

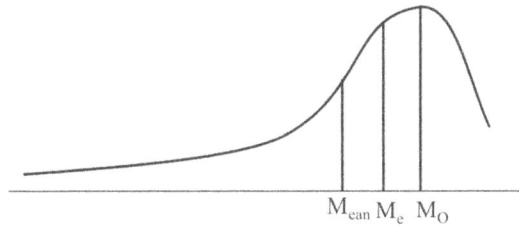

Fig. 3.2

In this figure, the distribution is positively skewed *Mean < Median < Mode*

For a symmetrical distribution, *Mean = Median = Mode*

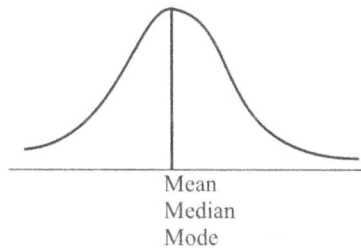

Fig. 3.3

In this figure, the distribution is symmetrical. *Mean = Median =* Mode

Measures of skewness

The degree of skewness is measured by its coefficient.

1. Pearsons first measure:

$$\text{Skewness} = \frac{Mean - \text{Mode}}{S \tan dard \text{ Deviation}}$$

2. Pearsons 2nd measure:

$$\text{Skewness} = \frac{3(\text{Mean} - \text{Median})}{\text{Standard Deviation}}$$

3. Bowley's measure:

$$SK = \frac{(Q_3 - Q_2) - (Q_2 - Q_1)}{(Q_3 - Q_2) + (Q_2 - Q_1)} = \frac{(Q_3 Q_1 - 2Q)}{Q_3 - Q_1}$$

where Q_1, Q_2, and Q_3 are 1st, 2nd & 3rd quintiles respectively.

3.9 Kurtosis

Kurtosis is the peakedness of the frequency curve. In two or more distributions having same average, dispersion and skewness, one may have high concentration of values near the mode, and in this case, frequency curve will show a sharper peak than the others. This characteristic of frequency distribution is known as kurtosis.

Kurtosis is measured by the coefficient β_2 is defined as

$$\beta_2 = \frac{u_4}{u_2^2}$$

where u_2 and u_4 are the 2nd and the 4th central months.

A distribution is said to be

Platy-kurtic if $\beta_2 < 3$

Meso-kurtic if $\beta_2 = 3$ and

Lepto-kurtic if $\beta_2 > 3$

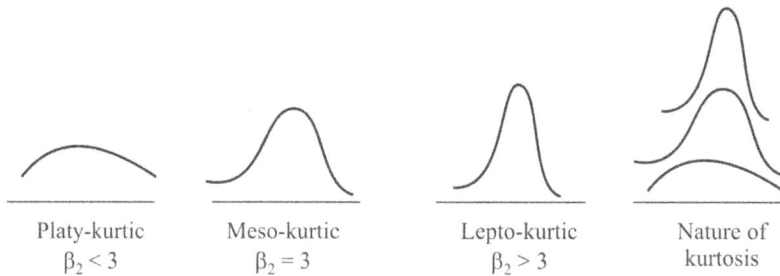

| Platy-kurtic | Meso-kurtic | Lepto-kurtic | Nature of |
| $\beta_2 < 3$ | $\beta_2 = 3$ | $\beta_2 > 3$ | kurtosis |

Examples

Example 1: Find the mean and the variance of uniform probability distribution given by

$$f(x) = \frac{1}{n} \quad \text{for } x = 1, 2, 3, \dots, n.$$

Sol: From the given data, the Probability distribution of the distribution is

X :	1	2	3	4	5	. . .	n
$P(X)$:	$\frac{1}{n}$	$\frac{1}{n}$	$\frac{1}{n}$	$\frac{1}{n}$	$\frac{1}{n}$	. . .	$\frac{1}{n}$

Mean of the random variable X is

$$E(X) = \sum_{i=1}^{n} x_i p_i(x_i)$$

$$= \frac{1}{n} + \frac{2}{n} + \frac{3}{n} + \ldots\ldots\ldots + \frac{n}{n}$$

$$= \frac{1}{n} \left[\frac{n(n+1)}{2} \right]$$

$$= \left[\frac{(n+1)}{2} \right]$$

Since variance $= E(X^2) - \{E(X)\}^2$ .....(3.12)

Now $$E(X^2) = \frac{1}{n} + \frac{4}{n} + \frac{9}{n} + \ldots\ldots\ldots + \frac{n^2}{n}$$

$$= \frac{1}{n} \left[\frac{n(n+1)(2n+1)}{6} \right] = \left[\frac{(n+1)(2n+1)}{6} \right]$$

From eq. (3.12),

$\therefore$ $$\text{Variance} = \left[\frac{(n+1)(2n+1)}{6} \right] - \left[\frac{(n+1)}{2} \right]^2$$

$$= \frac{n+1}{2} \left[\frac{2n+1}{3} - \frac{(n+1)}{2} \right]$$

$$= \frac{n+1}{2} \frac{(n-1)}{6}$$

$$= \frac{n^2-1}{12}$$

$\therefore$ Variance of the random variable is $\dfrac{n^2-1}{12}$

Example 2: A random variable X has the following distribution

X	-2	-1	0	1	2	3
$P(X)$	0.1	k	0.2	$2k$	0.3	k

Determine (i) k (ii) Mean (iii) Variance

Sol:

Since $\quad \sum P(x) = 1$

$\Rightarrow \qquad 0.1 + k + 0.2 + 2\,k + 0.3 + k = 1$

$\Rightarrow \qquad 0.6 + 4\,k = 1$

$\therefore \qquad k = 0.1$

Now the distribution of the random variable X is

X	–2	–1	0	1	2	3
$P(X)$	0.1	0.1	0.2	0.2	0.3	0.1

(ii)   Mean of the random variable X is

$$E(X) \;=\; \sum_{i=1}^{n} x_i p_i(x_i)$$

$$= (-2)*(0.1) + (-1)*(0.1) + (0)*(0.2) + (1)*(0.2) + (2)*(0.3) + (3)*(0.1)$$

$$= 0.8$$

(iii)   Variance of the random variable X is

$$E(X) \;=\; \sum_{i=1}^{n} x_i p_i(x_i)$$

$$= \sum x^2 P(x) - \{\sum x\, P(x)\}^2$$

$$= (-2)^2*(0.1) + (-1)^2*(0.1) + (0)^2*(0.2) + (1)^2*(0.2) +$$

$$(2)^2* (0.3) + (3)^2*(0.1) - \{0.8\}^2$$

$$= 2.16$$

***Example* 3:** Find the variance of the discrete random variable X whose probability distribution is given as

$$X \quad : \qquad 1 \quad\; 2 \quad\;\; 3 \quad\;\; 4 \quad\;\; 5$$

$$P(X=x): \qquad 0.1 \quad 0.1 \quad 0.3 \quad 0.3 \quad 0.2$$

Sol: The variance of the random variable X is

$$Var\,(X) = \sigma_x^2 \;=\; E\!\left(X^2\right) - \left[E(X)\right]^2 \qquad\qquad .....(3.13)$$

Now $\quad E\!\left(X^2\right) = \sum x^2\, P(X=x)$

$$= 1^2 \times (0.1) + 2^2 \times (0.1) + 3^2 \times (0.3) + 4^2 \times (0.3) + 5^2 \times (0.2)$$
$$= 0.1 + 4 \times 0.1 + 9 \times 0.3 + 16 \times 0.3 + 25 \times 0.2$$
$$= 0.1 + 0.4 + 2.7 + 4.8 + 5$$
$$= 13$$

And $E(X) = 1 \times 0.1 + 2 \times 0.1 + 3 \times 0.3 + 4 \times 0.3 + 5 \times 0.2$
$$= 3.4$$

Now from eq. (3.13), $V(X) = 13 - (3.4)^2$
$$= 1.44$$

∴ The variance of the random variable X is 1.44

***Example* 4:** A random variable X has the following probability distribution

X :	0	1	2	3
$P(X=x)$:	$\dfrac{1}{8}$	$\dfrac{3}{8}$	$\dfrac{3}{8}$	$\dfrac{1}{8}$

Find the mean, variance and standard deviation of the distribution

Sol: The mean of the random variable X is

$$E(X) = 0 \times \frac{1}{8} + 1 \times \frac{3}{8} + 2 \times \frac{3}{8} + 3 \times \frac{1}{8}$$
$$= \frac{3}{8} + \frac{6}{8} + \frac{3}{8}$$
$$= \frac{12}{8}$$
$$= \frac{3}{2}$$
$$= 1.5$$
$$V(X) = E(X^2) - [E(X^2)]$$

Since $E(X^2) = \sum x^2 p(x) = 0^2 \frac{1}{8} + 1^2 \frac{3}{8} + 2^2 \frac{3}{8} + 3^3 \frac{1}{8}$
$$= \frac{3}{8} + \frac{12}{8} \frac{9}{8}$$
$$= \frac{24}{8}$$
$$= 3$$

$$\therefore \quad V(X) = E(X^2) - [E(X^2)]$$

$$= 3 - (1.5)^2$$

$$= 3 - 2.25$$

$$= 0.75$$

And standard deviation of $X = \sigma_x = + \sqrt{V(X)}$

$$= + \sqrt{0.75}$$

$$= + 0.87$$

Example 5: An unbiased coin is tossed four times. If X denotes the number of heads from the distribution of X by writing down all the possible outcomes and hence calculate the expected value of variance of the random variable X

Sol: Let the random variable X represents the number of heads when tossing a coin four times. The its range 0, 1, 2, 3 and 4 and their respective probabilities are

$$\therefore P(X=0) = \frac{4C0}{24} = \frac{1}{16} \; ; \qquad\qquad P(X=1) = \frac{4C1}{24} = \frac{4}{16} = \frac{1}{4}$$

$$P(X=2) = \frac{4C2}{24} = \frac{6}{16} = \frac{3}{8} \; ; \qquad P(X=3) = \frac{4C3}{24} = \frac{1}{4} \quad \text{and} \quad P(X=4) = \frac{1}{16}$$

These probabilities represented in tabular form as

$X = x$:	0	1	2	3	4
$P(X=x)$:	$\frac{1}{16}$	$\frac{1}{4}$	$\frac{3}{8}$	$\frac{1}{4}$	$\frac{1}{16}$

Now the expected value of X is

$$E(X) = \sum x_i p_i$$

$$= 0.\frac{1}{16} + 1.\frac{1}{4} + 2.\frac{3}{8} + 3.\frac{1}{4} + 4.\frac{1}{16}$$

$$= \frac{1}{4} + \frac{6}{8} + \frac{3}{4} + \frac{1}{4}$$

$$= \frac{2+6+6+2}{8}$$

$$= \frac{16}{8}$$

$$= 2$$

And

$$E(X^2) = \sum x^2 p_i$$

$$= 0^2 \cdot \frac{1}{16} + 1^2 \cdot \frac{1}{4} + 2^2 \cdot \frac{3}{8} + 3^2 \cdot \frac{1}{4} + 4^2 \cdot \frac{1}{16}$$

$$= \frac{1}{4} + \frac{12}{8} + \frac{9}{4} + 1$$

$$= \frac{2 + 12 + 18 + 8}{8}$$

$$= \frac{40}{8} = 5$$

Now

$$V(X) = E(X^2) - [E(X)]^2$$

$$= 5 - 2^2$$

$$= 1$$

∴ Variance of the random variable X is 1.

***Example* 6:** Calculate expectation and variance of X. If the probability distribution of the random variable X is given by

x	−1	0	1	2	3
f	0.3	0.1	0.1	0.3	0.2

Sol: Since the expectation or mean of the random variable X is

$$E(X) = \sum_{i=1}^{n} x_i p_i(x_i)$$

$$= -1 * 0.3 + 0 * 0.1 + 1 * 0.1 + 2 * 0.3 + 3 * 0.2$$

$$= 1$$

For Variance of the random variable X is

$$V(x) = E(x^2) - \{E(x)\}^2$$

$$= \sum x^2 P(X=x) - \{\sum x P(X=x)\}^2$$

$$= \{(-1)^2 * 0.3 + 0^2 * 0.1 + 1^2 * 0.1 + 2^2 * 0.3 + 3^2 * 0.2\} - 1^2$$

$$= 3.4 - 1$$

$$= 2.4$$

***Example* 7:** Let X denote the minimum of two numbers that appear when a pair of dice is thrown once. Determine the (i) discrete probability distribution (ii) Expectation and (iii) Variance

***Sol*:** When two dice are thrown then the sample space S is

$$S = \begin{Bmatrix} (1,1),(1,2),(1,3),(1,4),(1,5),(1,6); & (2,1),(2,2),(2,3),(2,4),(2,5),(2,6) \\ (3,1),(3,2),(3,3),(3,4),(3,5),(3,6); & (4,1),(4,2),(4,3),(4,4),(4,5),(4,6) \\ (5,1),(5,2),(5,3),(5,4),(5,5),(5,6); & (6,1),(6,2),(6,3),(6,4),(6,5),(6,6)\} \end{Bmatrix}$$

From the given data, the random variable X represents the minimum of its numbers on the two dice. Then

$X = 1 = $ Min $[(1, 1),(1, 2),(2, 1),(1, 3),(3, 1) (1, 4),(4, 1), (1, 5),(5, 1), (1, 6),(6, 1)]$

$\therefore \qquad\qquad P(X = 1) = P(1) = \dfrac{11}{36}$

$X = 2 = $ Min $[(2, 2),(2, 3),(3, 2) (2, 4),(4, 2), (2, 5),(5, 2), (2, 6),(6, 2)]$

$\therefore \qquad\qquad P(X = 2) = P(2) = \dfrac{9}{36}$

$X = 3 = $ Min $[(3, 3), (3, 4),(4, 3), (3, 5),(5, 3), (3, 6),(6, 3)] \quad \therefore P(X = 3) = P(3) = \dfrac{7}{36}$

$X = 4 = $ Min $[(4, 4), (4, 5),(5, 4), (4, 6),(6, 4)] \qquad\qquad \therefore P(X = 4) = P(4) = \dfrac{5}{36}$

$X = 5 = $ Min $[(5, 5), (5, 6), (6, 5)] \qquad\qquad\qquad \therefore P(X = 5) = P(5) = \dfrac{3}{36}$

$X = 6 = $ Min $[(6, 6)] \qquad\qquad\qquad\qquad \therefore P(X = 6) = P(6) = \dfrac{1}{36}$

$\therefore$ Probability distribution of the random variable X is

$X = x$	1	2	3	4	5	6
$P(X = x)$	11/36	9/36	7/36	5/36	3/36	1/36

(ii)   ***Mean*** of the random variable X is

$$E(X) = \sum_{i=1}^{n} x_i p_i(x_i)$$

$$= 1 \times \frac{11}{36} + 2 \times \frac{9}{36} + 3 \times \frac{7}{36} + 4 \times \frac{5}{36} + 5 \times \frac{3}{36} + 6 \times \frac{1}{36}$$

$$= \frac{91}{36}$$

$$= 2.53$$

(iii) **Variance** of the random variable X is

$$V(X) = E(X^2) - \{E(X)\}^2$$

$$= \sum x^2 P(x) - \{\sum x\, P(x)\}^2$$

$$= 1^2 \times \frac{11}{36} + 2^2 \times \frac{9}{36} + 3^2 \times \frac{7}{36} + 4^2 \times \frac{5}{36}$$

$$+ 5^2 \times \frac{3}{36} + 6^2 \times \frac{1}{36} - \{2.53\}^2$$

$$= 1.97$$

Example 8: From the following table compute (i) $E(X)$ (ii) $E(2X \pm 3)$ (iii) $E(4X + 5)$ (iv) $E(X^2)$ (v) $V(X)$ and (vi) Var $E(2X \pm 3)$

$X = x$:	-3	-2	-1	0	1	2	3
$P(X = x)$:	0.05	0.1	0.3	0	0.3	0.15	0.1

Sol:

(i) Since $E(X) = \sum x^2 p(X = x_i)$

$$= -3 \times 0.05 + (-2) \times 0.1 + (-1) \times 0.3 + 0 \times 0 + 1 \times 0.3 + 2 \times (0.15) + 3(0.1)$$

$$= 0.25$$

(ii) $\quad E(2X \pm 3) = 2.\, E(X) \pm 3$

$$= 2(0.25) \pm 3$$

$$= 0.5 \pm 3$$

$$= -2.5, 3.5$$

(iii) $\quad E(4X + 5) = 4.\, E(X) + 5$

$$= 4\,(0.25) + 5$$

$$= 1 + 5$$

$$= 6$$

(iv) $\quad E(X^2) = (-3)^2\,(0.05) + (-2)^2\,(0.1) + (-1)^2\,(0.3) +$

$$0 \times 0 + 1^2\,(0.3) + 2^2\,(0.15) + 3^2\,(0.1)$$

$$= 2.95$$

(v) $\quad Var(X) = E(X^2) - [E(X)]^2$

$$= 2.95 - (0.25)$$

$$= 2.89$$

(vi) $Var(2X \pm 3) = 4.V(X)$

$$= 4 \times 2.89$$

$$= 11.56$$

$$= 11.56 \qquad \qquad \because V(aX + b) = a^2 V(X)$$

Example 9: The mean and variance of a binomial distribution are 4 and 3 respectably find $p(X \geq 1)$

Sol: From the given date mean $np = 4$ and variance $npq = 3$

$$\therefore q = \frac{npq}{np} = \frac{3}{4} \text{ This implies } p = 1 - q = 1 - (3/4) = 1/4$$

$$\therefore np = 4 \Rightarrow n.\frac{1}{4} = 4 \Rightarrow n = 16.$$

$$\therefore n = 16, \ p = 1/4 \ q = 3/4.$$

Now probability distribution of Binomial distribution is

$$P(X = x) = {}^nC_x \, p^x \, q^{n-x} \qquad x = 0, 1, 2, \ldots \ldots .n.$$

$$p(X \geq 1) = 1 - p(X < 1)$$

$$= 1 - p(X = 0)$$

$$= 1 - {}^{16}C_0 \left(\frac{1}{4}\right)^0 .\left(\frac{3}{4}\right)^{16-0}$$

$$= 1 - \left(\frac{3}{5}\right)^{16}$$

$$= 0.9899.$$

Example 10: Over a 10 minute period a counter records an average of 1.3 gamma particles per millisecond coming from a radio active substance. To a good approximation, the distribution of the count, X of gamma particles during the next millisecond is Poisson. Determine (i) λ (ii) The probability of one or more gamma particles and (iii) The variance.

Sol: From the given data, on an average gamma particle per millisecond coming from a radio active substance is 1.3.

(i) $\lambda = 1.3$

Probability function of Poisson distribution is

$$p(X = x) = \frac{e^{-\lambda}.\lambda^x}{\angle x} \qquad x = 0, 1, 2 \ldots$$

(ii) The probability that one or more gamma particles is

$$p(X \geq 1) = 1 - p(X < 1) = 1 - p(X = 0)$$

$$= 1 - \frac{e^{-2} . 2^0}{\angle 0} = 1 - e^{-1.3}$$

$$= 1 - 0.2725$$

$$= 0.7275$$

(iii) For the Poisson distribution, mean = variance

$$\therefore \text{ variance = mean} = \lambda = 1.3$$

***Example* 11:** Determine the variance of the geometric distribution whose probability function is

$$P(X = k) = q^{k-1} p$$

Sol:

Since $\quad V(X) = E(X^2) - [E(X)]^2 \qquad \qquad \qquad \text{.....(3.14)}$

Consider $\quad E(X^2) = \sum x^2 p(X = x)$

$$= \sum k^2 p(X = k)$$

$$= \sum [k(k-1) + k] p(X = k)$$

$$= \sum k(k-1) p(X = k) + \sum k p(X = k)$$

$$= \sum k(k-1) p(X = k) + E(X)$$

Now take $\quad E(X^2) = \sum_{k=2}^{\infty} k(k-1) p(X = k) \qquad \qquad \text{.....(3.15)}$

$$= \sum_{k=2}^{\infty} k(k-1) q^{k-1} p = p \sum_{k=2}^{\infty} k(k-1) q^{k-1}$$

since if $k = 1$, the term is zero

$$= p [2q + 2.3 \, q^2 + 3.4 \, q^3 + \text{----}]$$

$$= 2 p q [1 + 3q + 6q^2 + \text{----}]$$

$$= 2\,pq\,\frac{1}{(1-q)^3}$$

$$= \frac{2\,pq}{p^3}$$

$$= \frac{2q}{p^2} \qquad\qquad .....(3.16)$$

Hence from (3.14), (3.15) and (3.16)

$$V(X) = \frac{2q}{p^2} + \frac{1}{p} - \left(\frac{1}{p}\right)^2$$

$$= \frac{2q}{p^2} + \frac{1}{p} - \frac{1}{p^2}$$

$$= \frac{q}{p^2}$$

Hence the variance of geometric distribution is $\dfrac{q}{p^2}$

Example 12: Let one copy of a magazine out of 10 copies bears a special prize following geometric random distribution. Determine the mean and variance.

Sol: From the given data, the probability of getting a special prize is $p = \dfrac{1}{10}$ then $q = \dfrac{9}{10}$.

The probability distribution of geometric distribution is

$$P(X = k) = q^{K-1}\,p$$

The mean of the geometric distribution is $= \dfrac{1}{p} = \dfrac{1}{\frac{1}{10}} = 10$

The variance of the geometric distribution is $\dfrac{q}{p^2} = \dfrac{\frac{9}{10}}{\left(\frac{1}{10}\right)^2}$

$$= \frac{9}{10} \times 10^2$$

$$= 90$$

$\therefore$ The mean $= 10$ and variance $= 90$

Example 13: A wire if length a units is divided into two parts. If the first part is of length X find $E(X)$ and $Var(X)$.

Sol: Since the position of the point of division are equally likely, X is uniformly distributed in $(0, a)$.

Since for the Uniform distribution $f(x) = \dfrac{1}{b-a}$ $a < x < b$

Now for the sum pdf $f(x) = \dfrac{1}{a-0} = \dfrac{1}{a}$ $0 < x < a$

Now $E(X) = \displaystyle\int_{\infty}^{\infty} xf(x)dx = \int_{0}^{a} x.\dfrac{1}{a}\left(\dfrac{x^2}{2}\right)_{0}^{a}$

$= \dfrac{1}{a}.\dfrac{a^2}{2}$

$= \dfrac{a}{2}$

And $E(X^2) = \displaystyle\int_{-\infty}^{\infty} x^2 f(x)dx$

$= \displaystyle\int_{0}^{a} x^2.\dfrac{1}{a}dx$

$= \dfrac{1}{a}\left(\dfrac{x^3}{3}\right)_{0}^{a} \dfrac{1}{a}.\dfrac{a^3}{3}$

$= \dfrac{a^2}{3}$

Now $V(X) = E(X^2) - [E(X)]^2$

$= \dfrac{a^2}{3} - \left(\dfrac{a}{2}\right)^2$

$= \dfrac{a^2}{3} - \dfrac{a^2}{4}$

$= \dfrac{4a^2 - 3a^2}{12}$

$= \dfrac{a^2}{12}$

***Example* 14:** Determine the mean and variance of the uniform distribution whose density is

$$f(x) = \begin{cases} \dfrac{1}{b-a} & \text{if } a < x < b \\ 0 & \text{Otherwise} \end{cases}$$

Sol:

The rth moment about the origin is given by

$$\mu_r' = E\left(X^r\right)$$

$$= \int\limits_{-\infty}^{\infty} x^r . f(x)$$

$$= \int\limits_{a}^{b} x^r . \frac{1}{b-a} dx$$

$$= \frac{1}{b-a} \int\limits_{a}^{b} x^r dx$$

$$= \frac{1}{b-a} . \left[\frac{x^{r+1}}{r+1} \right]_a^b$$

$$= \frac{1}{b-a} \frac{b^{r+1} - a^{r+1}}{r+1}$$

$$\mu_r' = \frac{b^{r+1} - a^{r+1}}{(b-a)(r+1)} \qquad\qquad .....(3.17)$$

Put $r = 1$ we get the mean of the uniform distribution

$$\mu_r' = \frac{b^2 - a^2}{(b-a)2} = \frac{(b+a)(b-a)}{(b-a)2} = \frac{a+b}{2}$$

∴ The mean of the uniform distribution is $E(x) = \dfrac{a+b}{2}$

Put $r = 2$ then

$$\mu_2' = \frac{b^3 - a^3}{(b-a)\times 3} = \frac{(b-a)\left(b^2 + ab + a^2\right)}{(b-a)\times 3} = \frac{a^2 + ab + b^2}{3}$$

∴ The variance of the uniform distribution is

$$\mu_2 = \mu_2^1 - \left(\mu_1^{/}\right)^2$$

$$= \frac{a^2 + ab + b^2}{3} - \left(\frac{a+b}{2}\right)^2$$

$$= \frac{a^2 + ab + b^2}{3} - \frac{a^2 + 2ab + b^2}{4}$$

$$= \frac{4a^2 + 4ab + b^2 - 3a^2 - 6ab - 3b^2}{12}$$

$$= \frac{a^2 - 2ab + b^2}{12}$$

$$= \frac{(a-b)^2}{12}$$

$$= \frac{(b-a)^2}{12}$$

∴ Variance of the uniform distribution is $V(X) = \dfrac{(b-a)^2}{12}$

***Example* 15:** Determine the mean and variance of the exponentially distributed random variable whose density is

$$f(x) = \begin{cases} \dfrac{1}{b}e^{-(x-a)/b} & \text{if } x > a \\ 0 & \text{if } x < a \end{cases}$$

***Sol*:** The mean of the random variable X is

$$E(X) = \int_{-\infty}^{\infty} x^2 f(x)dx$$

$$= \int_{-\infty}^{a} x.odx + \int_{a}^{\infty} x.\frac{1}{b}e^{-(x+a)/b}dx$$

$$= \frac{1}{b}\int_{a}^{\infty} x.e^{-(x-a)/b}dx$$

put $x - a = t$ then $x = a + t$, $dx = dt$ and

when $x = a$ then $t = 0$ also when $x = \infty$ then $t = \infty$

$$E(X) = \frac{1}{b} \int_a^\infty (a+t)e^{-t/b} dt$$

$$= \frac{1}{b}\left[a \int e^{-t/b} dt + \int_0^\infty t^{-t/b} dt \right]$$

$$= \frac{1}{b}\left\{ a\left[\frac{e^{-t/b}}{-1/b} \right]_0^\infty + \left\{ t.\frac{e^{-t/b}}{-1/b} - 1.\frac{e^{-t/b}}{(-1/b)^2} \right\}_0^\infty \right\}$$

$$= \frac{1}{b}\left(a c0 - \frac{1}{-1/b} \right) + \left\{ 0 - b^2 \left((0-1) \right) \right\}$$

$$= \frac{1}{b}\left[+ ab + b^2 \right]$$

$$= \frac{1}{b}.b(a+b)$$

$$= a + b$$

Or

$$E(X) = \frac{1}{b} \int_a^\infty x.e^{-x/b}.e^{a/b} dx$$

$$= \frac{1}{b} e^{a/b} . \int_a^\infty xe^{-x/b} dx$$

$$= \frac{1}{b} e^{a/b} . \left\{ x.\frac{e^{-x/b}}{-1/b} - 1.\frac{e^{-x/b}}{(-1/b)^2} \right\}_a^\infty$$

$$= \frac{1}{b} e^{a/b} \left[-bxe^{-x/b} - b^2 e^{-x/b} \right]_a^\infty$$

$$= \frac{1}{b} e^{a/b} \left[\left(0 + abe^{-a/b} \right) - \left(0 - b^2 e^{-a/b} \right) \right]$$

$$= \frac{1}{b} e^{a/b} \left[abe^{-a/b} + b^2 e^{-a/b} \right]$$

And
$$E(X^2) = \int_a^\infty x^2 f(x) dx$$

$$= \int_a^\infty x^2 . \frac{1}{b} e^{-(x-a)/b} dx$$

$$= \int_a^\infty \frac{1}{b} x^2 . e^{(a-x)/b} dx$$

$$= \frac{1}{b} e^{a/b} \int_a^\infty x^2 e^{-x/b} dx$$

$$= \frac{1}{b} e^{a/b} \left\{ x^2 . \frac{e^{-x/b}}{-1/b} - 2x . \frac{e^{-x/b}}{(-1/b)^2} + 2 . \frac{e^{-x/b}}{(-1/b)^3} \right\}_a^\infty$$

$$= \frac{1}{b} e^{a/b} \left\{ -bx^2 \ e^{-x/b} - 2b^2 \ x \ e^{-x/b} - 2b^3 \ e^{-x/b} \right\}_a^\infty$$

$$= \frac{1}{b} e^{a/b} \left\{ -\left(0 - a^2 b e^{-a/b} \right) - 2b^2 \left(0 - ae^{-x/b} \right) - 2b^3 \left(0 - e^{-a/b} \right) \right\}$$

$$= \frac{1}{b} e^{a/b} \left\{ a^2 b e^{-a/b} + 2ab^2 e^{-a/b} + 2b^3 e^{-a/b} \right\}$$

$$= \frac{1}{b} e^{a/b} . b e^{-a/b} \left(a^2 + 2ab + 2b^2 \right)$$

$$= a^2 + 2ab + 2b^2$$

Now $Var\ (X)\ = E(X^2) - [E(X)]^2$
$$= a^2 + 2ab + 2b^2 - (a+b)^2$$
$$= a^2 + 2ab + 2b^2 - a^2 - b^2 - 2ab$$
$$= b^2$$

$\therefore Var(x) = b^2$

***Example* 16:** For the Raleigh density $f(x) = \dfrac{x}{\alpha^2} e^{-x^2/2\alpha^2}$, $0 < x < \infty$; find the nth moment about origin and deduce the value of mean and variance

Sol:

Given $f(x) = \dfrac{x}{\alpha^2} e^{-x^2/2\alpha^2}$ $0 < x < \infty$

The nth moment about origin is

$$E\left(X^n \right) = \int_0^\infty x^n \frac{x}{\alpha^2} . e^{-x^2/2\alpha^2} dx$$

$$= \frac{1}{\alpha^2} \cdot \int_0^\infty x^n \cdot x e^{-x^2/2\alpha^2} \, dx$$

$$= \frac{1}{\alpha^2} \cdot \int_0^\infty x^n \cdot e^{-x^2/2\alpha^2} \, dx$$

Put $\dfrac{x^2}{2\alpha^2} = t \Rightarrow x^2 = 2\alpha^2 t$ and $2x \, dx = 2\alpha^2 \, dt$ also $x = \left(2\alpha^2 t\right)^{1/2}$ and the limits are unaltered

$$= \frac{1}{\alpha^2} \cdot \int_0^\infty \left(2\alpha^2 t\right)^{n/2} \cdot e^{-t} \, dt$$

$$= \int_0^\infty \left(2\alpha^2\right)^{n/2} \cdot t^{n/2} e^{-t} \, dt$$

$$= 2^{n/2} \cdot \alpha^n \int_0^\infty t^{\left(\frac{n}{2}+1\right)-1} e^{-t} \, dt$$

$$= 2^{n/2} \cdot \alpha^n \ \Gamma\left(\frac{n}{2}+1\right)$$

$$\therefore \ E\left(X^n\right) = 2^{n/2} \cdot \alpha^n \ \Gamma\left(\frac{n}{2}+1\right) \text{ is the } n^{th} \text{ moment about origin.}$$

Put $n = 1$ then

$$E(X) = 2^{1/2} \cdot \alpha \ \Gamma\left(\frac{1}{2}+1\right)$$

$$= \sqrt{2} \cdot \alpha \ \Gamma\left(\frac{3}{2}\right)$$

$$= \sqrt{2} \cdot \alpha \frac{1}{2} \Gamma\left(\frac{1}{2}\right)$$

$$= \sqrt{2} \cdot \alpha \frac{1}{2} \sqrt{\pi} \qquad \because \Gamma\left(\frac{1}{2}\right) \sqrt{\pi}$$

$$= \sqrt{\frac{\pi}{2}} \ \alpha$$

∴ The mean of the given Raleigh density is $\sqrt{\dfrac{\pi}{2}}\,\alpha$

Put $n = 2$ then

$$E\left(X^n\right) = 2^{2/2}.\alpha^2\ \Gamma\left(\frac{2}{2}+1\right)$$

$$= 2.\alpha^2\Gamma(2)$$

$$= 2.\alpha^2.1.\Gamma(1)$$

$$= 2\alpha^2 \qquad\qquad \because \Gamma(1)=1$$

Therefore the variance of the Raleigh density function is

$$V(X) = E(X^2) - [E(X)]^2$$

$$= 2\alpha^2 - \left(\sqrt{\frac{\pi}{2}}\,\alpha\right)^2$$

$$= 2\alpha^2 - \frac{\pi}{2}\alpha^2$$

$$= \left(2 - \frac{\pi}{2}\right)\alpha^2$$

***Example* 17:** If the random variable X follows $N(0, 2)$ and $Y = 3X^2$, find the mean and variance of Y.

Sol: Given that $X \sim N(0, 2)$

i.e., X follows normal distribution with zero mean and standard deviation 2.

∴ Mean of the normal distribution is 0

i.e., $E(X) = 0$ and variance of the normal distribution is 4

∴ $V(X) = E(X^2) - [E(X)]^2 = 4$

⇒ $E(X^2) = 4$ $\because E(X) = 0$

Also given $Y = 3\,X^2$

The mean of Y is

$$E(Y) = E(3X^2)$$

$$= 3.\,E(X^2)$$

$$= 3 \times 4$$

$$= 12$$

And the variance of Y is

$$V(Y) = E(Y^2) - [E(Y)]^2 \qquad \rightarrow \qquad \qquad .....(3.18)$$

Since $E(Y^2) = E(9X^4)$

$$= 9E(X^4) = 9 \times 48 = 432$$

Since for the normal distribution $N(0.\sigma)$

$$E(X^{2r}) = \frac{(2r)! \sigma^{2r}}{2^r . r!}$$

$\therefore$ $E(X^4) = E(X^{2 \times 2})$

$$= \frac{(2.2)! \ 4^2}{2^2 . 2!}$$

$$= \frac{4! \ 16}{4 \times 2}$$

$$= 48$$

$\therefore$ From eq. (3.18)

$$V(Y) = 432 - 12^2$$

$$= 432 - 144$$

$$= 288$$

Example 18: Find the mean and variance of the uniform distribution

Sol: Since the rth moment about the origin is given by

$$\mu_r' = E(X^r)$$

$$= \int_{-\infty}^{\infty} x^r . f(x)$$

$$= \int_a^b x^r . \frac{1}{b-a} dx$$

$$= \frac{1}{b-a} \int_a^b x^r dx$$

$$= \frac{1}{b-a} \left[\frac{x^{r+1}}{r+1} \right]_a^b$$

$$= \frac{1}{b-a} \ \frac{b^{r+1} - a^{r+1}}{r+1}$$

$$\mu'_r = \frac{b^{r+1} - a^{r+1}}{(b-a)(r+1)} \qquad \qquad \text{.....(3.19)}$$

Put r = 1 we get the mean of the uniform distribution

$$\mu'_r = \frac{b^2 - a^2}{(b-a)2} = \frac{(b+a)(b-a)}{(b-a)2} = \frac{a+b}{2}$$

∴ The mean of the uniform distribution is E (x) = $\dfrac{a+b}{2}$

Put r = 2 then

$$\mu^1_2 = \frac{b^3 - a^3}{(b-a) \times 3} = \frac{(b-a)(b^2 + ab + a^2)}{(b-a) \times 3} = \frac{a^2 + ab + b^2}{3}$$

∴ The variance of the uniform distribution is

$$\begin{aligned}
\mu_2 &= \mu^1_2 - \left(\mu'_1\right)^2 \\
&= \frac{a^2 + ab + b^2}{3} - \left(\frac{a+b}{2}\right)^2 \\
&= \frac{a^2 + ab + b^2}{3} - \frac{a^2 + 2ab + b^2}{4} \\
&= \frac{4a^2 + 4ab + b^2 - 3a^2 - 6ab - 3b^2}{12} \\
&= \frac{a^2 - 2ab + b^2}{12} \\
&= \frac{(a-b)^2}{12} \\
&= \frac{(b-a)^2}{12}
\end{aligned}$$

∴ Variance of the uniform distribution is V(x) = $\dfrac{(b-a)^2}{12}$

***Example* 19:** If X is uniformly distributed with mean 1 and variance 4/3. Find $P(X < 0)$

Sol: Given that the random variable X is uniformly distributed

The p d f of uniform distribution is

$$f(x) = \frac{1}{b-a} \quad a < x < b$$

Also given the mean and variance of uniform distribution is 1 and 4/3

i.e., $E(X) = \dfrac{a+b}{2} = 1$ and $var(X) = \dfrac{(b-a)^2}{12} = 4/3$

$\Rightarrow$ $a+b=2$ and

$$(b-a)^2 = \frac{4}{3} \times 12 = 16 \text{ this implies } (b-a) = \pm\, 4$$

$$a+b=2 \text{ and } (b-a) = \pm\, 4$$

From these two equations:

If $(b-a) = -4$ and $a+b=2$ solving these two equations we get b = −1 and a = 3. Here a > b. But a < b.

If $(b-a) = 4$ and $a+b=2$ solving these two equations we get b = 3 and a = −1. Here a < b. It satisfies the condition a < b. So $(b-a) = 4$

Now $f(x) = \dfrac{1}{b-a} \quad -1 < x < 3$

$$= \frac{1}{3+1}$$

$$= \frac{1}{4}$$

$$p(X < 0) = \int_{-\infty}^{0} f(x)\, dx$$

$$= \int_{-1}^{0} \frac{1}{4}\, dx$$

$$= \frac{1}{4} \int_{-1}^{0} dx$$

$$= \frac{1}{4} (x)_{-1}^{0}$$

$$= \frac{1}{4}$$

$$\therefore p(X < 0) = \frac{1}{4}$$

***Example* 20:** Find the Variance of the Gamma distribution

Sol: Since variance of the Gamma distribution is

$$V(X) = E(X^2) - \{E(X)\}^2 \qquad \qquad .....(3.19)$$
$$E(X^2) =$$

$$= \int_0^\infty x^2 \frac{\beta}{\Gamma(\alpha)} (\beta x)^{\alpha-1} e^{-\beta x} dx.$$

$$= \frac{\beta}{\Gamma(\alpha)} \int_0^\infty x^2 (\beta x)^{\alpha-1} e^{\beta x} dx$$

Put t $= \beta x$ then $x = \dfrac{t}{p}$ and dx $= \dfrac{1}{\beta}$, the limits are unaltered.

$$E(X^2) = \frac{\beta}{\Gamma(\alpha)} \int_0^\infty \left(\frac{t}{\beta}\right)^2 (t)^{\alpha-1} e^{-t} \frac{dt}{\beta}$$

$$= \frac{\beta}{\Gamma(\alpha)} \int_0^\infty e^{-t} t^{\alpha+1} \, dt$$

$$= \frac{1}{\beta^2 \Gamma(\alpha)} \int_0^\infty e^{-t} t^{(\alpha+2)-1} \, dt.$$

$$= \frac{1}{\beta^2 \Gamma(\alpha)} \Gamma(\alpha + 2)$$

$$= \frac{(\alpha+1)(\alpha)\Gamma(\alpha)}{\beta^2 \Gamma(\alpha)}$$

$$= \frac{\alpha(\alpha+1)}{\beta^2}$$

$$\therefore \ V(X) = E(X^2) - [E(X)]^2$$

$$= \frac{\alpha(\alpha+1)}{\beta^2} - \left(\frac{\alpha}{\beta}\right)^2$$

$$= \frac{\alpha^2 + \alpha - \alpha^2}{\beta^2}$$

$$= \frac{\alpha}{\beta^2}$$

$$\therefore Var(X) = \frac{\alpha}{\beta^2} \text{ and mean} = \frac{\alpha}{\beta}$$

Example 21: Find the Variance of the Normal Distribution

Sol: Since $Var(X) = E(X^2) - [E(X)]^2 = E(X - \mu)^2$ .....(3.20)

Now

$$E(X - \mu)^2 = \int_{-\infty}^{\infty} (x - \mu)^2 f(x)\, dx = \int_{-\infty}^{\infty} (x - \mu)^2 \frac{1}{\sigma\sqrt{2\pi}} e^{-\frac{1}{2}\left(\frac{x-\mu}{\sigma}\right)^2} dx$$

Put $\dfrac{x - \mu}{\sigma} = z$ then $dx = \sigma\, dz$ also limits unaltered

$$= \frac{1}{\sigma\sqrt{2\pi}} \int_{-\infty}^{\infty} \sigma^2 z^2 e^{-\frac{1}{2}z^2} \sigma\, dz = \frac{\sigma^2}{\sqrt{2\pi}} \int_0^{\infty} z^2 e^{-\frac{1}{2}z^2}\, dz$$

$\because z^2 e^{-\frac{1}{2}z^2}$ is an even function

Put $\dfrac{z^2}{2} = t$ then $z\,dz = dt$ and $z = \sqrt{2t}$ also limits unaltered

$$= \frac{2\sigma^2}{\sqrt{2\pi}} \int_0^{\infty} 2t\, e^{-t^2} \frac{dt}{\sqrt{2t}} = \frac{2\sigma^2}{\sqrt{2\pi}} \int_0^{\infty} \sqrt{2t}\, e^{-t^2}\, dt$$

$$= \frac{2\sigma^2}{\sqrt{\pi}} \int_0^{\infty} t^{\frac{1}{2}} e^{-t^2}\, dt = \frac{2\sigma^2}{\sqrt{\pi}} \int_0^{\infty} t^{\frac{3}{2}-1} e^{-t^2}\, dt$$

$$= \frac{2\sigma^2}{\sqrt{\pi}} \Gamma\left(\frac{3}{2}\right) \qquad\qquad \because \int_0^{\infty} t^{\frac{3}{2}-1} e^{-t^2}\, dt = \Gamma\left(\frac{3}{2}\right)$$

$$= \frac{2\sigma^2}{\sqrt{\pi}} \frac{1}{2}\Gamma\left(\frac{1}{2}\right)$$

$$= \frac{2\sigma^2}{\sqrt{\pi}} \frac{1}{2}\sqrt{\pi} \qquad\qquad \because \Gamma\left(\frac{1}{2}\right) = \sqrt{\pi}$$

$$= \sigma^2$$

∴ Variance of the Normal Distribution is 'σ^2'

Example 22: Suppose the weights of 800 male students are normally distributed with mean $\mu = 140$ pounds and standard deviation 10 pounds. Find the number of students whose weights are (i) between 138 and 148 pounds and (ii) more than 152 pounds.

Sol: From the given data N = 800, $\mu = 140$ $\sigma = 10$

The standard normal variety $Z = \dfrac{x - \mu}{\sigma} \Rightarrow Z = \dfrac{x - 140}{10}$

(i) The probability that the students weights between 138 and 148 pounds:

i.e., $P(138 < X < 148)$:

When x = 138, then $Z = \dfrac{138 - 140}{10} = -0.2$ (say z_1)

When x = 148, then $Z = \dfrac{148 - 140}{10} = 0.8$ (say z_2)

∴ $P(138 < X < 148) = P(-0.2 < Z < 0.8) = F(0.8) - F(-0.2)$

$= 0.7881 - 0.4207$

$= 0.3674$

∴ The number of students whose weights are between 138 and 148 pounds

$= N. P (138 < X < 148)$

$= 800 \times 0.3674$

≈ 294

(ii) The probability that the students weights more than 152 pounds:

i.e., $P(X > 152)$:

When x = 152, then $Z = \dfrac{152 - 140}{10} = 1.2$

$P(X > 152) = P(Z > 1.2) = 1 - P(Z \leq 1.2) = 1 - F(1.2)$

$= 1 - 0.8849$

$= 0.1151$

∴ The number of students whose weights are more than 152 pounds

$= N. P (X > 152)$

$= 800 \times 0.1151$

≈ 92

Example 23: In a distribution exactly normal 7% of the items are under 35 and 89% are under 63. What are the mean and standard deviation of the distribution?

Sol: From the given data

$$P (X \leq 35) = \frac{7}{100} = 0.07 \qquad\qquad .....(3.21)$$

and $P(X \leq 63) = \dfrac{89}{100} = 0.89$ .....(3.22)

The standard normal variate $Z = \dfrac{x - \mu}{\sigma}$

When x = 35,   then $Z = \dfrac{35 - \mu}{\sigma}$

When x = 63,   then $Z = \dfrac{63 - \mu}{\sigma}$

From eq. (3.22), $P(X \leq 35) = 0.07$

$\Rightarrow$ $P(Z \leq \dfrac{35 - \mu}{\sigma}) = 0.07$

$\Rightarrow$ $\dfrac{35 - \mu}{\sigma} = -1.48$ (from tables)

$\Rightarrow$ $35 - \mu = -1.48\,\sigma \;\Rightarrow\; \mu - 1.48\sigma = 35$ .....(3.23)

From eq. (3.23), P (X< 63) = 0.89

$\Rightarrow$ $P (Z < \dfrac{63 - \mu}{\sigma}) = 0.89$

$\Rightarrow$ $\dfrac{63 - \mu}{\sigma} = 1.23 \;\Rightarrow\; 63 - \mu = 1.23\,\sigma \;\Rightarrow\; \mu + 1.23\sigma = 63$ .....(3.24)

From eqns (3.23) and (3.24)

         $\mu = 50.3$ and $\sigma = 10.3$

$\therefore$ The mean $\mu = 50.3$ and standard deviation $\sigma = 10.3$

Example 24: The marks obtained in statistics in a certain examination found to be normally distributed. If 15% of the students $\geq$ 60 marks, 40% < 30 marks. Find the mean and standard deviation.

Sol: From the given data

       $P(X \geq 60) = \dfrac{15}{100} = 0.15$ .....(3.25)

and $P(X < 30) = \dfrac{40}{100} = 0.4$ .....(3.26)

The standard normal variety $Z = \dfrac{x - \mu}{\sigma}$

When x = 60, then $Z = \dfrac{60 - \mu}{\sigma}$

When x = 30, then $Z = \dfrac{30 - \mu}{\sigma}$

From eq. (3.25), $P(X \geq 60) = 0.15$

$\Rightarrow$ $P(Z > \dfrac{60 - \mu}{\sigma}) = 0.15$

$\Rightarrow$ $1 - P(Z \leq \dfrac{60 - \mu}{\sigma}) = 0.15$ $\Rightarrow$ $P(Z \leq \dfrac{60 - \mu}{\sigma}) = 0.85$

$\Rightarrow$ $\dfrac{60 - \mu}{\sigma} = 1.04$ $\Rightarrow$ $60 - \mu = 1.04\,\sigma$ $\Rightarrow$ $\mu + 1.04\sigma = 60$ .....(3.27)

(From tables)

From eq. (3.27), $P(X < 30) = 0.4$

$\Rightarrow$ $P(Z < \dfrac{30 - \mu}{\sigma}) = 0.4$

$\Rightarrow$ $\dfrac{30 - \mu}{\sigma} = -0.25$ $\Rightarrow$ $30 - \mu = -0.25\,\sigma$ $\Rightarrow$ $\mu - 0.25\sigma = 30$ .....(3.28)

From eqns (3.27) and (3.28)

$\mu = 35.8$ and $\sigma = 23.2$

$\therefore$ The mean $\mu = 35.8$ and standard deviation $\sigma = 23.2$

Example 25: A sales tax officer has reported that the average sales of the 500 business that he has to deal with during a ear is Rs.36, 000/- with standard deviation of Rs. 10,000/- Assuming that the sales in these business are normally distributed find (i) the number of business as the sales of which are more than Rs. 40,000/- (ii) The percentage of business the sales of which are likely to range between Rs. 30,000 and Rs. 40,000.

Sol: From the given data N = 500, $\mu = 36{,}000$ and $\sigma = 10{,}000$

The standard normal variate $Z = \dfrac{x - \mu}{\sigma}$ $\Rightarrow$ $Z = \dfrac{x - 36{,}000}{10{,}000}$

(i) $P(X > 40{,}000)$:

When x = 40,000 , then $Z = \dfrac{40{,}00036{,}00}{10{,}000} = 0.4$ (say z_1)

$P(X > 40{,}000) = P(Z > 0.4) = 1 - P(Z \leq 0.4) = 1 - F(0.4)$

$= 1 - 0.6554$

$= 0.3446$

∴ The number of business as the sales which are more than Rs. 40,000/- are

$$= N. \ P(X > 40,000)$$

$$= 500 \times 0.3446 = 172.3$$

$$\approx 172$$

(ii) $P(30,000 < X < 40,000)$:

When x = 30,000, then $Z = \dfrac{30,000 - 36,000}{10,000} = -0.6$ (say z_1)

When x = 40,000, then $Z = \dfrac{40,000 - 36,000}{10,000} = 0.4$ (say z_1)

∴ $P(30,000 < X < 40,000) = P(-0.6 < Z < 0.4) = F(0.4) - F(-0.6)$

$$= 0.6554 - 0.2743$$

$$= 0.3811$$

∴ The number of business as the sales which are likely to range between 30,000 and 40,000 are

$$= N. \ P(30,000 < X < 40,000)$$

$$= 500 \times 0.3811 = 190.55$$

$$\approx 191$$

Out of 500 business 191 having sales between 30,000 and 40,000 for 100 business 38.2

∴ The required percentage is 38.2 %

***Example* 26:** The men and standard deviation of normal variate are 8 and 4 respectively. Find

(i) $P(5 \le X \le 10)$ (ii) $P(X \ge 5)$

Sol: From the given data , $\mu = 8$ and $\sigma = 4$

Let the random variable 'X' represents the mass of students

The standard normal variate $Z = \dfrac{x - \mu}{\sigma} \Rightarrow Z = \dfrac{x - 8}{4}$

(i) $P(5 \le X \le 10)$:

When x = 5, then $Z = \dfrac{5 - 8}{4} = -0.75$ (say z_1)

When x =10, then $Z = \dfrac{10-8}{4} = 0.5$ (say z_1)

$$P(5 \leq X \leq 10) = P(-0.75 < Z < 0.5) = F(0.5) - F(-0.75)$$
$$= 0.6915 - 0.2266$$
$$= 0.4649$$

(ii) $P(X \geq 5)$:

When x = 5 , then $Z = \dfrac{5-8}{4} = -0.75$ (say z_1)

$$P(X \geq 5) = P(Z > -0.75) = 1 - P(Z \leq -0.75) = 1 - F(-0.75)$$
$$= 1 - 0.2266$$
$$= 0.7734$$

Example **27:** If X is a normal variate with mean 30 and standard deviation 5. Find (i) $P(26 \leq X \leq 40)$ (ii) $P(X \geq 4\,5)$

Sol:

From the given data, $\mu = 30$ and $\sigma = 5$

Let the random variable X represents the mass of students

The standard normal variate $Z = \dfrac{x-\mu}{\sigma}$ $\Rightarrow$ $Z = \dfrac{x-30}{5}$

(i) $P(26 \leq X \leq 40)$:

When x = 26 , then $Z = \dfrac{x-30}{5} = -0.8$ (say z_1)

When x = 40 , then $Z = \dfrac{40-30}{5} = 2$ (say z_1)

$$P(26 \leq X \leq 40) = P(-0.8 < Z < 2) = F(2) - F(-0.8)$$
$$= 0.9772 - 0.2119$$
$$= 0.7653$$

(ii) $P(X \geq 45)$:

When x = 45, then $Z = \dfrac{45-30}{5} = 3$

$$P(X \geq 45) = P(Z > 3) = 1 - P(Z \leq 3) = 1 - F(3)$$
$$= 1 - 0.9987$$
$$= 0.0013$$

***Example* 28:** The marks obtained in mathematics by 1000 students is normally distributed with mean 78% and standard deviation 11%. Determine (i) How many students got marks above 90% (ii) What was the highest mark obtained by the lowest 10% of the students   (iii) With in what limits did the middle of 90% of the students lie

Sol: From the given data, the mean of the normal distribution is μ =78 and standard deviation σ = 11 and N = 1000

The standard normal variate $Z = \dfrac{x-\mu}{\sigma}$ i.e., $Z = \dfrac{x-78}{11}$

Let the random variable X represents the marks obtained by the students

(i)   The number of students got marks above 90% is :

Now $P(X > 90)$:

When x =90   then $Z = \dfrac{90-78}{11} = 1.091$

$$P(X > 90) = P(Z > 1.091) = 1 - P(Z \leq 1.091)$$

$$= 1 - F(1.091) = 1 - 0.8621$$

$$= 0.1379$$

$\therefore$ The number of students got marks above 90% is

$$= N.P(X < 90)$$

$$= 1000 \times 0.1379 = 137.9$$

$$\approx 138$$

(ii)   The highest mark obtained by the lowest 10% of the students is:

Let x_1 be the highest mark obtained by the lowest 10% of the students.

$$P(X \leq x_1) = \dfrac{10}{100} = 0.1$$

When x = x_1 then $Z = \dfrac{x_1 - 78}{11} = z_1$ (say)

$$P(X \leq x_1) = P(Z \leq z_1) = 0.1$$

$\Rightarrow$ $P(Z \leq z_1) = 0.1$ $\Rightarrow$ $z_1 = -1.28$

$\therefore$ $\dfrac{x_1 - 78}{11} = -1.28$ $\Rightarrow$ $x_1 - 78 = -14.08$

$\Rightarrow$ $x_1 = 63.92 \approx 64$

$\therefore$ The highest mark obtained by the lowest 10% of the students is 64%

(iii)   The middle of 90% of the students marks lies between:

Let the middle of 90% of the students lie between x_1 and x_2

Where x_1 be the lower limit and x_2 be the upper limit

$$P(X > x_1) = 0.9$$

When $X = x_1$ then $Z = \dfrac{x_1 - 78}{11} = z_1$ (say)

$$P(X > x_1) = P(Z > z_1) = 0.9$$

$\Rightarrow$ $1 - P(Z \leq z_1) = 0.9$

$\Rightarrow$ $P(Z \leq z_1) = 0.1 \Rightarrow z_1 = -1.28$

$\therefore$ $\dfrac{x_1 - 78}{11} = -1.28 \Rightarrow x_1 - 78 = -14.08$

$\Rightarrow$ $x_1 = 63.92 \approx 64$

Now x_2 be the upper limit

$$P(X < x_2) = 0.9$$

When $X = x_2$ then $Z = \dfrac{x_2 - 78}{11} = z_2$ (say)

$$P(X < x_2) = P(Z < z_2) = 0.9$$

$\Rightarrow$ $P(Z \leq z_2) = 0.9 \Rightarrow z_2 = 1.285$

$\therefore$ $\dfrac{x_2 - 78}{11} = 1.285$

$\Rightarrow$ $x_2 - 78 = 14.135 \Rightarrow x_2 = 92.135 \approx 92$

$\therefore$ The middle of 90% of the students marks lie between 64 and 92

3.10   Chebyshev's Theorem

Since the variance of the random variable X tells us, the variability of the observations about the mean. If a random variable has a smaller variance, we would expect most of the values to be scattered about the mean.

Chebyshev's inequality reveals a statement of a random variable having finite mean and variance, the upper bound for the amount of dispersion of the random variable from its mean can be fixed, i.e.., suppose that the random variable X having a mean μ and variance σ^2 then $|X - \mu|$ is the measure of the amount by which the value X differ from the mean value μ, in either direction.

Chebyshev's inequality

Statement: Suppose that X is a random variable (discrete or contiguous) having finite mean μ and variance σ^2, then if k is any positive number $P[|X - \mu| \geq k\sigma] \leq \dfrac{1}{k^2}$

Proof: Let X be a continuous random variable has the density function f(x), then variance is

$$\sigma^2 = E[X - \mu]^2$$

$$= \int_{-\infty}^{\infty} (X - \mu)^2 f(x)dx$$

$$= \int_{-\infty}^{\mu-k\sigma} (X - \mu)^2 f(x)dx + \int_{\mu-k\sigma}^{\mu+k\sigma} (X - \mu)^2 f(x)dx + \int_{\mu+k\sigma}^{\infty} (X - \mu)^2 f(x)dx$$

$$\geq \int_{-\infty}^{\mu-k\sigma} (X - \mu)^2 f(x)dx + \int_{\mu+k\sigma}^{\infty} (X - \mu)^2 f(x)dx$$

$\because \displaystyle\int_{\mu-k\sigma}^{\mu+k\sigma} (X - \mu)^2 f(x)dx \geq 0,$ Because $|X - \mu| \geq k\sigma$ then $X \geq \mu + k\sigma$ or $X \leq \mu - k\sigma$.

Also we have $(X - \mu)^2 \geq k^2\sigma^2$ in the remaining integrals.

$$\sigma^2 \geq \int_{-\infty}^{\mu-k\sigma} k^2\sigma^2 f(x)dx + \int_{\mu+k\sigma}^{\infty} k^2\sigma^2 f(x)dx$$

$$\geq k^2\sigma^2 \left[\int_{-\infty}^{\mu-k\sigma} f(x)dx + \int_{\mu+k\sigma}^{\infty} f(x)dx \right]$$

$$\frac{\sigma^2}{k^2\sigma^2} \geq \left[\int_{-\infty}^{\mu-k\sigma} f(x)dx + \int_{\mu+k\sigma}^{\infty} f(x)dx \right]$$

i.e., $$\left[\int_{-\infty}^{\mu-k\sigma} f(x)dx + \int_{\mu+k\sigma}^{\infty} f(x)dx \right] \leq \frac{1}{k^2}$$

$$P[|X - \mu| \geq k\sigma] \leq \frac{1}{k^2}$$

Hence $P[\mu-k\sigma < X < \mu+k\sigma] = \displaystyle\int_{\mu-k\sigma}^{\mu+k\sigma} f(x)dx \geq \left(1-\dfrac{1}{k^2}\right)$

Aliter: For the equation (3),

$$\sigma^2 \geq \int_{-\infty}^{\mu-k\sigma}(X-\mu)^2 f(x)dx + \int_{\mu+k\sigma}^{\infty}(X-\mu)^2 f(x)dx \qquad\qquad .....(3.29)$$

For the first integral $x \leq \mu - k\sigma$ this implies $x - \mu \leq -k\sigma$ .....(3.30)

For the second integral $x \geq \mu + k\sigma$ this implies $x - \mu \geq k\sigma$ .....(3.31)

From equations (3.30) and (3.31), $|x-\mu| \geq k\sigma$ i.e., $(x-\mu)^2 \geq k^2\sigma^2$ .....(3.32)

Substitute in eq. (3.29) then

$$\sigma^2 \geq \int_{-\infty}^{\mu-k\sigma} k^2\sigma^2 f(x)dx + \int_{\mu+k\sigma}^{\infty} k^2\sigma^2 f(x)dx$$

$$\geq k^2\sigma^2\left[\int_{-\infty}^{\mu-k\sigma} f(x)dx + \int_{\mu+k\sigma}^{\infty} f(x)dx\right]$$

$$\dfrac{\sigma^2}{k^2\sigma^2} \geq \left[\int_{-\infty}^{\mu-k\sigma} f(x)dx + \int_{\mu+k\sigma}^{\infty} f(x)dx\right]$$

i.e., $\left[\displaystyle\int_{-\infty}^{\mu-k\sigma} f(x)dx + \int_{\mu+k\sigma}^{\infty} f(x)dx\right] \leq \dfrac{1}{k^2}$

Hence $P[|X-\mu| \geq k\sigma] \leq \dfrac{1}{k^2}$

Note

1. Chebyshev's inequality be verified for discrete case also.

2. We observe that the Chebyshev's inequality states that regardless of the shape of the curve f(x), we have $P[(\mu-\epsilon) < X < (\mu+\epsilon)] \geq \left(1-\dfrac{\sigma^2}{\epsilon^2}\right)$ where $\epsilon = k\sigma > 0$

    i.e., $P[|X-\mu| \geq \epsilon] \leq \dfrac{\sigma^2}{\epsilon^2}$ i.e., the probability that X takes values in the interval $(\mu-\epsilon, \mu+\epsilon)$ centered as μ is close to 1 provided σ is most less than ϵ.

3. For k = 2, in the inequality, the random variable X has a probability of at least $1 - \dfrac{1}{2^2} = \dfrac{3}{4}$ of falling with in two standard deviations of the mean , i.e., $\dfrac{3}{4}$ or more of the observations of any distribution lie into the interval $\mu \pm 2\sigma$, similarly, the inequality states that at least $\dfrac{3}{8}$ of observations of any distribution fall in the interval $\mu \pm 3\sigma$

Examples

***Example* 1:** A random variable X has a mean $\mu = 8$, variance $\sigma^2 = 9$ and an unknown probability distribution. Find (i) $P(-4 < X < 20)$ and (ii) $P(|X - 8| \ge 6)$

Sol: From the given data, $\mu = 8$, $\sigma = 3$, to find

(i) $P(-4 < X < 20)$: We shall obtain the required probability by using Chebyschev's inequality

$$P[\mu - k\sigma < X < \mu + k\sigma] \ge \left(1 - \frac{1}{k^2}\right) \qquad .....(3.33)$$

Here $\mu - k\sigma = -4$ $\mu + k\sigma = 20$ this implies $k = 4$ since $\mu = 8$, $\sigma = 3$.

Now

$$\left(1 - \frac{1}{k^2}\right) = \left(1 - \frac{1}{4^2}\right) = 1 - \frac{1}{16} = \frac{15}{16} \qquad .....(3.34)$$

∴ $P(-4 < X < 20) \ge \dfrac{15}{16}$ from eq. (3.33) and (3.34)

(ii) $P(|X - 8| \ge 6)$: We shall obtain the required probability by using Chebyschev's inequality

$$P[|X - \mu| \ge k\sigma] \le \frac{1}{k^2} \qquad .....(3.35)$$

Here $k\sigma = 6$, ∴ $k = 2$ ∵ $\sigma = 3$

Hence $P[|X - 8| \ge 6] \le \dfrac{1}{4}$

***Example* 2:**   A random variable X has a mean 12 and variance 9. If the probability distribution is unknown, find $P(6 < X < 18)$

Sol: From the given data, $\mu = 12$, $\sigma = 3$, to find

$P(6 < X < 18)$: We shall obtain the required probability by using Chebyschev's inequality

$$P[\mu - k\sigma < X < \mu + k\sigma] \ge \left(1 - \frac{1}{k^2}\right) \qquad .....(3.36)$$

Here $\mu - k\sigma = 6$ $\mu + k\sigma = 18$ this implies $k = 2$ since $\mu = 12$, $\sigma = 3$.

Now

$$\left(1 - \frac{1}{k^2}\right) = \left(1 - \frac{1}{2^2}\right) = 1 - \frac{1}{4} = \frac{3}{4} \qquad .....(3.37)$$

$\therefore$ $P(6 < X < 18) \ge \dfrac{3}{4}$ from eq. (3.36) and (3.37)

Example3: For a geometric distribution, $P(x) = \dfrac{1}{2^x}$, x=1, 2, 3, . . . prove that by Chebyshev's inequality $P[|X - 2| \le 2] > \dfrac{1}{2}$ while the actual probability is $\dfrac{15}{16}$

Sol: Given geometric distribution, $P(x) = \dfrac{1}{2^x}$, x = 1, 2, 3, . . .   , the probability distribution is geometric distribution,

X :	1	2	3	4	. . .
$P(x)$:	$\dfrac{1}{2}$	$\dfrac{1}{2^2}$	$\dfrac{1}{2^3}$	$\dfrac{1}{2^4}$	. . .

Now $E(X) = \sum xP(x)$

$$= 1 \times \frac{1}{2} + 2 \times \frac{1}{2^2} + 3 \times \frac{1}{2^3} + \cdots$$

This is a series of A.G.P. Here $r = \dfrac{1}{2}$

Let $S = 1 \times \dfrac{1}{2} + 2 \times \dfrac{1}{2^2} + 3 \times \dfrac{1}{2^3} + \cdots \qquad .....(3.38)$

$$\frac{1}{2}S = 1 \times \frac{1}{2^2} + 2 \times \frac{1}{2^3} + 3 \times \frac{1}{2^4} + \cdots \qquad .....(3.39)$$

Subtract eq. (3.38) from (3.39), then we get

$$\frac{1}{2}S = 1 \times \frac{1}{2} + (2-1) \times \frac{1}{2^2} + (3-2) \times \frac{1}{2^3} + (4-3) \times \frac{1}{2^4} + \cdots \qquad .....(3.40)$$

i.e., $\quad \dfrac{1}{2}S = \dfrac{1}{2}+\dfrac{1}{2^2}+\dfrac{1}{2^3}+\dfrac{1}{2^4}+\cdots \quad = \dfrac{\dfrac{1}{2}}{1-\dfrac{1}{2}} \quad$ since the sum of infinite G.P is

$$\dfrac{a}{1-r} \quad \text{if } r<1$$

$$= 1$$

$\therefore \qquad S = 2$

Hence $\quad E(X)=2$

Also $\quad E(X^2) = \sum x^2 P(x)$

$$= 1^2 \times \dfrac{1}{2}+2^2 \times \dfrac{1}{2^2}+3^2 \times \dfrac{1}{2^3}+\cdots$$

$$= \dfrac{1}{2}\left[1+4\times\dfrac{1}{2}+9\times\dfrac{1}{2^2}+16\times\dfrac{1}{2^3}\cdots\right]$$

Let $\quad S = 1+4\times\dfrac{1}{2}+9\times\dfrac{1}{2^2}+16\times\dfrac{1}{2^3}\cdots$

$\qquad S = 1+4A+9A^2+16A^2+\cdots \qquad\qquad\qquad\qquad$ .....(3.41)

Put $A = \dfrac{1}{2}$ in the above equation

$$-3AS = -3A-12A^2-27A^3-48A^4\cdots \qquad\qquad \text{.....(3.42)}$$

$$-3A^2S = -3A^2-12A^3-27A^4-48A^5\cdots \qquad\qquad \text{.....(3.43)}$$

$$-A^3S = -A^3-4A^4\cdots \qquad\qquad\qquad\qquad \text{.....(3.44)}$$

Adding all these four equations

$$\left(1-3A+3A^2-A^3\right)S = 1+A$$

$$\left(1-A\right)^3 S = 1+A$$

$\therefore \qquad S = \left(1+A\right)\left(1-A\right)^{-3}$

$$= \left(1+\dfrac{1}{2}\right)\left(1-\dfrac{1}{2}\right)^{-3}$$

$$= \dfrac{3}{2}\times 2^3 = 12$$

$\therefore \qquad E(X^2) = \dfrac{1}{2}\times 12 = 6$

Variance of X is, $V(X) = E(X^2) - [E(X)]^2 = 6 - 2^2 = 2$

Therefore $\sigma = \sqrt{2}$. Hence the mean $\mu = 2$ and $\sigma = \sqrt{2}$

We shall find

$P(|X - 2| \leq 2)$: We shall obtain the required probability by using Chebyschev's inequality

$$P[|X - \mu| \leq k\sigma] \geq 1 - \frac{1}{k^2}$$

Here $k\sigma = 2$, $\therefore k = \sqrt{2}$ $\because \sigma = \sqrt{2}$

$$P(|X - 2| \leq 2) \geq 1 - \frac{1}{2}$$

Hence $P(|X - 2| \leq 2) \geq \frac{1}{2}$

***Example* 4:** A random variable X has density function $f(x) = \begin{cases} 2e^{-2x} & x \geq 0 \\ 0 & x < 1 \end{cases}$

(i) Find $P[|X - \mu| > 1]$ and ii) use Chebyshev's inequality to obtain an upper bound on $P[|X - \mu| > 1]$ and compare with the result in (i).

Sol: Given density function is $f(x) = \begin{cases} 2e^{-2x} & x \geq 0 \\ 0 & x < 1 \end{cases}$

We find the $P[|X - \mu| > 1]$ in the usual manner

Now $E(X) = \int\limits_{-\infty}^{+\infty} x f(x) dx$

$= \int\limits_{-\infty}^{0} x.0 \, dx + \int\limits_{0}^{\infty} x \, 2e^{-2x} \, dx$

$= 2\int\limits_{0}^{\infty} x e^{-2x} \, dx$

$= 2\left[x \frac{e^{-2X}}{-2} - \frac{e^{-2X}}{(-2)^2} \right]_{0}^{\infty}$

$= 2\left[0 - \frac{1}{4}(0 - 1) \right]$

$= \frac{1}{2}$

Therefore the mean of the random variable X is $\mu = \dfrac{1}{2}$

Consider

$$P\big[|X - \mu| < 1\big] = P\left[\left|X - \frac{1}{2}\right| < 1\right]$$

$$= P\left[-1 < X - \frac{1}{2} < 1\right]$$

$$= P\left[-\frac{1}{2} < X < \frac{3}{2}\right]$$

$$= \int_{0}^{\frac{3}{2}} f(x)\,dx$$

$$= \int_{0}^{\frac{3}{2}} 2e^{-2x}\,dx$$

$$= 2\int_{0}^{\frac{3}{2}} e^{-2x}\,dx$$

$$= 2\left[\frac{e^{-2X}}{-2}\right]_{0}^{\frac{3}{2}}$$

$$= -\left(e^{-2\times\frac{3}{2}} - 1\right)$$

$$= 1 - e^{-3}$$

$$\therefore \quad P\big[|X - \mu| < 1\big] = 1 - e^{-3}$$

Therefore $P\big[|X - \mu| \geq 1\big] = 1 - P\big[|X - \mu| < 1\big]$

$$= 1 - \left(1 - e^{-3}\right)$$

$$= e^{-3}$$

$$= 0.0498$$

Hence $P\big[|X - \mu| \geq 1\big] = 0.0498$

By using Chebyshev's inequality $P\big[|X - \mu| \geq 1\big]$:

Now $E(X^2) = \int\limits_{-\infty}^{+\infty} x^2 \, f(x) \, dx$

$= \int\limits_{-\infty}^{0} x^2 \, f(x) dx + \int\limits_{0}^{\infty} x^2 \, f(x) dx$

$= 0 + 2 \int\limits_{0}^{\infty} x^2 e^{-2x} dx$

$= 2\left[x^2 \left(\dfrac{e^{-2x}}{-2}\right) - 2x\left(\dfrac{e^{-2x}}{(-2)^2}\right) + 2\left(\dfrac{e^{-2x}}{(-2)^3}\right) - \right]_{0}^{\infty}$

$= 2\left[0 \ -0-\dfrac{1}{4}\,(0-1)- \right] = \dfrac{1}{2}$

$\therefore$ $V(X) = \sigma^2 = E(X^2) - \{E(X)\}^2$

$= \dfrac{1}{2} - \left(\dfrac{1}{2}\right)^2$

$= \dfrac{1}{4}$

Hence $\sigma = \dfrac{1}{2}$

Now Compare $P\big[|X - \mu| \geq 1\big]$ with $P\big[|X - \mu| \geq k\sigma\big] \leq \dfrac{1}{k^2}$

Here $k\sigma = 1$ this implies k = 2

$\therefore$ $P\big[|X - \mu| \geq 1\big] = \dfrac{1}{4} = 0.25$

Compare with (i), we see that bound furnished by Chebyshev's inequality is highly different. Chebyscheve's inequality provides estimates when it is impossible to obtain exact values.

Example 5: A random variable X has a mean 3 and variance 2. Use Chebyscheve's inequality to obtain an upper bound for i) $P\big[|X - 3| \geq 2\big]$ and ii) $P\big[|X - 3| \geq 1\big]$

Sol: From the given data, $\mu = 3$, $\sigma = \sqrt{2}$, to find

(i) $P\big[|X - 3| \geq 2\big]$: We shall compare with Chebyschev's inequality

$$P\big[|X - \mu| \ge k\sigma\big] \le \frac{1}{k^2} \qquad\qquad \text{.....(3.45)}$$

Here $k\sigma = 2$, $k = \sqrt{2}$ since $\sigma = \sqrt{2}$

$$\therefore P\big[|X - 3| \ge 2\big] \le \frac{1}{2} \qquad \text{Hence the upper bound for} \quad P\big[|X - 3| \ge 2\big] \quad \text{is} \quad \frac{1}{2}$$

(ii) $$P\big[|X - 3| \ge 1\big]$$

Here $k\sigma = 1$, $k = \frac{1}{\sqrt{2}}$ since $\sigma = \sqrt{2}$

$$P\big[|X - 3| \ge 1\big] \le 2$$

Hence the upper bound for $\quad P\big[|X - 3| \ge 1\big] \quad$ is 1 since probability never greater than 1.

Example 6: Two unbiased dice are thrown. Let X be a random variable that represents the sum of the numbers showing up. Prove that by Chebyshev's inequality $P\big[|X - 7| \ge 2\big] \le \frac{35}{24}$. Compare with the actual probability.

Sol: When two dice are thrown then the sample space S is

$$X(S) = \begin{cases} (1,1), (1,2), (1,3), (1,4), (1,5), (1,6); & (2,1), (2,2), (2,3), (2,4), (2,5), (2,6) \\ (3,1), (3,2), (3,3), (3,4), (3,5), (3,6); & (4,1), (4,2), (4,3), (4,4), (4,5), (4,6) \\ (5,1), (5,2), (5,3), (5,4), (5,5), (5,6); & (6,1), (6,2), (6,3), (6,4), (6,5), (6,6)\} \end{cases}$$

From the given data,

$S = \{2, 3, 4, 5, 6, 7, 8, 9, 10, 11, 12\}$ represents sum of the variables on the faces.

Now

$$P(2) = P(X = 2) = P(1,1) = \frac{1}{36}; \qquad\qquad P(3) = P(X = 3) = P\{(1,2)\,(2,1)\} = \frac{2}{36}$$

$$P(4) = P(X = 4) = P\{(1,3)\,(3,1)\,(2,2)\} = \frac{3}{36}$$

$$P(5) = P(X = 5) = P\{(1,4)\,(2,3)\,(3,2),(4,1)\} = \frac{4}{36} \quad \text{and so on}$$

$$P(12) = P(X = 12) = P\{(6,6)\} = \frac{1}{36}$$

∴ Probability distribution of the random variable X is

$X = x$	2	3	4	5	6	7	8	9	10	11	12
$P(X=x)$	$\dfrac{1}{36}$	$\dfrac{2}{36}$	$\dfrac{3}{36}$	$\dfrac{4}{36}$	$\dfrac{5}{36}$	$\dfrac{6}{36}$	$\dfrac{5}{36}$	$\dfrac{4}{36}$	$\dfrac{3}{36}$	$\dfrac{2}{36}$	$\dfrac{1}{36}$

Mean $\mu = \sum x\, P(x) = 2 * \dfrac{1}{36} + 3 * \dfrac{2}{36} + 4 * \dfrac{3}{36} + \ldots\ldots\ldots + 12 * \dfrac{1}{36}$

$$= \frac{252}{36} = 7$$

The variance of the random variable X is

$$\sigma^2 = E(X^2) - \{E(X)\}^2$$

Now

$$E(X^2) = \sum x^2\, P(x) = 2^2 * \frac{1}{36} + 3^2 * \frac{2}{36} + 4^2 * \frac{3}{36} + \ldots\ldots\ldots + 12^2 * \frac{1}{36}$$

$$= \frac{329}{6}$$

$$\sigma^2 = E(X^2) - \{E(X)\}^2$$

$$= \frac{329}{6} - 7^2$$

$$= \frac{35}{6}$$

Now by Chebyshev's inequality, we know that $P\big[|X - \mu| \ge \epsilon\big] \le \dfrac{\sigma^2}{\epsilon^2}$

Taking $\epsilon = 2$, $P\big[|X - 7| \ge 2\big] \le \dfrac{35/6}{4} = \dfrac{35}{24}$

Since by actual probability is given by

$$P\big[|X - 7| \ge 2\big] = 1 - P\big[|X - 7| < 2\big]$$

$$= 1 - P(5 < X < 9)$$

$$= 1 - \big[P(X = 6) + P(X = 7) + P(X = 8)\big]$$

$$= 1 - \left[\frac{5}{36} + \frac{6}{36} + \frac{5}{36}\right]$$

$$= 1 - \frac{16}{36}$$

$$= \frac{5}{9}$$

3.11   Moment Generating Function

Moment generating function is helpful to find the higher moments in easier manner. The moment generating function (m. g..f) of a random variable X about origin having the probability function $f(x)$ is denoted by $M_x(t)$ and is defined by

$$M_x(t) = E[e^{tx}] = \begin{cases} \displaystyle\sum_{x=0}^{\infty}[e^{tx}.p(X=x)] & \textit{If X is discrete Random Variable} \\ \\ \displaystyle\int_{-\infty}^{+\infty} e^{tx} f(x)dx & \textit{If X is Continuous Random Variable} \end{cases}$$

Now $M_x(t) = E[e^{tx}]$

$$= E\left[1 + \frac{tx}{1!} + \frac{(tx)^2}{2!} + \frac{(tx)^3}{3!} + \cdots + \frac{(tx)^r}{r!} + \cdots\right]$$

$$= 1 + tE[x] + \frac{t^2}{2!}E[x^2] + \frac{t^3}{3!}E[x3] + \cdots + \frac{t^r}{r!}E[x^r] + \cdots$$

$$= 1 + t\mu_1' + \frac{t^2}{2!}\mu_2' + \frac{t^3}{3!}\mu_3' + \cdots + \frac{t^r}{r!}\mu_r' + \cdots = \sum_{r=0}^{\infty}\frac{t^r}{r!}\mu_r' \qquad .....(3.46)$$

Where $\mu_r' = E[x^r]$ is the rth moment of X about the origin. Other words, the coefficient of $\dfrac{t^r}{r!}$ in $M_x(t)$ gives μ_r' (about the origin). Thus $M_x(t)$ generates moments, it is known as moment generating function.

Differentiating (1) with respect to t, r times then putting $t = 0$ we get

$$\frac{d^r}{dt^r}[M_x(t)] = \left[\frac{1}{r!}r!\mu_r' + t\mu_{r+1}' + \frac{t^2}{2!}\mu_{r+2}' + \cdots\right]_{t=0} = \mu_r'$$

$$\therefore \mu_r' = \frac{d^r}{dt^r}[M_x(t)] \qquad\qquad .....(3.47)$$

Note: The mean can be obtained form $M_x(t)$ as $E(X) = \mu_1' = \dfrac{d}{dt}[M_x(t)]$

In general, the moment generating function of X about the point $X=a$ is defined as

$M_x(t)$ (About $X=a$) $= E[e^{t(x-a)}]$

$$= E\left[1 + \frac{t(x-a)}{1!} + \frac{t^2(x-a)^2}{2!} + \frac{t^3(x-a)^3}{3!} + \cdots + \frac{t^r(x-a)^r}{r!} + \cdots\right]$$

$$= 1 + t\mu_1' + \frac{t^2}{2!}\mu_2' + \frac{t^3}{3!}\mu_3' + \cdots + \frac{t^r}{r!}\mu_r' + \cdots = \sum_{r=0}^{\infty} \frac{t^r}{r!}\mu_r' \qquad \ldots\ldots(3.48)$$

Where $\mu_r' = E\left[(x-a)^r\right]$ is the rth moment of X about the point $X=a$.

Properties of Moment Generating Function

1. If c is any constant, and X is a random variable then $M_{cx}(t) = M_x(ct)$

   Since $M_x(t) = E[e^{tx}]$

   $$M_{cx}(t) = E[e^{t.cx}]$$

   $$= E[e^{ct.x}]$$

   $$= M_x(ct)$$

   $\therefore$ $M_{cx}(t) = M_x(ct)$

   And $M_{-x}(t) = M_x(-t)$ this denotes that the random variable X has a density function symmetrical about the origin then the m.g.f is also symmetric about the origin.

2. Let $Y = aX + b$, where a and b are constants and X is at random variable then the moment generating function of Y is

   $$M_y(t) = E[e^{ty}] = E[e^{t(aX+b)}]$$

   $$= E[e^{taX} e^{tb}] = e^{tb} E[e^{at.X}]$$

   $$= e^{tb} M_X(at)$$

   $$= e^{tb} M_{aX}(t)$$

   $\therefore$ $M_{(aX+b)}(t) = e^{tb} M_X(at) = e^{tb} M_{aX}(t)$

3. Let X_1 and X_2 are twp random variables, then the m.g.f. of the sum of two random variables is equal to product of their m.g.fs

   $$M_{(X_1+X_2)}(t) = E[e^{t(X_1+X_2)}] = E[e^{t(aX+b)}]$$

   $$= E[e^{tX_1+tX_2}] = E[e^{tX_1} e^{tX_2}]$$

   $$= E[e^{tX_1}] \times E[e^{tX_2}]$$

   $$= M_{X_1}(t) \times M_{X_2}(t)$$

4. Even if the moment generating function exists for a random variable X it may have no moments.

5. A random variable can have its moment generating function and some for all moments, yet the moment generating function does not generate the moments.

6. A random variable X can have all or some moments, but moment generating function does not exist except perhaps at one point.

Examples

***Example* 1:** If a random variable X has the moment generating function $M_x(t) = \dfrac{1}{1-t}$, $t < 1$, determine the mean and variance of the random variable X

Sol: Given that the moment generating function of the random variable X is

$$M_x(t) = \frac{1}{1-t} = (1-t)^{-1}$$

$$= 1 + t + t^2 + t^3 + \cdots + t^r + \cdots \qquad \text{since } t < 1$$

This can be written as

$$Mx(t) = 1 + 1!\left(\frac{t}{1!}\right) + 2!\frac{t^2}{2!} + 3!\frac{t^3}{3!} + \cdots + r!\frac{t^r}{r!} + \cdots$$

$$= \sum_{r=0}^{\infty} \frac{t^r}{r!} \mu'_r$$

The coefficient of $\dfrac{t}{1!} = 1$, Therefore $E[x] = 1$. Hence the mean of the random variable X is 1.

The coefficient of $\dfrac{t^2}{2!} = 2!$, Therefore $E[x^2] = 2$

The variance of the random variable $X = E[x^2] - [E(x)]^2$

$$= 2 - 1 = 1$$

Hence Variance of the random variable X is 1.

***Example* 2:** If a random variable X has the moment generating function $M_x(t) = \dfrac{3}{3-t}$, $t < 1$, determine the standard deviation of the random variable X

Sol: Given that the moment generating function of the random variable X is

$$M_x(t) = \frac{3}{3-t} = \frac{3}{3\left(1 - \frac{t}{3}\right)} = \left(1 - \frac{t}{3}\right)^{-1}$$

$$= 1 + \frac{t}{3} + \frac{t^2}{3^2} + \frac{t^3}{3^3} + \cdots + \frac{t^r}{3^r} + \cdots \quad \text{since } t < 1$$

This can be written as

$$M_x(t) = 1 + \frac{1}{3}\left(\frac{t}{1!}\right) + \frac{2!}{3^2}\frac{t^2}{2!} + \frac{3!}{3^3}\frac{t^3}{3!} + \cdots + \frac{r!}{3^r}\frac{t^r}{r!} + \cdots$$

$$= \sum_{r=0}^{\infty} \frac{t^r}{r!} \mu_r'$$

The coefficient of $\dfrac{t}{1!}$ in $M_x(t) = \dfrac{1}{3}$, Therefore $E[X] = \dfrac{1}{3}$. Hence the mean of the random variable X is $\dfrac{1}{3}$.

The coefficient of $\dfrac{t^2}{2!}$ in $M_x(t) = \dfrac{2!}{3^2}$, Therefore $E[X^2] = \dfrac{2}{9}$

The variance of the random variable X is

$$V(X) = E[X^2] - [E(X)]^2$$

$$= \frac{2}{9} - \left(\frac{1}{3}\right)^2$$

$$= \frac{1}{9}$$

Hence the standard deviation of the random variable X is $+\dfrac{1}{3}$.

Example 3: Determine the moment generating function of the random variable whose moments are

$$\mu_r' = (r+1)! \, 2^r$$

Sol : Since $M_x(t) = 1 + t\mu_1' + \dfrac{t^2}{2!}\mu_2' + \dfrac{t^3}{3!}\mu_3' + \cdots + \dfrac{t^r}{r!}\mu_r' + \cdots = \displaystyle\sum_{r=0}^{\infty} \frac{t^r}{r!}\mu_r' \quad \ldots(3.49)$

Given that $\mu_r' = (r+1)! \, 2^r$

Then $M_x(t) = \sum\limits_{r=0}^{\infty} \dfrac{t^r}{r!}(r+1)!\,2^r$

$= \sum\limits_{r=0}^{\infty} t^r (r+1)\, 2^r$

$= \sum\limits_{r=0}^{\infty} (r+1)\,(2t)^r$

$= 1 + 2(2t) + 3(2t)^2 + 4(2t)^3 + \cdots$

$= (1-2t)^{-2}$

$= \dfrac{1}{(1-2t)^2}$

Example 4: The random variable X has the probability function $P(X = x) = q^{x-1}p$, $x = 1,2,3,\cdots$

Find the moment generating function and hence find the mean and variance of the random variable X, assume that $p + q = 1$

Sol: Since $M_x(t) = E[e^{tx}]$

$= \sum\limits_{x=0}^{\infty} [e^{tx}.p(X = x)]$ since X is discrete Random Variable

$= \sum\limits_{x=1}^{\infty} e^{tx} q^{x-1} p \quad = \dfrac{p}{q} \sum\limits_{x=1}^{\infty} (qe^t)^x$

$= \dfrac{p}{q}(qe^t) \sum\limits_{x=1}^{\infty} (qe^t)^{x-1}$

$= pe^t \left[1 + (qe^t) + (qe^t)^2 + \cdots \right]$

$= pe^t \left[1 - qe^t \right]^{-1}$

$\therefore \qquad M_x(t) = \dfrac{pe^t}{1 - qe^t}$

Since the mean of the random variable X is

$$E(X) = \mu_1' = \frac{d}{dt}[M_x(t)]_{t=0}$$

$$= \frac{d}{dt}\left[\frac{pe^t}{1-qe^t}\right]_{t=0} = \left[\frac{(1-qe^t)pe^t + pe^t(qe^t)}{(1-qe^t)^2}\right]_{t=0}$$

$$\mu_1' = \left[\frac{pe^t}{(1-qe^t)^2}\right]_{t=0}$$

$$= \frac{p}{(1-q)^2}$$

$$\therefore \quad E(X) = \frac{1}{p}$$

$$\therefore \quad E(X^2) = \mu_2' = \frac{d^2}{dt^2}[M_x(t)]_{t=0}$$

$$= \frac{d}{dt}\left[\frac{pe^t}{(1-qe^t)^2}\right]_{t=0}$$

$$= \left[\frac{(1-qe^t)^2 pe^t + pe^t 2(1-qe^t)qe^t}{(1-qe^t)^4}\right]_{t=0}$$

$$= \left[\frac{(1-q)^2 p + p2(1-q)q}{(1-q)^4}\right]$$

$$= \left[\frac{(1-q)p + p2q}{(1-q)^3}\right]$$

$$\mu_2' = \frac{(1+q)}{p^2}$$

Therefore the variance of the random variable X is

$$V(X) = E(X^2) - [E(X)]^2$$

$$= \frac{(1+q)}{p^2} - \left[\frac{1}{p}\right]^2 = \frac{q}{p^2}$$

Example 5: Find the moment generating function of Poisson distribution

Sol: If X is a Poisson variable, then the m.g.f is given by

$$M_x(t) = E[e^{tx}] = \sum_{x=0}^{\infty} [e^{tx} . p(X = x)]$$

$$= \sum_{x=0}^{\infty} e^{tx} \frac{e^{-\lambda} \lambda^x}{x} = e^{-\lambda} \sum_{x=0}^{\infty} \frac{(e^t \lambda)^x}{x!}$$

$$= e^{-\lambda} \left\{ 1 + \frac{\lambda e^{-t}}{1!} + \frac{(\lambda e^t)^2}{2!} + - - - - \right\}$$

$$e^{-\lambda} . e \lambda e^t = e^{\lambda(e^t - 1)}$$

$\therefore Mx(t) = e^{\lambda(e^t - 1)}$ is the m.g.f of Poisson distribution.

Example 6: Determine the moment Generating function of geometric distribution

Sol: Since $M_x(t) = E(e^{tx}) = \sum e^{tx} p(X = x)$

For geometric distribution, $P(X = k) = q^{k-1} p \; k = 1, 2 \ldots$

Now $M_x(t) = E(e^{tx})$

$$= \sum_{k=1}^{\infty} e^{tk} q^{k-1} p = p \sum_{k=1}^{\infty} e^{th} \frac{q^k}{q}$$

$$= \frac{p}{q} \sum_{k=1}^{q} (e^{tq})^k$$

$$= \frac{p}{q} \left[qe^t + (qe^t)^2 + (qe^t)^3 + \cdots \right]$$

$$= \frac{p}{q} qe^t \left[1 + (qe^t) + (qe^t)^2 + \cdots \right]$$

$$= pe^t \left[\frac{1}{1 - qe^t} \right]$$

$$= \frac{pe^t}{1 - qe^t}$$

Example 7: Determine the moment generating function of uniform distribution

Sol: The probability density functions of uniform distribution is

$$f(x) = \frac{1}{b - a} \qquad a < x < b$$

$$M_x(t) = E(e^{tx})$$

$$= \int_a^b e^{tx} f(x) dx$$

$$= \int_a^b e^{tx} \frac{1}{b-a} dx$$

$$= \frac{1}{b-a} \left(\frac{e^{tx}}{t} \right)_a^b$$

$$= \frac{1}{t(b-a)} \left(e^{bt} - e^{at} \right)$$

$$\therefore M_x(t) = \frac{e^{bt} - e^{at}}{t(b-a)}$$

Example 8: Find the moment generating function of the exponential distribution $f(x) = ae^{-ax}$ $0 \leq x < \infty$, $a > 0$. Hence find its mean and standard deviations.

Sol: Since

$$M_x(t) = E[e^{tx}]$$

$$= \int_{-\infty}^{+\infty} e^{tx} f(x) dx \quad \text{Here X is Continuous Random Variable}$$

$$= \int_0^{+\infty} e^{tx} ae^{-ax} dx$$

$$= a \int_0^{+\infty} e^{(t-a)x} dx$$

$$= a \left[\frac{e^{(t-a)x}}{(t-a)} \right]_0^{\infty}$$

$$= \frac{a}{(t-a)} \left[e^{-(a-t)x} \right]_0^{\infty}$$

$$= \frac{a}{(t-a)} [0-1] = \frac{a}{(t-a)}$$

$$= \frac{a}{a \left(1 - \frac{t}{a} \right)} = \left(1 - \frac{t}{a} \right)^1$$

$$= \left(1 + \frac{t}{a} + \frac{t^2}{a^2} + \frac{t^3}{a^3} + \cdots \right)$$

This can also be written as

$$M_x(t) = \left(1 + \frac{1!\, t}{a\, 1!} + \frac{2!\, t^2}{a^2\, 2!} + \frac{3!\, t^3}{a^3\, a^3} + \cdots \right)$$

μ_1' = The coefficient of $\dfrac{t}{1!}$ in $M_x(t)$ = $\dfrac{1}{a}$, Therefore $E[X] = \dfrac{1}{a}$. Hence the mean of

the random variable X is $\dfrac{1}{a}$.

μ_2' = The coefficient of $\dfrac{t^2}{2!}$ in $M_x(t) = \dfrac{2!}{a^2}$, Therefore $E[X^2] = \dfrac{2}{a^2}$

The variance of the random variable X is

$$V(X) = E[X^2] - [E(X)]^2 = \mu_2' - (\mu_1')^2$$
$$= \frac{2}{a^2} - \left(\frac{1}{a} \right)^2$$
$$= \frac{1}{a^2}$$

Hence the standard deviation of the random variable X is $+\dfrac{1}{a}$.

***Example* 9:** Obtain the moment generating function of the random variable X having pdf

$$f(x) = \begin{cases} x & 0 \le x < 1 \\ 2 - x & 1 \le x < 2 \\ 0 & \text{Otherwise} \end{cases}$$

Sol: Since $M_x(t) = E[e^{tx}]$

$$= \int_{-\infty}^{+\infty} e^{tx} f(x) dx \quad \text{Here } X \text{ is Continuous Random Variable}$$

$$= \int_0^1 e^{tx} x\, dx + \int_1^2 e^{tx} (2-x) dx$$

$$= \int\limits_{0}^{1} e^{tx} x \, dx + 2 \int\limits_{1}^{2} e^{tx} \, dx - \int\limits_{1}^{2} x e^{tx} \, dx$$

$$= \left[x \frac{e^{tx}}{t} - \frac{e^{tx}}{t^2} \right]_{0}^{1} + 2 \left(\frac{e^{tx}}{t} \right)_{1}^{2} - \left[x \frac{e^{tx}}{t} - \frac{e^{tx}}{t^2} \right]_{1}^{2}$$

$$= \left[\left(\frac{e^t}{t} - \frac{e^t}{t^2} \right) - \left(0 - \frac{1}{t^2} \right) \right] + 2 \left(\frac{e^{2t}}{t} - \frac{e^t}{t} \right) - \left[\left(\frac{e^t}{t} - \frac{e^t}{t^2} \right) - \left(0 - \frac{1}{t^2} \right) \right]$$

$$= \frac{e^t}{t} - \frac{e^t}{t^2} + \frac{1}{t^2} + 2 \frac{e^{2t}}{t} - 2 \frac{e^t}{t} - \left[2 \frac{e^{2t}}{t} - \frac{e^{2t}}{t^2} - \frac{e^t}{t} + \frac{e^t}{t^2} \right]$$

$$= + e^{2t} \left[\frac{2}{t} - \frac{2}{t} + \frac{1}{t^2} \right] + \frac{1}{t^2}$$

$$= e^t \left(-\frac{2}{t^2} \right) + e^{2t} \left(\frac{1}{t^2} \right) + \frac{1}{t^2}$$

$$= \frac{1}{t^2} \left(1 + e^{2t} - 2 e^t \right)$$

$$= \frac{1}{t^2} \left(1 - e^t \right)^2$$

$$= \left(\frac{e^t - 1}{t} \right)^2$$

Hence $M_x(t) = \left(\dfrac{e^t - 1}{t} \right)^2$

***Example* 10:** Determine the moment generating function of Poisson distribution

Sol: If X is a Poisson variable, then the m.g.f is given by

$$M_x(t) = E[e^{tx}] = \sum_{x=0}^{\infty} [e^{tx} . P(X = x)]$$

$$= \sum_{x=0}^{\infty} e^{tx} \frac{e^{-\lambda} \lambda^x}{x} = e^{-\lambda} \sum_{x=0}^{\infty} \frac{(e^t \lambda)^x}{x!}$$

$$= e^{-\lambda} \left\{ 1 + \frac{\lambda e^{-t}}{1!} + \frac{(\lambda e^t)^2}{2!} + - - - - \right\}$$

$$e^{-\lambda} . e \lambda e^t = e^{\lambda (e^t - 1)}$$

$\therefore M_x(t) = e^{\lambda (e^t - 1)}$ is the m.g.f of Poisson distribution

Example **11:** Determine the moment generating function of uniform distribution

Sol: The probability density functions of uniform distribution.

$$f(x) = \frac{1}{b-a} \qquad a < x < b$$

$$M_x(t) = E(e^{tx})$$

$$= \int_a^b e^{tx} f(x) dx$$

$$= \int_a^b e^{tx} \frac{1}{b-a} dx$$

$$= \frac{1}{b-a} \left(\frac{e^{tx}}{t} \right)_a^b$$

$$= \frac{1}{t(b-a)} \left(e^{bt} - e^{at} \right)$$

$$\therefore M_x(t) = \frac{e^{bt} - e^{at}}{t(b-a)}$$

Example **12:** Determine the moment generating function of Gamma distribution

Sol: Since $\qquad M_x(t) = E(e^{tx})$

$$= \int_0^\infty e^{tx} f(x) dx$$

$$= \int_0^\infty e^{tx} \frac{\beta}{\Gamma(\alpha)} (\beta x)^{\alpha-1} e^{-\beta x} dx$$

$$= \frac{\beta}{\Gamma(\alpha)} \int_0^\infty e^{-(\beta-t)x} (\beta x)^{\alpha-1} dx$$

Put $(\beta - t)x = y$ *then* $(\beta-t) dx = dy$ *and* $x = \dfrac{y}{\beta-t}$

$$\therefore M_x(t) = \frac{\beta}{\Gamma(\alpha)} \int_0^\infty e^{-y} \left(\frac{\beta y}{\beta-t} \right)^{\alpha-1} \frac{dy}{\beta-t}$$

$$= \left(\frac{\beta}{\beta-t} \right)^\alpha \frac{1}{\Gamma(\alpha)} \int_0^\infty e^{-y} y^{\alpha-1} dy$$

$$= \left(\frac{\beta}{\beta-t}\right)^{\alpha} \frac{1}{\Gamma(\alpha)} \Gamma(\alpha)$$

$$= \left(\frac{\beta}{\beta-t}\right)^{\alpha}$$

$\therefore$ The moment generating function of gamma distribution is $M_x(t) = \left(\frac{\beta}{\beta-t}\right)^{\alpha}$

3.12 Characteristic Function

In some situations, the m.g.f. does not exist for real values of t for some distributions, i.e.,

in some situations $E[e^{tx}] = \sum_{x=0}^{\infty}[e^{tx}.P(X=x)]$ or $\int_{-\infty}^{+\infty} e^{tx} f(x)dx$ does not converge

absolutely for real values of t. Then we define a function called characteristic function (c.f) as $E[e^{itx}]$ and is denoted by $\phi_X(t)$ which is useful to generate moments.

Therefore the characteristic function of a random variable X is defined as

$$\phi_X(t) = E[e^{itx}]$$

$$= \begin{cases} \sum_{x=0}^{\infty}[e^{itx}.p(X=x)] & \textit{If X is discrete Random Variable} \\ \int_{-\infty}^{+\infty} e^{itx} f(x)dx & \textit{If X is Continuous Random Variable} \end{cases}$$

The characteristic function is more advantages than m.g.f , since for some functions, the m.g.f. does not exist but the characteristic function always exists. Also there is a one to one correspondence between distribution function and its characteristic function of a random variable.

$$f_X(x) = \int_{-\infty}^{+\infty} e^{-itx} \phi_X(t)dt = \frac{1}{2\pi} \int_{-\infty}^{+\infty} e^{-itx} \phi_X(t)dt$$

This is known as the inverse Fourier Transform of $\phi_X(t)$

Properties

1. $\qquad\qquad \phi_X(0) = 1$

Since $\qquad\qquad \phi_X(t) = E[e^{itx}]$

$$\phi_X(0) = E[e^{i0x}]$$
$$= E[1]$$
$$= 1$$

Hence $\qquad \phi_X(0)=1$

Similarly we can prove the following properties

2. $\phi_X(0)$ The maximum magnitude of the characteristic function is unity. i.e., $|\phi_X(t)| \leq 1$

3. $\phi_X(t) = \phi_X(-t)$ if X is symmetric function.

4. $\phi_{aX+b}(t) = e^{ibt}\phi_X(at)$ where a and b are constants

5.   If X and Y are independent random variables then $\phi_{X+Y}(t) = \phi_X(t) \times \phi_X(t)$

6. $\phi_X(t)$ is a continuous function of t in $(-\infty, \infty)$

For any real value of t, then it is uniformly continuous for all t.

7. $\phi_X(t)$ and $\phi_X(-t)$ are complex conjugate function i.e., $\phi_X(t) = \overline{\phi_X(-t)}$

8.   The function $\varphi(t) = \log_e |\phi(t)|$ is called the second characteristic function.

9.   If X is a random variable with characteristic function $\phi_X(t)$ and if the rth moment exists, it can be determined by $\mu_r = (-i)^r \left[\dfrac{\partial^r}{\partial t^r}(\phi_X(t))\right]_{t=0}$

Examples

***Example* 1:** Obtain the characteristic function of the random variable

$$X = \begin{cases} a & \text{with probability } p \\ b & \text{withprobability } q = 1-p \end{cases}$$

Sol: Given that

$$\begin{array}{lcc} X: & a & b \\ P(X=x)]: & p & 1\text{--}p \end{array}$$

Since $\qquad \phi_X(t) = E[e^{itx}] = \displaystyle\sum_{x=0}^{\infty}[e^{itx}.P(X=x)]$

Here X is discrete Random Variable

$$= e^{ita}p + e^{itb}(1-p)$$

$$= e^{iat}p + e^{ibt} - e^{ibt}p$$

$$= e^{ibt} + p(e^{iat} - e^{ibt})$$

***Example* 2:** Find the characteristic function of the random variable X can assume the 1 and -1 with probability $\dfrac{1}{2}$ each. Also find the moment generating function.

Sol: Given that

X :	1	-1
$P(X=x)]$:	$\dfrac{1}{2}$	$\dfrac{1}{2}$

Since

$$\phi_X(t) = E[e^{itx}] = \sum_{x=0}^{\infty}[e^{itx}.P(X=x)]$$

$$= e^{it(1)}\frac{1}{2} + e^{it(-1)}\frac{1}{2}$$

$$= \frac{1}{2}\left(e^{it} + e^{-it}\right)$$

$$= \cos t$$

$$M_X(t) = E[e^{tx}] = \sum_{x=0}^{\infty}[e^{tx}.P(X=x)]$$

Here X is discrete Random Variable

$$= e^{t(1)}\frac{1}{2} + e^{t(-1)}\frac{1}{2}$$

$$= \frac{1}{2}\left(e^{t} + e^{-t}\right)$$

***Example* 3:** Determine the Characteristic Function of Poisson distribution

Sol: If X is a Poisson variate, then the characteristic function is given by

$$\varphi_X(t) = E\left[e^{itx}\right] = \sum e^{itx}.p(X=x)$$

$$= \sum_0^\infty e^{itx} \frac{e^{-\lambda}\lambda^x}{x!} = e^{-\lambda} \sum_0^\infty \frac{\left(\lambda e^{it}\right)^x}{x!}$$

$$= e^{-\lambda}.\left\{1 + \frac{\lambda e^{it}}{1!} + \frac{\left(\lambda e^{it}\right)^2}{2!} + \cdots\right\}$$

$$= e^{-\lambda} \ e^{\lambda e^{it}}$$

$$= e^{\lambda\left(e^{it}-1\right)}$$

$$\therefore \varphi_X(t) = e^{\lambda\left(e^{it}-1\right)} \quad \text{is the Characteristic function of Poisson distribution.}$$

Example 4: Find the characteristic function of the random variable X having the density function $f(x) = ce^{-a|x|}$, $-\infty < x < \infty$, $a > 0$ and c is constant

Sol: Since c is a constant. We also know that $\int\limits_{-\infty}^{+\infty} f(x)dx = 1$

$$\Rightarrow \qquad \int\limits_{-\infty}^{+\infty} ce^{-a|x|}dx = 1$$

$$\Rightarrow \qquad c\left[\int\limits_{-\infty}^{0} e^{-a(-x)}dx + \int\limits_{0}^{\infty} e^{-a(+x)}dx\right] = 1$$

$$\Rightarrow \qquad c\left[\int\limits_{-\infty}^{0} e^{ax}dx + \int\limits_{0}^{\infty} e^{-ax}dx\right] = 1$$

$$\Rightarrow \qquad c\left[\left(\frac{e^{ax}}{a}\right)_{-\infty}^{0} + \left(\frac{e^{-ax}}{-a}\right)_{0}^{\infty}\right] = 1$$

$$\Rightarrow \qquad c\left[\left(\frac{1-0}{a}\right) + \left(\frac{0-1}{-a}\right)\right] = 1$$

$$\Rightarrow \qquad c\left[\left(\frac{1}{a}\right) + \left(\frac{1}{a}\right)\right] = 1$$

$\Rightarrow$ $c\left(\dfrac{2}{a}\right) = 1$

$\Rightarrow$ $c = \dfrac{a}{2}$

Therefore the given density function $f(x) = \dfrac{a}{2}e^{-a|x|}$, $-\infty < x < \infty$,

Now the Characteristic function

$\varphi_X(t) = E[e^{itx}]$

$= \displaystyle\int_{-\infty}^{+\infty} e^{itx} f(x)dx$ Here X is Continuous Random Variable

$= \displaystyle\int_{-\infty}^{+\infty} e^{itx}\dfrac{a}{2}e^{-a|x|}dx = \dfrac{a}{2}\int_{-\infty}^{+\infty} e^{itx} e^{-a|x|}dx$

$= \dfrac{a}{2}\left[\displaystyle\int_{-\infty}^{0} e^{itx}e^{ax}dx + \int_{0}^{\infty} e^{itx}e^{-ax}dx \right]$

$= \dfrac{a}{2}\left[\displaystyle\int_{-\infty}^{0} e^{(it+a)x}dx + \int_{0}^{\infty} e^{(it-a)x}dx \right]$

$= \dfrac{a}{2}\left[\left(\dfrac{e^{(it+a)x}}{it+a}\right)_{-\infty}^{0} + \left(\dfrac{e^{(it-a)x}}{it-a}\right)_{0}^{\infty} \right]$

$= \dfrac{a}{2}\left[\left(\dfrac{1-0}{a+it}\right) + \left(\dfrac{0-1}{it-a}\right) \right]$

$= \dfrac{a}{2}\left[\left(\dfrac{1}{a+it}\right) + \left(\dfrac{1}{a-it}\right) \right]$

$= \dfrac{a}{2}\left[\left(\dfrac{a-it+a+it}{(a+it)(a-it)}\right) \right]$

$= \dfrac{a}{2}\left[\left(\dfrac{2a}{a^2+t^2}\right) \right]$

$\varphi_X(t) = \dfrac{a^2}{a^2+t^2}$

***Example* 5:** Determine the Characteristics function of the uniform distribution

Sol: $\varphi_x(t) = E\left(e^{itx}\right)$

$$= \int_a^b e^{itx} f(x)\, dx$$

$$= \frac{1}{b-a} \int_a^b e^{itx}\, dx$$

$$= \frac{1}{b-a} \cdot \frac{1}{it}\left(e^{itb} - e^{ita}\right)$$

$$= \frac{e^{itb} - e^{ita}}{it(b-a)}$$

***Example* 6:** Determine the Characteristics function of the Geometric distribution

Sol: Since $\varphi_x(t) = E(e^{itx}) = \sum e^{itx} P(X = x)$

For Geometric distribution, $p(X = k) = q^{k-1} p$ $k = 1, 2 \text{ ----}$

$$\phi_x(t) = \sum\left[e^{itx}\right] = \sum_{k=1}^{\infty} e^{itk} \cdot q^{k-1} p = \frac{p}{q} \cdot \sum_{k=1}^{q}\left(qe^{it}\right)^k$$

$$= \frac{p}{q}\left[qe^{it} + \left(qe^{it}\right)^2 + \left(qe^{it}\right)^3 + \text{-----}\right]$$

$$= \frac{p}{q} \cdot qe^{it}\left[1 + \left(qe^{it}\right) + \left(qe^{it}\right)^2 + \text{-----}\right]$$

$$= \frac{p}{q} \cdot pe^{it}\left[\frac{1}{1 - qe^{it}}\right] = \frac{pe^{it}}{1 - qe^{it}}$$

***Example* 7:** Find the characteristic function of the random variable X having the density function

$$f(x) = \begin{cases} \dfrac{1}{2a} & |x| < a \\ 0 & Otherwise \end{cases}$$

Sol: Now the Characteristic function

$$\varphi_X(t) = E[e^{itx}]$$

$$= \int_{-\infty}^{+\infty} e^{itx} f(x) dx \qquad \text{Here } X \text{ is Continuous Random Variable}$$

$$= \int_{-\infty}^{-a} e^{itx} .0.dx + \int_{-a}^{a} e^{itx} \frac{1}{2a} dx + \int_{a}^{+\infty} e^{itx} 0.dx$$

$$= \int_{-a}^{a} e^{itx} \frac{1}{2a} dx$$

$$= \frac{1}{2a} \int_{-a}^{a} e^{itx} dx$$

$$= \frac{1}{2a} \left(\frac{e^{itx}}{it} \right)_{-a}^{a} = \frac{1}{2a.it} \left(e^{ita} - e^{-ita} \right)$$

$$= \frac{1}{a.t} \left(\frac{e^{ita} - e^{-ita}}{2i} \right)$$

$$= \frac{1}{a.t} \sin at$$

Example 8: Obtain the characteristic function of a random variable X defined by the density function

$$f(x) = \begin{cases} 0 & x < a \\ 1 & 0 \le x \le 1 \\ 0 & x > 1 \end{cases}$$

Sol: Now the Characteristic function

$$\varphi_X(t) = E[e^{itx}]$$

$$= \int_{-\infty}^{+\infty} e^{itx} f(x) dx \qquad \text{Here } X \text{ is Continuous Random Variable}$$

$$= \int_{-\infty}^{0} e^{itx} .0.dx + \int_{0}^{1} e^{itx} .1.dx + \int_{1}^{+\infty} e^{itx} .0.dx$$

$$= \int_{0}^{1} e^{itx} .1.dx$$

$$= \left(\frac{e^{itx}}{it} \right)_{0}^{1}$$

$$= \frac{1}{.it}\left(e^{it} - 1\right)$$

$$= \frac{e^{it} - 1}{it}$$

Example 9: Show that the distribution function for which the characteristic function $e^{-|t|}$ has the density function is $f(x) = \dfrac{1}{\pi(1+x^2)}$, $-\infty < x < \infty$.

Sol: Since we know that $f_X(x) = \dfrac{1}{2\pi} \displaystyle\int_{-\infty}^{\infty} e^{-itx} \varphi_X(t)\, dt$

Given $\varphi_X(t) = e^{-|t|}$

Now $f_X(x) = \dfrac{1}{2\pi} \displaystyle\int_{-\infty}^{\infty} e^{-itx} e^{-|t|}\, dt$

$$= \frac{1}{2\pi}\left(\int_{-\infty}^{0} e^{-itx} e^{-(-t)}\, dt + \int_{0}^{\infty} e^{-itx} e^{-(t)}\, dt \right)$$

$$= \frac{1}{2\pi}\left(\int_{-\infty}^{0} e^{(1-ix)t}\, dt + \int_{0}^{\infty} e^{-(ix+1)t}\, dt \right)$$

$$= \frac{1}{2\pi}\left[\left(\frac{e^{(1-ix)t}}{1-ix} \right)_{-\infty}^{0} + \left(\frac{e^{-(1+ix)t}}{-(1+ix)} \right)_{0}^{\infty} \right]$$

$$= \frac{1}{2\pi}\left[\left(\frac{1-0}{1-ix} \right) + \left(\frac{0-1}{-(1+ix)} \right) \right]$$

$$= \frac{1}{2\pi}\left[\left(\frac{1}{1-ix} \right) + \left(\frac{1}{(1+ix)} \right) \right]$$

$$= \frac{1}{2\pi}\left[\left(\frac{(1+ix)+(1-ix)}{(1-ix)(1+ix)} \right) \right]$$

$$= \frac{1}{2\pi}\left[\left(\frac{2}{1+x^2} \right) \right]$$

$$= \frac{1}{\pi\left(1+x^2\right)}$$

Hence the distribution function for which the characteristic function $e^{-|t|}$ has the density function is $f(x) = \dfrac{1}{\pi(1+x^2)}, \quad -\infty < x < \infty$.

3.13 Transformations of a Random Variable

Change of one random variable in to another random variable through an operator is called a transformation. Let X be a random variable defined on S and g(.) be a function such that $Y = g(X)$ is also a random variable defined on S. The probability density function of a random variable X, $f_X(x)$ is known and the problem is to determine the density function $f_Y(y)$ and vice versa.

Let X be a random variable, the operator g can be linear, non-linear, etc., . There are many cases to consider.

If g(.) is any continuous function, then the distribution of $Y = g(x)$ is uniquely determined by that of X.

In this section we consider, mainly

1. X is continuous and g is also continuous and either monotonically increasing or decreasing with X
2. X is continuous and g is also continuous buy non-monotonic
3. X is discrete and g continuous.

If we wish we will also study other cases also.

3.14 Monotonic Transformations for a Continuous Random Variable

A transformation g is called monotonically increasing if $g(x_1) \le g(x_2)$ for $x_1 \le x_2$

A transformation g is called strictly monotonically increasing if $g(x_1) < g(x_2)$ for $x_1 < x_2$ e. g. Let g is continuous and differentiable at all points of x for which $f_X(x) \ne 0$ and let Y have a specific value y corresponding to the specific value x of X is $y = g(x)$ or $x = g^{-1}(y)$ where g^{-1} is the inverse transform of g. Then

$$P(Y \le y) = P(X \le x) \text{ Since one to one correspondence between } X \text{ any } Y.$$

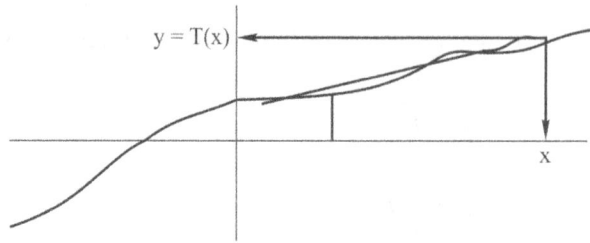

Fig. 3.3 Monotonically increasing

i.e., $F_Y(y) = P(Y \le y) = P(X \le x) = F_X(x)$

Similarly,

A transformation g is called monotonically decreasing if $g(x_1) \ge g(x_2)$ for $x_1 \le x_2$

A transformation g is called strictly monotonically decreasing if $g(x_1) > g(x_2)$ for $x_1 \le x_2$

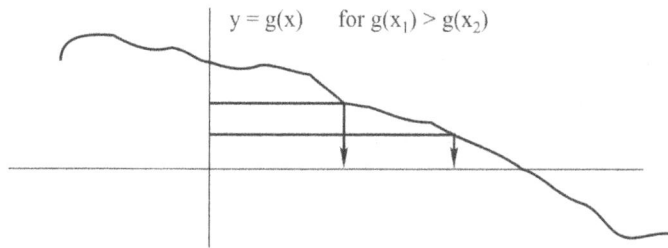

Fig. 3.4 Monotonically decreasing

Examples of Monotonic increasing and decreasing transformations

Consider a random variable X with known probability density function $f_X(x)$ and let $Y = g(x)$

(i) If $g(x)$ is a single valued function, then $f_X(x)\,dx = f_Y(y)\,dy$ or

$$f_Y(y) = f_X(x)\left|\frac{dx}{dy}\right| \qquad\qquad .....(3.50)$$

This is called one to one transformations, from figures (3.3) and (3.4).

(ii) If $g(x)$ is a multi valued function, then

$$f_Y(y) = f_X(x_1)\left|\frac{dx_1}{dy}\right| + f_X(x_2)\left|\frac{dx_2}{dy}\right| + \cdots \qquad\qquad .....(3.51)$$

This is called many to one correspondence.

If X is a continuous random variable with probability density function $f_X(x)$ and $g'(x)$ is positive then the random variable $Y = g(x)$ is continuous since $g'(x) > 0$, then g is strictly increasing function of X.

E.g., Let $y = \tan^{-1} x$

$$\Rightarrow \quad x = \tan y \quad \Rightarrow \quad \frac{dx}{dy} = \sec^2 y > 0$$

3.15   Non Monotonic Transformations of Continuous Random Variable

For a specific value of x, there corresponds a specific value y, is a single valued which is a monotonic transformation. A transformation may not be monotonic; there may be more than one interval of values of X corresponding to $Y \leq y$.

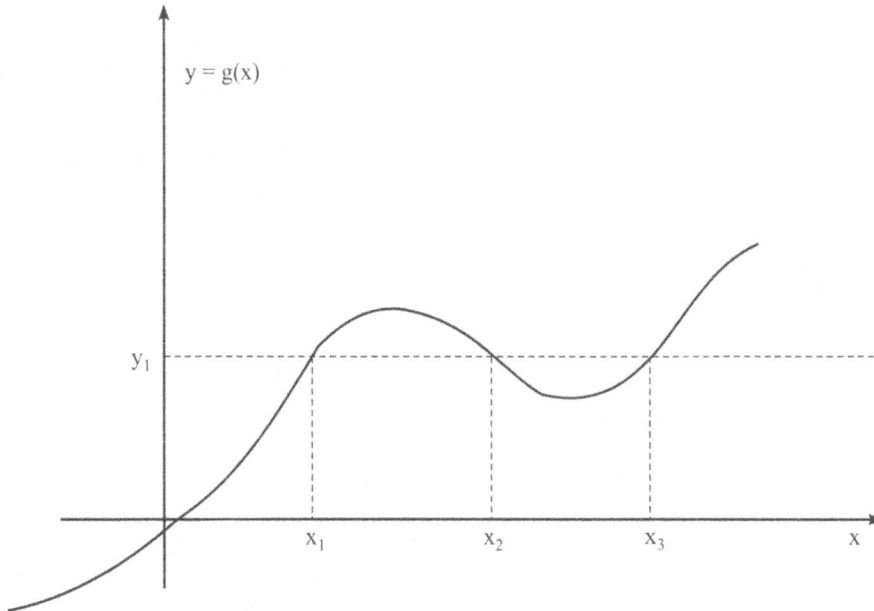

Fig. 3.5

From the figure, the values of Y, the event $Y \leq y$ corresponding to the event $X \leq x_1$ and $x_2 \leq X \leq x_3$ is called a non-monotonic transformation.

Then

$$P[Y \leq y] = P[x / Y \leq y]$$

$$= F_Y(y) \;=\; \int\limits_{(x/Y \le y)} f_X(x)dx \qquad\qquad .....(3.52)$$

To get the density function of Y, differentiate equation (3.52) then

$$\frac{d}{dy} F_Y(y) \;=\; \frac{d}{dy}\left[\int\limits_{(x/Y \le y)} f_X(x)dx\right] \qquad\qquad .....(3.53)$$

After simplification, we get the density function

$$f_Y(y) = \sum_n \frac{f_X(x)}{\left|\dfrac{d}{dx} g(x)\right|_{x=x_n}} \qquad\qquad .....(3.54)$$

Here the sum is taken as to include all the roots x_n, n = 1, 2, 3 . . . which are the real solutions of the equation $y = g(x)$ has no real roots for a given value of y, then $f_Y(y) = 0$

Theorem: suppose that X be a continuous random variable with p.d.f $f_X(x)$. Let $y = g(x)$ be strictly monotonic increasing or decreasing function of x. Also assume that $g(x)$ is continuous and differentiable for all values of x then the p.d.f of the random variable Y is given by

$$f_Y(y) = f_x(x)\frac{dx}{dy} \quad \text{where x is expressed inters of } y.$$

Proof: Here there are two cases:

(i)   Let $y = f(x)$ be strictly increasing function of x the $\dfrac{dx}{dy} > 0$

The distribution function of Y is given by

$$F_Y(y) = P(Y \le y) = P\big[(g(x) \le y)\big]$$

$$= P\big[x \le g^{-1}(y)\big] \qquad\qquad \because x = g^{-1}(y)$$

The inverse function is uniquely exists since $g(.)$ is strictly increasing

$$F_Y(y) = F_X\big[g^{-1}(y)\big]$$

$$= F_X(x) \qquad\qquad \because x = g^{-1}(y) \qquad .....(3.55)$$

Differentiating equation (1) w.r.t. y, then we get

$$\frac{d}{dy} F_y(y) = \frac{d}{dy} F_x(x)$$

$$\Rightarrow \qquad f_y(y) = \frac{d}{dy}\left[F_X(x)\frac{dx}{dy}\right] = f_X(x)\frac{dx}{dy}$$

(ii) Let $y = g(x)$ is strictly monotonic decreasing function of x then $\dfrac{dx}{dy} < 0$

Then the distribution function of Y is given by

$$F_Y(y) = P(Y \le y)$$
$$= P\left[(g(x) \le y)\right]$$
$$= P\left[x \ge g^{-1}(y)\right]$$
$$= 1 - P\left[x \le g^{-1}(y)\right] \qquad \because x = g^{-1}(y)$$
$$F_Y(y) = 1 - F_X\left[g^{-1}(y)\right]$$
$$= 1 - F_X(x) \qquad \because x = g^{-1}(y) \qquad .....(3.56)$$

The inverse function is uniquely exists.

Differentiate equation (2) w.r.t. y then we get

$$= \frac{d}{dy} F_y(y) = \frac{d}{dy}[1 - F(x)]$$

$$\Rightarrow \qquad f_y(y) = 0 - \frac{d}{dy}[F_X(x)]\frac{dx}{dy}$$

$$= -f_X(x)\frac{dx}{dy} \qquad .....(3.57)$$

Since in equation (3) y is a decreasing function of x. This implies x is a decreasing function of y

Since $\qquad \dfrac{dx}{dy} < 0$.

$$\therefore \qquad f_Y(y) = f_x(x)\left|\frac{dx}{dy}\right|.$$

3.16   Transformation of a Discrete Random Variable

Let X be a discrete random variable and $Y = g(x)$ is a continuous transformation then the probability density function and distribution functions are

$$f_X(x) = \sum_{n=1} P(x_n)\delta(x - x_n) \qquad .....(3.58)$$

and $\qquad F_X(x) = \sum_{n=1} P(x_n)u(x-x_n)$ $\qquad\qquad$ .....(3.59)

respectively. Here the sum is taken to include all the possible values of x_n, n = 1, 2, 3 . . .. of X.

If the transformation is monotonic, there is a one to one correspondence between X and Y so that a set $[y_n]$ corresponds to the set $[x_n]$ through the equation $y_n = g(x_n)$

Therefore $P(y_n) = P(x_n)$, thus

$$f_Y(y) = \sum_{n=1} P(y_n)\delta(y-y_n) \qquad\qquad .....(3.60)$$

and $\qquad F_Y(y) = \sum_{n=1} P(y_n)u(y-y_n)$ $\qquad\qquad$ .....(3.61)

If g is non-monotonic, the above procedure remains same except there exists the possibility that more than one value x_n corresponding to a value y_n. In such case, $P(y_n)$ will equal to the sum of the probabilities of the various values of x_n for which $y_n = g(x_n)$.

E.g., Let X be a discrete random variable which assumes the values -1, 0, 1, 2 with probabilities 0.1, 0.3, 0.4 and 0.2 respectively,

i.e., $\quad X = x_n$: $\qquad$ −1 $\qquad$ 0 $\qquad$ 1 $\qquad$ 2

$\qquad P(X = x_n)$: $\quad$ 0.1 $\quad$ 0.3 $\quad$ 0.4 $\quad$ 0.2

Then

$$f_X(x) = \sum_{n=1} P(x_n)\delta(x-x_n)$$

$$= P(x_1)\delta(x-x_1) + P(x_2)\delta(x-x_2) + P(x_3)\delta(x-x_3) + P(x_4)\delta(x-x_4)$$

$$= 0.1 \times \delta(x+1) + 0.3 \times \delta(x-0) + 0.4 \times \delta(x-1) + 0.2 \times \delta(x-2)$$

Let us assume X is transformed in to $\;Y = 2 - X^2 + \dfrac{X^3}{3}\;$ and find the density of Y.

The values of X map to respective values of Y given by $Y = \dfrac{2}{3},\ 2,\ \dfrac{4}{3}$ and $\dfrac{2}{3}$. The two

values x = −1 and x=2 map to one value $Y = \dfrac{2}{3}$. The $P\left(Y = \dfrac{2}{3}\right)$ is the sum of

probabilities of $P(X = -1)$ and $P(X = 2)$

Therefore $f_Y(y) = \sum_{n=1} P(y_n)\delta(y - y_n)$

$$= P(y_1)\delta(y - y_1) + P(y_2)\delta(y - y_2) + P(y_3)\delta(y - y_3) + P(y_4)\delta(y - y_4)$$

$$= 0.3 \times \delta(y - \frac{2}{3}) + 0.4 \times \delta(y - \frac{4}{3}) + 0.3 \times \delta(y - 2)$$

Quiz Questions

A random variable X has the following probability distribution is

X	1	2	3	4
$P(X = x)$	k	$2k$	$3k$	$4k$

And answer the following questions

1. The value of k is

    (a) $\dfrac{1}{10}$ (b) $\dfrac{3}{10}$ (c) $2\sqrt{3}$ (d)   none

2. The mean and variance of the above distribution is

    (a)   13,1   (b)   3,1   (c)   1, 3   (d)   none

3. Two persons are playing a game. Toss a coin, if head turns up he get one rupee, if tail turns up the losses one rupee. Now his expected gain is

    (a) $\dfrac{1}{4}$ (b)   1   (c)   0   (d)   none

4. A binomial distribution has mean 5 and stadard deviation is 2 then the number of trails is

    (a)   13   (b)   31   (c)   25   (d)   none

5. The standard deviation of binomial distribution $(q + p)^{16}$ is 2 the mean of the distribution is

    (a)   8   (b)   31   (c)   6   (d)   none

6. In a Binomial distribution, $9\,P(X = 4) = P(X = 2)$ if $n = 6$ then the probability p is

    (a) $\dfrac{1}{10}$ (b) $\dfrac{1}{2}$ (c) $\sqrt{3}$ (d)   none

7. For a Binomial distribution the mean is 10 and variance is 5 then the parameters are

   (a) $\dfrac{1}{10}, 20$ (b) $20, \dfrac{1}{2}$ (c) $1, \dfrac{1}{2}$ (d) none

8. Swetha has the probability $\dfrac{2}{3}$ of winning a game. If she plays 4 games, the probability that she winds at least one game is

   (a) $\dfrac{1}{81}$ (b) $\dfrac{80}{81}$ (c) $\dfrac{29}{81}$ (d) none

9. In a binomial distribution the mean is 4 and variance 3 then the distribution function is

   (a) $^{20}C_r \left(\dfrac{3}{4}\right)^{20-r} \left(\dfrac{1}{4}\right)^r$ (b) $^{16}C_r \left(\dfrac{3}{4}\right)^{16-r} \left(\dfrac{1}{4}\right)^r$

   (c) $^{20}C_r \left(\dfrac{2}{3}\right)^{20-r} \left(\dfrac{1}{3}\right)^r$ (d) $^{16}C_r \left(\dfrac{2}{3}\right)^{20-r} \left(\dfrac{1}{3}\right)^r$

10. In a large consignment of electric lamps 10% are defective. A random sample of 20 is taken for inspection. The probability it will have more than one defective is

    (a) $1 - \left(\dfrac{20 \times 9^{19}}{10^{20}}\right)$ (b) $1 - \left(\dfrac{9}{20}\right)^{20} - \left(\dfrac{20 \times 9^{19}}{10^{20}}\right)$

    (c) $1 - \left(\dfrac{9}{20}\right)^{20}$ (d) none

11. In five throws with a pair of dice, the probability of throwing doublets exactly two is

    (a) $^5C_1 \left(\dfrac{5}{6}\right)^3 \left(\dfrac{1}{6}\right)^2$ (b) $^5C_3 \left(\dfrac{5}{6}\right)^2 \left(\dfrac{1}{6}\right)^3$

    (c) $^5C_2 \left(\dfrac{5}{6}\right)^3 \left(\dfrac{1}{6}\right)^2$ (d) none

12. During war, one ship out of every 9 was sunk at an average in a making a voyage. The probability that exactly 3 out of a convoy of 6 ships would arrive safely is

    (a) $^6C_3 \left(\dfrac{1}{9}\right)^3 \left(\dfrac{8}{9}\right)^4$ (b) $\left(\dfrac{1}{9}\right)^3 \left(\dfrac{8}{9}\right)^3$

    (c) $^6C_1 \left(\dfrac{1}{9}\right)^3 \left(\dfrac{8}{3}\right)^4$ (d) none

13. The incidence of occupational disease in an industry is such that the workers have 20% chance of suffering from it. The probability that out of six workers chosen at random four or more will suffer from the disease is

(a) $\dfrac{3092}{3125}$ (b) $\dfrac{16}{25}$ (c) $\dfrac{53}{3125}$ (d)   None

14. An urn containing 100 chips, 10 are defective, what is the probability that out of a sample of 5 chips none is defective is

(a) 10^{-5} (b) $\left(\dfrac{1}{2}\right)^5$ (c) $\left(\dfrac{9}{10}\right)^5$ (d) $\left(\dfrac{9}{10}\right)$

15. In a binomical distribution, X is a binomial variate with mean 10 and variance 5 then $P(2 < X < 19)$ is

(a) $\displaystyle\sum_{x=3}^{18} {}^{20}C_x \left(\dfrac{1}{2}\right)^{20}$ (b) $\displaystyle\sum_{x=2}^{15} {}^{20}C_x \left(\dfrac{1}{2}\right)^{20}$

(c) $\displaystyle\sum_{x=3}^{14} {}^{20}C_x \left(\dfrac{1}{2}\right)^{20}$ (d)   none

16. In a poission distribution, the mean of the poission variate is m, then the sum of the terms in even places in this distribution is

(a) e^{-m} (b) $e^{-m}\cosh m$

(c) $e^{-m}\sinh m$ (d) $e^{-m}\coth m$

17. In a Poisson distribution $P(1) = P(2)$ then $P(0)$ is

(a) $\dfrac{1}{e^3}$ (b) $\dfrac{1}{e^2}$ (c) $\dfrac{1}{e}$ (d)   None

18. The chance of a traffic accident in a day in a street of a certain city is 0.001. If there are 1000 streets in the city, then out of 2000 days the days in which we expect no accidents in the whole city is

(a) $\dfrac{2000}{e}$ (b) $\dfrac{2000}{e^2}$ (c) $\dfrac{2000}{e^{10}}$ (d) $\dfrac{2000}{e^{20}}$

19. Let X has a Poisson distribution, if $P(X=0)= P(X=1)$ then $P(X=2)$ is

(a) $\dfrac{e}{2}$ (b) $\dfrac{e}{6}$ (c) $\dfrac{1}{6e}$ (d) $\dfrac{1}{2e}$

20. If a fixed volume v of blood contains on the average 20 red blood cells fro normal persons, then using Poisson distribution, we conclude that the probability that a specimen of volume v from a normal person will contain less than 15 red blood cells is

(a) $\sum\limits_{x=0}^{14} e^{-\left(\frac{v}{20}\right)} \dfrac{\left(\frac{v}{20}\right)^x}{x!}$

(b) $\sum\limits_{x=0}^{14} e^{-20} \dfrac{(20)^x}{x!}$

(c) $\sum\limits_{x=1}^{14} e^{-20} \dfrac{(20)^x}{x!}$

(d) $\sum\limits_{x=0}^{14} e^{-\left(\frac{20}{v}\right)} \dfrac{\left(\frac{20}{v}\right)^x}{x!}$

21. In a poission distribution the mean is λ and variance is σ^2 then the realation between them is

(a) $\lambda = \dfrac{1}{2\sigma^2}$ (b) $\sigma = \dfrac{1}{2\lambda}$ (c) $\lambda = \sigma^2$ (d) $\lambda^2 = \sigma^2$

22. The standard deviation of standard normal variate is

(a) 1   (b) 0   (c) 0.5   (d) none

23. If $X \sim N(\mu, \sigma^2)$ then the probability that random value of X will lie between $X = \mu$ and $X = x_1$ is

(a) $\int\limits_{\mu}^{x_1} e^{-\frac{1}{2}\left(\frac{x-\mu}{\sigma}\right)^2} dx$

(b) $\dfrac{1}{\sigma\sqrt{2\pi}} \int\limits_{\mu}^{x_1} e^{-\frac{1}{2}\left(\frac{x-\mu}{\sigma}\right)^2} dx$

(c) $\int\limits_{x}^{x_1} e^{-\frac{1}{2}\left(\frac{x-\mu}{\sigma}\right)^2} dx$

(d) none

24. If $f(x) = \begin{cases} \dfrac{1}{b-a} & a < x < b \\ 0 & otherwise \end{cases}$ then the mean of the random variable X is

(a) $\dfrac{b+a}{3}$ (b) e^{-2} (c) $\dfrac{b+a}{2}$ (d) none

25. If $f(x) = \begin{cases} \dfrac{1}{b-a} & a < x < b \\ 0 & otherwise \end{cases}$ then the variance of the random variable X is

(a) $\dfrac{(b-a)^2}{2}$ (b) e^{-2} (c) $\dfrac{(b+a)^2}{2}$ (d) none

26. In 64 sets of 5 tosses of an unbiased coin the expected number of cases in which we get 4 heads and one tail is

(a) $^{64}C_5 \dfrac{1}{2^5}$ (b) $^{64}C_4 \dfrac{1}{2}$ (c) 10   (d) none

Answers

1. (a)   2. (b)   3. (c)   4. (c)   5. (a)   6. (b)

7. (b)   8. (b)   9. (b)   10. (b)   11. (c)   12. (a)

13. (c)   14. (c)   15. (a)   16. (c)   17. (b)   18. (a)

19. (d)   20. (b)   21. (c)   22. (a)   23. (b)   24. (c)

25. (a)   26. (c)

Review Quiz Questions

1. The mean and standard deviation of two observations 3,4,6,7,9 . . . and 17, 22, 32, 37, 47, . . . are μ_1, σ_1; μ_2, σ_2 respectively then μ_2, σ_2 in terms of μ_1, σ_1 are

(a) $\mu_2 = 5\mu_1 + 2$ $\sigma_2^2 = 25 \sigma_1^2$ (b) $\mu_2 = \mu_1$, $\sigma_2^2 = 25 \sigma_1^2$

(c) $\mu_2 = \mu_1 + 2$, $\sigma_2^2 = \sigma_1^2$ (d) $\mu_1 = \mu_2$, $\sigma_1^2 = \sigma_2^2$

2. The standard deviation of 1,4,5,7,8 is σ_1 and standard deviation of 404, 416, 420 428, 432 is σ_2 then $\dfrac{\sigma_2}{\sigma_1} =$

(a)   3   (b)   4   (c)   5   (d)   None

3. For a symmetrical distribution skewness is

(a)   0   (b)   1   (c) -1 (d)   None

4. The limits for a Karl Pearson's coefficient of skewness are

(a)   ±4   (b)   ±2   (c)   ±3   (d)   None

5. The limits for Bowley's coefficient of skewness are

    (a)   ±1   (b)   ±2   (c)   ±3   (d)   None

6. For any frequency distribution coefficient of skewness is (numerical)

    (a)   greater than 1   (b)   less than 1

    (c)   equal to 1   (d)   none

7. The case of positive skewed distribution, the relation between mean, median and mode that hold is

    (a)   Median > Mean > Mode   (b)   Mean > Median > Mode

    (c)   Mean = Median = Mode   (d)   none

8. A random variable X has the following probability distribution the value of k

X :	1	2	3	4
$P(x)$:	k	$2k$	$3k$	$4k$

    (a)   10   (b) $\dfrac{1}{10}$ (c)   0   (d)   none

9. A random variable X has the following probability distribution the mean of the distribution is

X :	1	2	3	4
$P(x)$:	k	$2k$	$3k$	$4k$

    (a) $\dfrac{1}{3}$ (b)   1   (c)   3   (d)   none

10. A random variable X has the following probability distribution the variance of the distribution is

X :	1	2	3	4
$P(x)$:	k	$2k$	$3k$	$4k$

    (a)   1   (b)   0   (c)   10   (d)   none

11. A binomial distribution has mean 5 and standard deviation is 2 then the number of trails is

    (a)   25   (b)   30   (c)   20   (d)   none

12. The standard derivation of binomial distribution $(q+p)^{16}$ is 2 then the mean of the distribution is

    (a)   2   (b)   8   (c)   4   (d)   none

13. In a binomial distribution, $9\,P(X=4)=P(X=2)$. If $n=6$ then p is

    (a) $\dfrac{1}{3}$ (b) $\dfrac{1}{2}$ (c) $\dfrac{1}{\sqrt{10}}$ (d) 4

14. For a binomial distribution the mean is 10 variance is 5 then the parameters are

    (a) 10, 2   (b) $20,\dfrac{1}{2}$ (c) $5,\dfrac{1}{2}$ (d) none

15. The probability distribution function of Poisson distribution is

    (a) $\dfrac{e^{-\lambda}\lambda^{x}}{(x-1)\,!}$ (b) $\dfrac{e^{-\lambda}x}{\lambda^{x}}$ (c) $\dfrac{e^{-\lambda}\lambda^{x}}{x!}$ (d) none

16. For a Poisson distribution $p(1)=p(2)$ then $p(0)$ is

    (a) e (b) $\dfrac{1}{e}$ (c) $\dfrac{1}{e^{2}}$ (d) none

17. If X is a Poisson variate such that $P(X=1)=P(X=2)$ then $P(X=4)$ is

    (a) $\dfrac{2}{3}e^{-2}$ (b) $\dfrac{3}{2}e^{-2}$ (c) e^{-3} (d) none

18. If two regression coefficients are 0.8 and 0.2 then the correlation coefficient r is

    (a) 0.2   (b) 0.3   (c) 0.4   (d) none

19. The two regression equations $20x-9y-107=0 : 4x-5y+33=0$ and $\text{var}(X)=9$ then $\left(\bar{x},\bar{y}\right)$ is

    (a) (13, 17)   (b) (17, 13)   (c) (–17, –13)   (d) none

20. The two regression equations $20x-9y-107=0 : 4x-5y+33=0$ and $\text{var}(X)=9$ then σ_{y} is

    (a) 1   (b) 2   (c) 4   (d) none

21. If $r(X,Y)=1$ then the correlation between X and Y is called

    (a) perfect positive   (b) perfect negative

    (c) independent   (d) none

22. If $r(X,Y)=-1$ then the correlation between X and Y is called

    (a) perfect positive   (b) perfect negative

    (c) independent   (d) none

23. If $r(X,Y) = 0$ then X and Y are said to be

    (a)   perfect positive   (b)   perfect negative

    (c)   uncorrelated   (d)   none

24. If $r(X, Y) > 0$ then X and Y are said to be

    (a)   + correlation   (b)   –ve correlation

    (c)   uncorrelated   (d)   none

25. If $r(X, Y) < 0$ then X and Y are said to be

    (a)   + correlation   (b)   –ve correlation

    (c)   uncorrelated   (d)   none

26. The coefficient of correlation for a distribution in which standard deviation of X is 3 and standard deviation of y is 1.4 units and the covariance between y and x is 0.28 is

    (a)   0.67   (b)   0.2   (c)   0.6   (d)   none

27. The discrete distribution in which mean and variance are equal is

    (a)   normal   (b)   Poisson   (c)   Binomial   (d)   none

28. If the variance of a random variable X is 10 then the variance of $-2X$ is

    (a)   20   (b)   40   (c)   –20   (d)   none

29. In a Poisson distribution the probability $P(X = 0)$ is twice that of probability $P(X = 1)$. Then the mean of the distribution is

    (a) $\dfrac{1}{3}$ (b) $\dfrac{1}{2}$ (c) $\dfrac{1}{4}$ (d)   none

30. The discrete distribution in which the mean is greater than variance is

    (a)   Normal   (b)   Binomial   (c)   Poisson   (d)   none

31. A man takes a forward step with probability 0.4 and backward step with probability 0.6. The probability that at the end of eleven steps he is one step away from the starting point is

    (a) ${}^4C_5$ (b) ${}^4C_4$ (c) ${}^4C_3$ (d)   none

32. The variance of the standard normal variate is

    (a) $\dfrac{1}{2}$ (b) $\dfrac{1}{3}$ (c) 1   (d)   none

33. The probability that a bomb dropped from a plane strikes the target is $\frac{1}{5}$. The probability that out of 6 bombs dropped at least 2 bombs strikes the target approximately is

    (a)   0.44   (b)   0.34   (c)   0.54   (d)   none

34. The normal distribution is symmetric about X is equal to

    (a) μ (b)   0   (c)   1   (d)   none

Exercise Questions

1. Find the mean and variance of the Pascal's (Negative binomial) distribution given by $p(X = x) = {}^{n+k-1}C_k p^n q^k$ $k = 0, 1, 2 \ldots\ldots$

2. The mean and variance of binomial variable X with parameters n and p are 16 and 8. Find $P(x \geq 1)$ and $P(x > 2)$

3. 20% of items produced from a factory are defective. Find the probability that in a sample of 5 chosen at random

    (a) none is defective   (b)   one is defective   (c) $P(1 < x < 4)$

4. If a random variable has a Poisson distribution such that $P(1) = P(2)$. Find

    (i)   Mean of the distribution   (ii) $P(4)$ (iii) $P(x \geq 1)$

    (iv) $P(a < x < 4)$

5. Average number of accidents on only day on a national highway is 1.8. determine the probability that the number of accidents are

    (i)   At least one   (ii)   At most one

6. A distributor of bean seeds determines from extensive tests that 5% of large batch of seeds will not germinate. He sells the seeds in packets of 200 and guarantees 90% germination. Determine the probability that a particular packet will violate the guarantee.

7. In a normal distribution 7% of the items are under 35 and 89% are under 63. Find the mean and standard deviation of the distribution.

8. The marks obtained in mathematics by 1000 students is normally distribution with mean 78% and standard deviation 11% . Determine

    (i)   how many students got marks above 90%

    (ii) what was the highest mark obtained by the lowest 10% of the students

    (iii) within what limits did the middle of 90% of the students lie

9. If a Poisson distribution is such that $P(x = 1)3/2 = P(x = 3)$ find

   (i) $P(x \geq 1)$ (ii) $P(x = 3)$ (iii) $P(2 \leq x < 5)$

10. Given that the mean heights of students in a class is 158 cms with standard deviation of 20 cms. Find how many students heights lie between 150 cms and 170 cms, if there are 100 students in the class

11. If X is a normal variate with mean 30 and standard deviation 5. Find the probabilities that

   (i) $26 \leq x \leq 40$ (ii) $x \geq 45$

12. If X is a Poisson variety such that $3P(x = 4) = \frac{1}{2} P(x = 2) + P(x = 0)$. Find

   (i)   the mean of X (ii) $P(x \leq 2)$

13. If the variance of a Poisson variate is 3, then find the probability that

   (i) $x = 0$ (ii) $0 < x \leq 3$ (iii) $1 \leq x < 4$

14. Fit a binomial distribution to the following data.

X	0	1	2	3	4	5
f	2	14	20	34	22	8

15. In a binomial distribution consisting of 5 independent trials, probabilities of 1 and 2 success are 0.4096 and 0.2048 respectively. Find the parameter p of the distribution

Answers

1. Mean $E(X) = \dfrac{nq}{p}$ and $V(X) = \dfrac{nq}{p^2}$ 2.   0.9999

3. (a)   0.32768   (b)   0.4096   (c)   0.256

4. (i)   (2)   (ii)   (0.09022)   (iii)   (0.8647)   (iv)   (0.4511)

5. (i)   (0.8347)   (ii)   (0.4628)   6.   0.0016

7. 50.3, 10.332

8. (i)   (138)   (ii)   (64%)   (iii)   (60 to 96)

9. (i)   0.502,   (ii)   0.6474,   (iii)   0.7171   10.   38

11. (i)   0.7653,   (ii)   0.00135)   12.   (i)   2,   (ii)   0.498

13. (i)   0.04979,   (ii)   0.5974,   (iii)   0.5974

14. $(1.5 + 9.89 + 26 + 34.1 + 22.4 + 5.9$ i.e., 2, 10 26, 34, 22, 6)   15.   0.2

Review Exercise Questions

1. A merchant's file of 20 accounts contain 6 delinquent and 14 non-delinquent accounts. An auditor randomly selects 5 of these accounts for examination.

    (i)   What is the probability that the auditor finds exactly 2 delinquent accounts?

    (ii)   Find the expected number of delinquent accounts in the sample selected.

2. Briefly answer the following:

    (i)   Salient features of Poisson distribution

    (ii)   Importance of sample size

    (iii)   Binomial Distribution

3. A manufacturing process turns out articles that on the average 10% are defective. Compute the probability of 0, 1, 2 and 3 defective articles that might occur in a sample of 3 articles.

4. 30% of the bolts produced by a machine are defective. Deduce the probability distribution of the number of defectives in a sample of 5 bolts, chosen at random.

5. The average percentage of failures in a certain examination is 50. What is the probability that out of a group of 6 candidates, at least 4 passed in the examination?

6. The incidence of occupational disease in an industry is such that the workmen have a 20% chance of suffering from it. What is the probability that out of 6 workmen, 4 or more will contract the disease?

7. Suppose that half the population of a town are consumers of rice. 100 investigators are appointed to find out its truth. Each investigator interviews 10 individuals. How many investigators do you expect to report that three or less of the people interviewed are consumers of the rice?

8. Four coins are tossed simultaneously. What is the probability of getting (i) 2 heads and 2 tails (ii) at least two heads (iii) at least one head?

9. An oil exploration firm finds that 5% of the test wells it drills yield a deposit of natural gas. If it drills 6 wells, find the probability that at least one well will yield gas.

10. An accountant is to audit 24 accounts of a firm. Sixteen of these are of high-valued customers. If the accountant selects 4 of the accounts at random, what is the probability that he chooses at least one highly valued account?

11. 12% of the items produced by a machine are defective. What is the probability that out of a random sample of 20 items produced by the machine, 5 are defective?

12. The odds in favour of X winning a game against Y are 4 : 3. Find the probability of Y's winning 3 games out of 7 played.

13. On an average 2% of the population in an area suffers from *TB*. What is the probability that out of 5 persons chosen at random from this area, at least two suffer from *TB*?

14. In a multiple – choice examination, there are 20 questions. Each question has four alternative answers following it and students must select the one correct answer. Four marks are given for the correct answer and one mark is deducted for every wrong answer. A student must secure at least 50% of the maximum possible marks to pass the examination. Suppose the student has not studied at all so that he decides to select the answers to the questions on a random basis. What is the probability that he will pass in the examination?

15. An anti – aircraft battery had 3 out of 5 successes in shooting down the flying aircraft that came within the range. What is the chance that if 8 aircrafts came within range, not more than 2 got success?

16. Assuming that it is true that 2 in 10 industrial accidents are due to fatigue, find the probability that

    (i)   Exactly 2 of 8 industrial accidents will be due to fatigue.

    (ii) At least 2 of 8 industrial accidents will be due to fatigue.

17. If hens of a certain breed lay eggs on 5 days a week on an average, find how many days during a season of 100 days, a poultry keeper with 5 hens of this breed; will expect to receive at least 4 eggs?

18. If 5% of the electric bulbs manufactured by a company are defective, use Poisson distribution to find the probability that in a sample of 100 bulbs:

    (i) None is defective;   (ii)   5 bulbs will be defective. (Given: $e^{-5} = 0.007$)

19. Between the hours 2 *PM* and 4 *PM*, the average number of phone calls per minute coming through the switch board of a company is 2.35. Find the probability that during one particular minute, there will be at least 2 phone calls (Given: e – 2.35 = 0.095374).

20. A manufacturer of blades knows that 5% of his products are defective. If he sells blades in boxes of 100 and guarantees that not more than 10 blades will be defective, what is the probability (approximately) that a box will fail to meet the guaranteed quality?

21. It is known from past experience that in a certain plant, there are on the average 4 industrial accidents per month. Find the probability that in a given year, there will less than 4 accidents. Assume Poisson distribution (e – 4 = 0.0183)

22. Find the probability of at least 5 defective bolts will be found in a box of 200 bolts if it is known that 2 percent of such bolts are expected to be defective (assume Poisson Distribution and e-4 = 0.0183).

23. In a certain factory, turning out razor blades, there is a small chance $\dfrac{1}{500}$ for any blade to be defective. The blades are supplied in packets of 10. Use Poisson distribution to calculate the approximate number of packets containing no defective, one defective and two defective blades respectively in a consignment of 10,000 packets.

24. Five hundred television sets are inspected as they come off the production line and the number of defects per set is recorded below.

No. of defects(X)	:	0	1	2	3	4
No. of Sets	:	368	72	52	7	1

CHAPTER 4

Multiple Random Variables

4.1 Introduction

So far we have considered one dimensional random variable. In many engineering applications, it is necessary to make use of more than one variable, say two or more random variables.

Let us consider a sample space S and let X and Y are two random variables in it. Let x_i and y_j be different values assumed by the random variables X and Y respectively, then any ordered pair (x_i, y_j) is considered to be a random point in xy-plane.

4.2 Random Vector (Vector Random Variable)

The specified values of the random variables are called a random vector or a vector random variable and we call (X, Y) a 2dimensional random vector or a 2dimensional random variable.

Let (X, Y) be a 2dimensional random vector, then (X, Y) assumes a finite or a countably infinite number of values then (X, Y) is said to be 2dimensional discrete random variable.

Let (X, Y) be a 2dimensional random vector, then with each possible out come of (X, Y) say (x_i, y_j) associate a number $f(x_i, y_j)$ or $P(x_i, y_j)$ which has the value $P(X = x_i, Y = y_j)$

For example, A card is drawn from a pack of cards, it may be characterized according to its suit and to its denomination. Let the random variables X and Y represents that the values 1, 2, 3, 4 which corresponding to suit in some order say club, diamond, heart, spade and the values 1,2,3, . . . , 13 which corresponds to the denominations Ace, 2, 3, . . .10, J, Q, K respectively. Then the pair (X, Y) is a two dimensional random variable.

The probability of drawing a particular card will be denoted by $f(x_i, y_j)$ or $P(x_i, y_j)$ and the probability of getting each card is same, then the probability function of (x, y) is

$$f(x_i, y_j) = P(x_i, y_j) = \frac{1}{52} \qquad \begin{cases} 1 \le i \le 4 \\ 1 \le j \le 13 \end{cases}$$

In any experiment whose outcomes can be characterized by two (three) variables called bivariate (tri-variate) etc.

4.3   Joint Probability Distribution of (X, Y)

Let (X, Y) be a 2dimensional random variable, then (x_i, y_j) associated with a number $P(x_i, y_j)$ then the value $P(X = x_i, Y = y_j) = P_{ij}$

Table 4.1

Y / X	y_1	y_2	y_3	...	y_j	...	y_m	Total
x_1	P_{11}	P_{12}	P_{13}	...	P_{1j}	...	P_{1m}	$P_{1*} = f(x_1)$
x_2	P_{21}	P_{22}	P_{23}	...	P_{2J}	...	P_{2M}	$P_{2*} = f(x_2)$
x_3	P_{31}	P_{32}	P_{33}	...	P_{3J}	...	P_{3M}	$P_{3*} = f(x_3)$
. . .	. . .	. . .	. . .	. . .	. . .	. . .	. . .	. . .
x_i	P_{i1}	P_{i2}	P_{i3}	...	P_{ij}	...	P_{im}	$P_{i*} = f(x_i)$
. . .	. . .	. . .	. . .	. . .	. . .	. . .	. . .	. . .
x_n	P_{n1}	P_{n2}	P_{n3}	...	P_{nj}	...	P_{nl}	$P_{n*} = f(x_n)$
Total	$P_{*1} =$ $g(y_1)$	$P_{*2} =$ $g(y_2)$	$P_{*3} =$ $g(y_3)$	...	$P_{*j} =$ $g(y_j)$	...	P_{*m} $=$ $g(y_m)$	1 Grand Total

The joint probability distribution satisfies

$$p_{ij} = P(X = x_i, \ Y = y_j) = P(x_i, \ y_j) \geq 0 \text{ and } \sum_{i=1}^{n}\sum_{j=1}^{m} P(x_i, \ y_j) = 1$$

E.g., Three balls are drawn at random from a box containing 2 red, 3 white and 4 black balls. If the random variable X represents the number of red balls drawn and Y denotes the number of white balls drawn. To write the joint probability distribution of (X, Y)

Given that, there are 2 red, 3 white and 4 black balls.

There fore the total number of balls are 9.

The random variable X represents the number of red balls, here the range of X is 0, 1, 2. And the random variable Y represents the number of white balls, here the range of Y is 0, 1, 2, 3.

Three balls are drawn at random from 9 balls in $^9C_3$ ways

$P(X = 0, \ Y = 0)$: Probability of drawing no red, no white balls, this means that all the drawing balls are black balls, this can be done as

$$P(X = 0, \ Y = 0) = \frac{^2C_0 \, ^3C_0 \, ^4C_3}{^9C_3} = \frac{^4C_3}{^9C_3} = \frac{1}{21};$$

Similarly,

$$P(X = 0, \ Y = 1) = \frac{^2C_0 \, ^3C_1 \, ^4C_2}{^9C_3} = \frac{3 \times 6}{84} = \frac{3}{14};$$

$$P(X = 0, \ Y = 2) = \frac{^2C_0 \, ^3C_2 \, ^4C_1}{^9C_3} = = \frac{1 \times 3 \times 4}{84} = \frac{1}{7};$$

$$P(X = 0, \ Y = 3) = \frac{^2C_0 \, ^3C_3 \, ^4C_0}{^9C_3} = \frac{1 \times 1 \times 1}{84}$$

$$= \frac{1}{84}; \ P(X = 1, \ Y = 0) = \frac{^2C_1 \, ^3C_0 \, ^4C_2}{^9C_3} = \frac{2 \times \, ^4C_2}{^9C_3} = \frac{1}{7}$$

$$P(X = 1, \ Y = 1) = \frac{^2C_1 \, ^3C_1 \, ^4C_1}{^9C_3} \ x_1 < x_2 = \frac{2}{7};$$

$$P(X = 1, \ Y = 2) = \frac{^2C_1 \, ^3C_2 \, ^4C_0}{^9C_3} = \frac{2 \times 3 \times 1}{84} = \frac{1}{14}$$

$$P(X=2, Y=0) = \frac{^2C_2\,^3C_0\,^4C_1}{^9C_3} = \frac{1 \times 1 \times 4}{84} = \frac{1}{21};$$

$$P(X=2, Y=1) = \frac{^2C_2\,^3C_1\,^4C_0}{^9C_3} = \frac{1 \times 3 \times 1}{84} = \frac{1}{28}$$

$P(X=2, Y=2) = 0$ and $P(X=2, Y=3) = 0$ since we are drawing 3 balls only, not 4 balls or 5 ball

Now the joint probability distribution of (X,Y) represented in the following table as

Table 4.2

X \ Y	0	1	2	3
0	$\dfrac{1}{21}$	$\dfrac{3}{14}$	$\dfrac{1}{7}$	$\dfrac{1}{84}$
1	$\dfrac{1}{7}$	$\dfrac{2}{7}$	$\dfrac{1}{14}$	0
2	$\dfrac{1}{21}$	$\dfrac{1}{28}$	0	0

4.4 Marginal Distribution Functions (Marginal Probability distribution)

The Marginal Probability distribution $X = x_i$, $i=1, 2, 3 \ldots n$ is denoted by p_{i*} or $f(x_i)$ of $P(x_i, y_j)$ is the sum of the entries in the ith row in the table 1.1.

i.e., $P(X = x_i)$

$$P\{(X = x_i,\ Y = y_1)\ or\ (X = x_i,\ Y = y_2)\ or\ \ldots\ or\ (X = x_i,\ Y = y_m)\}$$

$$= p_{i1} + p_{i2} + p_{i3} + \ldots + p_{ij} + \ldots + p_{im}$$

$$= \sum_{j=1}^{m} P(x_i, y_j)$$

$$= p_{i*} = f(x_i) \qquad\qquad i = 1, 2, 3 \ldots n$$

is called marginal distribution function of $X = x_i$, $i = 1, 2, 3 \ldots n$. The set of $\{x_i, p_{i*}\}$ $i = 1,2,3..n$ is called the marginal probability distribution of $X = x_i$. Similarly we define $p_{*j} = \sum_{i=1}^{n} p_{ij} = P(Y = y_j)$ for $j = 1, 2, 3 \ldots m$. The set of $\{y_j, p_{*j}\}$ for $j = 1, 2,$ $3 \ldots m$ is called the marginal probability distribution of $Y = y_j$.

For the above example and table 1.1, we calculate the marginal probability distribution functions as

Now the marginal probabilities is represents as

The Marginal Probability distribution $X = x_i$, $i = 0, 1, 2$ is denoted by p_{i*} or $f(x_i)$ of $P(x_i, y_j)$ is the sum of the entries in the ith row

i.e., $P(X = x_i) =$

$P\{(X = x_i, \ Y = y_1) \ or \ (X = x_i, \ Y = y_2) \ or \ \ldots \ or \ (X = x_i, \ Y = y_m)\}$

$P(X = 0) = P(X = 0, Y = 0) + P(X = 0, Y = 1) +$

$P(X = 0, Y = 2) + P(X = 0, Y = 3)$

$\qquad = \dfrac{1}{21} + \dfrac{3}{14} + \dfrac{1}{7} + \dfrac{1}{84}$

$\qquad = \dfrac{35}{84} = p_{0*} = f(0)$

$P(X = 1) = P(X = 1, Y = 0) + P(X = 1, Y = 1) +$

$P(X = 1, Y = 2) + P(X = 1, Y = 3)$

$\qquad = \dfrac{1}{7} + \dfrac{2}{7} + \dfrac{1}{14} + 0$

$\qquad = \dfrac{1}{2} = p_{1*} = f(1)$

$P(X = 2) = P(X = 2, Y = 0) + P(X = 2, Y = 1) + P(X = 2, Y = 2) +$

$P(X = 2, Y = 3)$

$\qquad = \dfrac{1}{21} + \dfrac{1}{28} + 0 + 0$

$\qquad = \dfrac{1}{12} = p_{2*} = f(2)$

Similarly, the marginal probability distribution of $Y = y_i$'s are

$$P(Y = 0) = P(X = 0, Y = 0) + P(X = 1, Y = 0) + P(X = 2, Y = 0)$$

$$= \frac{1}{21} + \frac{1}{7} + \frac{1}{21} = \frac{5}{21} = p_{*0} = g(0)$$

$$P(Y = 1) = P(X = 0, Y = 1) + P(X = 1, Y = 1) + P(X = 2, Y = 1)$$

$$= \frac{3}{14} + \frac{2}{7} + \frac{1}{28} = \frac{15}{28} = p_{*1} = g(1)$$

$$P(Y = 2) = P(X = 0, Y = 2) + P(X = 1, Y = 2) + P(X = 2, Y = 2)$$

$$= \frac{1}{7} + \frac{1}{14} + 0 = \frac{3}{14} = p_{*2} = g(2)$$

$$P(Y = 3) = P(X = 0, Y = 3) + P(X = 1, Y = 3) + P(X = 2, Y = 3)$$

$$= \frac{1}{84} + 0 + 0 = \frac{1}{84} = p_{*3} = g(3)$$

These probabilities are represented in tabular form as

X \ Y	0	1	2	3	Totals
0	$\frac{1}{21}$	$\frac{3}{14}$	$\frac{1}{7}$	$\frac{1}{84}$	$\frac{35}{84} = p_{0*} = f(0)$
1	$\frac{1}{7}$	$\frac{2}{7}$	$\frac{1}{14}$	0	$\frac{1}{2} = p_{1*} = f(1)$
2	$\frac{1}{21}$	$\frac{1}{28}$	0	0	$\frac{1}{12} = p_{2*} = f(2)$
Totals	$\frac{5}{21} =$ $p_{*0} = g(0)$	$\frac{15}{28} =$ $p_{*1} = g(1)$	$\frac{3}{14} =$ $p_{*2} = g(2)$	$\frac{1}{84} =$ $p_{*3} = g(3)$	1

4.5   Joint Probability Density Function (for continuous case)

Let S be e sample space, X and Y be two random variables in it. The probability that X lies between $x - \dfrac{1}{2} dx$ and $x + \dfrac{1}{2} dx$ and at the same time Y lies between $y - \dfrac{1}{2} dy$ and $y + \dfrac{1}{2} dy$ is

$$P\left\{ x - \frac{1}{2} dx \le X \le x + \frac{1}{2} dx, \ y - \frac{1}{2} dy \le Y \le y + \frac{1}{2} dy \right\} = \int\limits_{x-\frac{1}{2}dx}^{x+\frac{1}{2}dx} \left\{ \int\limits_{y-\frac{1}{2}dy}^{y+\frac{1}{2}dy} f(x,y) dy \right\} .dx$$

Or we can also defined as $p(x_1 < X \le x_2, y_1 < Y \le y_2) = \displaystyle\int\limits_{y_1}^{y_2} \int\limits_{x_1}^{x_2} f(x,y) \, dx.dy$ And

$f(x,y)$ satisfies

(i) $f(x,y) \ge 0$ and $\displaystyle\int\limits_{-\infty}^{\infty} \int\limits_{-\infty}^{\infty} f(x,y) dy.dx = 1$ the above two are test for validity of some density function

4.6   Marginal Density Functions (Marginal Probability Density)

The Marginal Probability density function of X, is denoted by $f_x(x)$ and is defined as

$$f_x(x) = \int\limits_{-\infty}^{\infty} f(x,y) dy$$

Similarly $f_y(y) = \displaystyle\int\limits_{-\infty}^{\infty} f(x,y) dx$ is called the marginal density of Y.

4.7   Joint Distribution Function (Cumulative distribution function)

Let $F_X(x) = P(X \le x)$ and $F_Y(y) = P(Y \le y)$ are the probability distribution functions of two random variables X and Y respectively, these are also denoted by $F(x)$ and $F(y)$ respectively.

Now let (X, Y) be a two-dimensional random variable then the joint distribution function or cumulative distribution function is denoted by $F_{XY}(x, y)$ or $F(x, y)$ and is defined as

$$F_{XY}(x, y) = F(x, y)$$

$$= p(X \le x, Y \le y) = \begin{cases} \displaystyle\sum_{Y \le y}\sum_{X \le x} P(x_i, y_j) & \text{in discrete case} \\ \displaystyle\int_{-\infty}^{y}\int_{-\infty}^{x} f(x,y)dy.dx & \text{in continuous case} \end{cases}$$

The joint pdf of (X_1, X_2) is defined as $f(x_1, x_2) = \dfrac{\partial^2 F(x_1, x_2)}{\partial x_1 . \partial x_2}$

4.8 Properties of Distribution Function F(x, y)

1. $0 \le F_{XY}(x,y) \le 1$

2. $F_{XY}(+\infty, +\infty) = 1$

3. $F_{XY}(-\infty, -\infty) = F_{XY}(x, -\infty) = F_{XY}(-\infty, y) = 0$

4. $F_{XY}(x,y)$ is a non-decreasing function of x and y

5. $P(x_1 < X \le x_2, y_1 < Y \le y_2) = F_{XY}(x_2, y_2) - F_{XY}(x_2, y_1) - F_{XY}(x_1, y_2)$

   $+ F_{XY}(x_1, y_1) \ge 0$

This can be shown in fig.

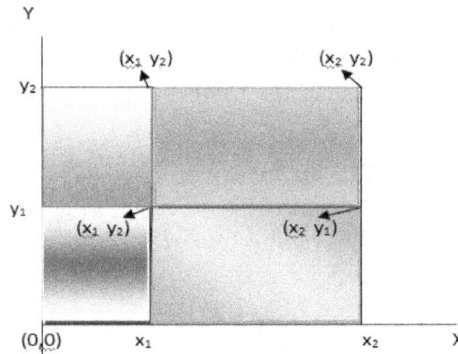

The cumulative distribution function of X alone (i.e., independent of Y) is denoted by $F_X(x)$ and is defined as

$$F_{XY}(x, -\infty) = F_X(x) = \int_{-\infty}^{\infty}\int_{-\infty}^{x} f(x,y)dx.dy$$

Then the probability density function of X (independent of Y) is

$$f_x(x) = \frac{d}{dx} F_x(x)$$

And the cumulative distribution function of Y alone (i.e., independent of X) is denoted by $F_y(y)$ and is defined as

$$F_{XY}(-\infty, y) = F_Y(y) = \int_{-\infty}^{\infty} \int_{-\infty}^{y} f(x, y)dx.dy$$

And the probability density function of Y (independent of X) is

$$f_y(y) = \frac{d}{dy} F_y(y)$$

If X and Y are independent, then

$$p(x \leq X \leq x+dx, y \leq Y \leq y+dy) = \left[f_x(x).dx \right]\left[f_y(y).dy \right]$$

This implies that

$$\Rightarrow \qquad f_{xy}(x,y) = f_x(x) . f_y(y)$$

i.e.,
$$p(x_1 \leq X \leq x_2, y_1 \leq Y \leq y_2) = \left[\int_{x_1}^{x_2} f_x(x).dx \right]\left[\int_{y_1}^{y_2} f(y).dy \right]$$

More generally, a vector $X : [X_1, X_2 \ldots \ldots X_n]$ whose components X_i are random variables is called a random vector $(X_1, X_2 \ldots \ldots X_n)$ assume all real values in some region R_n of the n dimensional space, the joint distribution of $(X_1, X_2 \ldots \ldots X_n)$ is defined as

$$F(x_1, x_2 \ldots x_n) = P[X_1 \leq x_1, X_2 \leq x_2 \ldots \ldots X_n \leq x_n]$$

The joint pdf of (X_1, X_2) is defined as $f(x_1, x_2) = \dfrac{\partial^2 F(x_1, x_2)}{\partial x_1 . \partial x_2}$

$$\therefore \qquad F(x_1, x_2) = \int_{-\infty}^{x_1} \int_{-\infty}^{x_2} f(x_1, x_2)dx_1.dx_2$$

Similarly, for extension,

The joint pdf of $(X_1, X_2 \ldots \ldots X_n)$ is defined as

$$f(x_1, x_2 \ldots x_n) = \frac{\partial^n F(x_1, x_2 \ldots x_n)}{\partial x_1 . \partial x_2 \ldots \partial x_n}$$

And satisfy the following conditions

(i) $f(x_1, x_2 \ldots x_n) \geq 0 \qquad \vee \; x_1, x_2 \ldots x_n$

(ii) $\quad \int\limits_{R_n} \int ....\int f(x_1,x_2....x_n)\, dx_1.dx_2....dx_n = 1$

(iii) $\quad p\,[(X_1, X_2 ... . . . X_n) \in D] = \int\limits_{D} \int ....\int f(x_1,x_2....x_n)\, dx_1.dx_2....dx_n = 1$

where D is the subset of the range space R_n.

4.9 Conditional Probability Distribution

The conditional probability distribution function of X given Y is defined as

$$p(X = x_i\,/\,Y = y_j) = \frac{p(X = x_i \cap Y = y_j)}{p(Y = y_j)} = \frac{p_{ij}}{p_{*j}}$$

The set of pairs $\left\{ x_j, \dfrac{p_{ij}}{p_{*j}} \right\}$, $i = 1, 2, 3 . . . .n$; is called the conditional probability distribution of X given $Y = y_j$

Similarly, the conditional probability distribution function of Y given X is defined as

$$p(Y = y_j\,/\,X = x_i) = \frac{p(X = x_i \cap Y = y_j)}{p(X = x_i)} = \frac{p_{ij}}{p_{i*}}$$

The set of pairs $\left\{ y_j, \dfrac{p_{ij}}{p_{i*}} \right\}$, $j = 1, 2, 3 . . . .m$; is called the conditional probability distribution of Y given $X = x_i$

4.10 Conditional Probability Density (continuous case)

The conditional probability density function of X given Y is defined as

$$P\left\{ x - \frac{1}{2}dx \le X \le x + \frac{1}{2}dx\,/\,y - \frac{1}{2}dy \le Y \le y + \frac{1}{2}dy \right\} = \frac{f(x,y)dx.dy}{f_y(y.)dy}$$

$$= \frac{f(x,y)}{f_y(y)}dx$$

This is denoted by $f(x\,/\,y)$.

Therefore the conditional probability density function of X given Y is

$$f(x\,/\,y) = \frac{f(x,y)}{f_y(y)}dx \qquad f_y(y) > 0$$

Similarly, the conditional probability density function of Y given X is defined as

$$P\left\{y-\frac{1}{2}dy \le Y \le y+\frac{1}{2}dy \,/\, x-\frac{1}{2}dx \le X \le x+\frac{1}{2}dx\right\} = \frac{f(x,y)dx.dy}{f_X(x)dx}$$

$$= \frac{f(x,y)}{f_X(x)}dy$$

This is denoted by $f(y/x)$.

Therefore the conditional probability density function of Y given X is

$$f(y/x) = \frac{f(x,y)}{f_x(x)}dy$$

Let A be an event $(X \le x)$ for the random variable X, If B is given, the conditional distribution function of the random variable X is

$$F_X(x/B) = P[X \le x/B] = \frac{P\big[(X \le x) \cap B\big]}{P[B]}$$

The event B can be defined from some characteristic of the physical experiment. It may be defined in terms of the random variable X or some other random variable other than X.

(a)   If the event B is defined in terms of X, let $B = (X \le b)$ where $-\infty < b < \infty$

$$\therefore F_X(x/X \le b) = P[X \le x/X \le b] = \frac{P\big[(X \le x) \cap X \le b\big]}{P[X \le b]}$$

(b)   If the event B is defined in terms of another variable Y, i.e., If the random variable X is conditioned by a second random variable Y, where $Y \le y$, is called *point conditioning*.

$$\therefore F_{X/Y}(x/y) = P[X \le x/Y \le y] = \frac{P\big[(X \le x) \cap (Y \le y)\big]}{P[Y \le y]}$$

provided $P[Y \le y] \ne 0$

This is also simply represented as $F_{X/Y}(x/y) = \dfrac{F_{X,Y}(x,y)}{F_Y(y)}$

If the random variable Y is defined in such a way that its value lies between two constants y_a and y_b.

i.e., between the interval y_a and y_b. ($y_a \le Y \le y_b$), this is called interval conditioning.

The corresponding conditional density function is given by

$$f(x_1 / x_2) = \frac{\partial\left[F_{X_1/X_2}(x_1 / x_2)\right]}{\partial x_1} = \frac{f_{XY}(x_1, x_2)}{f_Y(y)} \quad \text{provided } f_Y(y) > 0$$

Similarly, if X is given

$$f(x_2 / x_1) = \frac{\partial\left[F_{X_2/X_1}(x_2 / x_1)\right]}{\partial x_2} = \frac{f_{XY}(x_1, x_2)}{f_X(x)} \quad \text{provided } f_X(x) > 0$$

4.11 Properties of Conditional Distribution Function

The properties of conditional distribution function are similar to that of ordinary distribution function which is studied earlier.

Since $\qquad F(x) = p(X \le x)$

1. $F_X(-\infty/B) = p(X \le -\infty/B) = 0$
2. $F_X(+\infty/B) = p(X \le +\infty/B) = 1$
3. $0 \le F_X(x/B) \le 1$
4. If $x_1 < x_2$ then $F_X(x_1 / B) \le F_X(x_2 / B)$
5. $p(x_1 < X < x_2 / B) = F_X(x_2 / B) - F_X(x_1 / B)$

4.12 Properties of Conditional Density Function

The properties of conditional density function are similar to that of ordinary density function which is studied earlier.

1. $f_X(x/B) \ge 0 \qquad \vee\, x$

2. $\displaystyle\int_{-\infty}^{\infty} f_X(x/B)dx = 1$

3. $F_X(x/B) = \displaystyle\int_{-\infty}^{x} f_X(x/B)dx$

4. $p(x_1 < X < x_2 / B) = \displaystyle\int_{x_1}^{x_2} f_X(x/B)dx = F_X(x_2 / B) - F_X(x_1 / B)$

4.13 Statistical Independence (Independent Random Variables)

Let (X, Y) be a 2–dimensional discrete random variable, then (x_i, y_j) associated with a

number $P(x_i, y_j)$ such that $p(X = x_i / Y = y_j) = \dfrac{p(X = x_i \cap Y = y_j)}{p(Y = y_j)}$

$$= \frac{p_{ij}}{p_{*j}} = p_{i*} = p(X = x_i)$$

then the two random variables X and Y are said to be independent.

i.e., If X and Y are two independent random variables, then $P_{ij} = p_{i*} \times p_{*j}$

In the same manner, for a 2-dimensional continuous random variable, if X and Y are two independent random variables, then

$$f(x, y) = f_x(x) \times f_y(y)$$

Examples

***Example* 1:** The joint probabilities of two discrete random variables X and Y have

$P(X = 0, \ Y = 0) = \dfrac{2}{9}$ $P(X = 0, \ Y = 1) = \dfrac{1}{9}$, $P(X = 1, \ Y = 0) = \dfrac{1}{9}$ and

$P(X = 1, Y = 1) = \dfrac{5}{9}$. Test whether X and Y are independent.

Sol: The joint probability distribution is

X \ Y	0	1	Totals
0	$\dfrac{2}{9}$	$\dfrac{1}{9}$	$\dfrac{3}{9} = P(X = 0)$
1	$\dfrac{1}{9}$	$\dfrac{5}{9}$	$\dfrac{6}{9} = P(X = 1)$
Totals	$\dfrac{3}{9} = P(Y = 0)$	$\dfrac{6}{9}\ P(Y = 1)$	1

From the table $P(X = 0) = \dfrac{3}{9}$ and $P(X = 1) = \dfrac{6}{9}$

$$P(Y = 0) = \frac{3}{9} \text{ and } P(Y = 1) = \frac{6}{9}$$

Now consider $P(X = 0) \times P(Y = 0) = \dfrac{3}{9} \times \dfrac{3}{9} = \dfrac{1}{9}$

and $$P(X=0,\ Y=0)=\frac{2}{9}$$

Therefore $P(X=0,\ Y=0) \neq P(X=0).\ P(Y=0)$

X and Y are not independent

Example 2: For the following joint distribution of random variables X and Y, find $P(X \leq 1)$ and $P(Y \leq 3)$

X \ Y	1	2	3	4
0	$\frac{2}{32}$	$\frac{3}{32}$	$\frac{1}{32}$	$\frac{2}{32}$
1	$\frac{1}{16}$	$\frac{1}{16}$	$\frac{3}{8}$	$\frac{1}{8}$
2	$\frac{1}{32}$	$\frac{1}{32}$	$\frac{1}{64}$	$\frac{3}{64}$

Sol: From the given data,

$$
\begin{aligned}
P(X \leq 1) &= P(X=0) + P(X=1) \\
&= P(X=0,\ Y=1) + P(X=0,\ Y=2) + P(X=0,\ Y=3) + P(X=0,\ Y=4) \\
&\quad + P(X=1,\ Y=1) + P(X=1,Y=2) + P(X=1,Y=3) + P(X=1,Y=4) \\
&= P_{01} + P_{02} + P_{03} + +P_{04} + P_{11} + +P_{12} + P_{13} + P_{14} \\
&= \frac{2}{32} + \frac{3}{32} + \frac{1}{32} + \frac{2}{32} + \frac{1}{16} + \frac{1}{16} + \frac{3}{8} + \frac{1}{8} \\
&= \frac{7}{8}
\end{aligned}
$$

$$
\begin{aligned}
P(Y \leq 3) &= P(Y=1) + P(Y=2) + P(Y=3) \\
&= P(X=0,\ Y=1) + P(X=1,\ Y=1) + P(X=2,\ Y=1) + \\
&\quad P(X=0,\ Y=2) + P(X=1,\ Y=2) + P(X=2,\ Y=2) + \\
&\quad P(X=0,\ Y=3) + P(X=1,\ Y=3) + P(X=2,\ Y=3) \\
&= P_{01} + P_{11} + P_{21} + P_{02} + P_{12} + P_{22} + P_{03} + P_{13} + +P_{23} \\
&= \frac{2}{32} + \frac{1}{16} + \frac{1}{32} + \frac{3}{32} + \frac{1}{16} + \frac{1}{32} + \frac{1}{32} + \frac{3}{8} + \frac{1}{64} \\
&= \frac{49}{64}
\end{aligned}
$$

***Example* 3:** A_1, A_2 and A_3 are mutually exclusive events associated with a random experiment E_1. Events B_1, B_2 and B_3 are mutually exclusive events associated with a random experiment E_2. The joint probabilities of occurrence of these events and some marginal probabilities are listed in the table given below.

B $\quad$ A	$B1$	$B2$	$B3$
$A1$	$\dfrac{3}{36}$	*	$\dfrac{5}{36}$
$A2$	$\dfrac{5}{36}$	$\dfrac{4}{36}$	$\dfrac{5}{36}$
$A3$	*	$\dfrac{6}{36}$	*
$P(B_j)$	$\dfrac{12}{36}$	$\dfrac{14}{36}$	*

Find (a) the missing probabilities (*) in the table (b) $P(A_1/B_3)$ and $P(B_3/A_1)$ (c) Are the events A_2 and B_2 statistically independent.

Sol: From the given table

B $\quad$ A	B_1	B_2	B_3	$P(A_i)$
A_1	$\dfrac{3}{36}$	$\dfrac{4}{36}$	$\dfrac{5}{36}$	$\dfrac{12}{36}$
A_2	$\dfrac{5}{36}$	$\dfrac{4}{36}$	$\dfrac{5}{36}$	$\dfrac{14}{36}$
A_3	$\dfrac{4}{36}$	$\dfrac{6}{36}$	0	$\dfrac{10}{36}$
$P(B_j)$	$\dfrac{12}{36}$	$\dfrac{14}{36}$	$\dfrac{10}{36}$	1

(a) $P(B_1)$ can be obtained by adding all elements in the column headed by B_1 in the above joint probability matrix.

Since $\qquad P(A_1,B_1) + P(A_2,B_1) + P(A_3,B_1) = \dfrac{12}{36}$

$\Rightarrow \qquad \dfrac{3}{36} + \dfrac{5}{36} + P(A_3,B_1) = \dfrac{12}{36}$

$$\Rightarrow \qquad P(A_3, B_1) = \frac{12}{36} - \frac{8}{36} = \frac{4}{36}$$

So, the missing probability in the 3rd row, 1st column of above matrix is $\frac{4}{36}$.

Similarly we can obtained in the remaining 1st row , 2nd ; 3rd row , 3rd and 4^{rt} row , 3rd column of above matrix are $\frac{4}{36}$; 0 and $\frac{10}{36}$ respectively.

(b) Since $\qquad P(A_1/B_3) = P(A_1, B_3)/P(B_3)$

$$= \frac{5}{36} \div \frac{10}{36} = \frac{1}{2} \quad \text{and}$$

$$P(B_3/A_1) = P(A_1, B_3) /P(A_1)$$

$$= \frac{5}{36} \div \frac{12}{36} = \frac{5}{12}$$

(c) Now from the table

$$P(A_2, B_2) = \frac{4}{36}; \ P(A_2) = \frac{14}{36} \text{ and } P(B_2) = \frac{14}{36}$$

Since $P(A_2, B_2) \neq P(A_2).P(B_2)$

A_2 and B_2 are not statistically independent

***Example* 4:** For the following joint distribution of random variables X and Y, find
(a) $P(X \leq 1, Y \leq 3)$ (b) $P(X \leq 1/Y \leq 3)$ (c) $P(Y \leq 3/X \leq 1)$ (d) $P(X+Y \leq 4)$

Y X	1	2	3	4	5	6
0	0	0	$\frac{1}{32}$	$\frac{2}{32}$	$\frac{2}{32}$	$\frac{3}{32}$
1	$\frac{1}{16}$	$\frac{1}{16}$	$\frac{1}{8}$	$\frac{1}{8}$	$\frac{1}{8}$	$\frac{1}{8}$
2	$\frac{1}{32}$	$\frac{1}{32}$	$\frac{1}{64}$	$\frac{1}{64}$	0	$\frac{2}{64}$

Sol: From the given data,

(a) $P(X \leq 1, Y \leq 3) = \sum_{j=1}^{3} P(x=0, Y=j) + \sum_{j=1}^{3} P(x=1, Y=j) = \sum_{j=1}^{3} P_{0j} + \sum_{j=1}^{3} P_{1j}$

$$= P_{01} + P_{02} + P_{03} + P_{11} + P_{12} + P_{13}$$

$$= 0 + 0 + \frac{1}{32} + \frac{1}{16} + \frac{1}{16} + \frac{1}{8} = \frac{9}{32}$$

(b) $P(X \le 1 / Y \le 3)$:

since $P(X \le 1 / Y \le 3) = \dfrac{P(X \le 1, Y \le 3)}{P(Y \le 3)}$ Now first calculate P(Y≤3)

$$P(Y \le 3) = P(Y = 1) + P(Y = 2) + P(Y = 3)$$

$$= P(X = 0, Y = 1) + P(X = 1, Y = 1) + P(X = 2, Y = 1) +$$
$$P(X = 0, Y = 2) + P(X = 1, Y = 2) + P(X = 2, Y = 2) +$$
$$P(X = 0, Y = 3) + P(X = 1, Y = 3) + P(X = 2, Y = 3)$$

$$= P_{01} + P_{11} + P_{21} + P_{02} + P_{12} + + P_{22} + + + P_{03} + P_{13} + P_{23}$$

$$= 0 + \frac{1}{16} + \frac{1}{32} + 0 + \frac{1}{16} + \frac{1}{32} + \frac{1}{32} + \frac{1}{8} + \frac{1}{64} = \frac{23}{64}$$

Hence $P(X \le 1 / Y \le 3) = \dfrac{9}{32} \div \dfrac{23}{64} = \dfrac{18}{23}$

(c) $P(Y \le 3 / X \le 1)$:

Since $P(Y \le 3 / X \le 1) = \dfrac{P(X \le 1, Y \le 3)}{P(X \le 1)}$ Now first calculate P(X≤1)

$$P(X \le 1) = P(X = 0) + P(X = 1) = \sum_{j=1}^{6} P_{0j} + \sum_{j=1}^{6} P_{1j}$$

$$= P_{01} + P_{02} + P_{03} + P_{04} + P_{05} + P_{06} + P_{11} + P_{12} + P_{13} + P_{14} + P_{15} + P_{16}$$

$$= 0 + 0 + \frac{1}{32} + \frac{2}{32} + \frac{2}{32} + \frac{3}{32} + \frac{1}{16} + \frac{1}{16} + \frac{1}{8} + \frac{1}{8} + \frac{1}{8} + \frac{1}{8}$$

$$= \frac{28}{32}$$

Hence $P(Y \le 3 / X \le 1) = \dfrac{9}{32} \div \dfrac{28}{32} = \dfrac{9}{28}$

(d) $P(X + Y \le 4) = \displaystyle\sum_{j=1}^{4} P_{0j} + \sum_{j=1}^{3} P_{1j} + \sum_{j=1}^{2} P_{2j}$

$$= P_{01} + P_{02} + P_{03} + P_{04} + P_{11} + P_{12} + P_{13} + P_{21} + P_{22}$$

$$= 0 + 0 + \frac{1}{32} + \frac{2}{32} + \frac{1}{16} + \frac{1}{16} + \frac{1}{8} + \frac{1}{32} + \frac{1}{32} = \frac{13}{32}$$

***Example* 5:** The *i/p* to a communication channel is a random variable X and the *o/p* is another random variable Y. The joint probability mass function of X and Y are listed below.

Y / X	-1	0	1
-1	1/4	1/8	0
0	0	1/4	0
1	0	1/8	1/4

Determine (*i*) $P(Y = 1/X = 1)$ (*ii*) $P(X = 1/Y = 1)$

Sol: The marginal probability distribution of X and Y are

Y \ X	-1	0	1	Marginal probabilities of X
-1	$\dfrac{1}{4}$	$\dfrac{1}{8}$	0	$\dfrac{3}{8}$
0	0	$\dfrac{1}{4}$	0	$\dfrac{1}{4}$
1	0	$\dfrac{1}{8}$	$\dfrac{1}{4}$	$\dfrac{3}{8}$
Marginal probabilities of Y	$\dfrac{1}{4}$	$\dfrac{1}{2}$	$\dfrac{1}{4}$	1

(i)   Since $P(Y = 1 / X = 1) = \dfrac{P(X = 1, Y = 1)}{P(X = 1)}$

$$= \frac{1}{4} \div \frac{1}{4}$$

$$= 1$$

$\therefore$ $P(Y = 1 / X = 1) = 1$

(ii)   Since $P(X = 1 / Y = 1) = \dfrac{P(X = 1, Y = 1)}{P(Y = 1)}$

$$= \frac{1}{4} : \frac{3}{8}$$

$$= \frac{2}{3}$$

***Example* 6:** The joint probability mass function of (X,Y) is given by $P(x, y) = k(2x + 3y)$; $x = 0, 1, 2$ and $y = 1, 2, 3$. Find all the marginal and conditional probability distributions. Also find the probability distribution of $(X + Y)$

***Sol*:** From the given data, the joint probability distribution of (X, Y) is given

Y / X	1	2	3	Marginal probabilities of X
0	$3k$	$6k$	$9k$	$18k$
1	$5k$	$8k$	$11k$	$24k$
2	$7k$	$10k$	$13k$	$30k$
Marginal probabilities of Y	$15k$	$24k$	$33k$	$72k=1$

From the table $72k = 1 \Rightarrow k = \dfrac{1}{72}$

Hence the joint probability distribution of (X, Y) and the marginal distributions of X and Y are

Y / X	1	2	3	Marginal probabilities of X
0	$\dfrac{3}{72}$	$\dfrac{6}{72}$	$\dfrac{9}{72}$	$\dfrac{18}{72}$
1	$\dfrac{5}{72}$	$\dfrac{8}{72}$	$\dfrac{11}{72}$	$\dfrac{24}{72}$
2	$\dfrac{7}{72}$	$\dfrac{10}{72}$	$\dfrac{13}{72}$	$\dfrac{30}{72}$
Marginal probabilities of Y	$\dfrac{15}{72}$	$\dfrac{24}{72}$	$\dfrac{33}{72}$	1

Conditional distribution of $X = x_i$ given Y:

First we conditional distribution of $X = x_i$ given $Y = 1$ is

$$P(X = i / Y = 1) = \frac{P(X = i, Y = 1)}{P(Y = 1)} \qquad i = 0, 1, 2.$$

$$P(X = 0 / Y = 1) = \frac{P(X = 0, Y = 1)}{P(Y = 1)}$$

$$= \frac{3}{72} \div \frac{15}{72}$$

$$= \frac{1}{5}$$

$$P(X = 1 / Y = 1) = \frac{P(X = 1, Y = 1)}{P(Y = 1)}$$

$$= \frac{5}{72} \div \frac{15}{72}$$

$$= \frac{1}{3}$$

$$P(X = 2 / Y = 1) = \frac{P(X = 2, Y = 1)}{P(Y = 1)}$$

$$= \frac{7}{72} \div \frac{15}{72}$$

$$= \frac{7}{15}$$

Therefore the conditional distribution of X given $Y=1$ is given by

X	:	0	1	2	Total
Conditional probabilities	:	$\frac{1}{5}$	$\frac{1}{3}$	$\frac{7}{15}$	1

Similarly,

The conditional distribution of X given $Y=2$ is given by

X	:	0	1	2	Total
Conditional probabilities	:	$\frac{1}{4}$	$\frac{1}{3}$	$\frac{5}{2}$	1

The conditional distribution of X given $Y = 3$ is given by

X	:	0	1	2	Total	
Conditional probabilities	:	$\frac{3}{11}$	$\frac{1}{3}$	$\frac{1}{2}$	1	Check

The conditional distribution of Y = y_j given X:

The conditional distribution of Y given $X = 0$ is given by

Y	:	1	2	3	Total
Conditional probabilities	:	$\dfrac{1}{6}$	$\dfrac{1}{3}$	$\dfrac{1}{2}$	1

The conditional distribution of Y given $X = 1$ is given by

Y	:	1	2	3	Total
Conditional probabilities	:	$\dfrac{5}{24}$	$\dfrac{1}{3}$	$\dfrac{11}{24}$	1

The conditional distribution of Y given $X = 2$ is given by

Y	:	1	2	3	Total
Conditional probabilities	:	$\dfrac{7}{30}$	$\dfrac{1}{3}$	$\dfrac{13}{30}$	1

And the probability distribution of $(X+Y)$:

$X + Y$	:	*1*	*2*	*3*	*4*	*5*	
Total Probability	:	$\dfrac{3}{72}$	$\dfrac{11}{72}$	$\dfrac{24}{72}$	$\dfrac{21}{72}$	$\dfrac{13}{72}$	1

Since $X + Y = 1 = (0 + 1)$

$$= P_{01} = \frac{3}{72}$$

$X + Y = 2 = (0 + 2) + (1 + 1)$

$$= P_{02} + P_{11} = \frac{6}{72} + \frac{5}{72}$$

$$= \frac{11}{72}$$

$X + Y = 3 = (0 + 2) + (1 + 2) + (2 + 1)$

$$= P_{02} + P_{12} + P_{21}$$

$$= \frac{9}{72} + \frac{8}{72} + \frac{7}{72}$$

$$= \frac{24}{72} ;$$

$X + Y = 4 = (1 + 3) + (2 + 2)$

$$= P_{13} + P_{22}$$

$$= \frac{11}{72} + \frac{10}{72}$$

$$= \frac{21}{72}$$

$$X + Y = 5 = (2 + 3) = P_{23} = \frac{13}{72}$$

***Example* 7:** The following table represents the joint probability distribution of the discrete random variable X and Y.

Determine the

(a)   Marginal distribution of X and Y

(b)   Conditional distribution of X given $Y = 2$ and

(c)   The conditional distribution of Y given $X = 3$.

(d) $P(X \leq 2,\ Y = 2)$ (e) $P(Y \leq 3)$ (f) $P(X + Y \leq 4)$

Y X	1	2	3
0	$\frac{1}{12}$	$\frac{1}{6}$	0
1	0	$\frac{1}{9}$	$\frac{1}{5}$
2	$\frac{1}{18}$	$\frac{1}{4}$	$\frac{1}{15}$

Sol: From the given table, the marginal probability distribution of X and Y are

Y X	1	2	3	Marginal probabilities of X
0	$\frac{1}{12}$	$\frac{1}{6}$	0	$\frac{1}{4}$
1	0	$\frac{1}{9}$	$\frac{1}{5}$	$\frac{14}{45}$
2	$\frac{1}{18}$	$\frac{1}{4}$	$\frac{1}{15}$	$\frac{79}{180}$
Marginal probabilities of Y	$\frac{5}{36}$	$\frac{19}{36}$	$\frac{1}{3}$	1

Since the Marginal Probability distribution $X = x_i$, $i = 1, 2, 3$ is denoted by p_{i*} or $f(x_i)$ of $P(x_i, y_j)$ is the sum of the entries in the ith row

i.e., $\quad P(X = x_i)$

$$P\{(X = x_i, \ Y = y_1) \ or \ (X = x_i, \ Y = y_2) \ or \ \ldots \ or \ (X = x_i, \ Y = y_m)\}$$

$$P(X = 1) = P(X = 1, Y = 1) + P(X = 1, Y = 2) + P(X = 1, Y = 3)$$

$$= \frac{1}{12} + 0 + \frac{1}{8} = \frac{5}{36} = p_{1*} = f(1)$$

Similarly,

$$P(X = 2) = P(X = 2, Y = 1) + P(X = 2, Y = 2) + P(X = 2, Y = 3)$$

$$= \frac{1}{6} + \frac{1}{9} + \frac{1}{4} = \frac{19}{36} = p_{2*} = f(2)$$

$$P(X = 3) = P(X = 3, Y = 1) + P(X = 3, Y = 2) + P(X = 3, Y = 3)$$

$$= 0 + \frac{1}{5} + \frac{2}{15} = \frac{1}{3} = p_{3*} = f(2)$$

Similarly, the marginal probability distribution of $Y = y_i$'s are

$$P(Y = 1) = P(X = 1, Y = 1) + P(X = 2, Y = 1) + P(X = 3, Y = 1)$$

$$= \frac{1}{12} + \frac{1}{6} + 0 = \frac{3}{12} = \frac{1}{4} = p_{*1} = g(1)$$

$$P(Y = 2) = P(X = 1, Y = 2) + P(X = 2, Y = 2) + P(X = 3, Y = 2)$$

$$= 0 + \frac{1}{9} + \frac{1}{5} = \frac{14}{45} = p_{*2} = g(2)$$

$$P(Y = 3) = P(X = 1, Y = 3) + P(X = 2, Y = 3) + P(X = 3, Y = 3)$$

$$= \frac{1}{8} + \frac{1}{4} + \frac{2}{15} = \frac{79}{180} = p_{*3} = g(3)$$

Conditional distribution of $X = x_i$ given Y:

First we conditional distribution of $X = x_i$ given $Y = 2$ is

$$P(X = i / Y = 2) = \frac{P(X = i, Y = 2)}{P(Y = 2)} \qquad i = 1, 2, 3$$

$$P(X = 1 / Y = 2) = \frac{P(X = 1, Y = 2)}{P(Y = 2)} = 0 \div \frac{14}{45} = 0$$

$$P(X = 2 / Y = 2) = \frac{P(X = 2, Y = 2)}{P(Y = 2)} = \frac{1}{9} \div \frac{14}{45} = \frac{5}{14}$$

$$P(X = 3 / Y = 2) = \frac{P(X = 3, Y = 2)}{P(Y = 2)} = \frac{1}{5} \div \frac{14}{45} = \frac{9}{14}$$

Therefore the conditional distribution of X given $Y=2$ is given by

X :	1	2	3	Total
Conditional probabilities :	0	$\dfrac{5}{14}$	$\dfrac{9}{14}$	1

The conditional distribution of $Y = y_j$ given $X = 3$:

First we conditional distribution of $Y = y_j$ given $X = 3$ is

$$P(Y = j / X = 3) = \frac{P(X = 3, Y = j)}{P(X = 3)} \qquad j = 1, 2, 3$$

$$P(Y = 1 / X = 3) = \frac{P(X = 3, Y = 1)}{P(X = 3)} = 0 \div \frac{1}{3} = 0$$

$$P(Y = 2 / X = 3) = \frac{P(X = 3, Y = 2)}{P(X = 3)} = \frac{1}{5} \div \frac{1}{3} = \frac{3}{5}$$

$$P(Y = 3 / X = 3) = \frac{P(X = 3, Y = 3)}{P(X = 3)} = \frac{2}{15} \div \frac{1}{3} = \frac{2}{5}$$

The conditional distribution of Y given $X = 3$ is given by

Y :	1	2	3	Total
Conditional probabilities :	0	$\dfrac{3}{5}$	$\dfrac{2}{5}$	1

$$P(X \le 2,\ Y = 2) = P(X = 1, Y = 2) + P(X = 2, Y = 2) = 0 + \frac{1}{9} = \frac{1}{9}$$

$$P(Y \le 3) = P(Y = 1) + P(Y = 2) + P(Y = 3) = \frac{1}{4} + \frac{14}{45} + \frac{79}{180} = 1$$

$$P(X + Y \le 4) = P(X = 1, Y = 1) + P(X = 1, Y = 2) +$$

$$P(X = 2, Y = 1) + P(X = 2, Y = 2)$$

$$= \frac{1}{12} + 0 + \frac{1}{6} + \frac{1}{9} = \frac{13}{36}$$

***Example* 8:** Random variable X represents a binary communication system *i/p*, takes on one of two values *0* or *1* with probabilities $\frac{3}{4}$ and $\frac{1}{4}$ respectively. Due to errors caused by noise in the system, the random variable Y denotes the *o/p* differs from the input X occasionally. The behavior of the communication system is modeled by the conditional probabilities $P(Y = 1/X = 1) = \frac{3}{4}$ and $P(Y = 0/X = 0) = \frac{7}{8}$. Determine

(i) $P(Y = 1)$ (ii) $P(Y = 0)$ (iii) $P(X = 1/Y = 1)$

Sol: From the given data,

$$P(X = 0) = \frac{3}{4} \, ; \, P(X = 1) = \frac{1}{4} \, ; \; P(Y = 1/X = 1) = \frac{3}{4} \text{ and}$$

$$P(Y = 0/X = 0) = \frac{7}{8}$$

Y \\ X	0	1	Marginal probabilities of Y
0			
1		$\frac{3}{4}$	
Marginal probabilities of X	$\frac{3}{4}$	$\frac{1}{4}$	1

Since $P(Y = 1/X = 1) = \dfrac{P(X = 1, Y = 1)}{P(X = 1)} = \dfrac{3}{4}$ (given)

$\Rightarrow \qquad P(X = 1, Y = 1) = \dfrac{3}{4} \, P(X = 1)$

$$= \frac{3}{4} \times \frac{1}{4} = \frac{3}{16}$$

$\therefore \qquad P(X = 1, Y = 1) = \dfrac{3}{16}$

X Y	0	1	Marginal probabilities of Y
0		$\dfrac{1}{4}-\dfrac{3}{16}=\dfrac{1}{16}$	
1		$\dfrac{3}{16}$	
Marginal probabilities of X	$\dfrac{3}{4}$	$\dfrac{1}{4}$	1

Since $\qquad P(Y=0/X=0) = \dfrac{P(X=0,Y=0)}{P(X=0)} = \dfrac{7}{8}$ (given)

$\Rightarrow \qquad P(X=0,Y=0) = \dfrac{7}{8}\, P(X=0)$

$$= \dfrac{7}{8} \times \dfrac{3}{4} = \dfrac{21}{32}$$

$\therefore \quad P(X=0,Y=0) = \dfrac{21}{32}$

X Y	0	1	Marginal probabilities of Y
0	$\dfrac{21}{32}$	$\dfrac{1}{4}-\dfrac{3}{16}$ $=\dfrac{1}{16}$	$\dfrac{23}{32}$
1	$\dfrac{9}{32}-\dfrac{3}{16}=\dfrac{3}{32}$	$\dfrac{3}{16}$	$1-\dfrac{23}{32}=\dfrac{9}{32}$
Marginal probabilities of X	$\dfrac{3}{4}$	$\dfrac{1}{4}$	1

From the above table

(i) $P(Y=1) = \dfrac{9}{32}$

(ii) $P(Y = 0) = \dfrac{23}{32}$

(iii) $P(X = 1/Y = 1) = \dfrac{P(X = 1, Y = 1)}{P(Y = 1)} = \dfrac{3}{16} \div \dfrac{9}{32} = \dfrac{2}{3}$

Aliter:

Since $P(Y = 0/X = 1) = 1 - \dfrac{3}{4} = \dfrac{1}{4}$ and $P(Y = 1/X = 0) = 1 - \dfrac{7}{8} = \dfrac{1}{8}$

(i) Since $P(Y = 0/X = 1) = P(X = 1, Y = 0)/P(X = 1)$

$\Rightarrow$ $P(X = 1, Y = 0) = P(X = 1).P(Y = 0/X = 1)$

$$= \dfrac{1}{4} \dfrac{1}{4} = \dfrac{1}{16}$$

Since $P(Y = 0/X = 0) = P(X = 0 \ Y = 0)/P(X = 0)$

$\Rightarrow$ $P(X = 0, Y = 0) = P(X = 0).P(Y = 0/X = 0)$

$$= \dfrac{3}{4} \times \dfrac{7}{8} = \dfrac{21}{32}$$

$\therefore$ $P(Y = 0) = P(X = 0, Y = 0) + P(X = 1, Y = 0)$

$$= \dfrac{1}{16} + \dfrac{21}{32} = \dfrac{23}{32}$$

Since $P(Y = 1/X = 0) = P(X = 0, Y = 1)/P(X = 0)$

$\Rightarrow$ $P(X = 0, Y = 1) = P(X = 0).P(Y = 1/X = 0)$

$$= \dfrac{3}{4} \times \dfrac{1}{8} = \dfrac{3}{32}$$

$P(Y = 1/X = 1) = P(X = 1, Y = 1)/P(X = 1)$

$\Rightarrow$ $P(X = 1, Y = 1) = P(X = 1).P(Y = 1/X = 1)$

$$= \dfrac{1}{4} \times \dfrac{3}{4} = \dfrac{3}{16}$$

$\therefore$ $P(Y = 1) = P(X = 1, Y = 1) + P(X = 0, Y = 1)$

$$= \dfrac{3}{16} + \dfrac{3}{32} = \dfrac{9}{32}$$

(ii) $P(X = 1, Y = 1) = P(X = 1, Y = 1)/P(Y)$

$$= \frac{3}{16} \div \frac{3}{32} = \frac{2}{3}$$

Continuous case problems

Example 9: X and Y are two random variables having the joint density function

$$f(x, y) = e^{-(x+y)} \quad 0 < x < \infty ; \ 0 < y < \infty$$

Determine (i) $P(X > 1)$ (ii) $P(X < Y / X < 2Y)$ and (iii) $P(1 < X + Y < 2)$.

Sol: Give $f(x, y) = e^{-(x+y)} \quad 0 < x < \infty$

$$= e^{-x} - e^{-y}$$

$$= f_x(x) \ f_y(y) \quad 0 \le x < \infty, \ 0 \le y < \infty$$

$\Rightarrow$ X and Y are independent and

$$f_x(x) = e^{-x} \ \& \ f_y(y) = e^{-y} ; \quad y \ge 0; \quad x \ge 0$$

(i) $P(X > 1) = \int\limits_{1}^{\infty} f_x(x)dx = \int\limits_{1}^{\infty} e^{-x} \ dx = -(e^{-x})_1^{\infty} = \frac{1}{e}$

(ii) $P(X < Y / X < 2Y) = \dfrac{P(X < Y \cap X < 2Y)}{P(X < 2Y)} = \dfrac{P(X < Y)}{P(X < 2Y)}$ .....(4.1)

Now consider $P(X < Y) \quad = \int\limits_{0}^{\infty} \left[\int\limits_{0}^{y} f(x,y)dx \right] dy$

$$= \int\limits_{0}^{\infty} e^{-y} \left[\frac{e^{-x}}{-1} \right]_0^y dy$$

$$= -\int\limits_{0}^{\infty} e^{-y} \left(e^{-y} - 1 \right) dy$$

$$= -\left[\frac{e^{-2y}}{-2} + e^{-y} \right]_0^{\infty}$$

$$= 1 - \frac{1}{2} = \frac{1}{2}.$$

and $\qquad P(X < 2Y) = \int\limits_{0}^{\infty} \left[\int\limits_{0}^{2y} f(x, y)\, dx \right] dy$

$$= -\int\limits_{0}^{\infty} e^{-y}(e^{-2y} - 1)dy$$

$$= -\left[\frac{e^{-3y}}{-3} + e^{-y} \right]_{0}^{\infty}$$

$$= 1 - \frac{1}{3} = \frac{2}{3}$$

$\therefore$ From eq. (4.1),

$$P(X < Y / X < 2Y) = \frac{P(X < Y)}{P(X < 2Y)}$$

$$= \frac{1}{2} \div \frac{2}{3}$$

$$= \frac{3}{4}$$

(iii) $\qquad P(1 < X + Y < 2) = \iint\limits_{s} f(x, y)\, dx\, dy + \iint\limits_{\pi} f(x, y)\, dx\, dy$

$$= \int\limits_{0}^{1} \left[\int\limits_{1-x}^{2-x} f(x, y)\, dy \right] dx + \int\limits_{1}^{2} \left[\int\limits_{0}^{2-x} f(x, y)\, dy \right] dx$$

Since by change of order of integration

$$= \int\limits_{0}^{1} e^{-x} \left(\int\limits_{1-x}^{2-x} e^{-y}dy \right) dx + \int\limits_{1}^{2} e^{-x} \left(\int\limits_{0}^{2-x} e^{-y}dy \right) dx$$

$$= -\int\limits_{0}^{1} e^{-x} \left(e^{x-2} - e^{x-1} \right) dx - \int\limits_{1}^{2} e^{-x} \left(e^{x-2} - 1 \right) dx$$

$$= -\left(e^{-2} - e^{-1} \right) \int\limits_{0}^{1} 1\, dx - \int\limits_{1}^{2} \left(e^{-2} - e^{-x} \right) dx$$

$$= -\left(e^{-2} - e^{-1} \right) (x)_{0}^{1} - \left(e^{-2}x + e^{-x} \right)_{1}^{2}$$

$$= \frac{2}{e} - \frac{3}{e^2}$$

$$= \frac{2}{e} - \frac{3}{e^2}$$

$$= \frac{2e - 3}{e^2}$$

***Example* 10:** The joint probability function of two-dimensional random variables (X, Y) is

$$f(x, y) = \begin{cases} 2 & \text{if } 0 < x < 1, \quad 0 < y < x \\ 0 & \text{otherwise} \end{cases}$$

Determine

(i)   The Marginal Density function of X and Y

(ii)   The conditional density function of Y given $X=x$

(iii)   The conditional density function of X given $Y=y$ and

(iv)   Test for independence of X and Y

Sol: Since $f(x, y) \geq 0$ and $\displaystyle\int_{-\infty}^{\infty} \int_{-\infty}^{\infty} f(x, y) dy.dx = 1$ $\Rightarrow$ $\displaystyle\int_{0}^{1} \int_{0}^{x} f(x, y) dy.dx = 1$

$\Rightarrow$ $\displaystyle\int_{0}^{1} \int_{0}^{x} f(x, y) dy.dx = 1$ $\Rightarrow$ $\displaystyle\int_{0}^{1} \int_{0}^{x} 2 dy.dx = 1 \Rightarrow 2 \int_{0}^{1} \int_{0}^{x} dy.dx = 1$

$\Rightarrow$ $\displaystyle 2 \int_{0}^{1} x dx = 1$ $\Rightarrow 2 \left(\frac{x^2}{2} \right)_{0}^{1} = 1$ $\Rightarrow 2\frac{1}{2} = 1$ Hence verified.

(i)   The Marginal Density function of X is

$$f_x(x) = \int_{-\infty}^{\infty} f(x, y) dy = \int_{0}^{x} 2 dy = 2x \quad 0 < x < 1$$

Hence the marginal density function of X is $f_x(x) = 2x \quad 0 < x < 1$

The Marginal Density function of Y is

$$f_y(y) = \int_{-\infty}^{\infty} f(x, y) dx$$

$$= \int_{y}^{1} 2 dx$$

$$= 2 (x)_{y}^{1} = 2(y - 1) \qquad 0 < y < 1$$

$$\therefore f_y(y) = 2(y-1) \quad 0 < y < 1$$

(ii) Therefore the conditional probability density function of Y given X is

$$f(y/x) = \frac{f(x,y)}{f_x(x)} = \frac{2}{2x} = \frac{1}{x} \quad 0 < x < 1$$

(iii) Therefore the conditional probability density function of X given Y is

$$f(x/y) = \frac{f(x,y)}{f_y(y)} = \frac{2}{2(y-1)} = \frac{1}{y-1} \quad 0 < y < 1$$

(iv) Test for independence of X and Y:

$$f_x(x) \times f_y(y) = 2x \times 2(y-1)$$

$$= 4x(y-1) \neq 2 = f(x,y)$$

Therefore the random variables X and Y are not independent.

Example 11: The joint Probability density function of the random variables X and Y is

$$f(x,y) = \frac{1}{4}e^{-|x|-|y|} \quad \text{For} -\infty < X < \infty \text{ and } -\infty < Y < \infty$$

(a)   Are X and Y statistically independent variables and

(b)   Find $P(X \leq 1 \ and \ Y \leq 0)$

Sol: Since

$$f_x(x) = \int_{-\infty}^{\infty} f(x,y)dy$$

$$= \int_{-\infty}^{\infty} \frac{1}{4}e^{-|x|-|y|}dy$$

$$= \frac{1}{4}e^{-|x|}\left\{ \int_{-\infty}^{0} e^y dy + \int_{0}^{\infty} e^{-y} dy \right\}$$

$$= \frac{1}{4}e^{-|x|}(2)$$

$$= \frac{1}{2}e^{-|x|}$$

and $\qquad f_y(y) = \int\limits_{-\infty}^{\infty} f(x,y)dx \;\; = \int\limits_{-\infty}^{\infty} \frac{1}{4} e^{-|x|-|y|} dy$

$$= \frac{1}{4} e^{-|y|} \left\{ \int\limits_{-\infty}^{0} e^x dy + \int\limits_{0}^{\infty} e^{-x} dy \right\}$$

$$= \frac{1}{4} e^{-|y|} (2)$$

$$= \frac{1}{2} e^{-|y|}$$

Now $\qquad f_x(x) \times f_y(y) = \frac{1}{2} e^{-|x|} \times \frac{1}{2} e^{-|y|}$

$$= \frac{1}{4} e^{-|x|-|y|} = f(x,y)$$

Hence X and Y are independent.

(b) Since $\qquad P(X \le 1 \;and\; Y \le 0) \;=\; \int\limits_{-\infty}^{1} \int\limits_{-\infty}^{0} f(x,y)dxdy$

$$= \int\limits_{-\infty}^{1} \int\limits_{-\infty}^{0} \frac{1}{4} e^{-|x|-|y|} dxdy$$

$$= \frac{1}{4} \int\limits_{-\infty}^{1} \left\{ e^{-|x|} \left(\int\limits_{-\infty}^{0} e^y dy \right) \right\} dx$$

$$= (1/4) \int\limits_{-\infty}^{1} \left(e^{-|x|} \right).dx$$

$$= \frac{1}{4} \left\{ \int\limits_{-\infty}^{0} e^x dx + \int\limits_{0}^{1} e^{-x} dx \right\}$$

$$= \frac{1}{4} (2 - e^{-1})$$

$$= \frac{1}{2} - \frac{1}{4e}$$

***Example* 12:** Determine the constant b such that the function

$$f(x,y) = \begin{cases} 3xy & if\; 0 < x < 1, \quad 0 < y < b \\ 0 & otherwise \end{cases} \qquad \text{is a valid joint density function.}$$

Sol: Since $\qquad f(x,y) \geq 0$ and $\int\limits_{-\infty}^{\infty}\int\limits_{-\infty}^{\infty} f(x,y)dy.dx = 1$

Consider $\qquad \int\limits_{-\infty}^{\infty}\int\limits_{-\infty}^{\infty} f(x,y)dy.dx = 1$

This implies $\qquad \int\limits_{0}^{1}\int\limits_{0}^{b} f(x,y)dx.dy = 1$

$$= \int\limits_{0}^{1}\int\limits_{0}^{b} 3xy\,dx\,dy = 1$$

$$= \int\limits_{0}^{1} 3x\left[y^2/2\right]_0^b dx = 1$$

$$= (3/2)b^2 \int\limits_{0}^{1} x.dx = 1$$

$$\Rightarrow (3/2)b^2\left[x^2/2\right]_0^1 = 1 \quad \Rightarrow \quad \frac{3}{2}b^2\frac{1}{2} = 1$$

$$\therefore b = \frac{2}{\sqrt{3}}$$

Example 13: Two random variables X and Y have a joint probability density function

$$f(x,y) = \begin{cases} e^{-(x+y)} & \text{for } x \geq 0, \quad y \geq 0 \\ 0 & \text{otherwise} \end{cases}$$

Determine

(i)   Marginal density functions of X and Y

(ii) $P(X \leq 1 \text{ and } Y \leq 1)$ and   (iii) $P(X+Y \leq 1)$

Also prove that X and Y are independent.

Sol:

(i)   The Marginal Density function of X is

$$f_x(x) = \int\limits_{-\infty}^{\infty} f(x,y)dy = \int\limits_{0}^{\infty} e^{-(x+y)}dy = e^{-x}\int\limits_{0}^{\infty} e^{-y}dy = e^{-x} \qquad x \geq 0$$

Hence the marginal density function of X is $f_x(x) = e^{-x}$ $x \geq 0$

The Marginal Density function of Y is

$$f_y(y) = \int_{-\infty}^{\infty} f(x,y)dx$$

$$= \int_0^\infty e^{-(x+y)}dx = e^{-y}\int_0^\infty e^{-x}dx = e^{-y} \quad y \geq 0$$

$$\therefore f_y(y) = e^{-y} \quad y \geq 0$$

Test for independence of X and Y:

$$f_x(x) \times f_y(y) = e^{-x} \times e^{-y} = e^{-(x+y)} = f(x,y)$$

Therefore the random variables X and Y are independent

(ii) $$P(X \leq 1, Y \leq 1) = \int_0^1 \int_0^1 f(x,y)dxdy$$

$$= \int_0^1 \int_0^1 e^{-(x+y)}dxdy$$

$$= \int_0^1 e^{-x}\left(\int_0^1 e^{-y}dy\right)dx$$

$$= (1-e^{-1})\int_0^1 e^{-x}dx$$

$$= (1-e^{-1})^2$$

$$= \left(1-\frac{1}{e}\right)^2$$

Consider $$P((x+y) \leq 1) = \int_{x=0}^1 \int_{y=0}^{1-x} f(x,y)dx.dy$$

Since by change of order of integration, y in terms of x

$$= \int_0^1 e^{-x}\left(\int_0^{1-x} e^{-y}dy\right)dx$$

$$= \int_{x=0}^{1} e^{-x}[1 - e^{-(1-x)}]dx = \int_{x=0}^{1} (e^{-x} - e^{-1})dx$$

$$= [-e^{-x} - e^{-1}.x]_0^1$$

$$= (1 - 2e^{-1})$$

$$= \left(1 - \frac{2}{e}\right)^2$$

Or we can also change x in terms of y

$$P(x + y \leq 1) = \int_{y=0}^{1} \int_{x=0}^{1-y} f(x, y)dx.dy$$

$$= \int_{y=0}^{1} e^{-y}[\int_{x=0}^{1-y} e^{-x}dx]dy$$

$$= \int_{y=0}^{1} e^{-y}[1 - e^{-(1-y)}]dy$$

$$= \int_{y=0}^{1} (e^{-y} - e^{-1})dy$$

$$= (1 - 2e^{-1}) = \left(1 - \frac{2}{e}\right)^2$$

(ii) $P(X \leq 1 \; and \; Y \leq 1)$ and (iii) $P(X + Y \leq 1)$

***Example* 14:** The joint Probability density function of random variable X and Y is

$$f(x, y) = \frac{1}{2} \quad for \; \; 0 \leq x \leq y, \quad 0 \leq y \leq 2$$

Determine

(i)   Marginal density functions of X and Y

(ii) The conditional Probability density functions $f(x/y)$ and $f(y/x)$ and

(b) Test whether X and Y are independent.

Sol: Given that $f(x, y) = \frac{1}{2} \quad for \; \; 0 \leq x \leq y, \quad 0 \leq y \leq 2$

i.e., $f(x, y) = \frac{1}{2} \quad for \; \; 0 \leq x \leq y \leq 2$

(i) The Marginal density functions of X and Y

$$f_x(x) = \int_{-\infty}^{\infty} f(x,y)dy = \int_{x}^{2} \frac{1}{2}dy = \frac{1}{2}[y]_x^2 = \frac{1}{2}(2-x)$$

Hence the marginal density function of X is $f_x(x) = \frac{1}{2}(2-x)$

The Marginal Density function of Y is

$$f_y(y) = \int_{-\infty}^{\infty} f(x,y)dx$$

$$= \int_{0}^{y} \frac{1}{2}dx = \frac{1}{2}(x)_0^y = \frac{1}{2}y$$

$$\therefore f_y(y) = \frac{1}{2}y$$

Test for independence of X and Y:

Since $\qquad f_x(x) \times f_y(y) = \frac{1}{2}(2-x) \times \frac{1}{2}y$

$$= \frac{1}{4}(2-x)y \neq f(x,y)$$

Therefore the random variables X and Y are not independent

(ii) Therefore the conditional probability density function of X given Y is

$$f(x/y) = \frac{f(x,y)}{f_y(y)} = \frac{1}{2} \div \frac{1}{2}y = \frac{1}{y}$$

Therefore the conditional probability density function of Y given X is

$$f(y/x) = \frac{f(x,y)}{f_x(x)} = \frac{1}{2} \div \frac{1}{2}(2-x) = \frac{1}{2-x}$$

***Example* 15:** Let X and Y be jointly continuous random variables with joint Pdf

$$f(x,y) = \begin{cases} x^2 + \dfrac{xy}{3} & for \; 0 \le x \le 1, \quad 0 \le y \le 2 \\ 0 & otherwise \end{cases}$$

Are X and Y independent and Check $f(x/y)$ and $f(y/x)$ are Pdfs or not.

Sol: Given $f(x,y) = \begin{cases} x^2 + \dfrac{xy}{3} & for\ 0 \le x \le 1, \quad 0 \le y \le 2 \\ 0 & otherwise \end{cases}$

(i)   The Marginal density functions of X and Y

$$f_x(x) = \int\limits_{-\infty}^{\infty} f(x,y)dy = \int\limits_{0}^{2}\left(x^2 + \frac{xy}{3}\right)dy$$

$$= x^2\left[y\right]_0^2 + \frac{x}{3}\left[\frac{y^2}{2}\right]_0^2 = x^2[2] + \frac{x}{3}[2]$$

$$= 2x^2 + \frac{2x}{3}$$

The Marginal Density function of Y is

$$f_y(y) = \int\limits_{-\infty}^{\infty} f(x,y)dx$$

$$= \int\limits_{0}^{1}\left(x^2 + \frac{xy}{3}\right)dx$$

$$= \left[\frac{x^3}{3}\right]_0^1 + \frac{y}{3}\left[\frac{x^2}{2}\right]_0^1$$

$$= \frac{1}{3} + \frac{y}{6}$$

Now Test for independence of X and Y:

Since $f_x(x) \times f_y(y) = \left(2x^2 + \dfrac{2x}{3}\right) \times \left(\dfrac{1}{3} + \dfrac{y}{6}\right)$

$$\ne f(x,y)$$

Hence X and Y are not independent.

(ii) $f(x/y) = \dfrac{f(x,y)}{f_y(y)} = \left(x^2 + \dfrac{xy}{3}\right) \div \left(\dfrac{1}{3} + \dfrac{y}{6}\right) = \dfrac{\left(x^2 + \dfrac{xy}{3}\right)}{\dfrac{1}{3} + \dfrac{y}{6}}$

Now consider $\int\limits_{0}^{1} f(x/y)dx = \int\limits_{0}^{1}\dfrac{\left(x^2 + \dfrac{xy}{3}\right)}{\dfrac{1}{3} + \dfrac{y}{6}}dx = \dfrac{1}{\left(\dfrac{1}{3} + \dfrac{y}{6}\right)}\int\limits_{0}^{1}\left(x^2 + \dfrac{xy}{3}\right)dx$

$$= \frac{1}{\left(\dfrac{1}{3}+\dfrac{y}{6}\right)}\left(\frac{x^3}{3}+\frac{x^2 y}{6}\right)_0^1$$

$$= \frac{1}{\left(\dfrac{1}{3}+\dfrac{y}{6}\right)}\left(\frac{1}{3}+\frac{y}{6}\right)$$

$$= 1$$

Hence *f(x/y)* is a valid Probability density function.

$$\text{Consider } f(y \,/\, x)=\frac{f(x,y)}{f_x(x)} \;=\left(x^2+\frac{xy}{3}\right) \;\div\; \left(2x^2+\frac{2x}{3}\right) \;=\; \frac{\left(x^2+\dfrac{xy}{3}\right)}{2x^2+\dfrac{2x}{3}}$$

$$\int_0^2 f(y \,/\, x)dy \;=\; \int_0^2 \frac{\left(x^2+\dfrac{xy}{3}\right)}{2x^2+\dfrac{2x}{3}}dy \;=\; \frac{1}{\left(2x^2+\dfrac{2x}{3}\right)}\int_0^2\left(x^2+\frac{xy}{3}\right)dy =$$

$$= \frac{1}{\left(2x^2+\dfrac{2x}{3}\right)}\left[x^2\left(y\right)_0^2+\frac{x}{3}\left(\frac{y^2}{2}\right)_0^2\right]_0$$

$$= \frac{1}{\left(2x^2+\dfrac{2x}{3}\right)}\left(2x^2+\frac{2x}{3}\right)$$

$$= 1$$

Hence *f(y/x)* is a valid Pdf

***Example* 16:** The joint Pdf of (*X, Y*) is given to be

$$f(x,y) = \begin{cases} Ae^{-x-y} & for\ 0 \le x \le y, \quad 0 \le y < \infty \\ 0 & otherwise \end{cases}$$

(i)   Find *A*

(ii)   Find marginal density of *X* and *Y*

(iii)   Examine if *X* and *Y* are independent

(iv)   Find the conditional density function of *Y* given *X*=2.

Sol: For the probability density function $f(x, y) \geq 0$ and $\int\limits_{-\infty}^{\infty} \int\limits_{-\infty}^{\infty} f(x, y) dy. dx = 1$

(i) Consider $\int\limits_{-\infty}^{\infty} \int\limits_{-\infty}^{\infty} f(x, y) dy. dx = 1$

This implies $\int\limits_{0}^{y} \int\limits_{0}^{\infty} f(x, y) dx. dy = 1$

$\Rightarrow \qquad \int\limits_{0}^{y} \int\limits_{0}^{\infty} A e^{-x-y} dx. dy = 1$

$\Rightarrow \qquad A \int\limits_{0}^{\infty} e^{-y} \left(\int\limits_{0}^{y} e^{-x} dx \right) dy = 1 \qquad\qquad \Rightarrow \quad A \int\limits_{0}^{\infty} e^{-y} \left(\frac{e^{-x}}{-1} \right)_{0}^{y} . dy = 1$

$\Rightarrow \qquad A \int\limits_{0}^{\infty} e^{-y} \left(1 - e^{-y} \right). dy = 1 \qquad\qquad \Rightarrow \quad A \int\limits_{0}^{\infty} (e^{-y} - e^{-2y}). dy = 1$

$\Rightarrow \qquad A \left(\frac{e^{-y}}{-1} - \frac{e^{-2y}}{-2} \right)_{0}^{\infty} = 1 \qquad\qquad \Rightarrow \quad A \left(-\left(e^{-y}\right)_{0}^{\infty} + \frac{1}{2}\left(e^{-2y}\right)_{0}^{\infty} \right) = 1$

$\Rightarrow \qquad A \left(1 - \frac{1}{2} \right) = 1 \qquad\qquad\qquad\qquad \Rightarrow \quad A = 2$

The joint Pdf of (X, Y) is given to be

$$f(x, y) = \begin{cases} 2e^{-x-y} & \text{for } 0 \leq x \leq y, \quad 0 \leq y < \infty \\ 0 & \text{otherwise} \end{cases}$$

(ii) The Marginal density functions of X and Y

$$f_x(x) = \int\limits_{-\infty}^{\infty} f(x, y) dy = \int\limits_{x}^{\infty} 2e^{-x} e^{-y} dy = 2e^{-x} \left(\frac{e^{-y}}{-1} \right)_{x}^{\infty}$$

$$= -2e^{-x} \left(e^{-y} \right)_{x}^{\infty}$$

$$= -2e^{-x} \left(0 - e^{-x} \right)$$

$$= 2e^{-2x}$$

The Marginal Density function of Y is

$$f_y(y) = \int_{-\infty}^{\infty} f(x,y)dx = \int_{0}^{y} 2e^{-y}e^{-x}dx = 2e^{-y}\left(\frac{e^{-x}}{-1}\right)_0^y$$

$$= -2e^{-y}\left(e^{-x}\right)_0^y$$

$$= -2e^{-y}\left(e^{-y}-1\right)$$

$$= 2e^{-y}\left(1-e^{-y}\right)$$

(iii)   **To test X and Y are independent:**

$$f_x(x) \times f_y(y) = 2e^{-2x} \times 2e^{-y}\left(1-e^{-y}\right)$$

$$\neq f(x,y)$$

Therefore $f_x(x) \times f_y(y) \neq f(x,y)$

Hence X and Y are not independent:

(iv)   To find the conditional density function of Y given $X=2$.

$$f(Y/X=2) = \frac{f(X=2,y)}{f_x(X=2)}$$

$$= \frac{2e^{-2-y}}{2e^{-4}}$$

$$= e^{2-y}$$

***Example* 17:** The joint density function of random variables X and Y is $f(x, y) = 8xy$ for $0 < x < 1$ and $0 < y < x$. Find $f(y/x)$ and $f(x/y)$.

Sol: The Marginal density functions of X is

$$f_x(x) = \int_{-\infty}^{\infty} f(x,y)dy = \int_{0}^{x} 8xydy = 8x\left(\frac{y^2}{2}\right)_0^x = 4x^3$$

The Marginal Density function of Y is

$$f_y(y) = \int_{-\infty}^{\infty} f(x,y)dx$$

$$= \int\limits_{0}^{1} 8xy\,dx$$

$$= 8y\left(\frac{x^2}{2}\right)_{0}^{1}$$

$$= 4y$$

$$f(y/x) = \frac{f(x,y)}{f_x(x)}$$

$$= \frac{8xy}{4x^3}$$

$$= \frac{2y}{x^2}$$

And $$f(x/y) = \frac{f(x,y)}{f_y(y)}$$

$$= \frac{8xy}{4y}$$

$$= 2x$$

***Example* 18:** The joint probability density of the random variables X and Y is

$$f(x,y) = \begin{cases} ke^{-(x+y)} & \text{for } 0 \le x \le \infty, \quad 0 \le y < \infty \\ 0 & \text{otherwise} \end{cases}$$

(a) Find the value of the constant k.

(b) Find the probability density of X independently of Y.

(c) Find the probability $P(0 \le x \le 2, 2 \le y < 3)$ and

(d) Are the random variables dependent or independent

Sol: Given probability function is

$$f(x,y) = \begin{cases} ke^{-(x+y)} & \text{for } 0 \le x \le \infty, \quad 0 \le y < \infty \\ 0 & \text{otherwise} \end{cases}$$

(a) Since $\qquad \displaystyle\int\limits_{-\infty}^{\infty}\int\limits_{-\infty}^{\infty} f(x,y)dx.dy = 1$

This implies $\qquad \displaystyle\int\limits_{0}^{\infty}\int\limits_{0}^{\infty} f(x,y)dx.dy = 1$

$\Rightarrow \qquad \displaystyle\int\limits_{0}^{\infty}\int\limits_{0}^{\infty} ke^{-(x+y)}dx.dy = 1$

$\Rightarrow \qquad k\displaystyle\int\limits_{0}^{\infty}e^{-y}\left(\int\limits_{0}^{\infty}e^{-x}dx\right).dy = 1 \quad \Rightarrow \quad k\int\limits_{0}^{\infty}e^{-y}\left(\dfrac{e^{-x}}{-1}\right)_{0}^{\infty}.dy = 1$

$\Rightarrow \qquad k\displaystyle\int\limits_{0}^{\infty}e^{-y}\left(1-0\right).dy = 1 \qquad \Rightarrow \quad k\int\limits_{0}^{\infty}e^{-y}.dy = 1$

$\Rightarrow \qquad k\left(\dfrac{e^{-y}}{-1}\right)_{0}^{\infty} = 1 \qquad\qquad \Rightarrow \quad k\left(1-0\right) = 1$

$\Rightarrow \qquad k = 1$

The joint Pdf of (X, Y) is given to be

$$f(x,y) = \begin{cases} e^{-(x+y)} & for\ 0 \le x \le \infty, \quad 0 \le y < \infty \\ 0 & otherwise \end{cases}$$

(b) The probability density function of X independent of Y is or The Marginal density functions of X is

$$f_x(x) = \int\limits_{-\infty}^{\infty} f(x,y)dy$$

$$= \int\limits_{0}^{\infty} e^{-x}e^{-y}dy$$

$$= e^{-x}\left(\dfrac{e^{-y}}{-1}\right)_{0}^{\infty}$$

$$= -e^{-x}\left(e^{-y}\right)_{0}^{\infty}$$

$$= -e^{-x}\left(0-1\right)$$

$$= e^{-x}$$

(c) $P(0 \le x \le 2, 2 \le y < 3) = \int_0^2 \int_2^3 e^{-(x+y)} dx.dy$

$$= \int_2^3 e^{-y} \left(\int_0^2 e^{-x} dx \right) dy = \int_2^3 e^{-y} \left(\frac{e^{-x}}{-1} \right)_0^2 .dy$$

$$= \int_2^3 e^{-y} \left(1 - e^{-2} \right) .dy = \left(1 - e^{-2} \right) \int_2^3 e^{-y} .dy$$

$$= \left(1 - e^{-2} \right) \left(\frac{e^{-y}}{-1} \right)_2^3$$

$$= \left(1 - e^{-2} \right) \left(e^{-2} - e^{-3} \right)$$

(d) To test X and Y are independent:

Since $f_y(y) = \int_{-\infty}^{\infty} f(x, y) dx$

$$= \int_0^{\infty} e^{-x} e^{-y} dx = e^{-y} \left(\frac{e^{-x}}{-1} \right)_0^{\infty}$$

$$= -e^{-y} \left(e^{-x} \right)_0^{\infty}$$

$$= -e^{-y} (0 - 1)$$

$$= e^{-y}$$

$$f_x(x) \times f_y(y) = e^{-x} \times e^{-y}$$

$$= e^{-(x+y)} = f(x, y)$$

Therefore $f_x(x) \times f_y(y) \ne f(x, y)$

Hence X and Y are not independent

4.14   Sum of Two Random Variables

In any communication system, we observe that at the receiving end, in addition to the transmitted signal, noise also will get received. In order to analyze this signal plus voltage of noise receiver end. It will be more useful, the study of density function and distribution function of sum of two random variables. In this signal and noise can be considered as a combination of two independent random variables. i.e., X might represent a signal voltage and Y represents random noise at some time instant.

Let the random variables X and Y be two independent random variables whose sum is defined as

$$Z = X + Y$$

Now the probability distribution function of Z is given by

$$F_Z(z) = P(Z \leq z) = P(X + Y \leq z)$$

$$= \int\limits_{x=-\infty}^{\infty} \int\limits_{y=-\infty}^{z-x} f_{XY}(x,y)\,dxdy \qquad .....(4.2)$$

$$= \int\limits_{y=-\infty}^{\infty} \int\limits_{x=-\infty}^{z-y} f_{XY}(x,y)\,dxdy \qquad .....(4.3)$$

If X and Y are two independent random variables, then $f_{XY}(x,y) = f_X(x) \times f_Y(y)$

Therefore from equations (4.2) and (4.3),

$$F_Z(z) = \int\limits_{y=-\infty}^{\infty} \int\limits_{x=-\infty}^{z-y} f_{XY}(x,y)\,dxdy$$

$$= \int\limits_{y=-\infty}^{\infty} f_Y(y)\,dy \int\limits_{x=-\infty}^{z-y} f_X(x)\,dx \qquad \text{or} \qquad .....(4.4)$$

$$= \int\limits_{x=-\infty}^{\infty} f_X(x)\,dx \int\limits_{y=-\infty}^{z-x} f_Y(y)\,dy \qquad .....(4.5)$$

Then the probability density function of Z is

$$f_Z(z) = \frac{d}{dz}[F_Z(z)] = \int\limits_{x=-\infty}^{\infty} f_{XY}(x, z-x)\,dx \qquad .....(4.6)$$

Similarly,

$$f_Z(z) = \int\limits_{y=-\infty}^{\infty} f_{XY}(z-y, y)\,dy$$

These integrals express the density of Z in terms of $f_{XY}(x, z-x)$ or $f_{XY}(z-y, y)$. These densities are identified as a convolution integral.

Since X and Y are independent random variables,

$$f_Z(z) = \int\limits_{x=-\infty}^{\infty} f_{XY}(x, z-x)\,dx$$

$$= \int\limits_{x=-\infty}^{\infty} f_X(x).f_Y(z-x)\,dx$$

Similarly,

$$f_Z(z) = \int_{y=-\infty}^{\infty} f_{XY}(z-y, y) dy$$

$$= \int_{x=-\infty}^{\infty} f_Y(y) . f_X(z-y) dy$$

These equation are obtained by differentiating equations (1) and (2) by using Leibnitz's rule.

4.15   Sum of Several Random Variables

Since we have the density function of the sum of two independent random variables is the convolution of their individual density function.

$$f_Z(z) = f_X(x) * f_Y(y)$$

Similarly, the sum of the three random variables X_1, X_2 and X_3 can be expressed in terms of Z as $Z = X_1 + X_2 + X_3$, then the probability density fucntion of Z is

$$f_Z(z) = f_{X_1}(x_1) * f_{X_2}(x_2) * f_{X_3}(x_3)$$

This can be extended to a sum Z of n independent random variables X_1, $X_2, X_3 .. . X_n$ is

$$f_Z(z) = f_{X_1}(x_1) * f_{X_2}(x_2) * f_{X_3}(x_3) * \ldots * f_{X_n}(x_n)$$

The distribution function of Z is

$$F_Z(z) = \int_{-\infty}^{\infty} f_Z(z) \, dz$$

4.16   Central Limit Theorem

In this, the sum of a large number of independent random variables having the same probability density function having a mean and variance equals to n times the mean and variance of the original random variables and having a probability density function approaches Gaussian.

In any communication system, noise performance is theoretically studied by assuming noise as a random variable, in this study noise is treated as a Gaussian random variable because of central limit theorem.

4.17   Unequal Distribution

If X_1, X_2, $X_3 \ldots X_n$ be independently and identically distributed random variables with $E[X_i] = \mu$ and $V[X_i] = \sigma^2$, $i = 1, 2, 3, \cdots n$, then the random variable $S_n = X_1 + X_2 + X_3 + \ldots + X_n$ is asymptotically normal with mean $n\mu$ and variance $n\sigma^2$ under general conditions as $n \to \infty$

The central limit theorem guarantees only that the distribution of the sum of random variables become normal. It does not follow that the probability density is always Gaussian. In case of continuous random variables, there is no problem, but certain conditions imposed on the individual random variables (Cramer, 1946, Papoulis, 1965 and 1984) will guarantee that the density is Gaussian.

In case discrete random variable X_i, the sum S_n will also be discrete. So its density will contain impulses and is, therefore not gaussian even though the distribution approaches to gaussian.

4.18   Equal Distributions

If X_1, X_2, $X_3 \ldots X_n$ are independent and continuous random variables have the same distribution function and density function.

If X_1, X_2, $X_3 \ldots X_n$ be independently and identically distributed random variables with $E[X_i] = \mu$ and $V[X_i] = \sigma^2$, $i = 1, 2, 3, \cdots n$, then the random variable $S_n = X_1 + X_2 + X_3 + \ldots + X_n$ is asymptotically normal with mean $n\mu$ and variance $n\sigma^2$.

Proof: Since we have $\qquad S_n = X_1 + X_2 + X_3 + \ldots + X_n$

And each X_i has a mean μ and variance σ^2

$$\therefore E[S_n] = E[X_1] + E[X_2] + E[X_3] + \ldots + E[X_n]$$
$$= nE[X_i]$$
$$= n\mu$$

and

$$V[S_n] = V[X_1] + V[X_2] + V[X_3] + \ldots + V[X_n]$$

$\because \qquad X_1, X_2, X_3 \ldots X_n$ are independent

$$= nV[X_i]$$

$$= n\sigma^2$$

Now each $X_i \sim N(\mu, \sigma)$, $i = 1, 2, 3, \cdots n$ then

$$S_n = \sum X_i \sim N(n\mu, \sigma\sqrt{n}) \text{ then the standard normal variate}$$

$$Z = \frac{S_n - n\mu}{\sigma\sqrt{n}} \sim N(0,1)$$

The moment generating function of the standard normal variate Z is

$$M_Z(t) = E[e^{tZ}] = E\left[e^{t\left(\frac{S_n - n\mu}{\sigma\sqrt{n}}\right)} \right]$$

$$= E\left[e^{t\left(\frac{X_1 + X_2 + \cdots + X_n - (\mu + \mu + \cdots + \mu)}{\sigma\sqrt{n}}\right)} \right]$$

$$= E\left[e^{t\left\{ \left(\frac{X_1 - \mu}{\sigma\sqrt{n}}\right) + \left(\frac{X_2 - \mu}{\sigma\sqrt{n}}\right) + \cdots + \left(\frac{X_n - \mu}{\sigma\sqrt{n}}\right) \right\}} \right]$$

$$= E\left[e^{t\left(\frac{X_1 - \mu}{\sigma\sqrt{n}}\right)} . e^{t\left(\frac{X_2 - \mu}{\sigma\sqrt{n}}\right)} \cdots . e^{t\left(\frac{X_n - \mu}{\sigma\sqrt{n}}\right)} \right]$$

$$= \left\{ E\left[e^{t\left(\frac{X_1 - \mu}{\sigma\sqrt{n}}\right)} \right] \right\}^n \qquad \text{since } X_i\text{'s are independent}$$

Consider

$$E\left[e^{t\left(\frac{X_i - \mu}{\sigma\sqrt{n}}\right)} \right] = E\left[1 + \frac{t}{1!\sigma\sqrt{n}}(X_1 - \mu) + \frac{t^2}{2!\sigma^2 n}(X_1 - \mu)^2 + \cdots \right]$$

$$= 1 + \frac{t}{1!\sigma\sqrt{n}} \times 0 + \frac{t^2}{2!\sigma^2 n} \times \sigma^2 + \cdots$$

$$= 1 + \frac{t^2}{2n} + \cdots$$

$$\therefore \qquad M_Z(t) = \left\{ E\left[e^{t\left(\frac{X_1-\mu}{\sigma\sqrt{n}}\right)} \right] \right\}^n$$

$$= \left(1 + \frac{t^2}{2n} + \cdots \right)^n$$

As taking Limit $n \to \infty$,

$$\underset{n\to\infty}{Lt}\ M_Z(t) = \underset{n\to\infty}{Lt}\ \left(1 + \frac{t^2}{2n} + \cdots \right)^n$$

$$= e^{t^2/2}$$

This is same as the moment generating function of a standard normal variate. So Z is also a gaussian and central limit theorem is proved. This theorem is applicable event when the individual random variables are Gaussian or not.

Quiz Questions

1. Let (X, Y) be a 2dimensional random variable, then (x_i, y_j) associated with a number $P(x_i, y_j)$ then the value $P(X = x_i, Y = y_j)$ is

    (a) P_{ij} (b) P_{ji} (c) P_{i+j} (d)   none

2. The probability of drawing a particular card will be denoted by $f(x_i, y_j)$ or $P(x_i, y_j)$ and the probability of getting each card is same, then the probability function of (x, y) is

    $$f(x_i, y_j) = P(x_i, y_j) = \frac{1}{52} \qquad \begin{cases} 1 \le i \le 4 \\ 1 \le j \le -- \end{cases}$$

    The upper limit of j is

    (a)   10   (b)   12   (c)   13   (d)   none

3. The joint probability distribution satisfies $p_{ij} = P(X = x_i, Y = y_j) = P(x_i, y_j)$ is

    (a) < 0 (b) ≥ 0 (c) $= 0$ (d)   0

4. The Marginal Probability distribution $X = x_i$, $i=1, 2, 3 \ldots n$ is denoted by p_{i*} or $f(x_i)$ of $P(x_i, y_j)$ is

(a)   1   (b) $\displaystyle\sum_{j=1}^{m} P(x_i, y_j)$ (c) $\displaystyle\sum_{i=1}^{m} P(x_i, y_j)$ (d)   none

5.   Let S be e sample space, X and Y be two random variables in it. The probability that X lies between $x - \dfrac{1}{2}dx$ and $x + \dfrac{1}{2}dx$ and at the same time Y lies between $y - \dfrac{1}{2}dy$ and $y + \dfrac{1}{2}dy$ is denoted by

(a) $\displaystyle\int_{x-\frac{1}{2}dx}^{x+\frac{1}{2}dx} \left\{ \int_{y-\frac{1}{2}dy}^{y+\frac{1}{2}dy} f(x,y)dy \right\} .dx$ (b) $\displaystyle\sum_{j=1}^{m} P(x_i, y_j)$

(c) $\displaystyle\sum_{i=1}^{m} P(x_i, y_j)$ (d)   none

6.   The notation defined as $p(x_1 < X \le x_2, y_1 < Y \le y_2)$ it is denoted by

(a) $\displaystyle\int_{x_1}^{x_2} f(x,y)\, dx$ (b) $\displaystyle\int_{y_1}^{y_2}\int_{x_1}^{x_2} f(x,y)\, dx.dy$

(c) $\displaystyle\int_{y_1}^{y_2} f(x,y)\, .dy$ (d)   none

7.   For a continuous case, the Marginal Probability density function of X, is denoted by $f_x(x)$ and is defined as $f_x(x)$ is equal to

(a) $\displaystyle\int_{-\infty}^{\infty} f(x,y)dx$ (b) $f(x,y)$

(c) $\displaystyle\int_{-\infty}^{\infty} f(x,y)dy$ (d)   none

8.   Let (X, Y) be a two-dimensional random variable then the joint distribution function or is denoted by $F_{XY}(x, y)$ or $F(x, y)$ and is defined as

(a) $\begin{cases} \displaystyle\sum_{Y \le y}\sum_{X \le x} P(x_i, y_j)\, discrete\ case \\[2mm] \displaystyle\int_{-\infty}^{y}\int_{-\infty}^{x} f(x,y)dy.dx\ continuous\ case \end{cases}$

(b) $\begin{cases} \displaystyle\sum_{Y \le y}\sum_{X \le x} P(x_i, y_j) \text{ continuous case} \\ \\ \displaystyle\int_{-\infty}^{y}\int_{-\infty}^{x} f(x, y)dy.dx \text{ discrete case} \end{cases}$

(c) $\begin{cases} \displaystyle\sum_{Y \le y}\sum_{X \le x} P(x_i, y_j) \text{ continuous case} \\ \\ \displaystyle\int_{-\infty}^{y}\int_{-\infty}^{x} f(x, y)dy.dx \text{ continuous case} \end{cases}$

(d)   none

9.   The joint pdf of (X_1, X_2) is defined as $f(x_1, x_2)$ is

(a) $\dfrac{\partial F(x_1, x_2)}{\partial x_1 . \partial x_2}$

(b) $\dfrac{\partial^2 F(x_1, x_2)}{\partial x_1 . \partial x_2}$

(c) $\displaystyle\int_{-\infty}^{\infty} f(x, y)dy$

(d)   none

10.   Let X and Y be two random variables then the probability distribution function $F_{XY}(x, y)$ lies between

(a) $-1\&1$ (b) $1\&2$ (c) $0\&1$ (d)   none

11.   Let X and Y be two random variables then the probability distribution function $F(+\infty, +\infty)$ is

(a)   1   (b)   2   (c)   3   (d)   0

12.   Let X and Y be two random variables then the probability distribution function $F(x, -\infty)$ is

(a)   1   (b)   2   (c)   3   (d)   0

13.   The cumulative distribution function of X alone (i.e., independent of Y) is denoted by $F_X(x)$ and is defined as $F_X(x)$ is equal to

(a) $\displaystyle\int_{-\infty}^{\infty}\int_{-\infty}^{x} f(x, y)dx.dy$

(b) $\displaystyle\int_{-\infty}^{\infty}\int_{-\infty}^{y} f(x, y)dx.dy$

(c) $\displaystyle\int_{-\infty}^{y}\int_{-\infty}^{x} f(x, y)dx.dy$

(d)   0

14. Then the probability density function of X (independent of Y), $f_x(x)$ is equal to

   (a) $F_x(x)$ (b) $\dfrac{d}{dx}F_x(x)$ (c) $\int F_x(x)$ (d)   0

15. If X and Y are independent, then $f(x,y)$ is equal to

   (a) $f_x(x) \times f_Y(y)$ (b) $f_x(x) + f_Y(y)$

   (c) $f_x(x) - f_Y(y)$ (d)   0

16. The joint distribution function of (X_1, X_2) is denoted by $F(x_1, x_2)$ is equal to

   (a) $\displaystyle\int_{-\infty}^{\infty}\int_{-\infty}^{x_2} f(x_1,x_2)dx_1.dx_2$ (b) $\displaystyle\int_{-\infty}^{x_1}\int_{-\infty}^{x_2} f(x_1,x_2)dx_1.dx_2$

   (c) $\displaystyle\int_{-\infty}^{x_1}\int_{-\infty}^{\infty} f(x_1,x_2)dx_1.dx_2$ (d)   none

17. The conditional probability distribution function of X given Y is defined as $p(X = x_i \,/\, Y = y_j)$ Is

   (a) $\dfrac{p(X = x_i \cap Y = y_j)}{p(Y = y_j)}$ (b) $\dfrac{p(X = x_i \cap Y = y_j)}{p(X = x_i)}$

   (c) $\dfrac{p(X = x_i \cap Y = y_j)}{p(X = x_i)p(Y = y_j)}$ (d)   none

18. The conditional probability density function of X given Y is defined as

   $$P\left\{x - \frac{1}{2}dx \le X \le x + \frac{1}{2}dx \,/\, y - \frac{1}{2}dy \le Y \le y + \frac{1}{2}dy\right\}$$

   (a) $\dfrac{f(x,y)dx.dy}{f_X(x.)dx}$ (b) $\dfrac{f(x,y)dx.dy}{f_y(y.)dy}$

   (c) $\dfrac{f(x,y)dx.dy}{f_X(x.)dxf_y(y.)dy}$ (d)   none

19. Let A be an event $(X \le x)$ for the random variable X, If B is given, the conditional distribution function of the random variable X is $F_X(x/B) = P[X \le x/B]$ is

   (a) $\dfrac{P\left[(X \le x) \cap B\right]}{P[X \le x]}$ (b) $\dfrac{P\left[(X \le x) \cap B\right]}{P[B \le x]}$

   (c) $\dfrac{P\left[(X \le x) \cap B\right]}{P[B]}$ (d)   none

20. If the event B is defined in terms of another variable Y, i.e., If the random variable X is conditioned by a second random variable Y, where $Y \leq y$, is called a point conditioning. is denoted as $P[X \leq x / Y \leq y] = $ (provided $P[Y \leq y] \neq 0$)

(a) $\dfrac{P[(X \leq x) \cap (Y \leq y)]}{P[Y \leq y]}$

(b) $\dfrac{P[(X \leq x) \cap (Y \leq y)]}{P[X \leq x]}$

(c) $\dfrac{P[(X \leq x) \cap (Y \leq y)]}{P[(X \leq x)] P[Y \leq y]}$

(d) none

21. The point conditioning is simply represented as $F_{X/Y}(x / y)$ and is defined as

(a) $\dfrac{F_{X,Y}(x, y)}{F_X(y)}$

(b) $\dfrac{F_{X,Y}(x, y)}{F_Y(y)}$

(c) $\dfrac{F_{X,Y}(x, y)}{F_X(x)}$

(d) none

22. The conditional distribution function $F_X(-\infty/B) = p(X \leq -\infty/B)$

(a) 1 (b) ∞ (c) 0 (d) *none*

23. The conditional distribution function $F_X(x_1 / B) \leq F_X(x_2 / B)$ If

(a) $x_1 < x_2$ (b) $x_1 > x_2$ (c) $x_1 = x_2$ (d) none

24. The conditional density function $F_X(x / B)$ is equal to

(a) $\displaystyle\int_{-\infty}^{y} f_X(x / B) dx$

(b) $\displaystyle\int_{-\infty}^{x} f_X(x / B) dx$

(c) $\displaystyle\int_{-\infty}^{x} f_Y(y / B) dx$

(d) none

25. The conditional density function $p(x_1 < X < x_2 / B) = \displaystyle\int_{x_1}^{x_2} f_X(x / B) dx =$

(a) $F_X(x_2 / B) \times F_X(x_1 / B)$

(b) $F_X(x_2 / B) + F_X(x_1 / B)$

(c) $F_X(x_2 / B) - F_X(x_1 / B)$

(d) none

26. Let (X, Y) be a 2-dimensional discrete random variable, and the two random variables X and Y are independent, then $P_{ij} =$

(a) $P_{i*} \times P_{*j}$ (b) $P_{i*} + P_{*j}$ (c) $P_{i*} - P_{*j}$ (d) none

27. The joint probability function of two-dimensional random variables (X, Y) is

$$f(x, y) = \begin{cases} 2 & if \ 0 < x < 1, \quad 0 < y < x \\ 0 & otherwise \end{cases}$$

Then the Marginal Density function of X

(a)   0   (b)   x   (c)   2x–   (d)   none

28. The joint probability function of two-dimensional random variables (X, Y) is

$$f(x, y) = \begin{cases} 2 & if \ 0 < x < 1, \quad 0 < y < x \\ 0 & otherwise \end{cases}$$

Then the conditional density function of Y given $X=x$

(a) $\dfrac{1}{x}$ (b) $\dfrac{1}{y-1}$ (c)   2x_   (d)   none

29. The joint probability density of the random variables X and Y is   .

$$f(x, y) = \begin{cases} ke^{-(x+y)} & for \ 0 \le x \le \infty, \quad 0 \le y < \infty \\ 0 & otherwise \end{cases}$$

the value of the constant k is

(a)   1   (b)   0   (c)   0.5   (d)   none

30. The density function of the sum of two independent random variables is the convolution of their individual density function then $f_Z(z)$ is equal to

(a) $f_X(x) - f_Y(y)$ (b) $f_X(x) + f_Y(y)$

(c) $f_X(x) * f_Y(y)$ (d)   none

Answers

1. (a)	2. (c)	3. (b)	4. (b)	5. (a)
6. (b)	7. (c)	8. (a)	9. (b)	10. (c)
11. (a)	12. (d)	13. (a)	14. (b)	15. (a)
16. (b)	17. (a)	18. (b)	19. (c)	20. (a)
21. (b)	22. (c)	23. (a)	24. (b)	25. (c)
26. (a)	27. (c)	28. (a)	29. (a)	30. (c)

Review Quiz Questions

1. The joint probability distribution satisfies $\sum\limits_{i=1}^{n}\sum\limits_{j=1}^{m}P(x_i, y_j)$ is

    (a)   1   (b)   <1   (c)   >1   (d)   none

2. The Marginal Probability distribution $Y = y_j$, $j = 1, 2, 3$. m is denoted by p_{*j} or $f(y_j)$ of $P(x_i, y_j)$ is

    (a)   1   (b) $\sum\limits_{j=1}^{m}P(x_i, y_j)$

    (c) $\sum\limits_{i=1}^{n}P(x_i, y_j)$ (d)   none

3. Let S be e sample space, X and Y be two random variables in it. The probability that X lies between $x-\dfrac{1}{2}dx$ and $x+\dfrac{1}{2}dx$ and at the same time Y lies between $y-\dfrac{1}{2}dy$ and $y+\dfrac{1}{2}dy$ is denoted by

    (a) $P\left\{x-\dfrac{1}{2}dx \le X \le x+\dfrac{1}{2}dx, \ y-\dfrac{1}{2}dy \le Y \le y+\dfrac{1}{2}dy\right\}$

    (b) $P(x_i, y_j)$

    (c) $f(x_i, y_j)$

    (d)   none

4. Let X and Y are two random variables and $f(x, y)$ is a density function the it satisfies

    (a) $f(x,y) \ge 0$ (b) $\int\limits_{-\infty}^{\infty}\int\limits_{-\infty}^{\infty}f(x, y)dy.dx = 1$

    (c) $f(x,y) \ge 0; \int\limits_{-\infty}^{\infty}\int\limits_{-\infty}^{\infty}f(x, y)dy.dx = 1$ (d)   none

5. Let (X, Y) be a two dimensional random variable then the joint distribution function or is denoted by $F_{XY}(x, y)$ or $F(x, y)$ and is defined as

    (a) $p(X \le x, Y \le y)$ (b) $p(X \ge x, Y \le y)$

    (c) $p(X \le x, Y \ge y)$ (d)   none

6. Let X and Y be two random variables then the probability distribution function $F(-\infty, -\infty)$ is

    (a)   1   (b)   2   (c)   3   (d)   0

7. Let X and Y be two random variables then the probability distribution function $F(-\infty, y)$ is

    (a)   1   (b)   2   (c)   3   (d)   0

8. The cumulative distribution function of X alone (i.e., independent of Y) is denoted by $F_X(x)$ and is defined as $F_X(x)$ is equal to

    (a) $F_{XY}(x,-\infty)$ (b) $F_{XY}(y,-\infty)$

    (c) $F_{XY}(x,y)$ (d)   0

9. If X and Y are independent, then $p(x \le X \le x+dx, y \le Y \le y+dy)$ is equal to

    (a) $[f(x)dx][f(y)dy]$ (b) $[f(x)dy][f(y)dx]$

    (c) $F_{XY}(x,y)$ (d)   0

10. The joint pdf of $(X_1, X_2 \ldots \ldots X_n)$ is defined as

    (a) $f(x_1,x_2...x_n) = \dfrac{\partial F(x_1,x_2...x_n)}{\partial x_1 . \partial x_2...\partial x_n}$

    (b) $f(x_1,x_2...x_n) = \dfrac{\partial^n F(x_1,x_2...x_n)}{\partial x_1 . \partial x_2...\partial x_n}$

    (c) $f(x_1,x_2...x_n) = \dfrac{\partial^n F(x_1,x_2...x_n)}{(\partial x_1 . \partial x_2...\partial x_n)^n}$

    (d)   none

11. The conditional probability distribution function of X given Y is defined as $p(X = x_i / Y = y_j)$ is

    (a) $\dfrac{p_{ij}}{p_{j*}}$ (b) $\dfrac{p_{ij}}{p_{*j}}$ (c) $\dfrac{p_{ij}}{p_{*j}} p_{j*}$ (d)   none

12. The conditional probability distribution function of Y given X is defined as $p(Y = y_j / X = x_i)$ is

    (a) $\dfrac{p(X = x_i \cap Y = y_j)}{p(X = x_i)p(Y = y_j)}$ (b) $\dfrac{p(X = x_i \cap Y = y_j)}{p(Y = y_j)}$

    (c) $\dfrac{p(X = x_i \cap Y = y_j)}{p(X = x_i)}$ (d)   none

13. The conditional probability density function of Y given X is defined as

$$P\left\{y-\frac{1}{2}dy \le Y \le y+\frac{1}{2}dy \,/\, x-\frac{1}{2}dx \le X \le x+\frac{1}{2}dx\right\}$$

(a) $\dfrac{f(x,y)dx.dy}{f_X(x.)dx}$

(b) $\dfrac{f(x,y)dx.dy}{f_y(y.)dy}$

(c) $\dfrac{f(x,y)dx.dy}{f_X(x.)dxf_y(y.)dy}$

(d) none

14. If the event B is defined in terms of another variable Y, i.e., If the random variable X is conditioned by a second random variable Y, where $Y \le y$, is called

(a) point conditioning.

(b) interval conditioning

(c) mixed conditioning

(d) none

15. If the random variable Y is defined in such a way that its value lies between two constants y_a and y_b. is called

(a) point conditioning

(b) interval conditioning

(c) mixed conditioning

(d) none

16. The conditional distribution function $F_X(+\infty/B) = p(X \le +\infty/B)$

(a) 1

(b) ∞

(c) 0

(d) none

17. The conditional distribution function $p(x_1 < X < x_2 \,/\, B)$ If

(a) $F_X(x_2/B) + F_X(x_1/B)$

(b) $F_X(x_2/B) - F_X(x_1/B)$

(c) $F_X(x_2/B) \times F_X(x_1/B)$

(d) none

18. The conditional density function $\displaystyle\int_{-\infty}^{\infty} f_X(x/B)dx$ is equal to

(a) 1

(b) 0

(c) ∞

(d) none

19. For a 2-dimensional continuous random variable, if X and Y are two independent random variables, then $f(x,y) =$

(a) $f_x(x) \times f_y(y)$

(b) $f_x(x) + f_y(y)$

(c) $f_x(x) - f_y(y)$

(d) none

20.   The joint probability function of two-dimensional random variables (X, Y) is

$$f(x,y) = \begin{cases} 2 & \text{if } 0<x<1, \quad 0<y<x \\ 0 & \text{otherwise} \end{cases}$$

Then the Marginal Density function of Y

(a)   2x   (b)   x   (c) $2(y-1)$ _   (d)   none

21.   The joint probability function1 of two-dimensional random variables (X, Y) is

$$f(x,y) = \begin{cases} 2 & \text{if } 0<x<1, \quad 0<y<x \\ 0 & \text{otherwise} \end{cases}$$

Then the conditional density function of X given $Y=y$

(a) $\dfrac{1}{x}$ (b) $\dfrac{1}{y-1}$ (c) $2x$_   (d)   none

22.   The joint probability density of the random variables X and Y is   .

$$f(x,y) = \begin{cases} ke^{-(x+y)} & \text{for } 0 \le x \le \infty, \quad 0 \le y < \infty \\ 0 & \text{otherwise} \end{cases}$$

Then $P(0 \le x \le 2, 2 \le y < 3)$ is

(a)   1   (b)   0   (c)   0.5   (d)   none

23.   The sum of the three random variables X_1, X_2 and X_3 can be expressed in terms of Z as $Z = X_1 + X_2 + X_3$, then the probability density function of Z is $f_Z(z) =$

(a) $f_{X_1}(x_1) * f_{X_2}(x_2) * f_{X_3}(x_3)$ (b) $f_{X_1}(x_1) * f_{X_2}(x_2)$

(c) $f_{X_3}(x_3) * f_{X_2}(x_2)$ (d)   none

24.   The Marginal Probability density function of Y, is denoted by $f_y(y)$ and is defined as

$$f_y(y) =$$

(a) $\displaystyle\int_{-x}^{x} f(x,y)dx$ (b) $\displaystyle\int_{-\infty}^{\infty} f(x,y)dx$

(c) $\displaystyle\int_{-y}^{y} f(x,y)dx$ (d)   none

25. The cumulative distribution function of Y alone (i.e., independent of X) is denoted by $F_y(y)$ and is defined as $F_{XY}(-\infty, y) =$

    (a) $\displaystyle\int_{-\infty}^{\infty}\int_{-\infty}^{y} f(x, y)dx.dy$

    (b) $\displaystyle\int_{-\infty}^{\infty}\int_{-\infty}^{x} f(x, y)dx.dy$

    (c) $\displaystyle\int_{-\infty}^{\infty}\int_{-\infty}^{\infty} f(x, y)dx.dy$

    (d) none

26. The probability density function of Y (independent of X) is $f_y(y)$ is

    (a) $\dfrac{d}{dx}F_y(y)$

    (b) $\dfrac{d}{dy}F_Y(x)$

    (c) $\dfrac{d}{dy}F_y(y)$

    (d) none

Exercise Questions

1. The probability distribution function of X is

$X = x_i$	1	2	3
$P(X = x_i)$	$\dfrac{1}{3}$	$\dfrac{1}{6}$	$\dfrac{1}{2}$

   Find the third central moment

2. Two discrete random variables X and Y have joint pmf given by the following tables

YX	1	2	3
1	$\dfrac{1}{12}$	$\dfrac{1}{6}$	$\dfrac{1}{12}$
2	$\dfrac{1}{6}$	$\dfrac{1}{4}$	$\dfrac{1}{12}$
3	$\dfrac{1}{12}$	$\dfrac{1}{12}$	0

   Compute the probability of each of the following events

   (a) $X \leq 1\dfrac{1}{2}$ (b) XY is even   (c) Y is even given X is event

3. The noise voltage in an electric circuit can be modelled as a gaussian random v variable with zero mean and variance equal to 10^{-8}

   (i) What is the probabilyt that the value of lthe noise exceeds 10^{-4}

   (ii) What is the probability that the noise value is between -2×10^{-4} and 10^{-4}

4. A fair coin is tossed 10 times. Find the probability that the number of heads occuring is between 4 and 7 inclusive by sing i) Binomial distribution b) normal approximation to biomial distribution.

5. The number of newspapers that a certain delivery buy is able to sell in a day is found to be a numerical valued random phenomenon, with a probability fuction specified by the pmf p(.) given by $p(x) = \begin{cases} Ax & x = 1,2,3 \cdots 50 \\ A(100-x) & x = 51,52, \cdots 100 \\ 0 & Otherwise \end{cases}$

   Find (i) the vlaue of A that makes p(.) a pmf and sketch its graph (ii) What is the probability that the number of newspapers sold tomorrow is (a) more than 50 (b) less than 50 (c) equal to 50 (d) between 25 and 75 exclusive (e) an odd number (iii) let the events indicated in (ii) be denoted respectively by A, B, C, D and E, Find P(A/B), P(A/C), P(A/D) and P(C/D). Are A and B independent? Are A and D independent? Are C and D independent events?

6. The two random variables X and Y have

   $$P(X = 0, Y = 0) = \frac{2}{9}; \ P(X = 0, Y = 1) = \frac{1}{9}$$

   $P(X = 1, Y = 0) = \frac{1}{9}; \ P(X = 1, Y = 1) = \frac{5}{9}$ are X and Y are independent?

7. A certain binary system transmits two binary states $X = +1$, $X = -1$ with equal probabilities. Because of channel noise, the receiver makes recognition errors/ Also, as a result of path   distoration, the receiver may lose necessary signal strength to make any decision. Thus, there are three possible receiver states $Y = +1$, $Y = 0 Y = -1$ the corresponding loss of signal.

   Let $P(Y = 1 / X = +1) = 0.1; \quad P(Y = +1 / X = -1) = 0.2$

   $P(Y = 0 / X = +1) = P(Y = 0 / X = -1) = 0.05$

   Find   (a) $P(Y = +1); P(Y = -1); P(Y = 0)$ and

   (b) $P(X = +1 / Y = +1); P(X = -1 / Y = -1)$

8. The joint space for two random variables X and Y and corresponding probabilities are shown in table.

Find   (a) $F_{XY}(x, y)$

(b)   Marginal distribution function of X and Y

(c) $P(0.5 < X < 1.5)$

(d) $P(X \leq 1, Y \leq 2)$

X,Y	1,1	2,2	3,3	4,4
P	0.05	0.35	0.45	0.15

(e) $P(1 < X \leq 2, Y \leq 3)$

9.   The joint probability function of two discrete random variables X and Y is given by

$$f_{XY}(x, y) = \begin{cases} kxy & \text{for } x = 1,2,3 \text{ and } y = 1,2,3 \\ 0 & \text{otherwise} \end{cases}$$

Find (a) $P(1 \leq X \leq 2, Y \leq 2)$ $P(X \geq 2)$

10.   The joint probability density function of two random variables X and Y is

$$f_{XY}(x, y) = \begin{cases} \dfrac{1}{ab} & \text{for } 0 < x < a \text{ and } 0 < y < b \\ 0 & \text{otherwise} \end{cases}$$

Find $P(X + Y) \leq \dfrac{3a}{4})$ assuming $a < b$

11.   If X and Y are two independent random variables with $f_X(x) = \alpha.e^{-\alpha x}.u(x)$ and $f_Y(y) = \beta.e^{-\beta y}.u(y)$, then find the density of $Z = X + Y$ when (i) $\alpha \neq \beta$ and (ii) $\alpha = \beta$

12.   The joint probability density function of two random variables X and Y is

$$f_{XY}(x, y) = \begin{cases} A.e^{-x-y}\dfrac{1}{ab} & \text{for } 0 \leq x \leq y \text{ and } 0 < y < \infty \\ 0 & \text{otherwise} \end{cases}$$

Find the conditional density function Y given $X = 2$

13. The joint probability density function of two random variables X and Y is

$$f_{XY}(x,y) = \begin{cases} \dfrac{5}{16}.x^2 y & for \ \ 0 < y < x < 2 \\ 0 & otherwise \end{cases}$$

Find (i) the marginal density of X and Y and (ii) Are X and Y are independent?

14. The joint probability density function of two random variables X and Y is

$$f_{XY}(x,y) = \begin{cases} k(.x+y)^2 & for \ \ -2 < x < 2; \ -3 < y < 3 \\ 0 & otherwise \end{cases}$$

(i) For what value of K, the given $f_{XY}(x,y)$ is a valid density function?
(ii) Find the marginal density of X and Y

Answers

1. $\dfrac{7}{36}$ 2. (a) $\dfrac{1}{3}$ (b) $\dfrac{3}{4}$ (c) $\dfrac{1}{2}$

3. (i)   0.158   (ii)   0.818   4.   (i)   0.7734   (ii)   0.7718

5. (i) $A = \dfrac{1}{2500}$ (ii) $a)\,0.49$ $b)\,0.49$ $c)\,0.02$ $d)\dfrac{1850}{2500}$ $e)\,2\displaystyle\sum_{X=0}^{24}(2X+1)$

(iii) $P(A/B) = \dfrac{1}{2500}\displaystyle\sum_{X=51}^{100}(100-X)$

A and B are independent

$$P(A/C) = \dfrac{1}{2500}\displaystyle\sum_{X=51}^{100}(100-X)$$

$$P(A/D) = \dfrac{900}{1850}$$

A and D are not independent

$$P(C/D) = \dfrac{50}{1850}$$

C and D are not independent

6. X and Y are independent

7. (a) $P(Y=+1) = 0.525; P(Y=-1) = 0.425; P(Y=0) = 0.05$

(b) $P(X=+1/Y=+1) = 0.809; \ P(X=-1/Y=-1) = 0.882$

8. (a) $F_{XY}(x, y) = \begin{cases} 0.85 & \text{for } (X \le 3, Y \le 3) \\ 0.4 & \text{for } (X \le 2, Y \le 2) \\ 0.05 & \text{for } (X \le 1, Y \le 1) \end{cases}$

   (b) $P(X = i) = \begin{cases} 0.05 & \text{for } i = 1 \\ 0.35 & \text{for } i = 2 \\ 0.45 & \text{for } i = 3 \\ 0.15 & \text{for } i = 4 \end{cases}$

   (c) $P(Y = j) = \begin{cases} 0.05 & \text{for } j = 1 \\ 0.35 & \text{for } j = 2 \\ 0.45 & \text{for } j = 3 \\ 0.15 & \text{for } j = 4 \end{cases}$ (d)   0.05   (e)   0.35

9. (a) $\dfrac{1}{4}$ (b) $\dfrac{5}{6}$

10. $\dfrac{9a}{32b}$

11. (i) $\dfrac{\alpha\beta}{\alpha - \beta}\left[e^{-\beta Z} - e^{-\alpha Z}\right].u(Z)$ (ii) $\alpha^2.A.e^{-\alpha Z}.u(Z)$

12. e^{2-y}

13. $f_X(x) = \dfrac{5}{32}x^4$ and $f_Y(y) = \dfrac{5}{16}y\left(\dfrac{8 - y^3}{3}\right)$ b.X and Y are not independent

14. (i) $k = \dfrac{1}{104}$ (ii) $f_X(x) = \begin{cases} \dfrac{1}{104}(6x^2 + 18) & -2 < x < 2 \\ 0 & otherwise \end{cases}$

     And $f_Y(y) = \begin{cases} \dfrac{1}{104}(4y^2 + \dfrac{16}{3}) & -3 < y < 3 \\ 0 & otherwise \end{cases}$

Review Exercise Questions

1. For the following joint distribution of random variables X and Y, find $P(X \le 2)$ and $P(Y \le 4)$

X \ Y	1	2	3	4
0	$\dfrac{2}{32}$	$\dfrac{3}{32}$	$\dfrac{1}{32}$	$\dfrac{2}{32}$
1	$\dfrac{1}{16}$	$\dfrac{1}{16}$	$\dfrac{3}{8}$	$\dfrac{1}{8}$
2	$\dfrac{1}{32}$	$\dfrac{1}{32}$	$\dfrac{1}{64}$	$\dfrac{3}{64}$

2. A random variable X has the following distribution

X	0	1	2	3	4	5	6
P(X)	k	3k	5k	7k	9k	11k	13k

Find (i) $P(X < 5)$, $P(X \ge 6)$, $P(4 < X \le 7)$ and (ii) What will be the minimum value of x so that $P(X \le x) > 0.5$

3. A random variable X has the following probability distribution

x	1	2	3	4	5	6	7	8
f	k	2k	3k	4k	5k	6k	7k	8k

Find the value of (i) k (ii) $P(x \le 3)$ (iii) $P(3 \le x \le 6)$

4. A random variable has the following probability function

X	0	1	3	4	5	6	7
P(X)	0	k	2k	2k	3k	$2k^2$	$7k^2 + k$

Determine (i) k (ii) Evaluate $P(x < 5)$, $P(x \ge 5)$ (iii) $P(0 < x < 4)$

5. The joint probability mass function of (X, Y) is given by $P(x, y) = k(3x + 2y)$; $x = 0$, 1, 2 and $y = 1, 2, 3$. Find all the marginal and conditional probability distributions. Also find the probability distribution of $(X + Y)$

6. Random variable X represents a binary communication system i/p, takes on one of two values 0 or 1 with probabilities $\dfrac{1}{4}$ and $\dfrac{3}{4}$ respectively. Due to errors caused

by noise in the system, the random variable Y denotes the *o/p* differs from the input X occasionally. The behavior of the communication system is modeled by the conditional probabilities $P(Y = 1/X = 1) = \dfrac{1}{4}$ and $P(Y = 0/X = 0) = \dfrac{7}{8}$. Determine

(*i*) $P(Y = 1)$ (*ii*) $P(Y = 0)$ (*iii*) $P(X = 1/Y = 1)$

7. The following table represents the joint probability distribution of the discrete random variable X and Y.

Determine the

(a) Marginal distribution of X and Y

(b) Conditional distribution of X given $Y = 3$ and

(c) the conditional distribution of Y given $X=2$.

(d) $P(X \le 2, \ Y = 3)$

(e) $P(Y \le 3)$

(f) $P(X + \ Y \le 5)$

X \\ Y	1	2	3
0	$\dfrac{1}{12}$	$\dfrac{1}{6}$	0
1	0	$\dfrac{1}{9}$	$\dfrac{1}{5}$
2	$\dfrac{1}{18}$	$\dfrac{1}{4}$	$\dfrac{1}{15}$

8. For the following joint distribution of random variables X and Y, find (a) $P(X \le 2, Y \le 4)$ (b) $P(X \le 3/Y \le 3)$ (c) $P(Y \le 3/X \le 3)$ (d) $P(X + Y \le 5)$

X \\ Y	1	2	3	4	5	6
0	0	0	$\dfrac{1}{32}$	$\dfrac{2}{32}$	$\dfrac{2}{32}$	$\dfrac{3}{32}$
1	$\dfrac{1}{16}$	$\dfrac{1}{16}$	$\dfrac{1}{8}$	$\dfrac{1}{8}$	$\dfrac{1}{8}$	$\dfrac{1}{8}$
2	$\dfrac{1}{32}$	$\dfrac{1}{32}$	$\dfrac{1}{64}$	$\dfrac{1}{64}$	0	$\dfrac{2}{64}$

9. Is the function defined as follows a density

$$f(y) = \begin{cases} e^{-y} & \text{if } y \geq 0 \\ 0 & y < 0 \end{cases}$$

If so determine the probability that the variate having this density will fall in the interval (1, 2)? Find the cumulative probability function $F(3)$.

10. Find the constant k such that

$$f(y) = \begin{cases} k\, y^2 & \text{if } 0 \leq y \leq 3 \\ 0 & \text{else where} \end{cases}$$

is a probability density function

(i) Find the distribution function $F(y)$ (ii) $P(1 < Y \leq 2)$

11. The trouble shooting capability of an *IC* chip in a circuit is a random variable Y with distribution function given by

$$F(y) = \begin{cases} 0 & \text{for } y \leq 5 \\ 1 - \dfrac{9}{y^2} & \text{for } y > 5 \end{cases}$$

Where y denote the number of years. Find the probability that the chip will work properly (i) less than 10 years (ii) beyond 10 years   (iii) between 5 to 9 years.

12. Is the function defined by

$$f(x) = \begin{cases} 0 & x < 3 \\ \dfrac{3+x}{18} & \text{if } 2 \leq x \leq 5 \\ 0 & x > 5 \end{cases}$$

A probability density function? Find the probability that a variate having $f(x)$ as density function will fall in the interval $(2 \leq X \leq 3)$?

13. If probability density function

$$f(x) = \begin{cases} k\, x^2 & \text{in } 1 \leq x \leq 4 \\ 0 & \text{else where} \end{cases}$$

Find the value of k and find the probability between $x = \dfrac{3}{2}$ and $x = \dfrac{5}{2}$

14.   A continuous random variable Y has the distribution function

$$F(y) = \begin{cases} 0 & if \ y \leq 1 \\ k(y-1)^3 & if \ 1 \leq y \leq 4 \\ 1 & if \ y > 4 \end{cases}$$

Determine (i) $f(y)$ (ii) k

15.   The joint probability density of the random variables X and Y is

$$f(x,y) = \begin{cases} ke^{-(x+y)} & for \ 0 \leq x \leq \infty, \quad 0 \leq y < \infty \\ 0 & otherwise \end{cases}$$

(a) Find the value of the constant k.

(b) Find the probability density of X independently of Y.

(c) Find the probability $P(0 \leq x \leq 3, 3 \leq y < 5)$ and

Are the random variables independent?

16.   The joint density function of random variables X and Y is $f(x, y) = 8xy$ for $0 < y < 1$ and $0 < x < y$. Find $f(y/x)$ and $f(x/y)$.

17.   The joint Pdf of (X, Y) is given to be

$$f(x,y) = \begin{cases} Ae^{x-y} & for \ 0 \leq x \leq y, \quad 0 \leq y < \infty \\ 0 & otherwise \end{cases}$$

(i) Find A (ii)   Find marginal density of X and Y

(iii)   Examine if X and Y are independent

(iv)   Find the conditional density function of Y given $X=2$.

18.   Let X and Y be jointly continuous random variables with joint Pdf

$$f(x,y) = \begin{cases} y^2 + \dfrac{xy}{2} & for \ 0 \leq x \leq 1, \quad 0 \leq y \leq 2 \\ 0 & otherwise \end{cases}$$

Are X and Y independent and Check $f(x/y)$ and $f(y/x)$ are Pdf's or not.

19.   Determine the constant b such that the function

$$f(x,y) = \begin{cases} xy & if \ 0 < x < 1, \quad 0 < y < b \\ 0 & otherwise \end{cases}$$

is a valid joint density function.

20. The joint probability function of two-dimensional random variables (X, Y) is

$$f(x,y) = \begin{cases} 1 & if\ 0 < x < 1, \quad 0 < y < x \\ 0 & otherwise \end{cases}$$

Determine

(i) The Marginal Density function of X and Y

(ii) The conditional density function of Y given $X = x$

(iii) The conditional density function of X given $Y = y$ and

(iv) Test for independence of X and Y

Additional Questions

***Example* 1:** A random variable has the following probability function

X	0	1	3	4	5	6	7
$P(X)$	0	k	$2k$	$2k$	$3k$	$2k^2$	$7k^2 + k$

Determine

(a)   (i) k (ii) Evaluate $P(x < 6)$, $P(x \geq 6)$ (iii) $P(0 < x < 5)$

(b)   If X is a continuous random variable and $Y = aX + b$ prove that $E(Y) = a E(X) + b$ and $V(y) = a^2 V(X)$

Sol:

(a)   (i)   Since $\sum f(x) = 1$

$\Rightarrow$ $0 + k + 2k + 2k + 3k + 2k^2 + 7k^2 + k = 1$

$\Rightarrow$ $9k^2 + 9k - 1 = 0$ $\Rightarrow k = \dfrac{-3 \pm \sqrt{5}}{6}$

Note: Here k is negative for different values of X probability is negative , which is absurd. Hence this problem has insufficient data.

(b)   If X is a continuous random variable, then by definition

$$E[X] = \int_{-\infty}^{+\infty} x f(x) dx$$

Given

$$Y = a x + b$$

$$E(Y) = E\{a X + b\} = \int_{-\infty}^{+\infty} (aX + b) f(x) dx$$

$$= \int_{-\infty}^{+\infty} a x\, f(x) dx + b \int_{-\infty}^{+\infty} f(x) dx$$

$$= a \int_{-\infty}^{+\infty} x\, f(x) dx + b \int_{-\infty}^{+\infty} f(x) dx$$

$$\because \int_{-\infty}^{+\infty} f(x) dx = 1$$

$$= a E(X) + b$$

and

$$\because V(X) = E\left[X^2\right] - \left[E(X)\right]^2$$

$$\therefore V(Y) = E\left[Y^2\right] - \left[E(Y)\right]^2$$

$$= E\left[(aX+b)^2\right] - \left[E(aX+b))\right]^2$$

$$= E\left[a^2 X^2 + 2abX + b^2\right] - \left[aE(X) + b\right]^2$$

$$= a^2 E[X^2] + 2abE[X] + b^2 - \left\{a^2 E[X]^2 + 2abE[X] + b^2\right\}$$

$$= a^2 E[X^2] + 2abE[X] + b^2 - a^2 E[X]^2 - 2abE[X] - b^2$$

$$= a^2 \left\{E[X^2] - E[X]^2\right\}$$

$$= a^2 V(X)$$

$$\therefore V(y) = a^2 V(X)$$

Example 2: A continuous random variable X has the distribution function

$$F(x) = \begin{cases} 0 & if \ x \leq 1 \\ k(x-1)^4 & if \ 1 \leq x \leq 3 \\ 1 & if \ x > 3 \end{cases}$$

Determine (i) $f(x)$ (ii) k

Sol:

Since

$$f(x) = \frac{d}{dx} F(x),$$

$$\therefore f(x) = \begin{cases} 0 & if \ x \leq 1 \\ 4k(x-1)^3 & if \ 1 \leq x \leq 3 \\ 0 & if \ x > 3 \end{cases}$$

i.e.,

$$f(x) = \begin{cases} 4k(x-1)^3 & if \ 1 \leq x \leq 3 \\ 0 & else \ where \end{cases}$$

Since *f(x)* is a density function, $\int\limits_{-\infty}^{\infty} f(x)dx = 1$ and the range of x is 1 to 3

$$\int\limits_{-\infty}^{1} f(x).dx + \int\limits_{1}^{3} f(x)\,dx = 1$$

i.e., $$\int\limits_{-\infty}^{1} 0.dx + \int\limits_{1}^{3} 4k(x-1)^3\,dx = 1$$

$\Rightarrow$ $$4k \int\limits_{1}^{3} (x-1)^3\,dx = 1$$

$\Rightarrow$ $4k\left[\dfrac{(x-1)^4}{4}\right]_1^3 = 1$ $\Rightarrow$ $k\left[(x-1)^4\right]_1^3 = 1$

$\Rightarrow$ $k\left[(3-1)^4 - (1-1)^4\right] = 1 \Rightarrow k = \dfrac{1}{16}$

$$\therefore f(x) = \begin{cases} \dfrac{1}{4}(x-1)^3 & if \ \ 1 \le x \le 3 \\ 0 & else \ where \end{cases}$$

***Example* 3:** If probability density function

$$f(x) = \begin{cases} k\,x^3 & in \ \ 1 \le x \le 3 \\ 0 & else \ where \end{cases}$$

Find the value of *k* and find the probability between $x = \dfrac{1}{2}$ and $x = \dfrac{3}{2}$

Sol:

Since *f(x)* is a density function, $\int\limits_{-\infty}^{\infty} f(x)dx = 1$ and the range of *x* is 1 to 3

$$\int\limits_{1}^{3} f(x)\,dx = 1$$

i.e., $$\int\limits_{1}^{3} k.x^3\,dx = 1$$

$\Rightarrow \qquad k\left[\dfrac{x^4}{4}\right]_1^3 = 1$

$\Rightarrow \qquad \dfrac{k}{4}\left[x^4\right]_1^3 = 1$

$\Rightarrow \qquad \dfrac{k}{4}\left[3^4 - 1^4\right] = 1$

$\Rightarrow \qquad \dfrac{k}{4}[80] = 1$

$\therefore \qquad k = \dfrac{1}{20}$

$\therefore \qquad f(x) = \begin{cases} \dfrac{1}{20}x^3 & if \ \ 1 \le x \le 3 \\ 0 & else \ where \end{cases}$

$P\left(\dfrac{1}{2} \le x \le \dfrac{3}{2}\right):$

$P\left(\dfrac{1}{2} \le x \le \dfrac{3}{2}\right) = \dfrac{1}{20}\int_{\frac{1}{2}}^{\frac{3}{2}} x^3 \, dx = \dfrac{1}{20}\left[\dfrac{x^4}{4}\right]_{\frac{1}{2}}^{\frac{3}{2}}$

$= \dfrac{1}{80}\left[x^4\right]_{\frac{1}{2}}^{\frac{3}{2}} = \dfrac{1}{80}\left[\left(\dfrac{3}{2}\right)^4 - \left(\dfrac{1}{2}\right)^4\right] = \dfrac{1}{16}$

Example 4: A random variable X has the following probability distribution

x	1	2	3	4	5	6	7	8
f	k	2k	3k	4k	5k	6k	7k	8k

Find the value of (i) k (ii) $P(x \le 2)$ (iii) $P(2 \le x \le 5)$

Sol:

(i)   Since $\sum f(x) = 1$

$\Rightarrow \qquad k + 2k + 3k + 4k + 5k + 6k + 7k + 8k = 1$

$\Rightarrow \qquad 36k = 1 \quad \Rightarrow k = \dfrac{1}{36}$

$$\therefore \qquad k = \frac{1}{36}$$

Therefore the probability distribution of the random variable X is

x	1	2	3	4	5	6	7	8
f	$\dfrac{1}{36}$	$\dfrac{2}{36}$	$\dfrac{3}{36}$	$\dfrac{4}{36}$	$\dfrac{5}{36}$	$\dfrac{6}{36}$	$\dfrac{7}{36}$	$\dfrac{8}{36}$

Now

(ii) $\qquad P(X \le 2) = P(X = 1) + P(X = 2)$

$$= \frac{1}{36} + \frac{2}{36} = \frac{1}{12}$$

(iii) $\quad P(2 \le x \le 5) = P(X = 2) + P(X = 3) + P(X = 4) + P(X = 5)$

$$= \frac{2}{36} + \frac{3}{36} + \frac{4}{36} + \frac{5}{36} = \frac{7}{18}$$

Example 5: Find the constant k such that

$$f(x) = \begin{cases} k\,x^2 & if \ \ 0 \le x \le 3 \\ 0 & else \ where \end{cases}$$

is a probability density function

(i) Find the distribution function $F(x)$

(ii) $P(1 < X \le 2)$

Sol: Since $f(x)$ is a density function, $\displaystyle\int_{-\infty}^{\infty} f(x)dx = 1$ and the range of X is 0 to 3

$$\int_{0}^{3} f(x)\, dx = 1$$

i.e., $\qquad \displaystyle\int_{0}^{3} k.x^2\, dx = 1$

$\Rightarrow \qquad k\left[\dfrac{x^3}{3}\right]_{0}^{3} = 1$

$\Rightarrow \qquad \dfrac{k}{3}\left[x^3\right]_{0}^{3} = 1$

$$\Rightarrow \qquad \frac{k}{3}\left[3^3\right] = 1$$

$$\therefore \qquad k = \frac{1}{9}$$

$$\therefore \qquad f(x) = \begin{cases} \dfrac{1}{9}x^2 & if \;\; 0 \le x \le 3 \\ 0 & else\;where \end{cases}$$

(ii)   The distribution function F(x) is

$$F(x) = P(X \le x) = \int_{-\infty}^{x} f(x)dx = \int_{0}^{x}\frac{1}{9}x^2 dx = \frac{1}{9}\int_{0}^{x} x^2 dx$$

$$= \frac{1}{9}\left(\frac{x^3}{3}\right)_{0}^{x} = \frac{1}{9}\left(\frac{x^3}{3}-0\right)_{0}^{x} = \frac{x^3}{27}$$

$$\therefore \qquad F(x) = \begin{cases} 0 & if \;\; x < 0 \\ \dfrac{1}{27}x^3 & if \;\; 0 \le x \le 3 \\ 1 & x > 3 \end{cases}$$

Since the relation between the distribution function $F(x)$ and the density function $f(x)$ is

$$\frac{d}{dx}F(x) = f(x)$$

Example 6: Is the function defined as follows a density

$$f(x) = \begin{cases} e^{-x} & if \; x \ge 0 \\ 0 & x < 0 \end{cases}$$

If so determine the probability that the variate having this density will fall in the interval (1,2)? Find the cumulative probability function $F(2)$.

Sol: The probability density function $f(x)$ is

$$f(x) = \begin{cases} e^{-x} & if\, x \succ 0 \\ 0 & elsewhere \end{cases}$$

If it is density function, then it satisfies $\displaystyle\int_{-\infty}^{+\infty} f(x)\, dx = 1$

Now
$$\int\limits_{-\infty}^{+\infty} f(x)\, dx = \int\limits_{-\infty}^{0} f(x)\, dx + \int\limits_{0}^{\infty} f(x)\, dx$$

$$= \int\limits_{-\infty}^{0} 0\, dx + \int\limits_{0}^{\infty} e^{-x}\, dx$$

$$= -\left[e^{-x} \right]_{0}^{\infty}$$

$$= -\left[(0-1)\right] = 1$$

Hence $f(x) = \begin{cases} e^{-x} & if\, x > 0 \\ 0 & elsewhere \end{cases}$ is a density function

Now $P(1 < X < 2) = \int\limits_{1}^{2} f(x)\, dx$

$$= \int\limits_{1}^{2} e^{-x}\, dx$$

$$= -\left[e^{-x} \right]_{1}^{2} = -\{e^{-2} - e^{-1}\}$$

$$= e^{-1} - e^{-2}$$

$$= 0.3678 - 0.1353$$

$$= 0.2325$$

And the cumulative probability function $F(2)$ is

$$F(2) = P(X \le 2) = \int\limits_{0}^{2} f(x)\, dx$$

$$= \int\limits_{0}^{2} e^{-x}\, dx$$

$$= -\left[e^{-x} \right]_{0}^{2} = -\{e^{-2} - e^{0}\}$$

$$= 1 - e^{-2}$$

$$= 1 - 0.1353$$

$$= 0.8647$$

***Example* 7:** A random variable X has the following distribution

Find (i) $P(X < 4)$, $P(X \geq 5)$, $P(3 < X \leq 6)$ and (ii) What will be the minimum value of K so that $P(X \leq 2) > 0.3$

Sol: Since $\sum P(x) = 1$

$\Rightarrow$ $k + 3k + 5k + 7k + 9k + 11k + 13k = 1$

$\Rightarrow$ $49k = 1$

$\therefore$ $k = \dfrac{1}{49}$

The distribution of the random variable X is

$zi)$ $P(X < 4)$:

$P(X < 4) = P(X = 0) + P(X = 1) + P(X = 2) + P(X = 3)$

$= \dfrac{1}{49} + \dfrac{3}{49} + \dfrac{5}{49} + \dfrac{7}{49}$

$= \dfrac{16}{49}$

$= 0.3265$

(iii) $P(X \geq 5) = P(X = 5) + P(X = 6)$

$= \dfrac{11}{49} + \dfrac{13}{49}$

$= \dfrac{24}{49}$

$= 0.4898$

(ii) $P(3 < X \leq 6)$:

$P(3 < X \leq 6) = P(X = 4) + P(X = 5) + P(X = 6)$

$= \dfrac{9}{49} + \dfrac{11}{49} + \dfrac{13}{49}$

$= \dfrac{33}{49}$

$= 0.6734$

(iv)   The minimum value of X such that $P(X \leq x) > 0.3)$ is

The last row is $P(X \leq x)$

i.e., $P(X \le 0) = P(X=0)$

$$= \frac{1}{49} = 0.020$$

$P(X \le 1) = P(X=0) + P(X=1)$

$$= \frac{4}{49} = 0.081$$

$P(X \le 2) = P(X=0) + P(X=1) + P(X=2)$

$$= \frac{9}{49} = 0.1836$$

$P(X \le 3) = P(X=0) + P(X=1) + P(X=2) + P(X=3)$

$$= \frac{16}{49} = 0.3265$$

$P(X \le 4) = P(X=0) + P(X=1) + P(X=2) + P(X=3) + P(X=4)$

$$= \frac{25}{49} = 0.5102$$

$P(X \le 5) = P(X=0) + P(X=1) + P(X=2) + P(X=3) + P(X=4) + P(X=5)$

$$= \frac{36}{49} = 0.7346$$

$P(X \le 6) = P(X=0) + P(X=1) +$
$P(X=2) + P(X=3) + P(X=4) + P(X=5) + P(X=6)$

$$= \frac{49}{49} = 1$$

Therefore the minimum value of X such that $P(X \le x) > 0.3$ is 3

$\therefore$ $P(X \le 3) > 0.3$ is 3

Example 8: Is the function defined by

$$f(x) = \begin{cases} 0 & x > 2 \\ \dfrac{3+2x}{18} & if \ 2 \le x \le 4 \\ 0 & x > 4 \end{cases}$$

A probability density function? Find the probability that a variate having $f(x)$ as density function will fall in the interval $(2 \le X \le 3)$?

Sol: Given $f(x)$ is

$$f(x) = \begin{cases} 0 & x < 2 \\ \dfrac{3+2x}{18} & \text{if } 2 \le x \le 4 \\ 0 & x > 4 \end{cases}$$

If it is density function, then it satisfies $\displaystyle\int_{-\infty}^{+\infty} f(x)\,dx = 1$

Now $\displaystyle\int_{-\infty}^{+\infty} f(x)\,dx = \int_{-\infty}^{2} f(x)\,dx + \int_{2}^{4} f(x)\,dx + \int_{4}^{\infty} f(x)\,dx$

$$= \int_{-\infty}^{0} 0\,dx + \int_{2}^{4} \frac{3+2x}{18}\,dx + \int_{4}^{\infty} 0\,dx$$

$$= 0 + \frac{3}{18}[x]_2^4 + \frac{2}{18}\left[\frac{x^2}{2}\right]_2^4 + 0$$

$$= \frac{1}{6}[4-2] + \frac{1}{18}\left[4^2 - 2^2\right]$$

$$= \frac{1}{3} + \frac{2}{3} = 1 \quad \therefore \int_{-\infty}^{+\infty} f(x)\,dx = 1$$

Hence $f(x) = \begin{cases} 0 & x < 2 \\ \dfrac{3+2x}{18} & \text{if } 2 \le x \le 4 \\ 0 & x > 4 \end{cases}$ is a density function

Now $P(2 \le X \le 3) = \displaystyle\int_{2}^{3} f(x)\,dx$

$$= \int_{2}^{3} \frac{3+2x}{18}\,dx$$

$$= \frac{3}{18}[x]_2^3 + \frac{2}{18}\left[\frac{x^2}{2}\right]_2^3$$

$$= \frac{1}{6}[3-2] + \frac{1}{18}\left[3^2 - 2^2\right]$$

$$= \frac{1}{6} + \frac{5}{18} = \frac{8}{18} = \frac{4}{9} = 0.44$$

***Example* 9:** The trouble shooting capability of an *IC* chip in a circuit is a random variable *X* with distribution function given by

$$f(x) = \begin{cases} 0 & for \ x \le 3 \\ 1 - \dfrac{9}{x^2} & for \ x > 3 \end{cases}$$

Where *x* denote the number of years. Find the probability that the chip will work properly (i) less than 8 years (ii) beyond 8 years   (iii) between 5 to 7 years.

Sol: Given the distribution function *F(x)* is

$$f(x) = \begin{cases} 0 & for \ x \le 3 \\ 1 - \dfrac{9}{x^2} & for \ x > 3 \end{cases}$$

Since $f(x) = \dfrac{d}{dx} f(x),$

$$\therefore f(x) = \begin{cases} 0 & for \ x \le 3 \\ \dfrac{18}{x^3} & for \ x > 3 \end{cases}$$

(i)   The probability that the chip will work properly less than 8 years is

$$p(X \le 8) = \int_{3}^{8} f(x)dx$$

$$= \int_{3}^{8} \frac{18}{x^3}dx = 18\int_{3}^{8}\frac{1}{x^3}dx$$

$$= 18\left[\frac{x^{-2}}{-2}\right]_{3}^{8} = -9\left[\frac{1}{x^2}\right]_{3}^{8}$$

$$= -9\left[\frac{1}{8^2} - \frac{1}{3^2}\right] = -9\left[\frac{-55}{576}\right] = 0.86$$

(ii)   The probability that the chip will work properly beyond 8 years is

$$P(X \ge 8) = \int_{\infty}^{\infty} f(x)dx$$

$$= \int_8^\infty \frac{18}{x^3} dx = 18 \int_8^\infty \frac{1}{x^3} dx$$

$$= 18 \left[\frac{x^{-2}}{-2} \right]_3^8 = -9 \left[\frac{1}{x^2} \right]_8^\infty$$

$$= -9 \left[\frac{1}{\infty} - \frac{1}{8^2} \right] = -9 \left[\frac{-1}{64} \right] = 0.141$$

(iii)   The probability that the chip will work properly between 5 to 7 years is

$$P\big((5 < X < 7)\big) = \int_5^7 f(x) dx = \int_5^7 \frac{18}{x^3} dx = 18 \int_5^7 \frac{1}{x^3} dx$$

$$= 18 \left[\frac{x^{-2}}{-2} \right]_5^7 = -9 \left[\frac{1}{x^2} \right]_5^7 = -9 \left[\frac{1}{7^2} - \frac{1}{5^2} \right] = -9 \left[\frac{-24}{1225} \right]$$

$$= 0.1763$$

CHAPTER 5

Operations on Multiple Random Variables

5.1 Introduction

This unit is an extension of some of the topics in unit there statistical converges of a single random variable to multiple random variables. Here we discuss the concept of expectation extended to multiple random variables other operations also applied to there multiple random variables.

5.2 Expected Value of a Function of a Random Variable

Let X be a discrete random variable, assumes a discrete set of values $x_1, x_2, x_3, \cdots x_N$ with probabilities $P(x_1),\ P(x_2), P(x_3), \cdots P(x_N)$ respectively. Then the probability density function of X is

$$f_X(x) = \sum_{i=1}^{N} P(x_i)\delta(x - x_i)$$

Now the expected value of a real function g(.) of X is

$$\text{Mean} = E\big[g(x)\big] = \begin{cases} \int\limits_{-\infty}^{\infty} g(x) f_X(x)\, dx & \text{for } continuous \\ \sum g(x)\, P(x) & \text{for } discrete \end{cases}$$

is the expected value of a function of a random variable

Examples

***Example* 1:** If X and Y are independent random variables. The show that $E(XY) = E(X)$. $E(Y)$ and $E[g_1(x) . g_2(y)] = E[g_1(x)] . E[g_2(y)]$.

Sol: Since given that X and Y are independent random variables

$$E(XY) = \int\limits_{-\infty}^{\infty} \int\limits_{-\infty}^{\infty} xy \ f_{xy}(x,y)dxdy$$

$$= \int\limits_{-\infty}^{\infty} \int\limits_{-\infty}^{\infty} xy \ f_x(x) f_y(y) \ dx \ dy \qquad \therefore f_{xy}(x,y) = f_x(x).f_y(y).$$

$$= \int\limits_{-\infty}^{\infty} xy \ f_x(x)dx \int\limits_{-\infty}^{\infty} y. f_y(y)dy$$

$$= E(X).E(Y)$$

Also

$$E\big[g_1(x).g_2(y)\big] = \int\limits_{-\infty}^{\infty} \int\limits_{-\infty}^{\infty} g_1(y).g_2(x).f_{xy}(x,y)dxdy$$

$$= \int\limits_{-\infty}^{\infty} g_1(x) \ g_2(y).f_x(x)f_y(y)dxdy$$

$$= \int\limits_{-\infty}^{\infty} g_1(x) f_x(x)dx \int\limits_{-\infty}^{\infty} g_2(x).f_y(y)dy$$

$$= E[g_1(X)].E[g_2(Y)]$$

Similarly, we can also show that the above statement is true for X and Y are discrete.

***Example* 2:** Let X and Y be independent random variables, is each having density function

$$f_x(x) = \begin{cases} 2e^{-2x} & for \ x \ge 0 \\ 0 & else \ where \end{cases}$$

Find (a) $E(X + Y)$ (b) $E(X^2 + Y^2)$ (c) $E(XY)$.

Sol: Given $$f_x(x) = \begin{cases} 2e^{-2x} & for \ x \ge 0 \\ 0 & else \ where \end{cases}$$

Now $$E(X + Y) = \int\limits_{-\infty}^{\infty} \int\limits_{-\infty}^{\infty} (x+y)f_{xy}(x,y)dxdy$$

$$= \int\limits_{-\infty}^{\infty} \int\limits_{-\infty}^{\infty} (x+y)f_x(x) \ f_y(y) \ dxdy$$

Since $f_{xy}(x, y) = f_x(x) f_y(y)$, also $f_x(x) = 2e^{-ex}$ and $f_y(y) = 2e^{-ey}$

$$= \int_0^\infty \int_0^\infty x \ f_x(x) f_y(y) dx dy + \int_0^\infty \int_0^\infty y \ f_x(x) \ f_y(y) dx dy$$

$$= \int_0^\infty \int_0^\infty x \ 2e^{-2x} . 2e^{-2y} dx dy + \int_0^\infty \int_0^\infty y \ 2e^{-2y} . 2e^{-2x} \ dx dy .$$

$$= \int_0^\infty 2x \ e^{-2x} dx . \int_0^\infty 2e^{-2y} \ dy + \int_0^\infty 2y \ e^{-2y} dy . \int_0^\infty 2e^{-2x} dx.$$

Since we have $\int_0^\infty 2e^{-2x} dx = \int_0^\infty 2e^{-2y} dy = 1$

$$\therefore \quad E(X + Y) = 2 \int_0^\infty xe^{-2x} dx + 2 \int_0^\infty ye^{-2y} dy .$$

$$= 2 \left[x . \frac{e^{-2x}}{-2} - 1 . \frac{e^{-2x}}{(-2)^2} + y . \frac{e^{-2y}}{(-2)} - -1 . \frac{e^{-2y}}{(-2)^2} \right]_0^\infty$$

$$= 2 \left[(0-0) \ \frac{1}{4}(0-1) + (0-0) - \frac{1}{4}(0-1) \right]$$

$$= 2 \left(\frac{1}{4} + \frac{1}{4} \right)$$

$$= 1$$

$$E(X^2 + Y^2) = \int_{-\infty}^\infty \int_{-\infty}^\infty (x^2 + y^2) . 2e^{-2x} . 2e^{-2y} dx dy$$

$$= \int_{-\infty}^\infty \int_{-\infty}^\infty 2x^2 e^{-2x} dx dy + \int_{-\infty}^\infty \int_{-\infty}^\infty 2y^2 e^{-2x} . 2e^{-2y} \ dx dy .$$

$$= 2 \int_0^\infty x^2 e^{-2x} \ dx + 2 \int_0^\infty y_2 \ e^{-2y} dy$$

$$= 1$$

$$E(XY) = \int_0^\infty \int_0^\infty x \ y \ 2e^{-x} \ 2e^{-xy} \ dy$$

$$= \int_0^\infty 2x \ e^{-2x} dx \times \int_0^\infty 2y \ e^{-2y} dy$$

$$= 4\left[\left\{x.\frac{e^{-2x}}{-2} - 1.\frac{e^{-2x}}{(-2)^2}\right\} \times \left\{y.\frac{e^{-2y}}{(-2)} - -1.\frac{e^{-2y}}{(-2)^2}\right\}\right]_0^\infty$$

$$= 4\left[\left\{(0-0) - \frac{1}{4}(0-1)\right\} \times \left\{(0-0) - \frac{1}{4}(0-1)\right\}\right]$$

$$= 4\left(\frac{1}{4} \times \frac{1}{4}\right)$$

$$= 4\left(\frac{1}{16}\right)$$

$$= \frac{1}{4}$$

Example 3: Let X and Y have joint density function

$$f_{xy}(x,y) = \begin{cases} x+y & for\ 0 \le x \le 1; 0 \le y \le 1 \\ 0 & otherwise \end{cases}$$

Find the conditional expectation of a) Y given X and b) X given Y

Sol: Given that the joint density function of X and Y is

$$f_{xy}(x,y) = \begin{cases} x+y & for\ 0 \le x \le 1; 0 \le y \le 1 \\ 0 & otherwise \end{cases}$$

(a) $E(Y/X)$:

Since $E(Y/X) = \displaystyle\int_{-\infty}^{\infty} y.f(y/x)dy$

Also since $f(y/x) = \dfrac{f_{xy}(x,y)}{f_x(x)}$

and $f_x(x) = \displaystyle\int_0^1 f_{xy}(x/y)dy$

$$= \int_0^1 (x+y)dy$$

$$= \left(xy + \frac{y^2}{2}\right)_0^1$$

$$= x + \frac{1}{2}$$

$$\therefore f(y/x) = \frac{x+y}{x+\frac{1}{2}}.$$

Now $E(Y/X)$ $= \int\limits_0^1 y \cdot \frac{x+y}{x+\frac{1}{2}} dy$

$$= \frac{1}{x+\frac{1}{2}} \int\limits_0^1 y(x+y)dy$$

$$= \frac{1}{x+\frac{1}{2}} \int\limits_0^1 (xy+y^2)dy.$$

$$= \frac{1}{x+\frac{1}{2}} \left(\frac{xy^2}{2} + \frac{y^3}{3} \right)\Bigg|_0^1$$

$$= \frac{2}{2x+1} \left(\frac{x}{2} + \frac{1}{3} \right)$$

$$= \frac{2}{2x+1} \left(\frac{3x+2}{6} \right)$$

$$= \frac{3x+2}{3(2x+1)}$$

Therefore $E(Y/X) = \dfrac{3x+2}{3(2x+1)}$

And now

(b) $E(X/Y)$:

$$E(X/Y) = \int\limits_0^\infty x f(x/y)dx$$

since $\qquad f(x/y) = \dfrac{f_{xy}(x,y)}{f_y(y)}$

And $\qquad f_y(y) = \displaystyle\int_0^1 f_{xy}(x,y)dx$

$$= \int_0^1 (x+y)dx$$

$$= \left(\frac{x^2}{2} + xy\right)\Bigg|_0^1$$

$$= y + \frac{1}{2}$$

Proceeding in the above manner, after simplification, we get

$$E(x/y) = \frac{3y+2}{3(2y+1)}.$$

5.3 Joint Moments about the Origin

Let X and Y be two random variables is related as

$$g(X,Y) = X^i Y^j \quad i,j = 1,2,\cdots$$

The $(i+j)^{th}$ order joint moment of two dimensional random variable defined as

$$\begin{aligned}
\mu_{ij} \quad &= E\big[g(X,Y)\big] \\
&= E\big[X^i Y^j\big] \\
&= \begin{cases} \displaystyle\sum\sum x^i y^j f_{XY}(x,y) & \text{if } X \text{ and } Y \text{ discrete} \\ \displaystyle\int_{-\infty}^{\infty}\int_{-\infty}^{\infty} x^i y^j f_{XY}(x,y)dxdy & \text{if } X \text{ and } Y \text{ continuous} \end{cases}
\end{aligned}$$

$\mu_{0j}{}' = E\big[X^0 Y^j\big] = E\big[Y^j\big]$ this is the Jth order moment of Y and $\mu_{i0}{}'^{th}$ $= E\big[X^i Y^0\big] = E\big[X^i\big]$ is the ith order moment of X.

The first order moments $\mu_{01}{}'$ and $\mu_{10}{}'^{th}$ are the expected values of Y and X respectively, and are the coordinates of the "centre of gravity" of the function $f_{XY}(x,y)$.

The second order moments $\mu_{11}' = E[XY]$ is called the correlation between X and Y and is denoted by R_{XY}.

$$R_{XY} = \mu_{11}' = E[XY]$$

$$= \int\limits_{-\infty}^{\infty} \int\limits_{-\infty}^{\infty} xyf_{XY}(x,y)dxdy \quad \text{if } X \text{ and } Y \text{ continuous}$$

If X and Y are independent then we say that they are uncorrelated then

$$R_{XY} = \mu_{11}' = E[X] \times E[Y]$$

$$\therefore f_{XY}(x,y) = f_X(x) \times f_Y(y) \ X \text{ and } Y \text{ are independent}$$

Extension

For N random variables $X_1, X_2 \cdots X_N$ then the $(n_1 + n_2 + \cdots + n_N)^{th}$ order joint moment

$$\mu_{n_1, n_2 \cdots n_N}' = E\left[X_1^{n_1} X_2^{n_2} \cdots X_N^{n_N}\right]$$

$$= \int\limits_{-\infty}^{\infty} \int\limits_{-\infty}^{\infty} \cdots \int\limits_{-\infty}^{\infty} X_1^{n_1} X_2^{n_2} \cdots X_N^{n_N} f(x_1, x_2 \cdots x_N)dx_1 dx_2 \cdots dx_N$$

where $n_1, n_2 \cdots n_N$ are all integers, i.e., $n_1, n_2 \cdots n_N = 0, 1, 2 \cdots$

The second order moments are

$$\mu_{20}' = E\left[X^2 Y^0\right] = E\left[X^2\right]$$

$$\mu_{02}' = E\left[X^0 Y^2\right] = E\left[Y^2\right]$$

It can be shown that the ratio of μ_{11}' and the square root of the product of second order moments μ_{20}' and μ_{02}' i.e., $\dfrac{\mu_{11}'}{\sqrt{\mu_{20}' \mu_{02}'}}$ gives the correlation coefficient and here μ_{11}' is the covariance between X and Y.

5.4   Joint Central Moments

Let X and Y be two random variables, then the $(i + j)^{th}$ order central moments of two dimensional random variable is given by

$$\mu_{ij} = E\left[\left(X - \overline{X}\right)^i \left(Y - \overline{Y}\right)^j\right]$$

$$= \begin{cases} \sum\sum (x-\bar{x})^i (y-\bar{y})^j f_{XY}(x,y) & \text{if } X \text{ and } Y \text{ discrete} \\[2ex] \int\limits_{-\infty}^{\infty}\int\limits_{-\infty}^{\infty} (x-\bar{x})^i (y-\bar{y})^j f_{XY}(x,y)dxdy & \text{if } X \text{ and } Y \text{ continuous} \end{cases}$$

The second order central moments are

$$\mu_{20} = E\left[\left(X-\bar{X}\right)^2 \left(Y-\bar{Y}\right)^0\right] = E\left[\left(X-\bar{X}\right)^2\right] \qquad .....(5.1)$$

This is the variance of the random variables $X = \sigma_x^2$

$$\mu_{02} = E\left[\left(X-\bar{X}\right)^0 \left(Y-\bar{Y}\right)^2\right] = E\left[\left(Y-\bar{Y}\right)^2\right] \qquad .....(5.2)$$

This is the variance of the random variables $Y = \sigma_y^2$

$$\mu_{11} = E\left[\left(X-\bar{X}\right)\left(Y-\bar{Y}\right)\right] = \int\limits_{-\infty}^{\infty}\int\limits_{-\infty}^{\infty}(x-\bar{x})(y-\bar{y})f_{XY}(x,y)dxdy \qquad .....(5.3)$$

This is called the covariance between the random variables X and Y

Hence $\quad C_{XY} = \mu_{11} = E\left[\left(X-\bar{X}\right)\left(Y-\bar{Y}\right)\right]$

Since from equation (1)

$$\mu_{20} = E\left[\left(X-\bar{X}\right)^2\right]$$
$$= E\left[\left(X^2 - 2X\bar{X} + \bar{X}^2\right)\right]$$
$$= \left[E\left(X^2\right) - 2\bar{X}E(X) + E\left(\bar{X}\right)^2\right]$$
$$= \left[E\left(X^2\right) - 2\bar{X}^2 + \bar{X}^2\right]$$
$$= \left[E\left(X^2\right) - \bar{X}^2\right]$$
$$= \left[E\left(X^2\right) - \{E(X)\}^2\right]$$
$$= \mu_{20}' - \mu_{10}' \qquad .....(5.4)$$
$$= \sigma_x^2$$

Similarly from equation (5.2)

$$\mu_{20} = \mu_{02}{}' - \mu_{01}{}'$$ .....(5.5)

And from equation (5.3)

$$\mu_{11} = C_{XY} = R_{XY} - E(X)\,E(Y)$$ .....(5.6)

Since

$$\mu_{11} = E\left[\left(X - \overline{X}\right)\left(Y - \overline{Y}\right)\right]$$

$$= E\left[\left(XY - X\overline{Y} - \overline{X}Y + \overline{X}\overline{Y}\right)\right]$$

$$= \left[E(XY) - \overline{Y}E(X) - \overline{X}E(Y) + E\left(\overline{X}\overline{Y}\right)\right]$$

$$= \left[E(XY) - 2\overline{X}\overline{Y} + \overline{X}\overline{Y}\right]$$

$$= \left[E(XY) - \overline{X}\overline{Y}\right]$$

$$= \left[E(XY) - E(X)E(Y)\right]$$

$$= R_{XY} - E(X)E(Y)$$

$$= C_{XY}$$

If X and Y are independent then $E[XY] = E[X] \times E[Y]$

then $\qquad\qquad \mu_{11} = C_{XY} = 0$

If X and Y are orthogonal random variables then $C_{XY} = E(X)E(Y)$

If X and Y are correlated then the correlation coefficient between X and Y is denoted by ρ

$$\rho = \frac{Cov(X,Y)}{\sigma_x \sigma_y}$$

where

$$Cov(X,Y) = E[XY] - E[X].E[Y]$$

$$\sigma_x^2 = E\left[X^2\right] - \left[E(X)\right]^2 \quad \text{and} \quad \sigma_y^2 = E\left[Y^2\right] - \left[E(Y)\right]^2$$

The correlation coefficient ρ lies between $-1 \le \rho \le 1$

If $\rho = -1$ then the correlation is said to be perfect and is negative correlation

If $\rho = +1$ then the correlation is said to be perfect and is positive correlation

If $\rho = 0$ then the two variables X and Y are is said to be uncorrelated

Extension

For N random variables $X_1, X_2 \cdots X_N$ then the $(n_1 + n_2 + \cdots + n_N)^{th}$ order joint central moment

$$\mu_{m_1, n_2 \cdots n_N} = E\left[\left(X_1 - \overline{X}_1\right)^{n_1} \left(X_2 - \overline{X}_2\right)^{n_2} \cdots \left(X_N - \overline{X}_N\right)^{n_N}\right]$$

$$= \int_{-\infty}^{\infty}\int_{-\infty}^{\infty} \cdots \int_{-\infty}^{\infty} \left(X_1 - \overline{X}_1\right)^{n_1} \left(X_2 - \overline{X}_2\right)^{n_2} \cdots \left(X_N - \overline{X}_N\right)^{n_N} f(x_1, x_2 \cdots x_N) dx_1 dx_2 \cdots dx_N$$

where $n_1, n_2 \cdots n_N$ are all integers, i.e., $n_1, n_2 \cdots n_N = 0, 1, 2 \cdots$

5.5   Some Related Theorems

1. Let X and Y be two independent random variables, then Covariance between X and Y is zero.

   i.e., $Cov(X,Y)=0$

2. $$V(X+Y) = E\left[(X+Y)^2\right] - \left[E(X+Y)\right]^2$$

$$= E\left[\left(X^2 + Y^2 + 2XY\right)\right] - \left[E(X) + E(Y)\right]^2$$

$$= E\left[X^2\right] + E\left[Y^2\right] + 2E[XY] - [E(X)]^2 + [E(Y)]^2 - 2E(X)E(Y)$$

$$= E\left[X^2\right] - [E(X)]^2 + E\left[Y^2\right] - [E(Y)]^2 + 2\{E[XY] - E(X)E(Y)\}$$

$$= V(X) + V(Y) - 2Cov(X,Y)   \quad \ldots(1)$$

If X and Y be two independent random variables, then $V(X+Y) = V(X) + V(Y)$ since Covariance between X and Y is zero.

Similarly,

$$V(X-Y) = V(X) + V(Y)$$

If X and Y are independent random variables then

$$V(X+Y) = V(X-Y) = V(X) + V(Y)$$

Examples

***Example* 1:**   Let X and Y be independent random variables such that

$$X = \begin{cases} 1 & \text{with probability } \dfrac{1}{3} \\ 0 & \text{with probability } \dfrac{2}{3} \end{cases}$$

and

$$Y = \begin{cases} 2 & \text{with probability } \dfrac{3}{4} \\ -3 & \text{with probability } \dfrac{1}{4} \end{cases}$$

then find

(a) $E[3X + 2Y]$ (b) $E\left[2X^2 - Y^2\right]$

(c) $E[XY]$ and   (d) $E\left[X^2Y\right]$

Sol:

(a)   We shall find $E[3X + 2Y]$

Let $Z = 3X + 2Y$, then from the following table we find the value of z for various values of X and Y and the probability distribution of Z is

X	Y	$Z = 3X + 2Y$	$P(Z = z)$
1	2	7	$P(X = 1, Y = 2) = P(X = 1) * P(Y = 2) = \dfrac{1}{3} * \dfrac{3}{4} = \dfrac{1}{4}$
1	-3	-3	$P(X = 1, Y = -3) = P(X = 1) * P(Y = -3) = \dfrac{1}{3} * \dfrac{1}{4} = \dfrac{1}{12}$
0	2	4	$P(X = 0, Y = 2) = P(X = 0) * P(Y = 2) = \dfrac{2}{3} * \dfrac{3}{4} = \dfrac{1}{2}$
0	-3	-6	$P(X = 0, Y = -3) = P(X = 0) * P(Y = -3) = \dfrac{2}{3} * \dfrac{1}{4} = \dfrac{1}{6}$

$$E[Z] = \sum zP(Z = z) = 7 \times \frac{1}{4} + (-3) \times \frac{1}{12} + 4 \times \frac{1}{2} + (-6) \times \frac{1}{6} = \frac{5}{2}$$

Hence $E[3X + 2Y] = \dfrac{5}{2}$

(b)   We shall find $E\left[2X^2 - Y^2\right]$

Let $Z = 2X^2 - Y^2$, then from the following table we find the value of z for various values of X and Y and the probability distribution of Z is

X	Y	$Z = 2X^2 - Y^2$	$P(Z = z)$
1	2	-2	$P(X = 1, Y = 2) = P(X = 1) * P(Y = 2) = \dfrac{1}{3} * \dfrac{3}{4} = \dfrac{1}{4}$
1	-3	-7	$P(X = 1, Y = -3) = P(X = 1) * P(Y = -3) = \dfrac{1}{3} * \dfrac{1}{4} = \dfrac{1}{12}$
0	2	-4	$P(X = 0, Y = 2) = P(X = 0) * P(Y = 2) = \dfrac{2}{3} * \dfrac{3}{4} = \dfrac{1}{2}$
0	-3	-9	$P(X = 0, Y = -3) = P(X = 0) * P(Y = -3) = \dfrac{2}{3} * \dfrac{1}{4} = \dfrac{1}{6}$

Now $E[Z] = \sum zP(Z = z)$ $(-2) \times \dfrac{1}{4} + (-7) \times \dfrac{1}{12} + (-4) \times \dfrac{1}{2} + (-9) \times \dfrac{1}{6} = \dfrac{55}{12}$

Hence $E\left[2X^2 - Y^2\right] = \dfrac{55}{12}$

(c)   We shall find $E[XY]$

Let $Z = XY$, then from the following table we find the value of z for various values of X and Y and the probability distribution of Z is

X	Y	$Z = XY$	$P(Z = z)$
1	2	2	$P(X = 1, Y = 2) = P(X = 1) * P(Y = 2) = \dfrac{1}{3} * \dfrac{3}{4} = \dfrac{1}{4}$
1	-3	-3	$P(X = 1, Y = -3) = P(X = 1) * P(Y = -3) = \dfrac{1}{3} * \dfrac{1}{4} = \dfrac{1}{12}$
0	2	0	$P(X = 0, Y = 2) = P(X = 0) * P(Y = 2) = \dfrac{2}{3} * \dfrac{3}{4} = \dfrac{1}{2}$
0	-3	0	$P(X = 0, Y = -3) = P(X = 0) * P(Y = -3) = \dfrac{2}{3} * \dfrac{1}{4} = \dfrac{1}{6}$

Now $E[Z] = \sum z P(Z = z) \ = (2) \times \dfrac{1}{4} + (-3) \times \dfrac{1}{12} + (0) \times \dfrac{1}{2} + (0) \times \dfrac{1}{6} \ = \dfrac{1}{4}$

Hence $E[XY] = \dfrac{1}{4}$

(d)   We shall find $E\left[X^2Y\right]$

Let $Z = X^2Y$, then from the following table we find the value of z for various values of X and Y and the probability distribution of Z is

X	Y	$Z = X^2Y$	$P(Z = z)$
1	2	2	$P(X = 1, Y = 2) = P(X = 1) * P(Y = 2) = \dfrac{1}{3} * \dfrac{3}{4} = \dfrac{1}{4}$
1	−3	−3	$P(X = 1, Y = -3) = P(X = 1) * P(Y = -3) = \dfrac{1}{3} * \dfrac{1}{4} = \dfrac{1}{12}$
0	2	0	$P(X = 0, Y = 2) = P(X = 0) * P(Y = 2) = \dfrac{2}{3} * \dfrac{3}{4} = \dfrac{1}{2}$
0	−3	0	$P(X = 0, Y = -3) = P(X = 0) * P(Y = -3) = \dfrac{2}{3} * \dfrac{1}{4} = \dfrac{1}{6}$

Now $E[Z] = \sum z P(Z = z) \ = (2) \times \dfrac{1}{4} + (-3) \times \dfrac{1}{12} + (0) \times \dfrac{1}{2} + (0) \times \dfrac{1}{6} \ = \dfrac{1}{4}$

Hence $E\left[X^2Y\right] = \dfrac{1}{4}$

Example 2: Find the correlation coefficient between the two random variables X and Y if $E[X] = 4$, $E[Y] = 9$, $E[XY] = 100$, $E\left[X^2\right] = 81$ and $E\left[Y^2\right] = 256$

Sol: Since the correlation coefficient between the two random variables X and Y is

$$\rho = \frac{Cov(X,Y)}{\sigma_x \sigma_y}$$

where

$$Cov(X,Y) = E[XY] - E[X].E[Y]$$

$$\sigma_x^2 = E\left[X^2\right] - [E(X)]^2 \quad \text{and} \quad \sigma_y^2 = E\left[Y^2\right] - [E(Y)]^2$$

Now $Cov(X,Y) = 100 - 4 \times 9 = 64$

$$\sigma_x^2 = 81 - 16 = 65$$

$$\sigma_y^2 = 256 - 81 = 175$$

Therefore the correlation coefficient between X and Y is

$$\rho = \frac{64}{\sqrt{65}\sqrt{175}}$$

$$= \frac{64}{106.65}$$

$$= 0.6$$

Example 3: Find the first two moments about the mean for the random variable X defined by

$$X = \begin{cases} -2 & \text{with probability } \dfrac{1}{3} \\[2mm] 3 & \text{with probability } \dfrac{1}{2} \\[2mm] 1 & \text{with probability } \dfrac{1}{6} \end{cases}$$

Sol: Since $E[X] = \sum xP(X = x) = (-2) \times \dfrac{1}{3} + (3) \times \dfrac{1}{2} + (1) \times \dfrac{1}{6} = 1 \ (= \mu_1')$

The first moment about mean

$$\mu_1 = E\left[X - \mu_1'\right]$$

$$= E[X] - \mu_1'$$

$$= \mu_1' - \mu_1'$$

$$= 1 - 1 = 0$$

Therefore the first moment about mean is zero

The second moment about mean $\mu_2' = E\left[X^2\right] = (-2)^2 \times \dfrac{1}{3} + (3)^2 \times \dfrac{1}{2} + (1)^2 \times \dfrac{1}{6}$

$$= 6$$

Now $\qquad \mu_2 = \mu_2' - \left(\mu_1'\right)^2 = 6 - 1 = 5$

***Example* 4:** Let X and Y be two independently and identically distributed random variables with mean 2 and 4 and their second order moments are 8 and 25 respectively. Find the mean, second order moment and variance of the random variable $Z = 3X - Y$

Sol: Since X and Y are independent also given that

$$E[X] = 2, \quad E[Y] = 4, \ E\left[X^2\right] = 8 \text{ and } E\left[Y^2\right] = 25$$

The mean of the random variable $Z = 3X - Y$ is

$$E[Z] = E[3X - Y]$$
$$= 3E[X] - E[Y]$$
$$= 3 \times 2 - 4 = 2$$

Therefore $E[3X - Y] = 2$

The second moment of the random variable $Z = 3X - Y$ is

$$E\left[Z^2\right] = E\left[(3X - Y)^2\right]$$
$$= E\left[9X^2 - 6XY + Y^2\right]$$
$$= 9E\left[X^2\right] - 6E[XY] + E\left[Y^2\right] \text{ since } X \text{ and } Y \text{ are independent}$$
$$= 9 \times 8 - 6 \times 2 \times 4 + 25$$
$$= 49$$

Variance of $Z = E\left[Z^2\right] - [E(Z)]^2$
$$= 49 - 4$$
$$= 45$$

Therefore $V[3X - Y] = 45$

***Example* 5:** If X and Y are independent random variables having each zero mean and variances 25 and 9 respectively. Find the correlation coefficient between $(X+Y)$ and $(X-Y)$

Sol: Given that $E[X] = E[Y] = 0$; $E\left[X^2\right] = 25$; $E\left[Y^2\right] = 9$

$$\sigma_x^{\ 2} = V(X) = E\left[X^2\right] - [E(X)]^2 = E\left[X^2\right] = 25$$

$$\sigma_y^{\ 2} = V(Y) = E\left[Y^2\right] - [E(Y)]^2 = E\left[Y^2\right] = 9$$

Since the correlation coefficient between the two random variables X and Y is

$$\rho = \frac{Cov(X,Y)}{\sigma_x \sigma_y}$$

where

$$Cov(X,Y) = E[XY] - E[X].E[Y]$$

$$\sigma_x^2 = V(X) = E\left[X^2\right] - [E(X)]^2 \quad \text{and} \quad \sigma_y^2 = V(Y) = E\left[Y^2\right] - [E(Y)]^2$$

Similarly,

The correlation coefficient between the two random variables $X+Y$ and $X-Y$ is

$$\rho = \frac{Cov(X+Y, X-Y)}{\sigma_{x+y} \sigma_{x-y}}$$

where

$$
\begin{aligned}
Cov(X+Y, X-Y) \;&= E[(X+Y)(X-Y)] - E[(X+Y)]E[(X-Y)] \\
&= E\left[X^2\right] - E\left[Y^2\right] - \{E[X]+E[Y]\}\{E[X]-E[Y]\} \\
&= E\left[X^2\right] - E\left[Y^2\right] \qquad \because E[X] = E[Y] = 0 \\
&= 25 - 9 \\
&= 16
\end{aligned}
$$

Now

$$
\begin{aligned}
V(X+Y) &= E\left[(X+Y)^2\right] - [E(X+Y)]^2 \\
&= E\left[X^2 + Y^2 + 2XY\right] - \{E[X]+E[Y]\}^2 \\
&= E\left[X^2\right] + E\left[Y^2\right] + 2E[X]E[Y] \\
&\hspace{6cm} \because E[X] = E[Y] = 0 \\
&= E\left[X^2\right] + E\left[Y^2\right] \\
&= 25 + 9 \\
&= 34
\end{aligned}
$$

Similarly,

$$V(X-Y) = E\left[(X-Y)^2\right] - \left[E(X-Y)\right]^2$$

$$= E\left[X^2 + Y^2 - 2XY\right] - \left\{E[X] - E[Y]\right\}^2$$

$$= E\left[X^2\right] + E\left[Y^2\right] - 2E[X]E[Y] \qquad \because E[X] = E[Y] = 0$$

$$= E\left[X^2\right] + E\left[Y^2\right]$$

$$= 25 + 9$$

$$= 34$$

Therefore the correlation coefficient between X and Y is

$$\rho = \frac{16}{\sqrt{34}\sqrt{34}}$$

$$= \frac{16}{34}$$

$$= \frac{8}{17}$$

Hence the correlation coefficient between $X+Y$ and $X-Y$ is $\dfrac{8}{17}$

***Example* 6:** If X and Y are independent random variables having means 1 and 2 and variances 4 and 1 respectively. Their correlation coefficient is 0.4. Let U and V are defined as $U = -X + 2Y$ and $V = X + 3Y$. Find the *(i)* means of U and V *(ii)* Variances of U and V *(iii)* Correlation between U and V and *(iv)* Correlation coefficient between U and V.

Sol: Given that $E[X] = 1$; $E[Y] = 2$; $V(X) = \sigma_x^2 = 4$; $V(Y) = \sigma_y^2 = 1$ and $\rho_{xy} = 0.4$

Since the correlation coefficient between the two random variables X and Y is

$$\rho = \frac{Cov(X,Y)}{\sigma_x \sigma_y}$$

where

$$Cov(X,Y) = E[XY] - E[X].E[Y]$$

$$\sigma_x^2 = V(X) = E\left[X^2\right] - \left[E(X)\right]^2 \quad \text{and} \quad \sigma_y^2 = V(Y) = E\left[Y^2\right] - \left[E(Y)\right]^2$$

Also given $U = -X + 2Y$ and $V = X + 3Y$.

(i) Mean of U is

$$E[U] = E[-X + 2Y]$$
$$= -E[X] + 2E[Y]$$
$$= -1 + 2 \times 2$$
$$= 3$$

Mean of V is

$$E[V] = E[X + 3Y]$$
$$= E[X] + 3E[Y]$$
$$= 1 + 3 \times 2$$
$$= 7$$

(ii) Now variance of U is

$$V(U) = V(-X + 2Y) = V(X) + 4V(Y) + 4Cov(X,Y) \quad .....(5.7)$$

$$\therefore \qquad \rho_{xy} = 0.4 = \frac{Cov(X,Y)}{\sigma_x \sigma_y}$$

$$\Rightarrow \qquad Cov(X,Y) = 0.4\sigma_x \sigma_y$$
$$= 0.4 \times 2 \times 1$$
$$= 0.8 \qquad\qquad\qquad .....(5.8)$$

From equation (5.7) and (5.8)

$$V(U) = V(-X + 2Y)$$
$$= 4 + (4 \times 1) + (4 \times 0.8) \quad \text{since } V(X) = 4;\ V(Y) = 1$$
$$= 11.2 \qquad\qquad\qquad .....(5.9)$$

Similarly Variance of V is

$$V(V) = V(X + 3Y)$$
$$= V(X) + 9V(Y) + 6Cov(X,Y)$$
$$= 4 + (9 \times 1) + (6 \times 0.8) \ \because\ V(X) = 4;\ V(Y) = 1$$

and

$$Cov(X,Y) = 0.8$$
$$= 17.8 \qquad\qquad\qquad .....(5.10)$$

Since

$$E[X] = 1;\ E[Y] = 2;\ V(X) = \sigma_x^2 = 4;\ V(Y) = \sigma_y^2 = 1$$

these are $\sigma_x^2 = V(X) = E\left[X^2\right] - \left[E(X)\right]^2$

$\Rightarrow \qquad 4 = E\left[X^2\right] - 1$

$\Rightarrow \qquad E\left[X^2\right] = 4 + 1 = 5$

Similarly, $\sigma_y^2 = V(Y) = E\left[Y^2\right] - \left[E(Y)\right]^2$

$\Rightarrow \qquad 1 = E\left[Y^2\right] - 4$

$\Rightarrow \qquad E\left[Y^2\right] = 1 + 4 = 5$

The correlation coefficient between U and V is

$$\rho_{u,v} = \frac{Cov(U,V)}{\sigma_u \sigma_v}$$

i.e., the correlation coefficient between the two random variables $-X + 2Y$ and $X + 3Y$ is

$$\rho = \frac{Cov(-X + 2Y, X + 3Y)}{\sigma_{-x+2y}\sigma_{x+3y}} \qquad\qquad .....(5.11)$$

where

$$Cov(-X + 2Y, X + 3Y)$$
$$= E\left[(-X + 2Y)(X + 3Y)\right] - E\left[(-X + 2Y)\right]E\left[(X + 3Y)\right]$$
$$= E\left[(-X^2 - XY + 6Y^2)\right] - 3 \times 7$$
$$= -E(X^2) - E(XY) + 6E(Y^2) - 21$$
$$= -5 - (1 \times 2) + (6 \times 5) - 21$$
$$= 2$$

Therefore from equations (5.9), (5.10) and (5.11) the correlation coefficient is

$$\rho = \frac{2}{\sqrt{11.2}\sqrt{17.8}} = \frac{2}{\sqrt{199.36}} = \frac{2}{14.12} = 0.14$$

***Example* 7:** Let X be the score on the first die and Y be the score on the second die when two dice are thrown.   Let Z denote the maximum of X and Y, i.e., $Z = \text{Max}(X,Y)$. Write down i) the joint distribution of Z and X and ii) mean and variance of Z and covariance of (Z,X)

Sol: When two dice are thrown then the sample space S is

$$S = \left\{ \begin{array}{l} (1,1), (1,2), (1,3), (1,4), (1,5), (1,6); \quad (2,1), (2,2), (2,3), (2,4), (2,5), (2,6) \\ (3,1), (3,2), (3,3), (3,4), (3,5), (3,6); \quad (4,1), (4,2), (4,3), (4,4), (4,5), (4,6) \\ (5,1), (5,2), (5,3), (5,4), (5,5), (5,6); \quad (6,1), (6,2), (6,3), (6,4), (6,5), (6,6) \} \end{array} \right\}$$

From the given data, $Z = \text{Max}(X,Y)$

$Z = Max[(1,1)] = \max(1,1) = 1;$

$\therefore \quad P(Z = 1) = P(1) = \dfrac{1}{36}$

Similarly,

$Z = \text{Max } [(2,1),(2,2),(1,2)] = 2$

$P(Z = 2) = P(2) = \dfrac{3}{36}$

$Z = \text{Max } [(1,3),(3,1),(2,3),(3,2),(3,3)] = 3$

$P(Z = 3) = P(3) = \dfrac{5}{36}$

$Z = \text{Max } [(1,4),(4,1),(2,4),(4,2),(3,4),(4,3),(4,4)] = 4$

$P(Z = 4) = P(4) = \dfrac{7}{36}$

$Z = \text{Max } [(1,5),(5,1),(2,5),(5,2),(3,5),(5,3),(4,5),(5,4),(5,5)] = 5$

$P(Z = 5) = P(5) = \dfrac{9}{36}$

$Z = \text{Max } [(1,6),(6,1),(2,6),(6,2),(3,6),(6,3),(4,6),(6,4),(5,6),(6,5),(6,6)] = 6$

$\therefore \quad P(Z = 6) = P(6)$

$\quad = \dfrac{11}{36}$

Hence the Probability distribution of the random variable Z is

$Z = z$	1	2	3	4	5	6
$P(Z = z)$	1/36	3/36	5/36	7/36	9/36	11/36

Now $E[Z] = \sum z P(Z = z)$

$\quad = 1 \times \dfrac{1}{36} + 2 \times \dfrac{3}{36} + 3 \times \dfrac{5}{36} + 4 \times \dfrac{7}{36} + 5 \times \dfrac{9}{36} + 6 \times \dfrac{11}{36}$

$\quad = \dfrac{161}{36}$

$$E\left[Z^2\right] = \sum z^2 P(Z = z)$$

$$= 1^2 \times \frac{1}{36} + 2^2 \times \frac{3}{36} + 3^2 \times \frac{5}{36} + 4^2 \times \frac{7}{36} + 5^2 \times \frac{9}{36} + 6^2 \times \frac{11}{36}$$

$$= \frac{791}{36}$$

$$V\left[Z\right] = E\left[Z^2\right] - \left[E(Z)\right]^2$$

$$= \frac{791}{36} - \left(\frac{161}{36}\right)^2$$

$$= \frac{2555}{1296}$$

Now the joint Probability of X and Z is

X \ Z	1	2	3	4	5	6	Marginal Totals
1	$\frac{1}{36}$	$\frac{1}{36}$	$\frac{1}{36}$	$\frac{1}{36}$	$\frac{1}{36}$	$\frac{1}{36}$	$\frac{6}{36}$
2	0	$\frac{2}{36}$	$\frac{1}{36}$	$\frac{1}{36}$	$\frac{1}{36}$	$\frac{1}{36}$	$\frac{6}{36}$
3	0	0	$\frac{3}{36}$	$\frac{1}{36}$	$\frac{1}{36}$	$\frac{1}{36}$	$\frac{6}{36}$
4	0	0	0	$\frac{4}{36}$	$\frac{1}{36}$	$\frac{1}{36}$	$\frac{6}{36}$
5	0	0	0	0	$\frac{5}{36}$	$\frac{1}{36}$	$\frac{6}{36}$
6	0	0	0	0	0	$\frac{6}{36}$	$\frac{6}{36}$
Marginal Totals	$\frac{1}{36}$	$\frac{3}{36}$	$\frac{5}{36}$	$\frac{7}{36}$	$\frac{9}{36}$	$\frac{11}{36}$	1

These probabilities are obtained as

$$Z = \text{Max}[(1,1)] = 1;$$

$$\therefore \qquad P(X = 1, Z = 1) = \frac{1}{36}$$

Similarly,

$$Z = \text{Max} [(2,1),(2,2),(1,2)] = 2$$

$\therefore$ $P(X = 1, Z = 2) = \dfrac{1}{36}$ and $P(X = 2, Z = 2) = \dfrac{2}{36}$

$$Z = \text{Max} [(1,3),(3,1),(2,3),(3,2),(3,3)] = 3$$

$\therefore$ $P(X = 1, Z = 3) = \dfrac{1}{36}$; $P(X = 2, Z = 3) = \dfrac{1}{36}$ and $P(X = 3, Z = 3) = \dfrac{3}{36}$

$$Z = \text{Max} [(1,4),(4,1),(2,4),(4,2),(3,4),(4,3),(4,4)] = 4$$

$\therefore$ $P(X = 1, Z = 4) = \dfrac{1}{36}$; $P(X = 2, Z = 4) = \dfrac{1}{36}$;

$P(X = 3, Z = 4) = \dfrac{1}{36}$ and $P(X=4, Z=4) = \dfrac{4}{36}$

$$Z = \text{Max}[(1,5),(5,1),(2,5),(5,2),(3,5),(5,3),(4,5),(5,4),(5,5)] = 5$$

$\therefore$ $P(X = 1, Z = 5) = \dfrac{1}{36}$; $P(X = 2, Z = 5) = \dfrac{1}{36}$;

$P(X = 3, Z = 5) = \dfrac{1}{36}$; $P(X = 4, Z = 5) = \dfrac{1}{36}$ and $P(X = 5, Z = 5) = \dfrac{5}{36}$

$$Z = \text{Max} [(1,6),(6,1),(2,6),(6,2),(3,6),(6,3),(4,6),(6,4),(5,6),(6,5),(6,6)] = 6$$

$\therefore$ $P(X = 1, Z = 6) = \dfrac{1}{36}$; $P(X = 2, Z = 6) = \dfrac{1}{36}$

$P(X = 3, Z = 6) = \dfrac{1}{36}$; $P(X = 4, Z = 6) = \dfrac{1}{36}$

$P(X = 5, Z = 6) = \dfrac{1}{36}$ and $P(X = 6, Z = 6) = \dfrac{6}{36}$

Now $E[X] = \sum x P(X = x)$

$$= 1 \times \frac{6}{36} + 2 \times \frac{6}{36} + 3 \times \frac{6}{36} + 4 \times \frac{6}{36} + 5 \times \frac{6}{36} + 6 \times \frac{6}{36}$$

$$= \frac{126}{36}$$

$$E[XZ] = \sum xz P(x, z)$$

$$= \left\{ 1 \times \frac{1}{36} + 2 \times \frac{1}{36} + 3 \times \frac{1}{36} + 4 \times \frac{1}{36} + 5 \times \frac{1}{36} + 6 \times \frac{1}{36} \right\}$$

$$+\left\{4\times\frac{2}{36}+6\times\frac{1}{36}+8\times\frac{1}{36}+10\times\frac{1}{36}+12\times\frac{1}{36}\right\}$$

$$+\left\{9\times\frac{3}{36}+12\times\frac{1}{36}+15\times\frac{1}{36}+18\times\frac{1}{36}\right\}$$

$$+\left\{16\times\frac{4}{36}+20\times\frac{1}{36}+24\times\frac{1}{36}\right\}$$

$$+\left\{25\times\frac{5}{36}+30\times\frac{1}{36}\right\}+\left\{36\times\frac{6}{36}\right\}$$

$$=\frac{616}{36}$$

$$Cov(X,Z) = E[XZ] - E[X]E[Z]$$

$$=\frac{616}{36}-\frac{126}{36}\times\frac{161}{36}$$

$$=\frac{315}{216}$$

Example 8: Let X and Y are independent random variables whose moments are $\mu_{10} = 2$, $\mu_{20} = 14$, $\mu_{02} = 12$ and $\mu_{11} = -6$. Find the moment μ_{22}.

Sol: Given the random variables X and Y are independent. Then

$$f_{xy}(x, y) = f_x(x) \cdot f_y(y)$$

and also given $\mu_{10} = 2 = E(x)$

$$\mu_{20} = 14 = E(x^2)$$

$$\mu_{02} = 12 = E(y^2) \text{ and } \mu_{11} = -6 = E(xy)$$

Since x and y are independent

$$\therefore \quad E(xy) = E(x). E(y) = -6$$

$$\Rightarrow 2. E(y) = -6$$

$$\Rightarrow E(y) = -3$$

And $\mu_{22} = E[(x-\bar{x})(y-\bar{y})^2]$

$$= \int_{-\infty}^{\infty}\int_{-\infty}^{\infty}(x-\bar{x})^2(y-\bar{y})^2 f_{xy}(x, y) \, dxdy$$

$$\int_{-\infty}^{\infty}(x-\bar{x})^2 f_x(x) \, dx \cdot \int_{-\infty}^{\infty}(y-\bar{y})^2 f_y(y) \, dy$$

∴ $f_{xy}(x, y) = f_x(x) \cdot f_y(y)$

$$E\left[\left(x - \bar{x}\right)^2\right] \cdot E\left[\left(y - \bar{y}\right)^2\right]$$

$$= \{E(x^2) - [E(x)^2]\} \, \{E(y^2) - [E(y)^2]\}$$

$$= [14 - (2)^2] \, [12 - (-3)^2]$$

$$= 10 \times 3$$

$$= 30.$$

∴ $\mu_{22} = 30$

∴   The moment $\mu_{22} = 30$

Example 9: Let X_1, X_2 and X_3 are three independent random variables whose mean values are $X_1 = 3$, $X_2 = 6$ and $X_3 = -2$. Find the mean values of the following

(a) $g(X_1, X_2 X_3) = X_1 + 3X_2 + 4X_3$ (b) $g(X_1, X_2 X_3) = X_1 X_2 X_3$

(c) $g(X_1, X_2 X_3) = -2X_1 X_2 - 3X_1 X_3 + 4X_2 X_3$ (d) $g(X_1, X_2 X_3) = X_1 + X_2 + X_3$

Sol: Given that, $E(X_1) = 3$, $E(X_2) = 6$ and $E(X_3) = -2$

(a) $E[g(X_1, X_2 X_3)] = E[X_1 + 3X_2 + 4X_3]$

$$= E[X_1] + 3E[X_2] + 4E[X_3]$$

$$= 3 + 3 \times 6 + 4 \times (-2)$$

$$= 13$$

(b) $E[g(X_1, X_2 X_3)] = E[X_1 X_2 X_3]$

$$= E[X_1] E[X_2] E[X_3]$$

$$= 3 \times 6 \times (-2)$$

$$= -36$$

(c) $E[g(X_1, X_2 X_3)] = E[-2X_1 X_2 - 3X_1 X_3 + 4X_2 X_3]$

$$= (-2) E[X_1] E[X_2] - 3E[X_1] E[X_3] + 4E[X_2] E[X_3]$$

$$= (-2) \times 3 \times 6 - 3 \times 3 \times (-2) + 4 \times 6 \times (-2)$$

$$= -66$$

(d) $E[g(X_1, X_2 X_3)] = E[X_1 + X_2 + X_3]$

$$= E[X_1] + E[X_2] + E[X_3]$$

$$= 3 + 6 + (-2)$$

$$= 7$$

***Example* 10:** Let X and Y be two random variables each taking three values -1, 0 and 1 and having the joint probability distribution

X \ Y	−1	0	1	Total
−1	0	0.1	0.1	0.2
0	0.2	0.2	0.2	0.6
1	0	0.1	0.1	0.2
Total	0.2	0.4	0.4	1

(i) Show that X and Y have different expectations (ii) Show that X and Y uncorrelated (iii) Find variance of X and Y (iv) Find the conditional distribution of X given $Y = 0$ (v) Find $V(Y/X = -1)$

Sol: Given that

X \ Y	−1	0	1	Total
−1	0	0.1	0.1	0.2
0	0.2	0.2	0.2	0.6
1	0	0.1	0.1	0.2
Total	0.2	0.4	0.4	1

(i)   Now

$$E[X] = \sum xP(X = x)$$
$$= -1 \times (0.2) + 0 \times (0.4) + 1 \times (0.4)$$
$$= 0.2$$
$$E[Y] = \sum yP(Y = y)$$
$$= -1 \times (0.2) + 0 \times (0.6) + 1 \times (0.2)$$
$$= 0$$

(ii)   The correlation coefficient between X and Y is

$$\rho_{x,y} = \frac{Cov(X,Y)}{\sigma_x \sigma_y}$$

$$E[XY] = \sum xyP(x, y)$$
$$= \{(-1) \times (-1) \times 0 + (-1) \times 0.1 \times 0 + (-1) \times 0.1 \times 1\} +$$
$$\{0 \times 0.2 \times -1 + 0 \times 0.2 \times 0 + 0 \times 0.2 \times 1\} +$$
$$\{1 \times 0 \times (-1) + 1 \times 0.1 \times 0 + 1 \times 0.1 \times 1\}$$
$$= 0$$

$$Cov(X,Y) = E[XY] - E[X]E[Y]$$

$$= 0 - 0$$

$$= 0$$

Therefore from equation (1) correlation between X and Y is zero. Then X and Y are uncorrelated.

(iii) Variance of X:

$$V[X] = E\left[X^2\right] - [E(X)]^2$$

$$E\left[X^2\right] = (-1)^2 \times (0.2) + 0^2 \times (0.4) + 1^2 \times (0.4)$$

$$= 0.6$$

Hence $V[X] = 0.6 - (0.2)^2$

$$= 0.56$$

Variance of Y: $V[Y] = E\left[Y^2\right] - [E(Y)]^2$

$$E\left[Y^2\right] = (-1)^2 \times (0.2) + 0^2 \times (0.6) + 1^2 \times (0.2)$$

$$= 0.4$$

Therefore,

$$V[Y] = 0.4 - (0)^2$$

$$= 0.4$$

Hence $V[Y] = 0.4$

(iv) $P(X = -1/Y = 0) = \dfrac{P(X = -1 \cap Y = 0)}{P(Y = 0)}$

$$= \frac{0.2}{0.6}$$

$$= \frac{1}{3}$$

$$P(X = 0/Y = 0) = \frac{P(X = 0 \cap Y = 0)}{P(Y = 0)}$$

$$= \frac{0.2}{0.6}$$

$$= \frac{1}{3}$$

$$P(X=1/Y=0) = \frac{P(X=1 \cap Y=0)}{P(Y=0)}$$

$$= \frac{0.2}{0.6}$$

$$= \frac{1}{3}$$

(v) $V(Y/X=-1) = E(Y/X=-1)^2 - \left[E(Y/X=-1)\right]^2$ .....(5.13)

Now $E\left[Y/X=-1\right] = \sum yP(Y=y/X=-1)$

$$= -1 \times 0 + 0 \times (0.2) + 1 \times 0$$

$$= 0$$ .....(5.14)

$$E\left[Y/X=-1\right]^2 = \sum y^2 P(Y=y/X=-1)$$

$$= (-1)^2 \times 0 + 0^2 \times (0.2) + 1^2 \times 0$$

$$= 0$$ .....(5.15)

Therefore from equation (5.13), (5.14) and (5.15)

$$V(Y/X=-1) = 0-0$$

$$= 0$$

5.6 Joint Moment Generating Function

Let X_1 and X_2 be two independent random variables with joint probability density function $f_{X_1 X_2}(x_1, x_2)$ and their joint moment generating function is defined as

$$M_X(t) = M_{X_1 X_2}(t_1, t_2) = E\left[e^{t_1 x_1 + t_2 x_1}\right]$$

$$= \begin{cases} \displaystyle\sum_i^\infty \sum_j^\infty e^{t_1 x_1 + t_2 x_1} P\left[x_1(i), x_2(j)\right] & \text{If } X_1 \text{ and } X_2 \text{ discrete} \\ \displaystyle\int_{-\infty}^{\infty}\int_{-\infty}^{\infty} e^{t_1 x_1 + t_2 x_1} f_{X_1 X_2}(x_1, x_2) dx_1 dx_2 & \text{If } X_1 \text{ and } X_2 \text{ continuous} \end{cases}$$

It has similar properties as in the univariate case

Now

$$M_{X_1 X_2}(t_1, t_2) = E\left[e^{t_1 x_1 + t_2 x_1}\right]$$

$$= E\left[1+\left(t_1x_1+t_2x_1\right)+\frac{\left(t_1x_1+t_2x_1\right)^2}{2!}+\frac{\left(t_1x_1+t_2x_1\right)^3}{3!}+\cdots+\frac{\left(t_1x_1+t_2x_1\right)^r}{r!}+\cdots\right]$$

$$=1+E\left(t_1x_1+t_2x_1\right)+\frac{1}{2!}E\left(t_1x_1+t_2x_1\right)^2+\cdots+\frac{1}{r!}E\left(t_1x_1+t_2x_1\right)^r+\cdots$$

Now

$$E[X_1]=\mu_1'=\left[\frac{\partial}{\partial t_1}\{M(t_1,t_2)\}\right]_{t_1=0,t_2=0}=\frac{\partial}{\partial t_1}\{M(0,0)\}$$

$$E[X_2]=\mu_2'=\left[\frac{\partial}{\partial t_2}\{M(t_1,t_2)\}\right]_{t_1=0,t_2=0}=\frac{\partial}{\partial t_2}\{M(0,0)\}$$

$$E\left[X_1^2\right]=\left[\frac{\partial^2}{\partial t_1^2}\{M(t_1,t_2)\}\right]_{t_1=0,t_2=0}=\frac{\partial^2}{\partial t_1^2}\{M(0,0)\}$$

and in general

$$E\left[X_1^r\right]=\left[\frac{\partial^r}{\partial t_1^r}\{M(t_1,t_2)\}\right]_{t_1=0,t_2=0}=\frac{\partial^r}{\partial t_1^r}\{M(0,0)\}$$

$$E\left[X_2^2\right]=\left[\frac{\partial^2}{\partial t_2^2}\{M(t_1,t_2)\}\right]_{t_1=0,t_2=0}=\frac{\partial^2}{\partial t_2^2}\{M(0,0)\}$$

and in general

$$E\left[X_2^r\right]=\left[\frac{\partial^r}{\partial t_2^r}\{M(t_1,t_2)\}\right]_{t_1=0,t_2=0}=\frac{\partial^r}{\partial t_2^r}\{M(0,0)\}$$

$$E[X_1X_2]=\left[\frac{\partial^2}{\partial t_1\partial t_2}\{M(t_1,t_2)\}\right]_{t_1=0,t_2=0}=\frac{\partial^2}{\partial t_1\partial t_2}\{M(0,0)\}$$

and in general

$$E\left[X_1^rX_2^s\right]=\left[\frac{\partial^{r+s}}{\partial t_1^r\partial t_2^s}\{M(t_1,t_2)\}\right]_{t_1=0,t_2=0}=\frac{\partial^{r+s}}{\partial t_1^r\partial t_2^s}\{M(0,0)\}$$

Note: If the moment generating function is finite in an open rectangle containing (0,0) then the function completely determines the distribution of X_1 and X_2.

5.7   Properties of Joint Moment Generating Function

1. $M_{X_1, X_2}(t_1, 0) = M_{X_1}(t_1)$

2. $M_{X_1, X_2}(0, t_2) = M_{X_2}(t_2)$

3. $M_{X_1, X_2}(t_1, 0) \ M_{X_1+X_2}(t)$

4. If X_1 and X_2 are independent random variables then

$M_{X_1 X_2}(t_1, t_2) = M_{X_1}(t_1) \times M_{X_2}(t_2)$ for (t_1, t_2) in a rectangle about $(0,0)$

This gives that if X_1 and X_2 are independent, so $g_1(x) = e^{x_1 t_1}$ and $g_2(x) = e^{x_2 t_2}$

Extension

If $X_1, X_2, X_3 \cdots X_n$ are random variables defined on the sample space then their joint moment generating function is defined as the function $R^n \to R$, is

$$M_{X_1 X_2 \cdots X_n}(t_1, t_2 \cdots t_n) = E\left[e^{t_1 x_1 + t_2 x_1 \cdots + t_n x_n} \right]$$

As above the moment generating function uniquely determines the joint distribution of the random variable $X_1, X_2, X_3 \cdots X_n$ provided that this function is finite on some open disk around the origin .

For independence,

$$M_{X_1 X_2 \cdots X_n}(t_1, t_2 \cdots t_n) = M_{X_1}(t_1).M_{X_2}(t_2) \cdots M_{X_n}(t_n)$$

***Example* 1**: Suppose that X and Y are independent standard normal random variables and define $U = X + Y$ and $V = X - Y$. Find the moment generating function of U and V.

Sol: Since the moment generating function of x_1 and x_2 is

$$M_{X_1 X_2}(t_1, t_2) = E\left[e^{t_1 x_1 + t_2 x_1} \right]$$

Similarly the moment generating function of U and V is

$$M_{UV}(t_1, t_2) = E\left[e^{t_1 u + t_2 v} \right]$$
$$= E\left[e^{t_1(X+Y)+t_2(X-Y)} \right]$$
$$= E\left[e^{t_1(X+Y)} e^{t_2(X-Y)} \right]$$
$$= E\left[e^{(t_1+t_2)X} e^{(t_1-t_2)Y} \right]$$
$$= E\left[e^{(t_1+t_2)X} \right] E\left[e^{(t_1-t_2)Y} \right]$$

$$M_{UV}(t_1, t_2) = M_X(t_1 + t_2)M_Y(t_1 - t_2)$$

or $$M_{UV}(t_1, t_2) = e^{(t_1+t_2)^2/2} \cdot e^{(t_1-t_2)^2/2} = e^{2t_1^2/2}e^{2t_2^2/2}$$

We observe that the joint moment generating function of two independent random variables with mean zero and variance 2. This shows that U and V are independent.

***Example* 2**: Find the joint moment generating function for the two random variables having joint density is

$$f_{XY}(x, y) = \frac{y}{\sqrt{2\pi}} e^{-y - \frac{1}{2}(x-y)^2}$$

Sol:

5.8   Joint Characteristic Function

Let (X_1, X_2) be a two dimensional random variable, then the characteristic function of two random variables X_1 and X_2 is defined as

$$\varphi_X(t) = \varphi_{X_1 X_2}(t_1, t_2) = E\left[e^{it_1 x_1 + it_2 x_1}\right] \quad \text{where } i^2 = -1, \text{ and } t_1 \text{ and } t_2 \text{ are real}$$

this can also be written as

$$\varphi_{X_1 X_2}(w_1, w_2) = E\left[e^{jw_1 x_1 + jw_2 x_1}\right] \text{where } j^2 = -1, \text{ and } w_1 \text{ and } w_2 \text{ are real}$$

$$= \begin{cases} \sum\limits_{i}^{\infty}\sum\limits_{j}^{\infty} e^{it_1 x_1 + it_2 x_1} P[x_1(i), x_2(j)] & \textit{If } X_1 \textit{ and } X_2 \textit{ discrete} \\ \int\limits_{-\infty}^{\infty}\int\limits_{-\infty}^{\infty} e^{it_1 x_1 + it_2 x_1} f_{X_1 X_2}(x_1, x_2) dx_1 dx_2 & \textit{If } X_1 \textit{ and } X_2 \textit{ continuous} \end{cases}$$

It has similar properties as in the univariate case

5.9   Properties of Joint Characteristic Function

1. $\varphi_{X_1, X_2}(0,0) = 1$

2. $E[X_1] = \dfrac{1}{i}\left[\dfrac{\partial}{\partial t_1}\{\varphi(t_1, t_2)\}\right]_{t_1=0, t_2=0}$

3. $E[X_2] = \dfrac{1}{i}\left[\dfrac{\partial}{\partial t_2}\{\varphi(t_1, t_2)\}\right]_{t_1=0, t_2=0}$

4. $E\left[X_1^2\right] = \dfrac{1}{i^2}\left[\dfrac{\partial^2}{\partial t_1^2}\left\{\varphi(t_1,t_2)\right\}\right]_{t_1=0,t_2=0}$

and in general

$E\left[X_1^r\right] = \dfrac{1}{i^r}\left[\dfrac{\partial^r}{\partial t_1^r}\left\{\varphi(t_1,t_2)\right\}\right]_{t_1=0,t_2=0}$

5. $E\left[X_2^2\right] = \dfrac{1}{i^2}\left[\dfrac{\partial^2}{\partial t_2^2}\left\{\varphi(t_1,t_2)\right\}\right]_{t_1=0,t_2=0}$

And in general

$E\left[X_2^r\right] = \dfrac{1}{i^r}\left[\dfrac{\partial^r}{\partial t_2^r}\left\{\varphi(t_1,t_2)\right\}\right]_{t_1=0,t_2=0}$

6. $E\left[X_1 X_2\right] = \dfrac{1}{i^2}\left[\dfrac{\partial^2}{\partial t_1 \partial t_2}\left\{\varphi(t_1,t_2)\right\}\right]_{t_1=0,t_2=0}$

And in general

$E\left[X_1^r X_2^s\right] = \dfrac{1}{i^{r+s}}\left[\dfrac{\partial^{r+s}}{\partial t_1^r \partial t_2^s}\left\{\varphi(t_1,t_2)\right\}\right]_{t_1=0,t_2=0}$

7. $\varphi_{X_1}(t_1) = \varphi(t_1,0)$ and $\varphi_{X_2}(t_2) = \varphi(0,t_2)$

8. If X and Y are independent then $\varphi_{X_1 X_2}(t_1,t_2) = \varphi_{X_1}(t_1) \times \varphi_{X_2}(t_2)$

Theorem

The necessary and sufficient condition for X and Y is equal to the product of their individual characteristic functions

Proof:

Necessary condition

Let X and Y are independent, then by definition

$$\varphi_{X_1 X_2}(t_1,t_2) = E\left[e^{it_1 x_1 + it_2 x_2}\right]$$

$$= E\left[e^{it_1 x_1} . e^{it_2 x_2}\right]$$

$$= E\left[e^{it_1 x_1}\right]E\left[e^{it_2 x_2}\right]$$

$$= \varphi_{X_1}(t_1) \times \varphi_{X_2}(t_2)$$

$\therefore \qquad \varphi_{X_1 X_2}(t_1,t_2) = \varphi_{X_1}(t_1) \times \varphi_{X_2}(t_2)$

Sufficient condition

Let $f_{X_1X_2}(x_1, x_2)$ be joint probability density function of X and Y and $f_{X_1}(x_1)$ and $f_{X_2}(x_2)$ be the marginal probability density functions of X and Y respectively. Then by definition (for continuous random variables)

$$\phi_{X_1}(t_1) = \int_{-\infty}^{\infty} e^{it_1 x_1} f_{X_1}(x_1) dx_1$$

and $$\phi_{X_2}(t_2) = \int_{-\infty}^{\infty} e^{it_2 x_2} f_{X_2}(x_2) dx_2$$

$$\phi_{X_1}(t_1) \times \phi_{X_2}(t_2) = \int_{-\infty}^{\infty} e^{it_1 x_1} f_{X_1}(x_1) dx_1 \times \int_{-\infty}^{\infty} e^{it_2 x_2} f_{X_2}(x_2) dx_2$$

$$= \int_{-\infty}^{\infty} e^{it_1 x_1} f_{X_1}(x_1) dx_1 \times \int_{-\infty}^{\infty} e^{it_2 x_2} f_{X_2}(x_2) dx_2$$

$$= \int_{-\infty}^{\infty}\int_{-\infty}^{\infty} e^{it_1 x_1 + it_2 x_2} f_{X_1}(x_1) f_{X_2}(x_2) dx_1 dx_2$$

$$= \int_{-\infty}^{\infty}\int_{-\infty}^{\infty} e^{it_1 x_1 + it_2 x_2} f_{X_1 X_2}(x_1, x_2) dx_1 dx_2$$

$$= \phi_{X_1 X_2}(t_1, t_2)$$

$\therefore$ $$\phi_{X_1}(t_1) \times \phi_{X_2}(t_2) = \phi_{X_1 X_2}(t_1, t_2)$$

Hence this implies X and Y are independent

Note: For discrete random variables, the same result is obtained by using summation instead of integration.

Extension

This result can be extend to n variables $X_1, X_2, X_3 \cdots X_n$ then their joint characteristic function is

$$\phi_{X_1 X_2 \cdots X_n}(t_1, t_2 \cdots t_n) = \phi_{X_1}(t_1).\phi_{X_2}(t_2)\cdots\phi_{X_n}(t_n)$$

Example 1: Two random variables X and Y have the joint characteristic function $\phi_{XY}(w_1, w_2) = e^{-w_1^2 - 8w_2^2}$ then show that the random variables X and Y are both zero mean random variables and also that they are uncorrelated.

Sol: Since we have

$$E[X] = \frac{1}{i}\left[\frac{\partial}{\partial w_1}\{\varphi(t_1,t_2)\}\right]_{t_1=0,t_2=0}$$
       Since by the property

$$= \frac{1}{i}\left[\frac{\partial}{\partial w_1}\left\{e^{-w_1^2-8w_2^2}\right\}\right]_{w_1=0,w_2=0}$$

$$= \frac{1}{i}\left[e^{-w_1^2-8w_2^2}(-4w_1)\right]_{w_1=0,w_2=0}$$

$$= 0$$

Therefore the random variable X has zero mean

$$E[Y] = \frac{1}{i}\left[\frac{\partial}{\partial w_2}\{\varphi(t_1,t_2)\}\right]_{t_1=0,t_2=0}$$

$$= \frac{1}{i}\left[\frac{\partial}{\partial w_2}\left\{e^{-w_1^2-8w_2^2}\right\}\right]_{w_1=0,w_2=0}$$

$$= \frac{1}{i}\left[e^{-w_1^2-8w_2^2}(-16w_2)\right]_{w_1=0,w_2=0} = 0$$

Here also the random variable Y has zero mean

$$E[XY] = \frac{1}{i^2}\left[\frac{\partial^2}{\partial w_1 \partial w_2}\left\{e^{-w_1^2-8w_2^2}\right\}\right]_{w_1=0,w_2=0}$$

$$= \frac{1}{i^2}\left[\frac{\partial}{\partial w_1}\left\{e^{-w_1^2-8w_2^2}(-16w_2)\right\}\right]_{w_1=0,w_2=0} = 0$$

Since the correlation coefficient between X and Y is

$$\rho = \frac{Cov(X,Y)}{\sigma_x \sigma_y}$$

But $Cov(X,Y) = E[XY] - E[X].E[Y] = 0 \ \because E[XY] = 0 \ E[X] = 0 \ E[Y] = 0$

Therefore the correlation coefficient is zero. Hence these two random variables are uncorrelated.

Example 2: Find the joint characteristic function of the discrete random variable X and Y then their joint probability mass function is

$$f_{XY}(x,y) = \begin{cases} \dfrac{1}{6} & \text{for } (x,y) = (1,1) \\[6pt] \dfrac{1}{6} & \text{for } (x,y) = (-1,0) \\[6pt] \dfrac{1}{6} & \text{for } (x,y) = (1,0) \\[6pt] \dfrac{1}{6} & \text{for } (x,y) = (-1-,1) \\[6pt] \dfrac{1}{3} & \text{for } (x,y) = (0,0) \\[6pt] 0 & \text{otherwise} \end{cases}$$

Sol: Since the joint characteristic function is

$$\phi_{X_1 X_2}(t_1, t_2) = E\left[e^{it_1 x_1 + it_2 x_2}\right]$$

$$= \sum_{i}^{\infty}\sum_{j}^{\infty} e^{it_1 x_1 + it_2 x_2} P\left[x_1(i), x_2(j)\right]$$

$$= e^{i(t_1+t_2)}\frac{1}{6} + e^{it_1}\frac{1}{6} + e^{-it_1}\frac{1}{6} + e^{-i(t_1+t_2)}\frac{1}{6} + \frac{1}{3}$$

$$= \frac{1}{3} + \frac{1}{6}\left[\left(e^{it_1} + e^{-it_1}\right) + \left(e^{i(t_1+t_2)} + e^{-i(t_1+t_2)}\right)\right]$$

$$= \frac{1}{3} + \frac{1}{6}\left[2\frac{\left(e^{it_1} + e^{-it_1}\right)}{2} + 2\frac{\left(e^{i(t_1+t_2)} + e^{-i(t_1+t_2)}\right)}{2}\right]$$

$$= \frac{1}{3} + \frac{1}{3}\left[\cos t_1 + \cos(t_1 + t_1)\right]$$

Therefore the joint characteristic function of the discrete random variables X and Y is

$$\phi_{X_1 X_2}(t_1, t_2) = \frac{1}{3} + \frac{1}{3}\left[\cos t_1 + \cos(t_1 + t_1)\right]$$

Since the correlation coefficient between X and Y is

$$\rho = \frac{Cov(X,Y)}{\sigma_x \sigma_y}$$

where $Cov(X,Y) = E[XY] - E[X].E[Y]$ .....(5.16)

Now $E[XY] = \dfrac{1}{i^2}\left[\dfrac{\partial^2}{\partial t_1 \partial t_2}\{\varphi(t_1,t_2)\}\right]_{t_1=0,t_2=0}$

$= \dfrac{1}{i^2}\left[\dfrac{\partial^2}{\partial t_1 \partial t_2}\left\{\dfrac{1}{3}+\dfrac{1}{3}[\cos t_1 + \cos(t_1+t_1)]\right\}\right]_{t_1=0,t_2=0}$

$= \dfrac{1}{i^2}\left[\dfrac{\partial}{\partial t_2}\left\{\dfrac{-\sin t_1}{3}-\dfrac{\sin(t_1+t_1)}{3}\right\}\right]_{t_1=0,t_2=0}$

$= \dfrac{1}{i^2}\left[\dfrac{\partial}{\partial t_2}\left\{\dfrac{-1}{3}\sin(t_1+t_1)\right\}\right]_{t_1=0,t_2=0}$

$= \left[\dfrac{1}{3}\cos(t_1+t_1)\right]_{t_1=0,t_2=0}$

$= \dfrac{1}{3}$

∴ $E[XY] = \dfrac{1}{3}$

The joint probability distribution is

X \ Y	−1	0	1	$f_X(x)$
−1	$\dfrac{1}{6}$	$\dfrac{1}{6}$	0	$\dfrac{2}{6}$
0	0	$\dfrac{1}{3}$	0	$\dfrac{1}{3}$
1	0	$\dfrac{1}{6}$	$\dfrac{1}{6}$	$\dfrac{2}{6}$
$f_Y(y)$	$\dfrac{1}{6}$	$\dfrac{4}{6}$	$\dfrac{1}{6}$	1

Now $E[X] = \sum xP(X=x)$

$= (-1)\times\dfrac{2}{6}+0\times\dfrac{1}{3}+1\times\dfrac{2}{6}$

$= 0$

$$E[Y] = \sum yP(Y = y)$$

$$= (-1) \times \frac{1}{6} + 0 \times \frac{4}{6} + 1 \times \frac{1}{6}$$

$$= 0$$

$$E[X^2] = \sum x^2 P(X = x)$$

$$= (-1)^2 \times \frac{2}{6} + 0^2 \times \frac{1}{3} + 1^2 \times \frac{2}{6}$$

$$= \frac{2}{3}$$

$$E[Y^2] = \sum y^2 P(Y = y)$$

$$= (-1)^2 \times \frac{1}{6} + 0^2 \times \frac{4}{6} + 1^2 \times \frac{1}{6}$$

$$= \frac{1}{3}$$

$$V(X) = \sigma_x^2 = E[X^2] - [E(X)]^2$$

$$= \frac{2}{3} - 0 = \frac{2}{3}$$

and

$$V(Y) = \sigma_y^2 = E[Y^2] - [E(Y)]^2$$

$$= \frac{1}{3} - 0$$

$$= \frac{1}{3}$$

$$Cov(X,Y) = E[XY] - E[X].E[Y]$$

$$= \frac{1}{3} - 0$$

$$= \frac{1}{3}$$

Therefore the correlation coefficient between X and Y is

$$\rho = \frac{Cov(X,Y)}{\sigma_x \sigma_y}$$

$$= \frac{1/3}{\sqrt{\frac{2}{3}}\sqrt{\frac{1}{3}}}$$

$$= \frac{1}{\sqrt{2}}$$

Hence the correlation coefficient between X and Y is $\dfrac{1}{\sqrt{2}}$

Example 3: Let $X_1, X_2, X_3 \cdots X_n$ are independently and identically distributed random variables as

$$X_K = \begin{cases} 1 & \text{with probability } \dfrac{1}{2} \\ -1 & \text{with probability } \dfrac{1}{2} \end{cases}$$

Find the characteristic function of the random variable $\quad Y = \dfrac{X_1 + X_2 + \cdots + X_n}{\sqrt{n}}$

Sol: Since the characteristic function,   *Check the solution*

$$\phi_Y(t) = E\left[e^{ity}\right]$$

$$= E\left[e^{it\frac{X_1+X_2+\cdots+X_n}{\sqrt{n}}}\right]$$

$$= E\left[e^{it\frac{X_1}{\sqrt{n}}}\right] E\left[e^{it\frac{X_2}{\sqrt{n}}}\right] \cdots E\left[e^{it\frac{X_n}{\sqrt{n}}}\right] \qquad \qquad \dots(5.17)$$

$$= \left\{ E\left[e^{it\frac{X_k}{\sqrt{n}}}\right] \right\}^n$$

Since $X_1, X_2, X_3 \cdots X_n$ are independently and identically distributed random variables

Consider

$$E\left[e^{it\frac{X_k}{\sqrt{n}}}\right] = \sum_{k=1}^{n} e^{it\frac{X_k}{\sqrt{n}}} P(X = x_k) \qquad \because E[X] = \sum xP(X = x)$$

$$= e^{it\frac{(+1)}{\sqrt{n}}}\left(\frac{1}{2}\right) + e^{it\frac{(-1)}{\sqrt{n}}}\left(\frac{1}{2}\right) = \left(\frac{1}{2}\right)\left[e^{\frac{it}{\sqrt{n}}} + e^{-\frac{it}{\sqrt{n}}}\right]$$

$$= \cos\frac{t}{\sqrt{n}}$$

$$\therefore \qquad \phi_Y(t) = \left\{E\left[e^{it\frac{X_k}{\sqrt{n}}}\right]\right\}^n = \left(\cos\frac{t}{\sqrt{n}}\right)^n = \cos^n\left(\frac{t}{\sqrt{n}}\right)$$

Example 4: Find the joint moment generating function for the two random variables having joint density is

$$f_{XY}(x, y) = \frac{1}{2\pi}e^{-\frac{1}{2}(x^2+y^2)} \qquad -\infty < x, y < \infty$$

Sol: Since we have

$$\phi_{X_1X_2}(t_1, t_2) = E\left[e^{it_1x_1+it_2x_2}\right]$$

$$= \int_{-\infty}^{\infty}\int_{-\infty}^{\infty} e^{it_1x_1+it_2x_2} f_{X_1X_2}(x_1, x_2)dx_1dx_2$$

$$\phi_{XY}(t_1, t_2) = \int_{-\infty}^{\infty}\int_{-\infty}^{\infty} e^{it_1x+it_2y} f_{XY}(x, y)dxdy$$

$$= \int_{-\infty}^{\infty}\int_{-\infty}^{\infty} e^{it_1x+it_2y} \frac{1}{2\pi}e^{-\frac{1}{2}(x^2+y^2)}dxdy$$

$$= \frac{1}{2\pi}\int_{-\infty}^{\infty}\int_{-\infty}^{\infty} e^{-\frac{x^2}{2}}e^{it_1x}e^{-\frac{y^2}{2}}e^{it_2y}dxdy$$

$$= \frac{1}{2\pi}\int_{-\infty}^{\infty}\int_{-\infty}^{\infty} e^{-\frac{x^2}{2}+it_1x}dx.e^{-\frac{y^2}{2}+it_2y}dy$$

$$= \frac{1}{2\pi}\int_{-\infty}^{\infty}\int_{-\infty}^{\infty} e^{-\frac{x^2-2it_1x}{2}}dx.e^{-\frac{y^2-2it_2y}{2}}dy$$

$$= \frac{1}{2\pi} \int_{-\infty}^{\infty} \int_{-\infty}^{\infty} e^{-\frac{x^2-2it_1x+(it_1)^2-(it_1)^2}{2}} dx.e^{-\frac{y^2-2it_2y+(it_2)^2-(it_2)^2}{2}} dy$$

$$= \frac{1}{2\pi} \int_{-\infty}^{\infty} \int_{-\infty}^{\infty} e^{-\frac{(x-it_1)^2}{2}} e^{\frac{(it_1)^2}{2}} dx.e^{-\frac{(y-it_2)^2}{2}}.e^{\frac{(it_2)^2}{2}} dy$$

$$= \frac{1}{2\pi} \int_{-\infty}^{\infty} \int_{-\infty}^{\infty} e^{-\frac{(x-it_1)^2}{2}} dx.e^{-\frac{(y-it_2)^2}{2}} e^{-\frac{t_1^2}{2}-\frac{t_2^2}{2}} dy$$

$$= e^{-\frac{t_1^2+t_2^2}{2}} \left[\frac{1}{\sqrt{2\pi}} \int_{-\infty}^{\infty} e^{-\frac{(x-it_1)^2}{2}} dx. \frac{1}{\sqrt{2\pi}} \int_{-\infty}^{\infty} .e^{-\frac{(y-it_2)^2}{2}} .dy \right]$$

Since the integrals $\dfrac{1}{\sqrt{2\pi}} \displaystyle\int_{-\infty}^{\infty} e^{-\frac{(x-it_1)^2}{2}} dx$ and $\dfrac{1}{\sqrt{2\pi}} \displaystyle\int_{-\infty}^{\infty} .e^{-\frac{(y-it_2)^2}{2}} .dy$ are the areas

enclosed by the density of Gaussian Random Variables with mean it_1 and it_2 and unit standard deviations, and these are

$$\frac{1}{\sqrt{2\pi}} \int_{-\infty}^{\infty} e^{-\frac{(x-it_1)^2}{2}} dx = 1 \text{ and } \frac{1}{\sqrt{2\pi}} \int_{-\infty}^{\infty} .e^{-\frac{(y-it_2)^2}{2}} .dy = 1$$

$$\therefore \qquad \phi_{XY}(t_1,t_2) = e^{-\frac{t_1^2+t_2^2}{2}}$$

5.10 Jointly Gaussian Random Variables

In many cases, we have to deal with two or three dimensional Gaussian random variables and it is useful that we particularize the n-dimensional general case to second order.

We studying the localizations of autonomous robots, the random vector X plays the role of the robot's location. Depending on the robots characteristics and on the operating environment, the location may be expressed as

(i)   A two dimensional vector with the position in a two dimensional environment

(ii) A three dimensional vector (two dimensional position and orientation) representing a mobile robot's location in an horizontal environment

(iii) A six dimensional vector (three positions and three orientations) in a underwater vehicle.

When characterizing a two dimensional laser scanner in a statistical frame work, ach range measurement is associated with a given pan angle corresponding the scanning mechanism. Therefore the pair (distance and angle)may be considered as a random vector whose statistical characterization depends on the physical principle of the sensor device.

The above examples refer quantities (e.g., robot position, senser measurements) that are not deterministic. To account for the associated uncertainties, we consider them as random vectors.

The univariate Gaussian distribution has the form $f(x) = \dfrac{1}{\sigma\sqrt{2\pi}} e^{-\frac{1}{2}\left(\frac{x-\mu}{\sigma}\right)^2}$

$-\infty \leq X \leq \infty$

If we have a random variable X which is distributed according to normal distribution with mean μ and standard deviation σ as $X \sim N(\mu, \sigma^2)$

The joint Gaussian distribution is a generalization of a single variate normal distribution.

In this section we consider two Gaussian (Normal) random variables joint probability and we extend for n Gaussian random variables. Gaussian random variables are very useful in every field of science and engineering.

5.11 n-Random Variable Case

A random vector $X = [X_1, X_2, X_3 \cdots X]^T \in R^n$ is a Gaussian if its probability density function is given by

$$f_{X_1 X_2 \cdots X_n}(x_1, x_2 \cdots x_n) = \frac{1}{\left[\sqrt{2\pi}\right]^n (|C|)^{\frac{1}{n}}} e^{-\frac{1}{2}\left[(X-\mu)^T C^{-1}(X-\mu)\right]} \qquad \text{.....(5.18)}$$

where n is the dimension of the random variable

$\mu = E[X]$ is the mean of the random vector X,

The matrix $X - \mu = \begin{bmatrix} X_1 - \mu_1 \\ X_1 - \mu_2 \\ \vdots \quad \vdots \quad \vdots \\ X_1 - \mu_n \end{bmatrix}$

Here $\mu_1,\ \mu_2 \cdots \mu_n$ are the means of the random variables $X_1,\ X_2, X_3 \cdots X_n$ respectively

$$C = \begin{bmatrix} c_{11} & c_{12} \cdots c_{1j} & \cdots & c_{1n} \\ c_{21} & c_{22} \cdots c_{2j} & \cdots & c_{2n} \\ \cdots & \cdots & & \cdots \\ c_{i1} & c_{i2} \cdots c_{ij} & \cdots & c_{in} \\ \cdots & \cdots & & \cdots \\ c_{n1} & c_{n2} \cdots c_{nj} & \cdots & c_{nn} \end{bmatrix}_{n \times n} \quad \text{is the covariance matrix}$$

where $c_{ij} = E\left[(x_i - \mu_i)(x_j - \mu_j)\right]$ $\qquad i = 1, 2, \cdots n, \ j = 1, 2, \cdots n$

$$C = \begin{bmatrix} E\left[(x_1 - \mu_1)^2\right] & E\left[(x_1 - \mu_1)(x_2 - \mu_2)\right] \cdots E\left[(x_1 - \mu_1)(x_j - \mu_j)\right] \cdots E\left[(x_1 - \mu_1)(x_n - \mu_n)\right] \\ E\left[(x_2 - \mu_2)(x_1 - \mu_1)\right] & E\left[(x_2 - \mu_2)^2\right] & \cdots E\left[(x_2 - \mu_2)(x_j - \mu_j)\right] \cdots E\left[(x_2 - \mu_2)(x_j - \mu_j)\right] \\ \cdots & \cdots & \cdots \\ E\left[(x_i - \mu_i)(x_1 - \mu_1)\right] & E\left[(x_i - \mu_i)(x_2 - \mu_2)\right] \cdots E\left[(x_i - \mu_i)(x_j - \mu_j)\right] \cdots E\left[(x_2 - \mu_2)(x_j - \mu_j)\right] \\ \cdots & \cdots & \cdots \\ E\left[(x_n - \mu_n)(x_1 - \mu_1)\right] & E\left[(x_n - \mu_n)(x_2 - \mu_2)\right] \cdots E\left[(x_n - \mu_n)(x_j - \mu_j)\right] \cdots & E\left[(x_n - \mu_n)^2\right] \end{bmatrix}_{n \times n}$$

The covariance matrix is symmetric matrix is a symmetric matrix. The diagonal elements of C are the variances of the random variables ($i=j$) and the general elements are the $c_{ij} = E\left[(x_i - \mu_i)(x_j - \mu_j)\right]$ $\qquad$ for $i \ne j$, $i = 1, 2, \cdots n$, $j = 1, 2, \cdots n$ represents the covariance of two random variables X_i and X_j.

Similarly, to the scalar case, the probability density function of a Gaussian random vector is completely characterized by its first and second moments.

5.12 Two Random Variables Case

A random vector $Z = [X \ Y]$ is a Gaussian if its probability density function is given by

$$f_{X,Y}(x,y) = \frac{1}{\left[\sqrt{2\pi}\right]^2 (|C|)^{\frac{1}{2}}} e^{\frac{1}{2}\left[(Z-\mu)^T C^{-1}(Z-\mu)\right]} \qquad \qquad .....(5.19)$$

Let X and Y be two Gaussian random variables, here n=2 in the general n-random variable case, follows i.e., $X \sim N(\mu_1, \sigma_x^2) \ Y \sim N(\mu_2, \sigma_y^2)$ then the matrix

$$Z - \mu = \begin{bmatrix} X - \mu_1 \\ Y - \mu_2 \end{bmatrix}$$

Here μ_1, μ_2 are the means of the random variables X and Y respectively and the covariance matrix

$$C = \begin{bmatrix} \sigma_x^2 & \rho\, \sigma_x \sigma_y \\ \rho\, \sigma_x \sigma_y & \sigma_y^2 \end{bmatrix}_{2 \times 2}$$

$\because \qquad \rho = \dfrac{Cov(X,Y)}{\sigma_x \sigma_y}$

$\Rightarrow \qquad Cov(X,Y) = \rho \sigma_x \sigma_y = E\big[(x - \mu_1)(y - \mu_2)\big] = c_{12} = c_{21}$

here $\qquad i = 1,2., \ \ j = 1,2.$

Now

$$|C| = \left| \begin{bmatrix} \sigma_x^2 & \rho\, \sigma_x \sigma_y \\ \rho\, \sigma_x \sigma_y & \sigma_y^2 \end{bmatrix} \right|$$

$$= \sigma_x^2 \sigma_y^2 - \rho^2 \sigma_x^2 \sigma_y^2$$

$$= (1 - \rho^2)\sigma_x^2 \sigma_y^2$$

$\therefore \qquad \sqrt{|C|} = \sqrt{(1 - \rho^2)}\, \sigma_x \sigma_y$

The inverse of the matrix C is

$$C^{-1} = \frac{adj\ C}{|C|} = \frac{1}{(1 - \rho^2)\sigma_x^2 \sigma_y^2} \begin{bmatrix} \sigma_y^2 & -\rho\, \sigma_x \sigma_y \\ -\rho\, \sigma_x \sigma_y & \sigma_x^2 \end{bmatrix}$$

Now consider $(Z - \mu)^T C^{-1}(Z - \mu)$,

$(Z - \mu)^T C^{-1}(Z - \mu)$

$$= \begin{bmatrix} X - \mu_1 & Y - \mu_2 \end{bmatrix} \frac{1}{(1 - \rho^2)\sigma_x^2 \sigma_y^2} \begin{bmatrix} \sigma_y^2 & -\rho\, \sigma_x \sigma_y \\ -\rho\, \sigma_x \sigma_y & \sigma_x^2 \end{bmatrix} \begin{bmatrix} X - \mu_1 \\ Y - \mu_2 \end{bmatrix}$$

$$= \frac{1}{(1 - \rho^2)\sigma_x^2 \sigma_y^2} \begin{bmatrix} X - \mu_1 & Y - \mu_2 \end{bmatrix} \begin{bmatrix} \sigma_y^2 & -\rho\, \sigma_x \sigma_y \\ -\rho\, \sigma_x \sigma_y & \sigma_x^2 \end{bmatrix} \begin{bmatrix} X - \mu_1 \\ Y - \mu_2 \end{bmatrix}$$

$$= \frac{1}{(1 - \rho^2)\sigma_x^2 \sigma_y^2} \begin{bmatrix} (X - \mu_1)\sigma_y^2 - \rho(Y - \mu_2)\sigma_x \sigma_y \\ -(X - \mu_1)\rho\, \sigma_x \sigma_y + (Y - \mu_2)\sigma_x^2 \end{bmatrix}^T \begin{bmatrix} X - \mu_1 \\ Y - \mu_2 \end{bmatrix}$$

$$= \frac{1}{(1-\rho^2)\sigma_x^2\sigma_y^2} \left[(X-\mu_1)^2\sigma_y^2 - 2\rho(X-\mu_1)(Y-\mu_2)\sigma_x\sigma_y + (Y-\mu_2)\sigma_x^2 \right]$$

$$= \frac{1}{(1-\rho^2)} \left[\frac{(X-\mu_1)^2}{\sigma_x^2} - 2\rho\frac{(X-\mu_1)(Y-\mu_2)}{\sigma_x\sigma_y} + \frac{(Y-\mu_2)^2}{\sigma_y^2} \right]$$

$$\therefore \quad (Z-\mu)^T C^{-1}(Z-\mu)$$

$$= \frac{1}{(1-\rho^2)} \left[\left(\frac{X-\mu_1}{\sigma_x}\right)^2 - 2\rho\left(\frac{X-\mu_1}{\sigma_x}\right)\left(\frac{Y-\mu_2}{\sigma_y}\right) + \left(\frac{Y-\mu_2}{\sigma_y}\right)^2 \right]$$

Therefore the equation of the n-dimensional case to two dimensional case is

$$f_{X,Y}(x,y) = \frac{1}{2\pi\sqrt{(1-\rho^2)}\sigma_x\sigma_y} \left\{ e^{-\frac{1}{2(1-\rho^2)}\left[\left(\frac{X-\mu_1}{\sigma_x}\right)^2 - 2\rho\left(\frac{X-\mu_1}{\sigma_x}\right)\left(\frac{Y-\mu_2}{\sigma_y}\right) + \left(\frac{Y-\mu_2}{\sigma_y}\right)^2\right]} \right.$$

Is the joint density of X and Y.

If X and Y are independent then the correlation coefficient $\rho = 0$, then the joint probability density function is

$$f_{X,Y}(x,y) = \frac{1}{2\pi\sigma_x\sigma_y} e^{-\frac{1}{2}\left[\left(\frac{X-\mu_1}{\sigma_x}\right)^2 + \left(\frac{Y-\mu_2}{\sigma_y}\right)^2\right]}$$

$$= \frac{1}{\sqrt{2\pi}\sigma_x} e^{-\frac{1}{2}\left(\frac{X-\mu_1}{\sigma_x}\right)^2} \frac{1}{\sqrt{2\pi}\sigma_x} e^{-\frac{1}{2}\left(\frac{Y-\mu_2}{\sigma_y}\right)^2}$$

$$- f_X(x) \times f_Y(y)$$

$$\therefore \quad f_{X,Y}(x,y) = f_X(x) \times f_Y(y)$$

5.13 Properties

1. In the bivariate Gaussian distribution, the random variables X and Y each follows normal distribution with means μ_1, μ_2 and variances σ_x^2, σ_y^2 respectively.

   i.e., $X \sim N(\mu_1, \sigma_x^2)$ $Y \sim N(\mu_2, \sigma_y^2)$

2. As in univariate normal distribution, the joint Gaussian distribution has $\int\limits_{x}\int\limits_{y} f_{X,Y}(x,y)dxdy = 1$ and $f_{X,Y}(x,y) \geq 0$

3. The joint Gaussian random variable occupies a surface in xy-plane

4. If the random variables X and Y are independent (then $\rho = 0$) then the surfaces of constant $f_{X,Y}(x,y)$ are concentric circles around the origin.

5. The curve of the two dimensional Gaussian curve is fig.

6. The bivariate Gaussian distribution has a maximum at the origin

7. If there is a positive correlation between the variables, the distribution is a stretched diagonally, forming elliptical isopleths with positive shaped major axes. If there is a negative correlation, the major axes have a negative slope.

8. When σ_x, σ_y and ρ is known then the joint probability density function is completely determined.

5.14 Transformations of Multiple Random Variables

Multivariate transformations are more powerful than the univariate transformations. They depend on the concept of Jacobian.

First we consider a single functional transformation of multiple random variables for finding the density of a random variable and then we extend for several functions of several random variables.

5.15 Random Variable Expressed as a Function of Two Random Variables

Let X_1, X_2 be two independent random variables whose densities are known. Now we wish to know the density of $Z = X_1 + X_2$.

To obtain the density of Z, the first method by using the properties of moment generating function, but this method is limited to certain conditions.

The second method, by using a bivariate transformation that deals easily to find the joint density of the two random variables.

The distribution of Z is the convolution of the distribution of X_1 and X_2. With same modifications, the distribution is to be calculate, it appears as a marginal distribution of a multivariate distribution that is built artificially for the sole purpose of integrating out all the variables but the variable of interest.

Let a random variable Z be expressed as a function of two independent random variables with densities $f_{X_1}(x_1)$ and $f_{X_2}(x_2)$ and Z=g(X_1, X_2). For finding the density of Z by introducing z, let z be a given number and R_z be the region of the $x_1 x_2$ - plane such that g($x_1 x_2$) $\leq z$

Now
$$F_Z(z) = P(Z \leq z)$$

$$= P[g(X_1, X_1) \leq z]$$

$$= P[(X_1, X_1) \in R_z]$$

$$= \int_{x_1} \int_{x_2} f_{X_1, X_2}(x_1, x_2) dx_1 dx_2$$

Now the density function of Z is

$$f(z) = \frac{d}{dz} F_Z(z)$$

$$= \frac{d}{dz} \left\{ \int_{x_1} \int_{x_2} f_{X_1, X_2}(x_1, x_2) dx_1 dx_2 \right\}$$

Now we extend for n random variables

Let $Z = g(X_1, X_2, \cdots X_n)$, we shall find the density function of Z

The distribution function of Z is

Now
$$F_Z(z) = P(Z \leq z)$$

$$= P[g(X_1, X_2, \cdots X_n) \leq z]$$

$$= P[(X_1, X_2, \cdots X_n) \in R_z]$$

$$= \int_{x_1} \int_{x_2} \cdots \int_{x_n} f_{X_1, X_2, \cdots, X_n}(x_1, x_2 \cdots x_n) dx_1 dx_2 \cdots dx_n$$

Now the density function of Z is

$$f(z) = \frac{d}{dz} F_Z(z)$$

$$= \frac{d}{dz} \left\{ \int_{x_1} \int_{x_2} \cdots \int_{x_n} f_{X_1, X_2, \cdots, X_n}(x_1, x_2 \cdots x_n) dx_1 dx_2 \cdots dx_n \right\}$$

5.16   Linear Transformations of Gaussian Random Variables

Let $f_{X_1,X_2}(x_1,x_2)$ be the joint density function of the continuous random variables X_1 and X_2 at (x_1,x_2). If the functions given by $Y_1 = g(x_1,x_2)$ and $Y_2 = g(x_1,x_2)$ are partially differentiable with respect to x_1 and x_2 represent a one to one transformation for all values with in the range of X_1 and X_2 for which $f_{X_1,X_2}(x_1,x_2) \neq 0$, then for these values of x_1 and x_2, and y_1 and y_2 can be uniquely solved for x_1 and x_2.

To given $x_1 = w_1(y_1,y_2)$ and $x_2 = w_2(y_1,y_2)$ and for corresponding values of y_1 and y_2, the joint probability density of $Y_1 = g_1(X_1,X_2)$ and $Y_2 = g_2(X_1,X_2)$ is

$$f_{Y_1,Y_2}(y_1,y_2) = f_{X_1,X_2}\left[w_1(y_1,y_2),w_2(y_1,y_2)\right].\left|j(x,y)\right|$$

where J is the Jacobean of the transformation and is defined as

$$j(x,y) = \begin{vmatrix} \dfrac{\partial x_1}{\partial y_1} & \dfrac{\partial x_1}{\partial y_2} \\ \dfrac{\partial x_2}{\partial y_1} & \dfrac{\partial x_2}{\partial y_2} \end{vmatrix} \text{ and all other point } f_{Y_1,Y_2}(y_1,y_2) = 0$$

Let X_1 and X_2 be two Gaussian random variables and have jointly Gaussian as $N(\mu_1,\mu_2,\sigma_x^2,\sigma_y^2)$

Let $Y_1 = a_{11}X_1 + a_{12}X_2$

And $Y_2 = a_{21}X_1 + a_{22}X_2$ .....(5.20)

Be new random variables also having jointly normal distribution. If $a_{11}a_{22} - a_{12}a_{21} \neq 0$ then the system of equations can be written as $Y = A X$

where $\quad A = \begin{bmatrix} a_{11} & a_{12} \\ a_{21} & a_{22} \end{bmatrix} \quad X = \begin{bmatrix} X_1 \\ X_2 \end{bmatrix} \quad \overline{X} = \begin{bmatrix} \overline{X_1} \\ \overline{X_2} \end{bmatrix} \text{ and } Y = \begin{bmatrix} Y_1 \\ Y_2 \end{bmatrix} \text{ then } \overline{Y} = \begin{bmatrix} \overline{Y_1} \\ \overline{Y_2} \end{bmatrix}$

Therefore the system of equations cab represent as

$$\begin{bmatrix} Y_1 \\ Y_2 \end{bmatrix} = \begin{bmatrix} a_{11} & a_{12} \\ a_{21} & a_{22} \end{bmatrix}\begin{bmatrix} X_1 \\ X_2 \end{bmatrix} \qquad .....(5.21)$$

This system of equations can also be represented as

$$[Y] = [A][X] \qquad .....(5.22)$$

and $$[Y - \overline{Y}] = [A][X - \overline{X}] \qquad .....(5.23)$$

$$[X] = [A]^{-1} [Y] \quad\quad\quad .....(5.24)$$

and
$$\left[X - \overline{X} \right] = [A]^{-1} \left[Y - \overline{Y} \right] \quad\quad .....(5.25)$$

Let
$$[A]^{-1} = \begin{bmatrix} a^{11} & a^{12} \\ a^{21} & a^{22} \end{bmatrix} \text{ from equation (5.24)}, [X] = [A]^{-1} [Y]$$

$$\begin{bmatrix} X_1 \\ X_2 \end{bmatrix} = \begin{bmatrix} a^{11} & a^{12} \\ a^{21} & a^{22} \end{bmatrix} \begin{bmatrix} Y_1 \\ Y_2 \end{bmatrix}$$

$$= \begin{bmatrix} a^{11}Y_1 + a^{12}Y_2 \\ a^{21}Y_1 + a^{22}Y_1 \end{bmatrix}$$

$$\therefore X_1 = a^{11}Y_1 + a^{12}Y_2$$

and
$$X_2 = a^{21}Y_1 + a^{22}Y_2 \quad\quad .....(5.26)$$

Differentiate these equations (5.26) partially with respect to Y_1 and Y_2 respectively

$$\frac{\partial X_1}{\partial Y_1} = a^{11}; \ \frac{\partial X_1}{\partial Y_2} = a^{12}; \ \frac{\partial X_2}{\partial Y_1} = a^{21} \text{ and } \frac{\partial X_2}{\partial Y_2} = a^{22} \quad\quad .....(5.27)$$

Similarly from

$$\therefore \quad\quad X_1 - \overline{X}_1 = a^{11}\left(Y_1 - \overline{Y}_1\right) + a^{12}\left(Y_2 - \overline{Y}_2\right)$$

and
$$X_2 - \overline{X}_2 = a^{21}\left(Y_1 - \overline{Y}_1\right) + a^{22}\left(Y_2 - \overline{Y}_2\right) \quad\quad .....(5.28)$$

The density of the new variables Y_1 and Y_2 is found by solving the right   hand side equation (5.28), by taking the Jacobian of the inverse transformation by using (8) and (9), we find that jacobian J equals to the determinant of the matrix $[A]^{-1}$

i.e.,
$$J = \left\| [A]^{-1} \right\| \Rightarrow \begin{vmatrix} \dfrac{\partial x_1}{\partial y_1} & \dfrac{\partial x_1}{\partial y_2} \\ \dfrac{\partial x_2}{\partial y_1} & \dfrac{\partial x_2}{\partial y_2} \end{vmatrix} = \left\| [A]^{-1} \right\|$$

Therefore $\quad |J| = \left\| [A]^{-1} \right\|$

Now consider the covariance by using equation,

$$C_{X_1X_2} = E\left[\left(X_1 - \overline{X}_1\right)\left(X_2 - \overline{X}_2\right)\right]$$

Since

$$C_{X_1X_2} = C_{X_2X_1}$$

$$C_{X_2X_1} = E\left[\left(X_2 - \overline{X}_2\right)\left(X_1 - \overline{X}_1\right)\right]$$

$$C_{X_1X_1} = E\left[\left(X_1 - \overline{X}_1\right)^2\right]$$

$$C_{X_2X_2} = E\left[\left(X_2 - \overline{X}_2\right)^2\right]$$

Now

$$C_{X_1X_2} = E\left[\left(X_1 - \overline{X}_1\right)\left(X_2 - \overline{X}_2\right)\right]$$

$$= E\left[\left\{a^{11}\left(Y_1 - \overline{Y}_1\right) + a^{12}\left(Y_2 - \overline{Y}_2\right)\right\}\left\{a^{21}\left(Y_1 - \overline{Y}_1\right) + a^{22}\left(Y_2 - \overline{Y}_2\right)\right\}\right]$$

$$= a^{11}a^{21}E\left[\left(Y_1 - \overline{Y}_1\right)^2\right] + a^{11}a^{22}E\left[\left(Y_1 - \overline{Y}_1\right)\left(Y_2 - \overline{Y}_2\right)\right] +$$

$$a^{12}a^{21}E\left[\left(Y_1 - \overline{Y}_1\right)\left(Y_2 - \overline{Y}_2\right)\right] + a^{12}a^{22}E\left[\left(Y_2 - \overline{Y}_2\right)^2\right]$$

Therefore

$$C_{X_1X_2} = C_{X_2X_1} = a^{11}a^{21}E\left[\left(Y_1 - \overline{Y}_1\right)^2\right] + a^{11}a^{22}E\left[\left(Y_1 - \overline{Y}_1\right)\left(Y_2 - \overline{Y}_2\right)\right] + 1$$

$$a^{12}a^{21}E\left[\left(Y_1 - \overline{Y}_1\right)\left(Y_2 - \overline{Y}_2\right)\right] + a^{12}a^{22}E\left[\left(Y_2 - \overline{Y}_2\right)^2\right]$$

$$C_{X_1X_1} = E\left[\left(X_1 - \overline{X}_1\right)^2\right] = E\left[\left\{a^{11}\left(Y_1 - \overline{Y}_1\right) + a^{12}\left(Y_2 - \overline{Y}_2\right)\right\}^2\right]$$

$$= E\left[\left\{a^{11}\left(Y_1 - \overline{Y}_1\right) + a^{12}\left(Y_2 - \overline{Y}_2\right)\right\} \times \left\{a^{11}\left(Y_1 - \overline{Y}_1\right) + a^{12}\left(Y_2 - \overline{Y}_2\right)\right\}\right]$$

$$= E\left[\left(a^{11}\right)^2\left(Y_1 - \overline{Y}_1\right)^2\right] + a^{11}a^{12}E\left[\left(Y_1 - \overline{Y}_1\right)\left(Y_2 - \overline{Y}_2\right)\right] +$$

$$a^{11}a^{12}E\left[\left(Y_1 - \overline{Y}_1\right)\left(Y_2 - \overline{Y}_2\right)\right] + \left(a^{12}\right)^2\left[\left(Y_2 - \overline{Y}_2\right)^2\right]$$

$$= \left(a^{11}\right)^2 E\left[\left(Y_1 - \overline{Y}_1\right)^2\right] + 2a^{11}a^{12}E\left[\left(Y_1 - \overline{Y}_1\right)\left(Y_2 - \overline{Y}_2\right)\right] + \left(a^{12}\right)^2\left[\left(Y_2 - \overline{Y}_2\right)^2\right]$$

$$= \left(a^{11}\right)^2 \sigma_{Y_1}^2 + 2a^{11}a^{12}Cov(Y_1, Y_2 + \left(a^{12}\right)^2 \sigma_{Y_2}^2$$

$$\therefore C_{X_1X_1} = \left(a^{11}\right)^2 \sigma_{Y_1}^2 + 2a^{11}a^{12}Cov(Y_1, Y_2 + \left(a^{12}\right)^2 \sigma_{Y_2}^2$$

Similarly,

$$C_{X_2X_2} = \left(a^{21}\right)^2 \sigma_{Y_1}^2 + 2a^{21}a^{22}Cov(Y_1,Y_2) + \left(a^{22}\right)^2 \sigma_{Y_2}^2$$

Now the covariance matrix

$$[C_X] = \begin{bmatrix} C_{X_1X_1} & C_{X_1X_2} \\ C_{X_2X_1} & C_{X_2X_2} \end{bmatrix}$$

$$= [A]^T \ [C_y] \ \left[[A]^T\right]^{-1}$$

Taking the inverse on both sides then

$$[C_X]^{-1} = [A]^T \ [C_y]^{-1} \ [A]$$

Taking the determinant on both sides then

$$\left|[C_X]^{-1}\right| = \left|[C_y]^{-1}\right|\left|[A]^T\right|^2 \qquad \qquad .....(5.29)$$

Since X_1 and X_2 are jointly Gaussian, their joint density is given by

$$f_{X_1,X_2}(x_1,x_2) = \frac{1}{2\pi\left(|C_X|\right)^{\frac{1}{2}}} e^{-\frac{1}{2}\left[(X-\overline{X})^T[C_X]^{-1}(X-\overline{X})\right]}$$

$$= \frac{\left|(C_X)^{-1}\right|^{\frac{1}{2}}}{2\pi} e^{-\frac{1}{2}\left[(X-\overline{X})^T[C_X]^{-1}(X-\overline{X})\right]} \qquad \qquad .....(5.30)$$

From equations (5.29) and (5.30), we get

$$f_{X_1,X_2}(x_1,x_2) = \frac{\left|[C_y]^{-1}\right|[A]^2\right|^{\frac{1}{2}}}{2\pi} e^{-\frac{1}{2}\left[(X-\overline{X})^T[A]^T[C_y]^{-1}[A](X-\overline{X})\right]} \qquad \qquad .....(5.31)$$

Now $f_{Y_1,Y_2}(y_1,y_2) = f_{X_1,X_2}(x_1,x_2) \ |J|$

$$= \frac{\left|(C_Y)^{-1}\right|}{2\pi} e^{-\frac{1}{2}\left[(Y-\overline{Y})^T[C_Y]^{-1}(Y-\overline{Y})\right]} \qquad \qquad .....(5.32)$$

Since $\left[Y-\overline{Y}\right]=[A]\left[X-\overline{X}\right]$ this implies $\left[Y-\overline{Y}\right]^T=\left[X-\overline{X}\right]^T[A]^T$

Equation (5.32) indicates the new random variables Y_1 and Y_2 are jointly Gaussian. Thus the linear transformation of Gaussian random variables gives Gaussian random variables.

Extension for n random variables

Let $X_1, X_2 \cdots X_n$ be two Gaussian random variables and have jointly Gaussian as $N(\mu_1, \mu_2 \cdots \mu_n, \sigma_{x_1}^2, \sigma_{x_2}^2 \cdots \sigma_{x_n}^2)$

Let
$$Y_1 = a_{11}X_1 + a_{12}X_2 + \cdots + a_{1n}X_n$$
$$Y_2 = a_{21}X_1 + a_{22}X_2 + \cdots + a_{2n}X_n$$
$$\vdots \qquad \vdots \qquad \vdots$$
$$Y_n = a_{n1}X_1 + a_{n2}X_2 + \cdots + a_{nn}X_n$$

Proceeding in the above two random variable case the density of the random variable Y is

$$f_{Y_1,Y_2\cdots Y_n}(y_1, y_2 \cdots y_n) = \frac{\left|(C_Y)^{-1}\right|^{\frac{1}{2}}}{(2\pi)^{\frac{n}{2}}} e^{-\frac{1}{2}\left[(Y-\bar{Y})^T [C_Y]^{-1}(Y-\bar{Y})\right]}$$

Examples

Example 1: Let X_1 and X_2 be two Gaussian random variables having zero means and variances $\sigma_1^2 = 9$ and $\sigma_2^2 = 16$. Their covariance $C_{X_2X_2}$ equals to 4. If X_1 and X_2 are linearly transformed to new variables Y_1 and Y_2 according to $Y_1 = X_1 - 2X_2$ and $Y_2 = 3X_1 + 4X_2$. Find the means and variances of Y_1 and Y_2 and covariance both Y_2 and Y_2.

Sol: Since given $\sigma_1^2 = \sigma_{x_1}^2 = 9$ and $\sigma_2^2 = \sigma_{x_2}^2 = 16$

Here the coefficient matrix

$$A = \begin{bmatrix} a_{11} & a_{12} \\ a_{21} & a_{22} \end{bmatrix} \text{ i.e., } A = \begin{bmatrix} 1 & -2 \\ 3 & 4 \end{bmatrix}$$

$$[C_X] = \begin{bmatrix} 9 & 4 \\ 4 & 16 \end{bmatrix}$$

Since X_1 and X_2 are zero mean and Gaussian, Y_1 and Y_2 will also be zero means and Gaussian $\bar{Y}_1 = \bar{Y}_2 = 0$

$$\left[C_y \right] = \left[A \right]^T \left[C_X \right] \left[\left[A \right]^T \right]^{-1}$$

$$= \begin{bmatrix} 1 & -2 \\ 3 & 4 \end{bmatrix} \begin{bmatrix} 9 & 4 \\ 4 & 16 \end{bmatrix} \begin{bmatrix} 1 & 3 \\ -2 & 4 \end{bmatrix}$$

$$= \begin{bmatrix} 1 \times 9 - 2 \times 4 & 1 \times 4 - 2 \times 16 \\ 3 \times 9 + 4 \times 4 & 3 \times 4 + 4 \times 16 \end{bmatrix} \begin{bmatrix} 1 & 3 \\ -2 & 4 \end{bmatrix}$$

$$= \begin{bmatrix} 1 & -28 \\ 43 & 76 \end{bmatrix} \begin{bmatrix} 1 & 3 \\ -2 & 4 \end{bmatrix}$$

$$= \begin{bmatrix} 1 \times 1 + 28 \times 2 & 1 \times 3 - 28 \times 4 \\ 43 \times 1 + 76 \times -2 & 43 \times 3 + 76 \times 4 \end{bmatrix}$$

$$= \begin{bmatrix} 57 & -109 \\ -109 & 433 \end{bmatrix}$$

$\sigma_{Y_1}^2 = 57$ and $\sigma_{Y_2}^2 = 433$ and $\mathrm{Cov}(Y_1, Y_2) = -109$

***Example* 2:** In a particular factory, the average salary for an employee is Rs. 1,20,000 per year. This year, management awards the bonuses to every employee as follows. A Dussarah bonus of Rs. 1000, and an incentive bonus is 5 percent of the employee's salary. Find the mean bonus received by employees?

Sol: The required linear transformation is $Y = mX + b$ (Where Y is the bonus variable, m is the slope (i.e., the multiplicative constant $= 0.05$), X is the salary variable and b is the additional bonus, since the management adopt the linear transformation to each employee's salary is $Y = mX + b$)

$$Y = 0.05 * X + 1000$$

$$Y = 0.05 * 1, 20,000 + 1500$$

$$= 6,000 + 1,000$$

$$= 7,000$$

Therefore the mean bonus received by employees is 7,000

Additional Examples

***Example* 1:** For two random variables X and Y

$$f_{XY}(xy) = 0.15\delta(x+1)\delta(y) + 0.1\delta(x)\delta(y) + 0.1\delta(x)\delta(y-2) +$$
$$0.4\delta(x-1)\delta(y+2) + 0.2\delta(x-1)\delta(y-1) + 0.05\delta(x-1)\delta(y-3)$$

Find

   (i)   The Cross Correlation between X and Y

   (ii)   The covariance between X and Y

   (iii) The correlation coefficient between X and Y

   (iv)   Are X and Y are uncorrelated or orthogonal

Sol: Given that

$$f_{XY}(xy) = 0.15\delta(x+1)\delta(y) + 0.1\delta(x)\delta(y) + 0.1\delta(x)\delta(y-2) +$$
$$0.4\delta(x-1)\delta(y+2) + 0.2\delta(x-1)\delta(y-1) + 0.05\delta(x-1)\delta(y-3)$$

i.e., the joint probability matrix of X and Y is

Y \ X	−2	0	1	2	3	Marginal totals
−1	0	0.15	0	0	0	0.15
0	0	0.1	0	0.1	0	0.2
1	0.4	0	0.2	0	0.05	0.65
Marginal totals	0.4	0.25	0.2	0.1	0.05	1.00

   Now

   (i)   The Cross Correlation between X and Y is

$$R_{XY} = E[XY] = \sum_i \sum_j x_i y_j P(x_i, y_j)$$

$$= (-1)\times(-2)\times0 + (-1)\times(0)\times0.15 + (-1)\times(1)\times0 + (-1)\times(2)\times0 + (-1)\times(3)\times0 +$$
$$(0)\times(-2)\times0 + (0)\times(0)\times0.1 + (0)\times(1)\times0 + (0)\times(2)\times0.1 + (0)\times(3)\times0 +$$
$$(1)\times(-2)\times0.4 + (1)\times(0)\times0 + (1)\times(1)\times0.2 + (1)\times(2)\times0 + (1)\times(3)\times0.05$$

$$= 0 + 0 + (-0.8 + 0.2 + 0.15)$$

$$= -0.45$$

   (ii)   The covariance between X and Y is

$$Cov(X,Y) = E[XY] - E[X]E[Y]$$

   Now

$$E[X] = \sum xP(X = x) = -1\times0.15 + 0\times0.2 + 1\times0.65 = 0.5$$

$$E[Y] = \sum yP(Y = y)$$

$$= -0.25 \ -2\times0.4 + 0\times0.25 + 1\times0.2 + 2\times0.1 + 3\times0.05$$

$$\therefore Cov(X,Y) = -0.45 - (0.5 \times -0.25)$$

$$= -0.45 + 0.125$$

$$= -0.325$$

(iii) The correlation coefficient between X and Y is

$$\rho_{x,y} = \frac{Cov(X,Y)}{\sigma_x \sigma_y}$$

(iv) Variance of X:

$$V[X] = E[X^2] - [E(X)]^2$$

$$E[X^2] = \sum x^2 P(X = x)$$

$$= (-1)^2 \times 0.15 + 0^2 \times 0.2 + 1^2 \times 0.65$$

$$= 0.8$$

Therefore, $V[X] = 0.8 - (0.5)^2$

$$= 0.55$$

Hence $\sigma_x = 0.74$

Variance of Y:

$$V[Y] = E[Y^2] - [E(Y)]^2$$

Now $E[Y^2] = \sum y^2 P(Y = y)$

$$= (-2)^2 \times 0.4 + 0^2 \times 0.25 + 1^2 \times 0.2 + 2^2 \times 0.1 + 3^2 \times 0.05$$

$$= 2.65$$

Therefore,

$$V[Y] = 2.65 - (0.25)^2$$

$$= 2.65 - 0.625$$

$$= 2.5875$$

Hence $\sigma_y = 1.61$

Therefore the correlation Coefficient between X and Y is

$$\rho_{x,y} = \frac{-0.325}{0.74 \times 1.61}$$

$$= -0.273$$

Hence the correlation coefficient between X and Y is -0.273

(iv) Are X and Y are uncorrelated or orthogonal

Here from the above (iii) statement the two variables are correlated.

Example 2: Verify Cauchy-Schwartz's inequality (Cosine inequality) for two random variables X and Y stated as

$$\left[E(XY)\right]^2 \le E\left[X^2\right]E\left[Y^2\right]$$

Sol: Let us consider the inequality $\left[E(X-aY)^2\right] \ge 0$

$\Rightarrow$ $\qquad E\left[X^2 - 2aXY + a^2Y^2\right] \ge 0$

$\Rightarrow$ $\qquad E\left[X^2\right] - 2aE\left[XY\right] + a^2\left[Y^2\right] \ge 0$ $\qquad\qquad$ .....(5.33)

$\Rightarrow$ $\qquad E\left[X^2\right] - a\left\{2E\left[XY\right] + a\left[Y^2\right]\right\} \ge 0$

$\Rightarrow$ $\qquad E\left[X^2\right] \ge 2aE\left[XY\right] - a^2\left[Y^2\right]$ $\qquad\qquad$ .....(5.34)

Consider $\dfrac{d}{da}\left\{E\left[X^2\right] - 2aE\left[XY\right] + a^2E\left[Y^2\right]\right\} = 0$

$\Rightarrow$ $\qquad -2E\left[XY\right] + 2aE\left[Y^2\right] = 0$

$\Rightarrow$ $\qquad a = \dfrac{E\left[XY\right]}{E\left[Y^2\right]}$

From equation (5.34) becomes

$\Rightarrow$ $\qquad E\left[X^2\right] - 2\dfrac{E\left[XY\right]}{E\left[Y^2\right]}E\left[XY\right] + \left(\dfrac{E\left[XY\right]}{E\left[Y^2\right]}\right)^2 E\left[Y^2\right] \ge 0$

$\Rightarrow$ $\qquad E\left[X^2\right] - 2\dfrac{\left\{E\left[XY\right]\right\}^2}{E\left[Y^2\right]} + \dfrac{\left\{E\left[XY\right]\right\}^2}{E\left[Y^2\right]} \ge 0$

$\Rightarrow$ $\qquad E\left[X^2\right] - \dfrac{\left\{E\left[XY\right]\right\}^2}{E\left[Y^2\right]} \ge 0$

$\Rightarrow \qquad E\left[X^2\right] \geq \dfrac{\left\{E[XY]\right\}^2}{E\left[Y^2\right]}$

$\Rightarrow \qquad E\left[X^2\right]E\left[Y^2\right] \geq \left\{E[XY]\right\}^2$

Hence $\qquad \left\{E[XY]\right\}^2 \leq E\left[X^2\right]E\left[Y^2\right]$

***Example* 3:** Consider random variables Y_1 and Y_2 are related to arbitrary random variables X and Y by the coordinate rotation $Y_1 = X \cos\theta + Y \sin\theta$ and $Y_2 = -X \sin s\theta + Y \cos\theta$

Find *Covariance of Y_1 and Y_2 and for what value of θ,* the random variables Y_1 and Y_2 uncorrelated

Sol: *Given that*

$$Y_1 = X \cos\theta + Y \sin\theta \text{ and } Y_2 = -X \sin s\theta + Y \cos\theta$$

Let the means of the random variables X and Y are $\overline{X}$ and $\overline{Y}$, then the means of Y_1 and Y_2 are $\overline{Y}_1 = \overline{X} \cos\theta + \overline{Y} \sin\theta$ and $\overline{Y}_2 = -\overline{X} \sin s\theta + \overline{Y} \cos\theta$

Covariance of Y_1 and Y_2 is

$$C_{Y_1 Y_2} = E\left[\left(Y_1 - \overline{Y}_1\right)\left(Y_2 - \overline{Y}_2\right)\right] \qquad \qquad ....(5.35)$$

$$= E\left[\left\{\left(X - \overline{X}\right)\cos\theta + \left(Y - \overline{Y}\right)\sin\theta\right\}\left\{-\left(X - \overline{X}\right)\sin\theta + \left(Y - \overline{Y}\right)\cos\theta\right\}\right]$$

$$= E\left[-\left(X - \overline{X}\right)^2 \cos\theta \sin\theta + \left(X - \overline{X}\right)\left(Y - \overline{Y}\right)\cos^2\theta - \left(X - \overline{X}\right)\left(Y - \overline{Y}\right)\sin^2\theta + \left(Y - \overline{Y}\right)^2 \cos\theta \sin\theta\right]$$

$$= -E\left[\left(X - \overline{X}\right)^2\right]\cos\theta \sin\theta + E\left[\left(X - \overline{X}\right)\left(Y - \overline{Y}\right)\right]\cos^2\theta - E\left[\left(X - \overline{X}\right)\left(Y - \overline{Y}\right)\right]\sin^2\theta + E\left[\left(Y - \overline{Y}\right)^2\right]\cos\theta \sin\theta$$

$$= -\sigma_x^2 \cos\theta \sin\theta + C_{XY} \cos^2\theta - C_{XY} \sin^2\theta + \sigma_y^2 \cos\theta \sin\theta$$

$$= \left(\sigma_y^2 - \sigma_x^2\right)\cos\theta \sin\theta + C_{XY}\left(\cos^2\theta - \sin^2\theta\right)$$

$$= \left(\sigma_y^2 - \sigma_x^2\right)\frac{1}{2}\sin 2\theta + C_{XY} \cos 2\theta$$

Since we know that, if ρ is the correlation coefficient between X and Y, then from equation (5.35)

$$C_{XY} = \rho\sigma_x\sigma_y \text{ since } \rho = \frac{Cov(X,Y)}{\sigma_x\sigma_y}$$

Therefore $\quad C_{Y_1 Y_2} = \left(\sigma_y^2 - \sigma_x^2\right)\frac{1}{2}\sin 2\theta + \rho\sigma_x\sigma_y \cos 2\theta \qquad \qquad ....(5.36)$

If Y_1 and Y_2 are uncorrelated then $C_{Y_1 Y_2} = 0$, then from equation (5.36) we get

$$\left(\sigma_y^2 - \sigma_x^2 \right) \frac{1}{2} \sin 2\theta + \rho \sigma_x \sigma_y \cos 2\theta = 0$$

$$\Rightarrow \tan 2\theta = \frac{\rho \sigma_x \sigma_y}{\sigma_y^2 - \sigma_x^2}$$

$$\Rightarrow \theta = \tan^{-1} \left(\frac{\rho \sigma_x \sigma_y}{\sigma_y^2 - \sigma_x^2} \right)$$

For this value of θ, the two random variables are uncorrelated.

***Example* 4:** Let the two random variables X and Y have the joint density function

$$f_{xy}(x,y) = \begin{cases} \dfrac{(x+y)^2}{40} & \text{for } -1 < x < 1; \ -3 \le y \le 3 \\ 0 & \text{Otherwise} \end{cases}$$

Find all the third order moments for X and Y

Sol: Given that the two random variables X and Y have the joint density function

$$f_{xy}(x,y) = \begin{cases} \dfrac{(x+y)^2}{40} & \text{for } -1 < x < 1; \ -3 \le y \le 3 \\ 0 & \text{Otherwise} \end{cases}$$

The $(i + j)^{th}$ order joint moment of two dimensional random variable about origin is defined as

$$\mu_{ij} = E\left[X^i Y^j \right]$$

$$= \begin{cases} \displaystyle\sum\sum x^i y^j f_{XY}(x,y) & \text{if } X \text{ and } Y \text{ discrete} \\ \displaystyle\int_{-\infty}^{\infty}\int_{-\infty}^{\infty} x^i y^j f_{XY}(x,y)\,dxdy & \text{if } X \text{ and } Y \text{ continuous} \end{cases}$$

The third order moments of X and Y are $\mu_{30}{}'$, $\mu_{03}{}'$, $\mu_{21}{}'$ and $\mu_{12}{}'$ Where

$$\mu_{30}{}' = E\left[X^3 Y^0 \right] = E\left[X^3 \right]$$

$$= \int_{-\infty}^{\infty}\int_{-\infty}^{\infty} x^3 f_{XY}(x,y)\,dxdy$$

$$= \int_{-3}^{3}\int_{-1}^{1}x^3\frac{(x+y)^2}{40}\,dxdy$$

$$= \frac{1}{40}\int_{-3}^{3}\int_{-1}^{1}x^3(x+y)^2\,dxdy$$

$$= \frac{1}{40}\int_{-3}^{3}\int_{-1}^{1}x^3(x^2+2xy+y^2)\,dxdy$$

$$= \frac{1}{40}\int_{-3}^{3}\int_{-1}^{1}(x^5+2x^4y+x^3y^2)\,dxdy$$

$$= \frac{1}{40}\int_{-3}^{3}\left[\frac{x^6}{6}+2\frac{x^5}{5}y+\frac{x^4}{4}y^2\right]_{-1}^{1}\,dy$$

$$= \frac{1}{40}\int_{-3}^{3}\left[\left(\frac{1}{6}-\frac{1}{6}\right)+2\left(\frac{1}{5}+\frac{1}{5}\right)y+\left(\frac{1}{4}-\frac{1}{4}\right)y^2\right]\,dy$$

$$= \frac{1}{40}\int_{-3}^{3}\frac{4}{5}y\,dy \quad = \frac{1}{50}\int_{-3}^{3}y\,dy$$

$$= \frac{1}{50}\left[\frac{y^2}{2}\right]_{-3}^{3} = \frac{1}{50}\left[\frac{9}{2}-\frac{9}{2}\right]$$

$$= 0$$

$$\mu_{03}' = E\left[X^0Y^3\right] = E\left[Y^3\right]$$

$$= \int_{-\infty}^{\infty}\int_{-\infty}^{\infty}y^3 f_{XY}(x,y)\,dxdy$$

$$= \int_{-3}^{3}\int_{-1}^{1}y^3\frac{(x+y)^2}{40}\,dxdy$$

$$= \frac{1}{40}\int_{-3}^{3}\int_{-1}^{1}y^3(x+y)^2\,dxdy$$

$$= \frac{1}{40}\int_{-3}^{3}\int_{-1}^{1}y^3(x^2+2xy+y^2)\,dxdy$$

$$= \frac{1}{40} \int_{-3}^{3} \int_{-1}^{1} \left(x^2 y^3 + 2xy^4 + y^5 \right) dx \, dy$$

$$= \frac{1}{40} \int_{-3}^{3} \left[\frac{x^3}{3} y^3 + 2 \frac{x^2}{2} y^4 + xy^5 \right]_{-1}^{1} dy$$

$$= \frac{1}{40} \int_{-3}^{3} \left[\left(\frac{1}{3} + \frac{1}{3} \right) y^3 + (1-1) y^4 + (1+1) y^5 \right] dy$$

$$= \frac{1}{40} \int_{-3}^{3} \left(\frac{2}{3} y^3 + 2y^5 \right) dy \quad = \frac{1}{40} \left(\frac{2}{3} \frac{y^4}{4} + 2 \frac{y^6}{6} \right)_{-3}^{3}$$

$$= \frac{1}{40} \left[\frac{1}{6} \left(3^4 - 3^4 \right) + \frac{1}{3} \left(3^6 - 3^6 \right) \right]_{-3}^{3} \quad = 0$$

$$\mu_{21}' = E\left[X^2 Y^1 \right] = E\left[X^2 Y \right]$$

$$= \int_{-\infty}^{\infty} \int_{-\infty}^{\infty} x^2 y f_{XY}(x, y) \, dx \, dy$$

$$= \int_{-3}^{3} \int_{-1}^{1} x^2 y \frac{(x+y)^2}{40} \, dx \, dy$$

$$= \frac{1}{40} \int_{-3}^{3} \int_{-1}^{1} x^2 y (x+y)^2 \, dx \, dy$$

$$= \frac{1}{40} \int_{-3}^{3} \int_{-1}^{1} x^2 y \left(x^2 + 2xy + y^2 \right) dx \, dy$$

$$= \frac{1}{40} \int_{-3}^{3} \int_{-1}^{1} \left(x^4 y + 2x^3 y^2 + x^2 y^3 \right) dx \, dy$$

$$= \frac{1}{40} \int_{-3}^{3} \left[\frac{x^5}{5} y + 2 \frac{x^4}{4} y^2 + \frac{x^3}{3} y^3 \right]_{-1}^{1} dy$$

$$= \frac{1}{40} \int_{-3}^{3} \left[\left(\frac{1}{6} + \frac{1}{6} \right) y + 2 \left(\frac{1}{4} - \frac{1}{4} \right) y^2 + \left(\frac{1}{3} + \frac{1}{3} \right) y^3 \right] dy$$

$$= \frac{1}{40} \int_{-3}^{3} \left(\frac{2}{6} y + \frac{2}{3} y^3 \right) dy \quad = \frac{1}{40} \left(\frac{2}{6} \frac{y^2}{2} + \frac{2}{3} \frac{y^4}{4} \right)_{-3}^{3}$$

$$= \frac{1}{40} \left[\frac{1}{6} \left(3^2 - 3^2 \right) + \frac{1}{6} \left(3^4 - 3^4 \right) \right] \quad = 0$$

$$\mu_{12}{}' = E\left[X^1Y^2\right] = E\left[XY^2\right]$$

$$= \int\limits_{-\infty}^{\infty}\int\limits_{-\infty}^{\infty} xy^2 f_{XY}(x,y)\,dxdy$$

$$= \int\limits_{-3}^{3}\int\limits_{-1}^{1} xy^2 \frac{(x+y)^2}{40}\,dxdy$$

$$= \frac{1}{40}\int\limits_{-3}^{3}\int\limits_{-1}^{1} xy^2 (x+y)^2\,dxdy$$

$$= \frac{1}{40}\int\limits_{-3}^{3}\int\limits_{-1}^{1} xy^2 \left(x^2 + 2xy + y^2\right)dxdy$$

$$= \frac{1}{40}\int\limits_{-3}^{3}\int\limits_{-1}^{1}\left(x^3 y^2 + 2x^2 y^3 + xy^4\right)dxdy$$

$$= \frac{1}{40}\int\limits_{-3}^{3}\left[\frac{x^4}{4}y^2 + 2\frac{x^3}{3}y^3 + \frac{x^2}{2}y^4\right]_{-1}^{1} dy$$

$$= \frac{1}{40}\int\limits_{-3}^{3}\left[\left(\frac{1}{4}-\frac{1}{4}\right)y^2 + 2\left(\frac{1}{3}+\frac{1}{3}\right)y^3 + \left(\frac{1}{2}-\frac{1}{2}\right)y^4\right]dy$$

$$= \frac{1}{40}\int\limits_{-3}^{3}\frac{4}{3}y^3 dy = \frac{1}{30}\left(\frac{y^4}{4}\right)_{-3}^{3}$$

$$= \frac{1}{30}\left[\frac{1}{4}\left(3^4 - 3^4\right)\right] = 0$$

Hence all the third order moments for X and Y are zeros.

***Example* 5:** Let Y be a random variable defined as the sum of N statistically independent random variables, i.e., $Y = X_1 + X_2 + \cdots + X_N$. If X_i, $i =1, 2, 3 \ldots N$ then find the probability density of Y

***Sol*:** Given that $Y = X_1 + X_2 + \cdots + X_N$ is the sum of N statistically independent random variables X_i, $i =1, 2, 3 \ldots N$ and each have finite mean μ and variance σ^2.

$$E[Y] = E[X_1] + E[X_n] + \cdots + E[X_N]$$
$$= \mu + \mu + \cdots + \mu = N\mu$$

Since all X_i's are statistically identically distributed. Similarly,

$$V[Y] = V[X_1] + V[X_n] + \cdots + V[X_N]$$
$$= \sigma^2 + \sigma^2 + \cdots + \sigma^2 = N\sigma^2$$

Since all X$_i$'s are statistically identically distributed.

Y is a normal variate with mean $N\mu$ and variance $N\sigma^2$

then the standard normal variate $\dfrac{Y - N\mu}{\sigma\sqrt{N}} \sim N(0, 1)$

Now the moment generating function of Y is

$$M_Y(t) = E\left[e^{ty}\right] = E\left[e^{t\frac{Y-N\mu}{\sigma\sqrt{N}}}\right]$$

$$= E\left[Exp\left\{t\left(\frac{Y - N\mu}{\sigma\sqrt{N}}\right)\right\}\right]$$

$$= E\left[Exp\left\{t\left(\frac{(X_1 + X_2 + \cdots + X_N) - N\mu}{\sigma\sqrt{N}}\right)\right\}\right]$$

$$= E\left[Exp\left\{t\left(\frac{(X_1 - \mu + X_2 - \mu + \cdots + X_N - \mu)}{\sigma\sqrt{N}}\right)\right\}\right]$$

$$= E\left[Exp\left\{t\frac{(X_1 - \mu)}{\sigma\sqrt{N}} + t\frac{(X_2 - \mu)}{\sigma\sqrt{N}} + \cdots + t\frac{(X_N - \mu)}{\sigma\sqrt{N}}\right\}\right]$$

$$= E\left[exp\left\{t\frac{(X_1 - \mu)}{\sigma\sqrt{N}}\right\}\right]E\left[exp\left\{t\frac{(X_2 - \mu)}{\sigma\sqrt{N}}\right\}\right]\cdots E\left[exp\left\{t\frac{(X_N - \mu)}{\sigma\sqrt{N}}\right\}\right]$$

$$= \left\{E\left[exp\left\{t\frac{(X_i - \mu)}{\sigma\sqrt{N}}\right\}\right]\right\}^N$$

Since all X$_i$'s are independent and identically distributed random variables

$$= \left[E\left\{1 + t\frac{(X_i - \mu)}{\sigma\sqrt{N}} + \frac{t^2}{2!}\left(\frac{X_i - \mu}{\sigma\sqrt{N}}\right)^2 + \cdots\right\}\right]^N$$

$$= \left[\left\{1 + \frac{t}{\sigma\sqrt{N}}E(X_i - \mu) + \frac{t^2}{2!\sigma^2 N}E(X_i - \mu)^2 + \cdots\right\}\right]^N$$

$$= \left[\left\{1 + \frac{t}{\sigma\sqrt{N}} \times 0 + \frac{t^2}{2!\sigma^2 N}\sigma^2 + \cdots\right\}\right]^N$$

$$= \left\{1 + \frac{t^2}{2!N} + \cdots\right\}^N$$

As Limit $N \to \infty$,

$$\underset{N\to\infty}{Lt}\ E\left[e^{ty}\right] = \underset{N\to\infty}{Lt}\ \left\{1+\frac{t^2}{2!N}+\cdots\right\}^N = e^{\frac{t^2}{2}}$$

This is same as Moment Generating Function of Standardized normal variate. Therefore Y is a Gaussian random variable.

Hence the density function of Y is $f_Y(y) = \frac{1}{\sigma\sqrt{2\pi}} e^{-\frac{1}{2}\left(\frac{y-\mu}{\sigma}\right)^2}$ $-\infty < Y < \infty$

Example 6: Show that the variance of a weighted sum of uncorrelated random variables equals the weighted sum of the variances of the random variables.

Sol: Let $W_1, W_2, \cdots W_N$ are the respective weights of the random variables $X_1, X_2, \cdots, X_N$

$$g(X_1, X_2, \cdots, X_N) = \sum_{i=1}^{N} W_i X_i$$

The Mean of the function g is

$$E\left[g(X_1, X_2, \cdots, X_N)\right] = E\left[\sum_{i=1}^{N} W_i X_i\right]$$

$$= \sum_{i=1}^{N} E\left[W_i X_i\right]$$

$$= \sum_{i=1}^{N}\left[\int_{-\infty}^{\infty}\int_{-\infty}^{\infty}\cdots\int_{-\infty}^{\infty} w_i x_i f_{X_1, X_2, \cdots X_N,}(x_1, x_2, \cdots x_N)\, dx_1 dx_2 \cdots dx_n\right] \qquad ....(5.37)$$

The marginal density of N random variables $X_1, X_2, \cdots, X_N$ the joint density function is found by integrating out all variables except k variables of interest $X_1, X_2, \cdots, X_N$ of the joint density function using this rule in equation (2), we get,

$$E\left[\sum_{i=1}^{N} W_i X_i\right] = \sum_{i=1}^{N}\left[\int_{-\infty}^{\infty} w_i x_i f_{X_i}(x_i)\, dx_i\right]$$

$$= \sum_{i=1}^{N} E\left[W_i X_i\right]$$

The mean value of a weighted sum of random variables equals the weighted sum of mean values.

Now assume that

$$g(X_1, X_2, \cdots, X_N) = X$$

Then $X = \sum_{i=1}^{N} W_i X_i$ and $\overline{X} = \sum_{i=1}^{N} W_i \overline{X}_i$

Now $X - \overline{X} = \sum_{i=1}^{N} W_i \left(X_i - \overline{X} \right)$

$$= \sum_{i=1}^{N} W_i X_i - \overline{X} \sum_{i=1}^{N} W_i$$

Variance of X is

$$\sigma_x^2 = E\left[\left(X - \overline{X} \right)^2 \right]$$

$$= E\left[Y^2 \right]$$

Example 7: Let X and Y be two random variables having their joint density function is

$$f_{XY}(x,y) = \begin{cases} \dfrac{xy}{9} & 0 < x < 2, 0 < y < 3 \\ 0 & Else\ where \end{cases}$$

Show that X and Y are statistically independent and also uncorrelated.

Sol: Given that the joint density function

$$f_{XY}(x,y) = \begin{cases} \dfrac{xy}{9} & 0 < x < 2, 0 < y < 3 \\ 0 & Else\ where \end{cases}$$

Since we know that

$$f_X(x) = \int_{-\infty}^{\infty} f_{XY}(x,y)\,dy$$

$$= \frac{x}{9} \int_0^3 y\,dy$$

$$= \frac{x}{9} \left[\frac{y^2}{2} \right]_0^3$$

$$= \frac{x}{2}$$

Therefore $f_X(x) = \begin{cases} \dfrac{x}{2} & 0 < x < 2 \\ 0 & Otherwise \end{cases}$

Since we know that $f_Y(y) = \displaystyle\int_{-\infty}^{\infty} f_{XY}(x,y)\,dy$

$$= \frac{y}{9}\int_0^2 x\,dy$$

$$= \frac{y}{9}\left[\frac{x^2}{2}\right]_0^2 = \frac{2y}{9}$$

Therefore $f_Y(y) = \begin{cases} \dfrac{2y}{9} & 0 < y < 3 \\ 0 & Otherwise \end{cases}$

Here we observe that

$$f_X(x) \times f_Y(y) \quad \frac{x}{2} \times \frac{2y}{9} = \frac{xy}{9} = f_{XY}(x,y)$$

Therefore the two random variables X and Y are independent

Also, If X and Y are independent random variables, then the covariance between X and Y is zero. This implies the correlation coefficient between the two variables X and Y is zero. Then we say that the two variables are uncorrelated.

***Example* 8:** Random variables Z and W are defined by $Z = X + aY$, and $W = X - aY$ where a is a real number. Determine a such that Z and W are orthogonal.

Now the joint Probability of X and Z is

X \ Z	1	2	3	4	5	6	Marginal Totals
1	$\dfrac{1}{36}$	$\dfrac{1}{36}$	$\dfrac{1}{36}$	$\dfrac{1}{36}$	$\dfrac{1}{36}$	$\dfrac{1}{36}$	$\dfrac{6}{36}$
2	0	$\dfrac{2}{36}$	$\dfrac{1}{36}$	$\dfrac{1}{36}$	$\dfrac{1}{36}$	$\dfrac{1}{36}$	$\dfrac{6}{36}$
3	0	0	$\dfrac{3}{36}$	$\dfrac{1}{36}$	$\dfrac{1}{36}$	$\dfrac{1}{36}$	$\dfrac{6}{36}$

Contd....

4	0	0	0	$\frac{4}{36}$	$\frac{1}{36}$	$\frac{1}{36}$	$\frac{6}{36}$
5	0	0	0	0	$\frac{5}{36}$	$\frac{1}{36}$	$\frac{6}{36}$
6	0	0	0	0	0	$\frac{6}{36}$	$\frac{6}{36}$
Marginal Totals	$\frac{1}{36}$	$\frac{3}{36}$	$\frac{5}{36}$	$\frac{7}{36}$	$\frac{9}{36}$	$\frac{11}{36}$	1

$$E[XZ] = \sum xzP(x,z)$$

$$= \left\{ 1 \times \frac{1}{36} + 2 \times \frac{1}{36} + 3 \times \frac{1}{36} + 4 \times \frac{1}{36} + 5 \times \frac{1}{36} + 6 \times \frac{1}{36} \right\}$$

$$+ \left\{ 4 \times \frac{2}{36} + 6 \times \frac{1}{36} + 8 \times \frac{1}{36} + 10 \times \frac{1}{36} + 12 \times \frac{1}{36} \right\}$$

$$+ \left\{ 9 \times \frac{3}{36} + 12 \times \frac{1}{36} + 15 \times \frac{1}{36} + 18 \times \frac{1}{36} \right\}$$

$$+ \left\{ 16 \times \frac{4}{36} + 20 \times \frac{1}{36} + 24 \times \frac{1}{36} \right\}$$

$$+ \left\{ 25 \times \frac{5}{36} + 30 \times \frac{1}{36} \right\} + \left\{ 36 \times \frac{6}{36} \right\}$$

$$= \frac{616}{36}$$

***Example* 9:** Two independent random variables X and Y are having their densities as $f_X(x) = e^{-x}u(x)$ and $f_Y(y) = e^{-y}u(y)$ $0 < x, y < 1$, find $P(X + Y \leq 1)$

Sol: Given that X and Y are independent random variables then $f_{XY}(xy) = f_X(x)f_Y(y)$

Now

$$P(X + Y \leq 1) = \int\limits_{-\infty}^{\infty}\int\limits_{-\infty}^{\infty} f_{XY}(xy)\,dxdy = \int\limits_{-\infty}^{\infty}\int\limits_{-\infty}^{\infty} f_X(x)f_y(y)\,dxdy = \int\limits_{-\infty}^{\infty} f_X(x)dx \int\limits_{-\infty}^{\infty} f_y(y)dy$$

$$= \int\limits_{0}^{1} e^{-x}u(x)dx \int\limits_{0}^{1-x} e^{-y}u(y)dy \int\limits_{0}^{1} e^{-x}u(x)dx \int\limits_{0}^{1-x} e^{-y}u(y)dy$$

Quiz Questions

1. Let X be a discrete random variable, assumes a discrete set of values $x_1, x_2, x_3, \cdots x_N$ with probabilities $P(x_1), P(x_2), P(x_3), \cdots P(x_N)$ respectively. Now the expected value of a real function g(.) of X is defined as $E[g(x)]$ is equal to

   (a) $\sum g(x) P(x)$

   (b) $\int\limits_{-\infty}^{\infty} g(x) f_X(x)\, dx$

   (c) $\begin{cases} \int\limits_{-\infty}^{\infty} g(x) f_X(x)\, dx \\ \sum g(x) P(x) \end{cases}$

   (d) none

2. Let X and Y be two independent random variables, is each having density function

   $$f_x(x) = \begin{cases} 2e^{-2x} & for\ x \geq 0 \\ 0 & else\ where \end{cases}$$

   Then $E(X + Y)$ is

   (a) 0   (b) 1   (c) ∞ (d) none

3. Let X and Y have joint density function

   $$f_{xy}(x, y) = \begin{cases} x + y & for\ 0 \leq x \leq 1; 0 \leq y \leq 1 \\ 0 & otherwise \end{cases}$$

   the conditional expectation of Y given X is

   (a) $\dfrac{3x+2}{(2x+1)}$ (b) $\dfrac{3x+2}{2(2x+1)}$ (c) $\dfrac{3x+2}{3(2x+1)}$ (d) none

4. Let X and Y be two discrete random variables is related as

   $$g(X,Y) = X^i Y^j \quad i, j = 1, 2, \cdots$$

   The $(i + j)^{th}$ order joint moment of two dimensional random variable μ_{ij} defined as

   (a) $\sum\sum x^i y^j f_{XY}(x, y)$

   (b) $\int\limits_{-\infty}^{\infty}\int\limits_{-\infty}^{\infty} x^i y^j f_{XY}(x, y)\, dx dy$

   (c) $= \begin{cases} \sum\sum x^i y^j f_{XY}(x, y) \\ \int\limits_{-\infty}^{\infty}\int\limits_{-\infty}^{\infty} x^i y^j f_{XY}(x, y)\, dx dy \end{cases}$

   (d) none

5. The first order moments $\mu_{10}{}^{/th}$ is the expected values of X is defined as

   (a) $E[X]$ (b) $E\left[X^i\right]$ (c) $E\left[X^j\right]$ (d)   none

6. The second order moments are called the ... between X and Y and is

   (a)   skewness   (b)   regression

   (c)   correlation   (d)   kurtosis

7. The correlation between X and Y and is denoted by

   (a) R_{XY}.   (b) C_{XY} (c) C_{YX} (d)   none

8. If X and Y are continuous random variables having their joint density function is $f_{XY}(x,y)$ then its correlation $R_{XY} = \mu_{11}'$ is

   (a) $\sum\sum x^i y^j f_{XY}(x,y)$ (b) $\int\limits_{-\infty}^{\infty}\int\limits_{-\infty}^{\infty} xy f_{XY}(x,y)\,dxdy$

   (c) $=\begin{cases}\sum\sum x^i y^j f_{XY}(x,y)\\ \int\limits_{-\infty}^{\infty}\int\limits_{-\infty}^{\infty} x^i y^j f_{XY}(x,y)\,dxdy\end{cases}$ (d)   none

9. If the two random variables X and Y are uncorrelated then $f_{XY}(x,y)$ is equal to

   (a) $f_X(x) + f_Y(y)$ (b) $f_X(x) - f_Y(y)$

   (c) $f_X(x) \times f_Y(y)$ (d)   none

10. The covariance between the random variables X and Y is denoted by $C_{XY} = \mu_{11}$ is defined as

   (a) $E\left[(X-\overline{X})(Y-\overline{Y})\right]$ (b) $E\left[(X+\overline{X})(Y+\overline{Y})\right]$

   (c) $E\left[(X+\overline{X})(Y-\overline{Y})\right]$ (d)   none

11. $\mu_{02}' - \mu_{01}' =$

   (a) μ_{02}' (b) μ_{20} (c) μ_{01}' (d)   none

12. If X and Y are independent then $\mu_{11} = C_{XY} =$

   (a)   1   (b) μ_{20} (c)   0   (d)   none

13. If X and Y are orthogonal random variables then C_{XY} is equal to

    (a) $E(X)E(Y)$ (b) $E(X)+E(Y)$

    (c) $E(X)-E(Y)$ (d)   none

14. If X and Y are correlated then the correlation coefficient between X and Y is denoted by ρ is

    (a) $\dfrac{Cov(X,Y)}{\sigma_x^2 \sigma_y}$ (b) $\dfrac{Cov(X,Y)}{\sigma_x \sigma_y}$

    (c) $\dfrac{Cov(X,Y)}{\sigma_x \sigma_y^2}$ (d)   none

15. If the correlation coefficient between X and Y, $\rho = \pm 1$ then the correlation is said to be

    (a)   +ve correlation   (b)   −ve correlation

    (c)   perfect correlation   (d)   none

16. Let X and Y be two independent random variables, then Covariance between X and Y is

    (a)   0   (b)   1   (c) ∞ (d)   none

17. $V(X+Y) =$

    (a) $V(X)+V(Y)+2Cov(X,Y)$ (b) $V(X)+V(Y)-2Cov(X,Y)$

    (c) $V(X)+V(Y)\div 2Cov(X,Y)$ (d)   none

18. If X and Y be two independent random variables, then $V(X+Y)$

    (a) $V(X)-V(Y)$ (b) $V(X)\times V(Y)$

    (c) $V(X)+V(Y)$ (d)   none

19. Let X and Y be independent random variables such that

$$X = \begin{cases} 1 & \text{with probability } \dfrac{1}{3} \\ 0 & \text{with probability } \dfrac{2}{3} \end{cases} \quad \text{and} \quad Y = \begin{cases} 2 & \text{with probability } \dfrac{3}{4} \\ -3 & \text{with probability } \dfrac{1}{4} \end{cases}$$

then the value of $E[3X+2Y]$ is

    (a) $\dfrac{5}{2}$ (b)   2   (c) $\dfrac{5}{4}$ (d) $\dfrac{3}{2}$

20. The correlation coefficient between the two random variables X and Y if $E[X] = 4$, $E[Y] = 9$, $E[XY] = 100$, $E[X^2] = 81$ and $E[Y^2] = 256$ is

    (a)   0.5   (b)   0.6   (c)   0.8   (d)   none

21. The random variable X defined by

$$X = \begin{cases} -2 & \text{with probability } \dfrac{1}{3} \\[2mm] 3 & \text{with probability } \dfrac{1}{2} \\[2mm] 1 & \text{with probability } \dfrac{1}{6} \end{cases}$$

    first moment about the mean for the random variable X is

    (a)   0.5   (b)   0.6   (c)   0   (d)   none

22. Let X and Y be two independently and identically distributed random variables with mean 2 and 4 and their second order moments are 8 and 25 respectively, the mean of the random variable $Z = 3X - Y$ is

    (a)   2   (b)   1   (c)   0   (d)   none

23. Let X_1, X_2 and X_3 are three independent random variables whose mean values are $X_1 = 3$, $X_2 = 6$ and $X_3 = -2$. Then the mean values $g(X_1, X_2 X_3) = X_1 + 3X_2 + 4X_3$

    (a)   12   (b)   13   (c)   10   (d)   none

24. Let X_1, X_2 and X_3 are three independent random variables whose mean values are $X_1 = 3$, $X_2 = 6$ and $X_3 = -2$. Then the mean value of $g(X_1, X_2 X_3) = -2X_1 X_2 - 3X_1 X_3 + 4X_2 X_3$

    (a)   32   (b)   –36   (c)   –66   (d)   none

25. Let X_1 and X_2 be two independent continuous random variables with joint probability density function $f_{X_1 X_2}(x_1, x_2)$ and their joint moment generating function is defined as $M_{X_1 X_2}(t_1, t_2)$ is equal to

    (a) $\displaystyle\int\limits_{-\infty}^{\infty}\int\limits_{-\infty}^{\infty} e^{t_1 x_1 + t_2 x_1} f_{X_1 X_2}(x_1, x_2)\, dx_1 dx_2$ (b) $\displaystyle\sum_i \sum_j e^{t_1 x_1 + t_2 x_1} P[x_1(i), x_2(j)]$

    (c) $\displaystyle\sum_i \sum_j e^{t_1 x_1 + t_2 x_1} P[x_1(i)]$ (d) $\displaystyle\int\limits_{-\infty}^{\infty}\int\limits_{-\infty}^{\infty} f_{X_1 X_2}(x_1, x_2)\, dx_1 dx_2$

26. If X_1 and X_2 are independent random variables then $M_{X_1X_2}(t_1,t_2)= \ldots$ for (t_1,t_2) in a rectangle about $(0,0)$

    (a) $M_{X_1}(t_1)-M_{X_2}(t_2)$ (b) $M_{X_1}(t_1)\times M_{X_2}(t_2)$

    (c) $M_{X_1}(t_1)+M_{X_2}(t_2)$ (d) none

27. Suppose that X and Y are independent standard normal random variables and define $U= X+Y$ and $V= X-Y$. Then the moment generating function of U and V is

    (a) $M_X(t_1+t_2)M_Y(t_1+t_2)$ (b) $M_X(t_1-t_2)M_Y(t_1+t_2)$

    (c) $M_X(t_1+t_2)M_Y(t_1-t_2)$ (d) $M_X(t_1-t_2)M_Y(t_1-t_2)$

28. Let (X_1,X_2) be a two dimensional discrete random variable, then the characteristic function of two random variables X_1 and X_2 is defined as $\phi_{X_1X_2}(t_1,t_2)$ is equal to

    (a) $\sum_i^\infty \sum_j^\infty e^{it_1x_1+it_2x_1}P[x_1(i),x_2(j)]$ (b) $\int_{-\infty}^\infty \int_{-\infty}^\infty e^{it_1x_1+it_2x_1}f_{X_1X_2}(x_1,x_2)dx_1dx_2$

    (c) $\sum_i^\infty \sum_j^\infty e^{t_1x_1+t_2x_1}P[x_1(i)]$ (d) $\int_{-\infty}^\infty \int_{-\infty}^\infty f_{X_1X_2}(x_1,x_2)dx_1dx_2$

29. If X and Y are independent then $\phi_{X_1X_2}(t_1,t_2)$ is equal to

    (a) $\phi_{X_1}(t_1)+\phi_{X_2}(t_2)$ (b) $\phi_{X_1}(t_1)\times \phi_{X_2}(t_2)$

    (c) $\phi_{X_1}(t_1)-\phi_{X_2}(t_2)$ (d) $\phi_{X_1}(t_1)\pm \phi_{X_2}(t_2)$

30. Two random variables X and Y have the joint characteristic function $\phi_{XY}(w_1,w_2)=e^{-w_1^2-8w_2^2}$ then the mean of the random variables X is

    (a) 2   (b) 1   (c) 0   (d) none

31. $\int_{-x}^x E\left(\dfrac{Y}{X=x}\right).f(x).dx$ is equal to

    (a) $E(X)$ (b) $E\left(\dfrac{Y}{X}\right)$ (c) $E(Y)$ (d) $f(x)$.

32. The Jacobian of the transformation $x = v;\ y = \dfrac{1}{2}(u - v)$

(a) $\dfrac{1}{2}$ (b) $-\dfrac{1}{2}$ (c) $\quad 1$ (d) $\quad 2$

33. If $X_1, X_2, \dots X_n$ are indentically distributed random variable such that $X_K = 1$ with prob. 1/2

     $= 2$ with prob. 1/3

     $= -1$ with prob. 1/6, then

     $E\left(X_1^2 + X_2^2 + \dots + X_n^2 \right)$

(a) $\quad n$ (b) $\quad 2n$ (c) $\quad n/2$ (d) $\quad n^2$

34. $E(X + Y) = E(X) + E(Y)$ for

(a) only independent X and Y (b) any X and Y

(c) orthogonal X and Y (d) uncorrelated X and Y

35. X and Y are continuous random variables with joint density

$$f X_1 X_2 (x_1 x_2) = e^{-(x_1 + x_2)} \quad \text{for} \quad x_1 \geq 0 \leq 1;\ x_2 \geq 0$$

$$= 0 \quad \text{elsewhere}$$

Then $P\left(\dfrac{x_1 > 2}{x_2 > 1} \right)$ is same as

(a) $P(X_1 > 2)$ (b) $P(X_2 > 1)$

(c) $P(X_1 > 2, X_2 > 1)$ (d) $P(X_1 \geq 2, X_2 \geq 1)$

36. X and Y are two statistically independent random variables. Then variance of the random variable $Z = 3X - Y$ is

(a) $9 \times var\,(X) - var(Y) + 6 \times Cov\,(X, Y)$

(b) $9 \times var\,(X) + var\,(Y)$

(c) $3 \times Var\,(X) + var\,(Y)$

(d) $3 \times Var\,(X) - var\,(Y)$

37. X and Y are Gaussian random variables with variances σ_X^2 and σ_y^2 . Then the random variables $V = X + kY$ and $W = X - kY$ are statically independent for K equal to

(a) $\sigma_x.\sigma_y$ (b) $\dfrac{\sigma_x}{\sigma_y}$ (c) $\dfrac{\sigma_y}{\sigma_{xy}}$ (d) $\sigma_x + \sigma_y$

38. X ad Y are jointly Gaussian random variables with same variance and $\rho_{XY} = -1$. The angle θ of a coordination rotation that generates new random variables that are statistically independent is

(a) $\pi/2$ (b) $-\pi/2$ (c) $-\pi/4$ (d) π

39. The random variables X and Y are transformed to get new random variables V and W as

$V = X \cos \theta + Y \operatorname{Sin} \theta$

$W = X \sin \theta - Y \cos \theta$

If $f(x,y) = \dfrac{1}{2\pi\sigma^2}\exp\left[\dfrac{(x^2 + y^2)}{2\sigma^2}\right]\, m$ then $f_{VW}(v,w)$

(a)   (b) $f_{XY}(x,y)$ (c) $f_{XY}(x,y)$ (d) $f_{XY}(x,y)$

40. X and Y are random variables transformed as $x=\dfrac{u}{v}$ and $y=v$. the jacobian of this transformation is

(a) v (b) $-v$ (c) $1/v$ (d) $1/v^2$

41. Two random variables X and Y with indentical moment generating function are . . . distributed.

(a)   Identically   (b)   differently

(c)   both   (d)   none

42. The average $\overline{g(X,Y)} = X^m Y^m$ are called . . .

(a)   moment of the order n

(b)   moments of the order m

(c)   joint moments of the order m, n

(d)   none

43. The joint moments of two statistically independent random variables is equal to . . .

  (a)   Sum of the individual moments   (b)   the product of the individual moments

  (c)   both   (d)   none

44. Whether or not X_1, X_2 are independent, $E(X_1 + X_2)$ is equal to . . .

  (a) $E(X_1) + E(X_2)$ (b) $E[X_1] E[X_2]$

  (c)   both   (d)   none

45. The covariance of two independent random variable is

  (a)   zero   (b)   one   (c)   two   (d)   none

46. Moment generating function of a random variable is used to generate moments about . . . of the random variable.

  (a)   Mean   (b)   origin   (c)   infinite   (d)   none

47. The second joint moment about origin of two random variables X and Y is given by . . .

  (a) $\int_{-\infty}^{\infty} \int_{-\infty}^{\infty} xy f(x,y) dx\, dy$ (b) $\int_{0}^{\infty} \int_{0}^{\infty} xy f(x,y) dx\, dy$

  (c) $\int_{-\infty}^{\infty} \int_{-\infty}^{\infty} f(x,y) dx\, dy$ (d)   none

48. Characteristic function is used to find moments about . . . of a random variable.

  (a)   origin   (b)   mean   (c)   both   (d)   none

49. If the PDF of a random variable is symmetric about the origin, then its moment generating function is . . .

  (a)   asymmetric   (b)   general

  (c)   symmetric   (d)   none

50. $\sigma_{XY} = E(XY) - \ldots$

  (a) $E[X] + E(Y)$ (b) $E[X] E[Y]$

  (c) $E[X]$ (d)   none

Answers

1. (a)	2. (b)	3. (c)	4. (a)	5. (b)	6. (c)
7. (a)	8. (b)	9. (c)	10. (a)	11. (b)	12. (c)
13. (a)	14. (b)	15. (c)	16. (a)	17. (b)	18. (c)
19. (a)	20. (b)	21. (c)	22. (a)	23. (b)	24. (c)

25. (a)	26. (b)	27. (c)	28. (a)	29. (b)	30. (c)
31. (c)	32. (b)	33. (b)	34. (b)	35. (a)	36. (b)
37. (b)	38. (c)	39. (b)	40. (c)	41. (a)	42. (c)
43. (b)	44. (b)	45. (a)	46. (b)	47. (a)	48. (a)
49. (c)	50.				

Review Quiz Questions

1. Let X be a continuous random variable, defined in the interval assumes a discrete set of values $-\infty < X < \infty$ with probability density function $f_X(x)$. Now the expected value of a real function g(.) of X is defined as $E[g(x)]$ is equal to

(a) $\sum g(x) P(x)$

(b) $\int\limits_{-\infty}^{\infty} g(x) f_X(x)\, dx$

(c) $\begin{cases} \int\limits_{-\infty}^{\infty} g(x) f_X(x)\, dx \\ \sum g(x) P(x) \end{cases}$

(d) none

2. Let X and Y be two independent random variables, is each having density function

$$f_x(x) = \begin{cases} 2e^{-2x} & for\ \ x \geq 0 \\ 0 & else\ where \end{cases}$$

Then $E(X^2 + Y^2)$ is

(a)   0   (b)   1   (c)   ∞   (d)   none

3. Let X and Y have joint density function

$$f_{xy}(x, y) = \begin{cases} x+y & for\ \ 0 < x < 1; 0 < y \leq 1 \\ 0 & otherwise \end{cases}$$

the conditional expectation of X given Y is

(a) $\dfrac{3y+2}{(2y+1)}$ (b) $\dfrac{3y+2}{2(2y+1)}$ (c) $\dfrac{3y+2}{3(2y+1)}$ (d)   none

4. Let X and Y be two continuous random variables is related as

$$g(X,Y) = X^i Y^j \quad i, j = 1, 2, \cdots$$

The $(i+j)^{th}$ order joint moment of two dimensional continuous random variable μ_{ij} defined as

(a) $\sum\sum x^i y^j f_{XY}(x,y)$

(b) $\int\limits_{-\infty}^{\infty}\int\limits_{-\infty}^{\infty} x^i y^j f_{XY}(x,y)dxdy$

(c) $= \begin{cases} \sum\sum x^i y^j f_{XY}(x,y) \\ \int\limits_{-\infty}^{\infty}\int\limits_{-\infty}^{\infty} x^i y^j f_{XY}(x,y)dxdy \end{cases}$

(d)   none

5. The first order moments $\mu_{0j}{}'$ is the expected values of Y is defined as

(a) $E[Y]$ (b) $E[Y^j]$ (c) $E[Y^i]$ (d)   none

6. $E\left[\left(X-\overline{X}\right)^2\right]$ is denoted as

(a) μ_{20} (b) $E[Y^j]$ (c) μ_{02} (d)   none

7. If $C_{XY}=E(X)E(Y)$ then the random variables X and Y are said to be

(a) orthogonal   (b) correlated (c)   congruent   (d)   none

8. If the correlation coefficient between X and Y, is zero i.e., $\rho=0$ then the the two variables are said to be

(a)   correlatied   (b)   uncorrelated

(c)   +vely orrelatied   (d)   none

9. If $Cov(X,Y)=0$ then the two random variables X and Y be

(a)   independent   (b)   neither independent nor dependent

(c)   either independent or dependent   (d)   none

10. If X and Y be two independent random variables, then $V(X–Y)$

(a) $V(X)-V(Y)$ (b) $V(X)\times V(Y)$

(c) $V(X)+V(Y)$ (d)   none

11. Let X and Y be independent random variables such that

$$X = \begin{cases} 1 & \text{with probability } \dfrac{1}{3} \\ 0 & \text{with probability } \dfrac{2}{3} \end{cases} \quad \text{and} \quad Y = \begin{cases} 2 & \text{with probability } \dfrac{3}{4} \\ -3 & \text{with probability } \dfrac{1}{4} \end{cases}$$

then the value of $E\left[2X^2 - Y^2\right]$ is

(a) $\dfrac{55}{12}$ (b) $\dfrac{53}{12}$ (c) $\dfrac{52}{3}$ and   (d) $\dfrac{5}{12}$

12. The random variable X defined by

$$X = \begin{cases} -2 & \text{with probability } \dfrac{1}{3} \\ 3 & \text{with probability } \dfrac{1}{2} \\ 1 & \text{with probability } \dfrac{1}{6} \end{cases}$$

first second moment about the mean for the random variable X is

(a)   0.5   (b)   6   (c)   0   (d)   none

13. The random variable X defined by

$$X = \begin{cases} -2 & \text{with probability } \dfrac{1}{3} \\ 3 & \text{with probability } \dfrac{1}{2} \\ 1 & \text{with probability } \dfrac{1}{6} \end{cases}$$

The variance of the random variable X is

(a)   5   (b)   6   (c)   0   (d)   none

14. Let X and Y be two independently and identically distributed random variables with mean 2 and 4 and their second order moments are 8 and 25 respectively, second order moment of the random variable $Z = 3X - Y$

(a)   49   (b)   94   (c)   0   (d)   none

15. Let X and Y be two independently and identically distributed random variables with mean 2 and 4 and their second order moments are 8 and 25 respectively, variance of the random variable $Z = 3X - Y$

(a)   54   (b)   45   (c)   0   (d)   none

16. Let X and Y are independent random variables whose moments are $\mu_{10} = 2$, $\mu_{20} = 14$, $\mu_{02} = 12$ and $\mu_{11} = -6$. Then the moment μ_{22} .is

(a)   5   (b)   6   (c)   0   (d)   none

17. Let X_1, X_2 and X_3 are three independent random variables whose mean values are $X_1 = 3$, $X_2 = 6$ and $X_3 = -2$. Then the mean values $g(X_1, X_2 X_3) = X_1 X_2 X_3$

(a)   32   (b)   -36   (c)   36   (d)   none

18. Let X_1, X_2 and X_3 are three independent random variables whose mean values are $X_1 = 3$, $X_2 = 6$ and $X_3 = -2$. Then the mean value $g(X_1, X_2 X_3) = X_1 + X_2 + X_3$

(a)   2   (b)   7   (c)   0   (d)   none

19. Let X_1 and X_2 be two independent discrete random variables with joint probability density function $f_{X_1 X_2}(x_1, x_2)$ and their joint moment generating function is defined as $M_{X_1 X_2}(t_1, t_2)$ is equal to

(a) $\displaystyle\int_{-\infty}^{\infty}\int_{-\infty}^{\infty} e^{t_1 x_1 + t_2 x_1} f_{X_1 X_2}(x_1, x_2) dx_1 dx_2$ (b) $\displaystyle\sum_i \sum_j e^{t_1 x_1 + t_2 x_1} P[x_1(i), x_2(j)]$

(c) $\displaystyle\sum_i \sum_j e^{t_1 x_1 + t_2 x_1} P[x_1(i)]$ (d) $\displaystyle\int_{-\infty}^{\infty}\int_{-\infty}^{\infty} f_{X_1 X_2}(x_1, x_2) dx_1 dx_2$

20. The two random variables X and Y having joint density is

$$f_{XY}(x, y) = \frac{y}{\sqrt{2\pi}} e^{-y - \frac{1}{2}(x-y)^2}$$

The joint moment generating function for the two random variables is

(a) $\dfrac{3y+2}{(2y+1)}$ (b) $\dfrac{3y+2}{2(2y+1)}$ (c) $\dfrac{3y+2}{3(2y+1)}$ (d)   none

21. Let (X_1, X_2) be a two dimensional continuous random variable, then the characteristic function of two random variables X_1 and X_2 is defined as $\phi_{X_1 X_2}(t_1, t_2)$ is equal to

(a) $\displaystyle\sum_i \sum_j e^{it_1 x_1 + it_2 x_1} P[x_1(i), x_2(j)]$ (b) $\displaystyle\int_{-\infty}^{\infty}\int_{-\infty}^{\infty} e^{it_1 x_1 + it_2 x_1} f_{X_1 X_2}(x_1, x_2) dx_1 dx_2$

(c) $\displaystyle\sum_i \sum_j e^{t_1 x_1 + t_2 x_1} P[x_1(i)]$ (d) $\displaystyle\int_{-\infty}^{\infty}\int_{-\infty}^{\infty} f_{X_1 X_2}(x_1, x_2) dx_1 dx_2$

22. Two random variables X and Y have the joint characteristic function $\phi_{XY}(w_1, w_2) = e^{-w_1^2 - 8w_2^2}$ then the mean of the random variables Y is

(a)   2   (b)   1   (c)   0   (d)   none

23. The Jacobian of the transformation $x = v; \ y = \dfrac{1}{2}(u + v)$ is

(a) $\dfrac{1}{2}$ (b) $-\dfrac{1}{2}$ (c)   1   (d)   none

24. If $X_1, X_2, \ldots X_n$ are indentically distributed random variable such that $X_K = 1$ with prob. 1/3

$$= 2 \quad \text{with prob. } 1/2$$
$$= -1 \quad \text{with prob. } 1/6, \text{ then}$$

$E\left(X_1^2 + X_2^2 + \ldots + X_n^2\right)$ is

(a) n (b) $2n$ (c) $n/2$ (d)   *none*

25. $E(X+Y) = E(X) + E(Y)$ for

(a)   only independent X and Y (b)   any X and Y

(c)   orthogonal X and Y (d)   uncorrelated X and Y

26. X and Y are continuous random variables with joint density

$$fX_1 X_2 (x_1, x_2) = e^{-2(x_1 + x_2)} \quad \text{for} \ \ 0 \le x_1 \le 1; \ x_2 \ge 0 = 0 \text{ elsewhere}$$

Then $PP\left(\dfrac{X_1 > 2}{X_2 > 1}\right)$ is same as

(a) $P(X_1 > 2)$ (b) $P(X_2 > 1)$ (c) $P(X_1 > 2, X_2 > 1)$ (d)   none

27. X and Y are two statistically independent random variables. Then variance of the random variable $Z = 2X - 3Y$ is

   (a) $9 \times var\,(X) - var(Y) + 6 \times Cov\,(X, Y)$

   (b) $9 \times var\,(X) + var\,(Y)$

   (c) $4 \times Var\,(X) - 9var\,(Y)$

   (d) $3 \times Var\,(X) - var\,(Y)$

28. X and Y are Gaussian random variables with variances σ_x^2 and σ_y^2 . Then the random variables $V = X - kY$ and $W = X + kY$ are statically independent for K equal to

   (a) $\sigma_x \cdot \sigma_y$ (b) $\dfrac{\sigma_x}{\sigma_y}$ (c) $\dfrac{\sigma_y}{\sigma_{xy}}$ (d)   none

29. X ad Y are jointly Gaussian random variables with same variance and $\rho_{XY} = +1$.

   The angle θ of a coordination rotation that generates new random variables that are statistically independent is

   (a) $\pi\!/\!2$ (b) $-\pi\!/\!2$ (c) $-\pi\!/\!4$ (d) π

30. The random variables X and Y are transformed to get new random variables V and W as $V = X\cos\theta - Y\sin\theta$

   $W = X\sin\theta + Y\cos\theta$

   If $f(x,y) = \dfrac{1}{2\pi\sigma^2}\,exp\left[\dfrac{(x^2 + y^2)}{2\sigma^2}\right]$ m then $f_{VW}(v,w)$ is

   (a) $-f_{Xy}(x,y)$ (b) $f_{XY}(x,y)$ (c) $f_{XY}(x,y)$ (d)   none

31. X and Y are random variables transformed as $x = \dfrac{u}{v}$ and $y = v$. the jacobian of this transformation is

   (a) v (b) $-v$ (c) $1/v$ (d) $1/v^2$

32. The average $\overline{g(X,Y)} = X^n Y^n$ are called . . .

   (a)   moment of the order n   (b)   moments of the order m

   (c)   joint moments of the order m, n   (d)   none

33. Whether or not X_1, X_2 are independent, $E(X_1 + X_2)$ is equal to ----------------------
    (a) $E(X_1) + E(X_2)$ (b) $E[X_1]E[X_2]$ (c)   both   (d)   none

34. The second joint moment about origin of two random variables X and Y is given by ------------

    (a) $\int_{-\infty}^{\infty} \int_{-\infty}^{\infty} xy \, f(x, y) \, dx \, dy$ (b) $\int_{0}^{\infty} \int_{0}^{\infty} xy \, f(x, y) \, dx \, dy$

    (c) $\int_{-\infty}^{\infty} \int_{-\infty}^{\infty} f(x, y) \, dx \, dy$ (d)   none

35. Characteristic function is used to find moments about -------------------- of a random variable.
    (a)   origin   (b)   mean   (c)   both   (d)   none

36. $\sigma_{XY} = E(XY) -$ -------------------------
    (a) $E[X] + E(Y)$ (b) $E[X] E[Y]$
    (c) $E[X]$ (d)   none

Exercise Questions

1. Let X and Y be random variable having joint density function

   $fXY (x, y) = x + y$ for $0 \leq x \leq 1, 0 \leq y \leq 1$

   $= 0$ otherwise

   Find (a) $Var (X)$ (b) $Var (Y)$ (c) $Cov (X,Y)$ (d) P_{XY}

2. Find the conditional variance of X given Y, where X and Y are random variables with joint density

   $fXY(x, y) = 1$ for $0 \leq x \leq 1, 0 \leq y \leq 1$

   $= 0$ otherwise

3. X and Y are random variables with density

   $fXY (x, y) = kxy$ for $0 \leq x \leq 1, 0 \leq y \leq 1$

   $= 0$ otherwise.

   Where k is a constant. Find $E(X^2 + Y^2)$

4. Let X be random variable with $E (X) = 3$ and $var (X) = 2$. Verify that the random variable X and the random variable $Y = - 6X + 22$ are orthogonal.

5. Find the correlation coefficient between X and Y from the following data.

X	1	2	3	4	5	6	7	8	9
Y	9	8	10	12	11	13	14	16	15

6. If X, Y and Z are three independent random variable of same mean and variance, verify that mean square value of $(Y-Z)$ is twice the variance of X.

7. Two random variable X_1 and X_2 have variance k and 2 respectively. A random variable Y is defined as $Y = 3X_2 - X_1$. If $var(Y) = 25$, find k.

8. Let X and Y be the random variable defined as $X = \cos\theta$ and $Y = \sin\theta$, where θ is a uniform random variable over (0.2π). Show that X and Y are not independent, even though they are uncorrelated.

9. X and Y are the random variable with joint density function

$$fXY(x, y) = \frac{xy}{96} \text{ for } 0 \le x \le 4,\, 0 \le y \le 5$$

$$= 0 \quad \text{otherwise}$$

Find the density of the random variable $U = X + 2Y$

10. Let X and Y have joint density function

$$f_{XY}(x,y) = \frac{3}{4} + xy, \qquad\qquad \text{for} \quad 0 < x < 1, \quad 0 < y < 1$$

$$= 0 \qquad\qquad\qquad \text{otherwise}$$

Find (a) $f(y/x)$ (b) $P\dfrac{Y > 1/2}{X = 1/2}$

11. Let X and Y have joint density function

$$fXY(x, y) = e^{-(x+y)ti\theta} \quad \text{for} \quad 0 \ge 0;\, y \ge 1,$$

If $U = \dfrac{X}{Y}$ and $V = X + Y$, Find the joint density of U and V

12. X and Y are independent random variables with density e^{-x} for $x \ge 0$ and e^{-y} for $y = \ge 0$ respectively. Find joint density of $U = \dfrac{X}{X+Y}$ and $V = X + Y$

13. If X and Y are independent normal random variable with zero mean and variance σ^2, fins the joint density of $R = \sqrt{X^2 + Y^2}$ and $\theta = tan^{-1}\left(\dfrac{Y}{X}\right)$ where R and θ are the new random variables defind. Find the marginal density of R and θ also.

14. The joint density of two random variable X and y is

$$fXY(x, y) = 1 \text{ for } 0 \leq x \leq 1; 0 \leq y \leq 1 0 \leq x \leq 1; 0 \leq y \leq 1$$

$$= 0 \qquad\qquad \text{elsewhere}$$

Find the density function of $Z = XY$.

15. If X and Y are two uncorrelated random variable with same variance. If the random variables $U = X + kY$ and $V = X + \dfrac{\sigma_x}{\sigma_y}$, Y are uncorrelated, find k.

16. If X and Y are two discrete random variables with the joint probability matrix.

$$\begin{array}{c} X/Y \quad -1 \quad\ \ 0 \quad\ \ 1 \\[4pt] P(X, Y) = \begin{array}{c} -1 \\ 0 \\ 1 \end{array} \begin{bmatrix} 0 & 0.2 & 0 \\ 0.1 & 0.2 & 0.1 \\ 0.1 & 0.2 & 0.1 \end{bmatrix} \end{array}$$

(a)   Check whether X and Y are uncorrelated or not

(a)   Find var $\left(\dfrac{Y}{X = -1}\right)$

(b) 17.   The joint characteristic function of two random variable X and Y is given as

$$\emptyset XY\left((w_1, w_2) = exp - 3w_2^2 - 5w\dfrac{2}{2}\right)$$

Verify that X and Y are of zero mean and orthogonal.

18. Find the expression for density function of the random variable $Y = X_1 + X_2 + \ldots + X_n$, where all X_1's are

(i)   independent

(ii)   independent and identically distributed random variables, in terms of their characteristic function $\emptyset_{Xi}(w)$.

19. If X and Y are independent identically distributed normal random variable $N(0, 1)$, find

$$E\left(\sqrt{X^2 + Y^2}\right).$$

(Hint : random variable $Z = \left(\sqrt{X^2 + Y^2}\right)$ is a Rayleigh random variable)

20. Let Y be a random variable defined as the sum of N statistically independent random variables, i.e., $Y = X_1 + X_2 + \cdots + X_N$. If X_i, $i = 1, 2, 3 \ldots N$ then find the probability density of Y.

Answers

1. (a) $\dfrac{11}{144}$;   (b) $\dfrac{11}{144}$;   (c) $\dfrac{-1}{144}$;   (d) $\dfrac{-1}{11}$

2. $\dfrac{1}{12}$ 3.   1   5.   0.946   7.   7

9. g(u) $=\dfrac{(u-2)^2 (u+4)}{2304}$ for $2 < u < 6$

$=\dfrac{(3u-18)}{144}$ for $6 < u < 10$

$=\dfrac{348u - u^3 - 2128}{2304}$ for $10 < u < 14$

$= 0$ elsewhere

10. 10. (a) $=\dfrac{3+4xy}{3+2x}$ for $0 < y < 1 = 0$ otherwise

(b) $\dfrac{9}{16}$

11. $fUV(u,v) = \dfrac{ve^{-v}}{(1+u)^2}$ for $u \geq 0, u \geq v \geq 0$

12. $v \cdot e^{-v}$ for $0 \leq u \leq 1, v \geq 0$

13. $f(r) = \dfrac{r}{\sigma^2} \cdot e - \dfrac{r^2}{2\sigma^2}$ for $r \geq 0; 0 \leq \theta, \leq y \leq 2\pi$

$$f(r) = \frac{r}{\sigma^2}.e^{-\frac{r^2}{2\sigma^2}} \qquad \text{for} \quad r \geq 0$$

$$= 0 \qquad \qquad \text{elsewhere}$$

$$f\theta = \frac{1}{2\pi} \qquad \qquad \text{for } 0 \leq \theta \leq 2\pi$$

$$= 0 \qquad \qquad \text{elsewhere}$$

14. $fZ(z) = -ln_z \qquad \text{for } 0 < z < 1$

$$= 0 \qquad \qquad \text{otherwise}$$

15. $-\dfrac{\sigma_y}{\sigma_x}$

16. (a) uncorrelated (b) 0

18. (i) $fY(y) = \dfrac{1}{2\pi} \displaystyle\int_{-\infty}^{\infty} \left[\prod_{i=1}^{n} \varnothing X_i(w) \right] e^{-jwy} \, dw$

(ii) $fY(y) = \dfrac{1}{2\pi} \displaystyle\int_{-\infty}^{\infty} [\varnothing X_i(w)]^2 \, e^{-jwy} \, dw$

19. $\sqrt{\dfrac{\pi}{2}}$

Review Exercise Questions

1. Let X and Y be independent random variables, is each having density function

$$f_x(x) = \begin{cases} 4e^{-x} & \text{for } x \geq 0 \\ 0 & \text{else where} \end{cases}$$

Find (a) $E(X+Y)$ (b) $E(X^2 + Y^2)$ (c) $E(XY)$.

2. Let X and Y have joint density function

$$f_{xy}(x,y) = \begin{cases} 2x + 3y & \text{for } 0 \leq x \leq 2; \; 0 \leq y \leq 2 \\ 0 & \text{otherwise} \end{cases}$$

Find the conditional expectation of $a)$ Y given X and $b)$ X given Y

3. Let X and Y be independent random variables such that

$$X = \begin{cases} 2 & \text{with probability } \dfrac{2}{3} \\ 1 & \text{with probability } \dfrac{1}{3} \end{cases} \quad \text{and} \quad Y = \begin{cases} 2 & \text{with probability } \dfrac{1}{4} \\ -1 & \text{with probability } \dfrac{3}{4} \end{cases}$$

then find

(a) $E[2X + 3Y]$ (b) $E[X^2 - 3Y^2]$

(c) $E[XY]$ and   (d) $E[XY^2]$

4. Find the correlation coefficient between the two independent random variables X and Y if $E[X]=2$, $E[Y]=3$, $E[XY]=10$, $E[X^2]=9$ and $E[Y^2]=16$

5. Find the first two moments about the mean for the random variable X defined by X

$$X = \begin{cases} 2 & \text{with probability } \dfrac{1}{2} \\ 3 & \text{with probability } \dfrac{1}{3} \\ -1 & \text{with probability } \dfrac{1}{6} \end{cases}$$

6. Let X and Y be two independently and identically distributed random variables with mean 4 and 16 and their second order moments are 64 and 625 respectively. Find the mean, second order moment and variance of the random variable $Z = 9X\text{-}Y$

7. If X and Y are independent random variables having each zero mean and variances 5 and 3 respectively. Find the correlation coefficient between $(2X+3Y)$ and $(2X-3Y)$

8. If X and Y are independent random variables having means 10 and 20 and variances 40 and 10 respectively. Their correlation coefficient is 0.04. Let U and V are defined as $U = -2X+Y$ and $V = 3X + Y$. Find the *(i)* means of U and V *(ii)* Variances of U and V *(iii)* Correlation between U and V and *(iv)* Correlation coefficient between U and V.

9. Let X be the score on the first die and Y be the score on the second die when two dice are thrown. Let Z denote the maximum of X and Y, i.e., $Z = \text{Mim}(X,Y)$. Write down i) the joint distribution of Z and X and ii) mean and variance of Z and covariance of (Z, X)

10. Let X and Y are independent random variables whose moments are $\mu_{10} = 12$, $\mu_{20} = 24$, $\mu_{02} = 22$ and $\mu_{11} = -16$. Find the moment μ_{22}.

11. Let X_1, X_2 and X_3 are three independent random variables whose mean values are $X_1 = 31$, $X_2 = 61$ and $X_3 = -12$. Find the mean values of the following

12. Let X and Y be two random variables each taking three values $^-1$, 0 and 1 and having the joint probability distribution

X / Y	-2	-1	0	Total
-2	0	0.1	0.1	0.2
-1	0.2	0.2	0.2	0.6
0	0	0.1	0.1	0.2
Total	0.2	0.4	0.4	1

(i) Show that X and Y have different expectations (ii) Show that X and Y uncorrelated (iii) Find variance of X and Y (iv) Find the conditional distribution of X given $Y = -1$ (v) Find $V(Y/X = -2)$

13. Suppose that X and Y are independent standard normal random variables and define $U = 2X + 3Y$ and $V = 2X - 3Y$. Find the moment generating function of U and V.

14. Find the joint moment generating function for the two random variables having joint density is

$$f_{XY}(x, y) = \frac{2y}{\sqrt{2\pi}} e^{-2y - \frac{1}{2}(x-y)^2}$$

15. Two random variables X and Y have the joint characteristic function

$$\phi_{XY}(w_1, w_2) = e^{-2w_1{}^2 - w_2{}^2}$$ then show that the random variables X and Y are both zero mean random variables and also that they are uncorrelated.

16. Find the joint characteristic function of the discrete random variable X and Y then their joint probability mass function is

$$f_{XY}(x,y) = \begin{cases} \dfrac{1}{3} & \text{for } (x,y) = (1,1) \\[2mm] \dfrac{1}{3} & \text{for } (x,y) = (-1,0) \\[2mm] \dfrac{1}{3} & \text{for } (x,y) = (1,0) \\[2mm] \dfrac{1}{2} & \text{for } (x,y) = (-1-,1) \\[2mm] 0 & \text{otherwise} \end{cases}$$

17. Let $X_1, X_2, X_3 \cdots X_n$ are independently and identically distributed random variables as

$$X_K = \begin{cases} 1 & \text{with probability } \dfrac{1}{3} \\[2mm] -1 & \text{with probability } \dfrac{2}{3} \end{cases}$$

Find the characteristic function of the random variable $Y = \dfrac{X_1 + X_2 + \cdots + X_n}{\sqrt{n}}$

18. Find the joint moment generating function for the two random variables having joint density is

$$f_{XY}(x,y) = \dfrac{1}{2\pi} e^{-\frac{1}{2}(x^2 - y^2)} \qquad -\infty < x, y < \infty$$

19. Let X_1 and X_2 be two Gaussian random variables having zero means and variances $\sigma_1^2 = 3$ and $\sigma_2^2 = 4$. Their covariance $C_{X_2 X_2}$ equals to 2. If X_1 and X_2 are linearly transformed to new variables Y_1 and Y_2 according to $Y_1 = 2X_1 + X_2$ and $Y_2 = 2X_1 - 3X_2$. Find the means and variances of Y_1 and Y_2 and covariance both Y_1 and Y_2.

20. In a particular factory, the average salary for an employee is Rs. 2,40,000 per year. This year, management awards the bonuses to every employee as follows. A Dussarah bonus of Rs. 1500, and an incentive bonus is 7.5 percent of the employee's salary. Find the mean bonus received by employees?

21. For two random variables X and Y

$$f_{XY}(xy) = 0.1\delta(x-1)\delta(y) + 0.2\delta(x)\delta(y) + 0.3\delta(x)\delta(y+2) +$$
$$0.5\delta(x+1)\delta(y-2) + 0.3\delta(x+1)\delta(y+1) + 0.02\delta(x+1)\delta(y+3)$$

Find (i) The Cross Correlation between X and Y (ii) The covariance between X and Y (iii) The correlation coefficient between X and Y (iv) Are X and Y are uncorrelated or orthogonal

22. Consider random variables Y_1 and Y_2 are related to arbitrary random variables X and Y by the coordinate rotation

$$Y_1 = 2X \cos\theta + 3Y \sin\theta \text{ and } Y_2 = -2X \sin s\theta + 3Y \cos\theta$$

Find *Covariance of Y_1 and Y_2 and for what value of* θ, the random variables Y_1 and Y_2 uncorrelated

23. Let the two random variables X and Y have the joint density function

$$f_{xy}(x,y) = \begin{cases} \dfrac{(x-y)^2}{20} & for \ -1 < x < 2; \ -5 \le y \le 5 \\ 0 & O \ therwise \end{cases}$$

Find all the third order moments for X and Y

24. Let X and Y be two random variables having their joint density function is

$$f_{XY}(x,y) = \begin{cases} \dfrac{xy}{3} & 0 < x < 3, 0 < y < 4 \\ 0 & Else \ where \end{cases}$$

Test whether X and Y are statistically independent or not and also correlated or uncorrelated.

25. Random variables Z and W are defined by $Z = aX + Y$, and $W = aX - Y$ where a is a real number. Determine a such that Z and W are orthogonal.

26. Two independent random variables X and Y are having their densities as

$$f_X(x) = e^{-2x}u(x) \text{ and } f_Y(y) = e^{-2y}u(y) \ 0 < x, y < 1, \text{ find } P(X+Y \le 2)$$

27. Let X and Y be random variable having joint density function

$$fXY(x,y) = x - y \quad for \ 0 \le x \le 1, 0 \le y \le 1$$
$$= 0 \quad otherwise$$

Find (a) *Var (X)* (b) *Var (Y)* (c) *Cov (X,Y)* (d) P_{XY}

28. Find the conditional variance of X given Y, where X and Y are random variables with joint density

$$fXY(x,y) = 1 \text{ for } 0 \le x \le 1, 0 \le y \le 1$$

$$= 0 \quad \text{otherwise}$$

29. X and Y are random variables with density

$$fXY(x, y) = 2xy \quad \text{for } 0 \le x \le 1, 0 \le y \le 1$$

$$= 0 \quad \text{otherwise.}$$

Find $E(X^2+Y^2)$

30. Let X be random variable with $E(X) = 2$ and $var(X) = 3$. Verify that the random variable X and the random variable $Y = -4X + 18$ are orthogonal.

31. If X, Y and Z are three independent random variable of same mean and variance, test whether the mean square value of $(X - Z)$ is twice the variance of Y.

CHAPTER 6

Random Processes

6.1 Introduction

A random variable X is defined as a function of possible outcomes s of an experiment. The random process also known as stochastic process, is extension of random variable with inclusion of time. The random process is denoted by $X(t,s)$ where 's' represents the possible outcome and 't' denotes time.

Random process $X(t,s)$ represents a family or ensemble of time functions when t and s are variables.

If we consider random process for a fixed possible outcome $s = s_i$, when 't' is variable then random process $X(t, s_i) = X_i(t)$ represents a sample function ensemble member, or sometimes a realization of the process as shown in Fig. 6.1

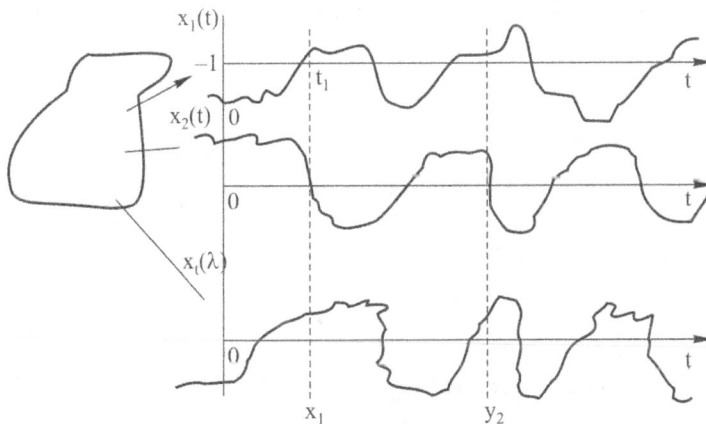

Fig. 6.1 Random process.

441

For a fixed sample point s_1, the random process represents a time function which is sample function of the random process for that particular fixed sample point. $= X_1(t)$ is the sample function for sample point s_1 as shown in Fig 6.1. Similarly we obtain other sample functions $X_2(t)$, $X_3(t)$, ... $X_n(t)$ where n represents very large integer value for the sample points $s_2, s_3, \ldots s_n$. Therefore random process $X(t, s)$ represents 'Ensemble' or family of all sample functions.

If we let t is fixed and s is variable, the random process represents a random variable $X_1(t) = X(t_1, s)$ is obtained from the random process, when time t is fixed at t_1. X_1 corresponds to a vertical slice through the sample functions at a time t_1 as shown in fig 6.1. Similarly $X_2 = X(t_2, s)$ is obtained when time t is fixed at t_2. In this way we obtain infinite number of random variables in a random process.

Therefore the random process X (t, s) becomes sample function X_i (t) when s is fixed $(s = s_i)$ and t is variable and the random process represents the random variable X_j when t is fixed $(t = t_j)$ and the sample point s is variable.

6.2   Classification of Random Processes

We classify the random process depending on the characteristics of time t and random variable $X = X(t)$ at time t where t and X having the values in the ranges $-\infty < t < \infty$ and $-\infty < X < \infty$

Continuous Random Process

If both X and t can have any of continuous of values, then the $X(t)$ is called a continuous random process as shown in Fig. 6.1. Example of this category of random process is Thermal noise generated by any realizable network.

Discrete Random Process

In this random process, the random variable having only discrete values while t is continuous as shown in Fig. 6.2

Fig 6.2 describes such a process derived by heavily limiting the sample functions shown in Fig 6.1. The sample function has only two discrete values that is positive level and negative level.

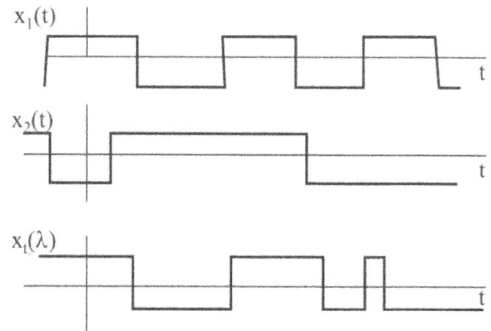

Fig. 6.2 A discrete random process.

6.3 Continuous Random Sequence

A random process for which X is continuous but time has only discrete values is called continuous random sequence. Such a sequence is obtained by periodically sampling the ensemble members of Fig. 6.1.

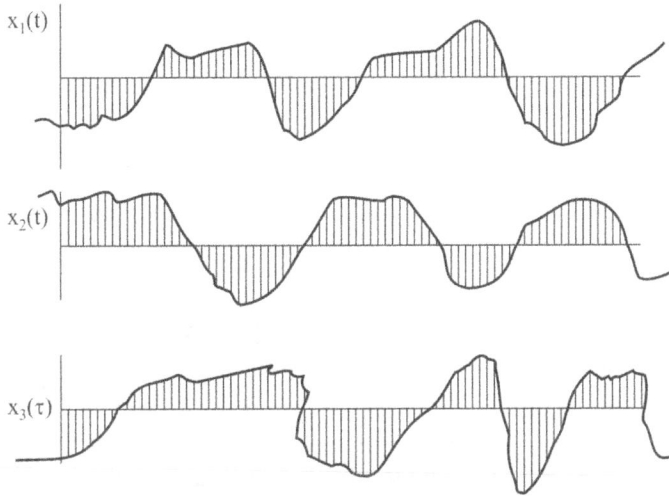

Fig. 6.3 A continuous random sequence

Random sequences is also called a discrete time (*DT*) random process since the continuous random sequence is defined at only discrete time values and usually denoted by $X(n)$ which are important in the analysis of various *DSP* systems.

6.4 Discrete Random Sequences

If both time t and random variable X are discrete in nature, then the random process is called discrete random sequence as shown in Fig. 6.4.

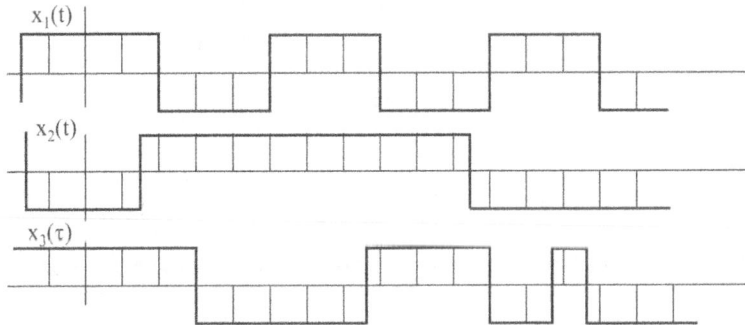

Fig. 6.4 Discrete random sequences formed by simplify the wave form of Fig. 6.2.

6.5   Deterministic Process

A process is called deterministic process if the future values of any sample function can be determined from past values.

6.6   Non Deterministic Random Process

If future values of any sample function can not be predicted from observed past values, the process is called Non deterministic Random Process.

6.7   Stationary – Non stationary Random Process

A random process becomes a random variable when time is fixed at some value. The random variable will possess statistical properties such as mean value, moments, variance which is related to density function.

   If two random variables are obtained from the random process for two different values of time these variables will possess statistical properties such as means, variances, joint moments etc.

   A random process X(t) is called stationary random process if all its statistical properties do not change with time. Other processes are called non-stationary.

6.8   Probability Distribution and Density Function

We define different orders of distributions as stated below

First order probability distribution for random variable $X_1 = X(t_1)$ is defined as

$$F_X(x_1, \ t_1) = \ p\{X(t_1) \le x_1\} \qquad \qquad .....(6.1)$$

for any real number x_1

   Second order joint distribution for two random variables $X_1 = X(t_1)$ and $X_2 = X(t_2)$ is defined as

$$F_X(x_1, x_2; \ t_1, t_2) = \ p\{X(t_1) \le x_1, \ X(t_2) \le x_2\} \qquad \qquad .....(6.2)$$

In a similar manner, the Nth order joint distribution function is defined as

$$F_X(x_1, ... x_N; \ t_1, ... t_N)$$

$$= p\{X(t_1) \le x_1, \ X(t_2) \le x_2, ... X(t_N) \le x_N\} \qquad \qquad .....(6.3)$$

   Probability density functions are obtained from appropriate derivatives of above three relationships

$$f_X(x_1, \ t_1) = \frac{d}{dx_1} F_X(x_1, \ t_1) \qquad \qquad .....(6.4)$$

$$f_X(x_1, x_2; t_1, t_2) = \frac{\partial^2}{\partial x_1 \partial x_2} F_X(x_1, x_2; t_1, t_2) \qquad \text{.....(6.5)}$$

$$f_X(x_1, \ldots x_N; t_1, \ldots t_N) = \frac{\partial^N}{\partial x_1 \ldots \partial x_N} F_X(x_1, \ldots x_N; t_1, \ldots t_N)$$

$$\text{.....(6.6)}$$

6.9 First Order Stationary Process

A random process is called stationary to order one if it is order *PDF* does not change with a shift in time origin

$$f_X(x_1, t_1) = f_X(x_1, t_1 + \Delta) \qquad \text{.....(6.7)}$$

must be valid for t_1 and real number Δ.

From equation (6.7), it is evident that $f_x(x_1; t_1)$ is independent of time t_1 and the process has mean value which is constant.

$$E[X(t)] = \overline{X} = \text{constant} \qquad \text{.....(6.8)}$$

6.10 Second Order Stationary Process

A process is said to be second order stationary process if its second order *PDF* satisfies the condition such as

$$f_X(x_1, x_2; t_1, t_2) = f_X(x_1, x_2; t_1 + \Delta, t_2 + \Delta) \qquad \text{.....(6.9)}$$

for all t_1, t_2 and Δ.

The auto correlation function of a second order stationary process,

$$R_{XX}(t_1, t_2) = E[X(t_1) X(t_2)] \qquad \text{.....(6.10)}$$

which is a function of t_1 and t_2

Use of eq. (6.9) in the evaluation of auto correlation we get

$$R_{XX}(t_1, t_1 + \tau) = E[X(t_1) X(t_1 + \tau)] = R_{XX}(\tau) \qquad \text{.....(6.11)}$$

where $\tau = t_2 - t_1$.

The most useful form is wide sense stationary process which satisfy the following two conditions

$$E[X(t)] = \overline{X} = \text{constant} \qquad \text{.....(6.12)}$$

$$E[X(t) \cdot X(t + \tau)] = R_{XX}(\tau) \qquad \text{.....(6.13)}$$

A process stationary to order two is clearly wide sense stationary. However the converse is not true

6.11   Statistical Independence

Two processes X(t) and Y(t) are statistically independent if joint density function must be factorable by two independent density functions. For any change of times (t_1, t_2, ... t_N, t_1^1, t_2^1, ... t_N^1)

$$f_{XY}\left(x_1,...x_N,y_1,...y_N; t_1, ..t_N, t_1', ..t_N'\right) = f_X\left(x_1,...x_N; t_1,..t_N\right) \times$$

$$f_Y\left(y_1,...y_N; t_1',.. t_N'\right) \qquad .....(6.14)$$

6.12   N th Order and Strict Sense Stationary

A random process is said to be stationary to order N if its Nth order *PDF* is invariant to time shift N i.e., if it satisfies the condition

$$f_X\left(x_1,...x_N; t_1,..t_N\right) = f_X\left(x_1,...x_N; t_1+\Delta,....t_N+\Delta\right) \qquad .....(6.15)$$

for all values of time $t_1...t_N$ and shift Δ

A process stationary to all orders $N = 1,2,...N$ is called strict sense stationary

6.13   Time Averages and Ergodicity

The mean value and autocorrelation can also be determined using time averages and are defined by

$$\overline{x} = \underset{T\to\infty}{Lt} \frac{1}{2T} \int_{-T}^{T} x(t)dt \qquad .....(6.16)$$

$$R_{xx}(\tau) = \underset{T\to\infty}{Lt} \frac{1}{2T} \int_{-T}^{T} x(t)x(t+\tau)dt \qquad .....(6.17)$$

the variation of last two integrals give two numbers for any sample functions. $\overline{x}$ and R_{xx} (τ) are actually random variables when all sample functions are considered.

By taking the expectation on both sides of equations (6.16) and (6.17) we obtain

$$E\left[\overline{x}\right] = \overline{X} \qquad .....(6.18)$$

$$E\left[R_{xx}(\tau)\right] = R_{XX}(\tau) \qquad .....(6.19)$$

We write the above equations as

$$\overline{x} = \overline{X} \qquad .....(6.20)$$

$$R_{xx}(\tau) = R_{XX}(\tau) \qquad \qquad .....(6.21)$$

by assuming that the random variables $\bar{x}$ and $R_{xx}(\tau)$ have zero variances.

The time averages $\bar{x}, R_{xx}(\tau)$ are equal to statistical averages $\bar{x}$ and $R_{XX}(\tau)$. Processes that satisfy this condition are called ergodic processes.

Two random processes are said to be jointly ergodic if they are individually ergodic and also satisfies the time cross correlation function.

$$R_{xy}(\tau) = \underset{T \to \infty}{Lt} \frac{1}{2T} \int_{-T}^{T} x(t)y(t+\tau)dt = R_{XY}(\tau) \qquad .....(6.22)$$

Which is equal to statistical cross correlation function.

6.14 Mean Ergodic Processes

The random process $X(t)$ becomes mean ergodic when statistical averages $\bar{x}$ are equal to time averages $\bar{x}$

$$E[X(t)] = \overline{X} = \bar{x} = \frac{1}{T} \int_{-T}^{T} X(t)dt \qquad .....(6.23)$$

If $E[X(t)] = \bar{x}$, then the random process is called mean ergodic process.

6.15 Correlation Ergodic Processes

A stationary process $X(t)$ with autocorrelation function $R_{x\,x}(\tau)$ is said to be autocorrelation ergodic or ergodic in the autocorrelation if and only if, for all 'τ'

$$R_{XX}(\tau) = \underset{T \to \infty}{Lt} \frac{1}{2T} \int_{-T}^{T} X(t)X(t+\tau)dt = R_{xx}(\tau) \qquad .....(6.24)$$

Two processes $X(t)$ and $Y(t)$ are called cross correlation ergodic or ergodic in the correlation sense if the time cross correlation function is equal to the statistical cross correlation function.

6.16 Correlation Functions

The correlation function between two random variables is a measure of the similarity between the variables. A common function is the expected value of the product of two random variables.

There are two types of correlation functions
1. Auto correlation function
2. Cross correlation function

6.17   Auto Correlation Function

The auto correlation function of a random process $X(t)$ is the correlation of $E[X_1, X_2]$ of two random variables $X_1 = X(t_1)$ and $X_2 = X(t_2)$ defined by the process at times t_1 and t_2.

The autocorrelation function of random process $R_{xx}(t_1, t_2)$ is mathematically expressed as

$$R_{XX}(t_1, t_2) = E[X(t_1) . X(t_2)]$$ .....(6.25)

Let $t_1 = t$, $t_2 = t + \tau$ where $\tau = t_2 - t_1$ is the time delay

$$R_{XX}(t, t + \tau) = E[X(t) . X(t + \tau)]$$ .....(6.26)

If $X(t)$ is at least wide sense stationary process(WSS), then $R_{xx}(t, t + \tau)$ must be a function of time difference

$$R_{XX}(\tau) = E[X(t) \times X(t + \tau)]$$ .....(6.27)

6.18   Properties of Auto Correlation Function

For WSS process the auto correlation function has the following properties

1. $R_{xx}(\tau)$ is bounded by its value at the origin

    i.e., $$|R_{XX}(\tau)| \le R_{XX}(0)$$ .....(6.28)

2. Auto correlation function has even symmetry

    i.e., $$R_{XX}(-\tau) = R_{XX}(\tau)$$ .....(6.29)

3. The mean value in the random process is obtained from $R_{xx}(\tau)$ by putting $\tau = 0$

    i.e., $$R_{XX}(0) = E[X^2(t)] = \overline{X}^2$$ .....(6.30)

4. If $X(t)$ has periodic component, then $R_{xx}(\tau)$ will have periodic component with same period

5. If the random process $X(t)$ has no periodic components and if $E[X(t)] = \overline{X} \ne 0$ then

    $$\underset{|T| \to \infty}{Lt} R_{XX}(\tau) = \overline{X}^2$$ .....(6.31)

6. If $X(t)$ is ergodic and has no periodic component, then

    $$\underset{|T| \to \infty}{Lt} R_{XX}(\tau) = 0$$ .....(6.32)

7. $R_{xx}(\tau)$ can not have an arbitrary shape

6.19 Cross Correlation Function and its Properties

Cross correlation deals with the correlation between two different random processes $X(t)$ and $Y(t)$ and is defined as

$$R_{XY}(t, t + \tau) = E[X(t)Y(t + \tau)] \quad\quad .....(6.33)$$

If $X(t)$ and $Y(t)$ are jointly wide sense stationary random process; the cross correlation $R_{XY}(t, t + \tau)$ is independent of absolute time. We can write $R_{XY}(t, t + \tau)$ as

$$R_{XY}(\tau) = E[X(t)Y(t + \tau)] \quad\quad .....(6.34)$$

The $X(t)$ and $Y(t)$ are said to be orthogonal processes if they satisfy the condition

$$R_{XY}(\tau) = 0 \quad\quad .....(6.35)$$

If the two processes are statistically independent, we can write R_{XY} as

$$R_{XY}(t, t + \tau) = E[X(t)] \; E[Y(t + \tau)] \quad\quad .....(6.36)$$

If, in addition to being independent, $X(t)$ and $Y(t)$ are wide sense stationary, we can write equation (6.4-13) as

$$R_{XY}(\tau) = \overline{X} \; \overline{Y} = E[X(t)] \; E[Y(t)] \quad\quad .....(6.37)$$

The properties of the correlation function are listed as below

1. $R_{XY}(-\tau) = R_{YX}(\tau)$ \quad\quad .....(6.38)

2. $|R_{XY}(\tau)| \le \sqrt{R_{XY}(0) \; R_{YY}(0)}$ \quad\quad .....(6.39)

3. $|R_{XY}(\tau)| \le \frac{1}{2}[R_{XX}(0) \; R_{YY}(0)]$ \quad\quad .....(6.40)

Property 1 describes the symmetry of $R_{XY}(\tau)$. Property 2 and 3 explains the bounds on the magnitude of $R_{XY}(\tau)$.

6.20 Covariance Functions

The auto covariance function is defined by

$$C_{XX}(t, t + \tau) = E[\{X(t) - E[X(t)]\} \; \{X(t + \tau) - E[X(t + \tau)]\}] \quad\quad .....(6.41)$$

We also write C_{XX} as

$$C_{XX}(t, t + \tau) = R_{XX}(t, t + \tau) - E[X(t) \; X(t + \tau)] \quad\quad .....(6.42)$$

The covariance between two random processes $X(t)$ and $Y(t)$ is called cross covariance of $X(t)$ and $Y(t)$ and is defined by

$$C_{XY}(t, t + \tau) = E[\{X(t) - E[X(t)]\} \; \{Y(t + \tau) - E[Y(t + \tau)]\}] \quad\quad .....(6.43)$$

We can also write C_{XY} as

$$C_{XY}(t, t + \tau) = R_{XY}(t, t + \tau) - E[X(t) \; Y(t + \tau)] \quad\quad .....(6.44)$$

Equations (6.42) and (6.43) becomes

$$C_{XX}(\tau) = R_{XX}(\tau) - \overline{X}^2 \qquad \qquad .....(6.45)$$

$$(\tau) = R_{XY}(\tau) - \overline{X}\ \overline{Y} = E[X(t)]\ E[Y(t)] \qquad \qquad .....(6.46)$$

For joint wide sense stationary random processes $X(t)$ and $Y(t)$.

If $\tau = 0$, equation (6.43) gives the variance of random process.

For wide sense stationary process, the variance is independent of time and is given by (6.46) with $\tau = 0$

$$\sigma_X^2 = E\left[\{X(t) - E[X(t)]\}^2\right] \qquad \qquad .....(6.47)$$

Two random processes $X(t)$ and $Y(t)$ are called Uncorrelated processes if

$$C_{XY}(t, t+T) = 0 \quad ....(6.4\text{-}25)$$

From equation (6.47), we get

$$R_{XY}(t, t+\tau) = E[X(t)]\ E[Y(t+\tau)] \qquad \qquad .....(6.48)$$

We conclude that the independent processes are uncorrected but the converse is not necessarily true.

6.21 Gaussian Random Process

N random variables $X_1 = X(t_1)$ .... $X_i = X(t_i)$ ... $X_N = X(t_N)$ are obtained by observing the continuous random process $X(t)$ at N time instants $t_1,...t_i... t_N$. These random variables are jointly Gaussian if they have a joint density as given by

$$f_X(x_1,...x_N; t_1,..t_N) = \frac{1}{\sqrt{(2\pi)^N |[C_X]|}}\ e^{-\frac{1}{2}\left(\frac{x-\overline{X}}{C_X}\right)^t} \qquad \qquad .....(6.49)$$

Where the matrices are given by

$$[x - \overline{X}] = \begin{bmatrix} x_1 - \overline{X}_1 \\ x_2 - \overline{X}_2 \\ .... \\ .... \\ x_N - \overline{X}_N \end{bmatrix} \qquad \qquad .....(6.50)$$

$$[C_X] = \begin{bmatrix} C_{11} & C_{12} & \ldots\ldots C_{1N} \\ C_{21} & C_{22} & \ldots\ldots C_{2N} \\ \ldots & \ldots & \ldots & \ldots \\ \ldots & \ldots & \ldots & \ldots \\ C_{N1} & C_{N2} & \ldots\ldots C_{NN} \end{bmatrix}$$ .....(6.51)

Then the process is called Gaussian Process

The mean values X_i of $X(t_i)$ are given by

$$\overline{X}_i = E[X_i] = E[X(t_i)]$$ .....(6.52)

Elements of $[C_X]$, called the covariance matrix of the N random variables, are given by

$$C_{ij} = E[(X_i - \overline{X}_i)(X_j - \overline{X}_j)] = \begin{cases} \sigma^2_{X_i} & \text{if } i = j \\ C_{X_i X_i} & \text{if } i \neq j \end{cases}$$ .....(6.53)

Which is the auto covariance of X_i and X_j

Substitution of mean and covariance values in equation (6.5-1) completely describes the Gaussian random process.

Alternative specification of random process is evident by expanding the equation (6.53), i.e

$$C_{XX}(t_i, t_k) = R_{XX}(t_i, t_k) = E[X(t_i)] \; E[X(t_k)]$$ ....(6.54)

And by using the mean and auto correlation functions

If the Gaussian process is not stationary, the mean and auto covariance $C_{x\,x}$ will depend on the absolute time. The mean will be constant for wide sense stationary processes while the auto correlation and auto covariance functions will depend only time differences.

$$C_{XX}(t_i, t_k) = C_{XX}(t_k - t_i)$$ .....(6.55)

$$R_{XX}(t_i, t_k) = R_{XX}(t_k - t_i)$$ .....(6.56)

A wide sense stationary Gaussian process is also strictly stationary random process.

6.22 Poisson Random Process

Poisson process describes the number of times that some event that has occurred as a function of time where time events occurs at random times.

To describe the Poisson process, let $X(t)$ represents the number of event occurrences with time, then $X(t)$ has integer valued non decreasing sample functions as shown in Fig. 6.5(a) for the random occurrence time of Fig. 6.5(b).

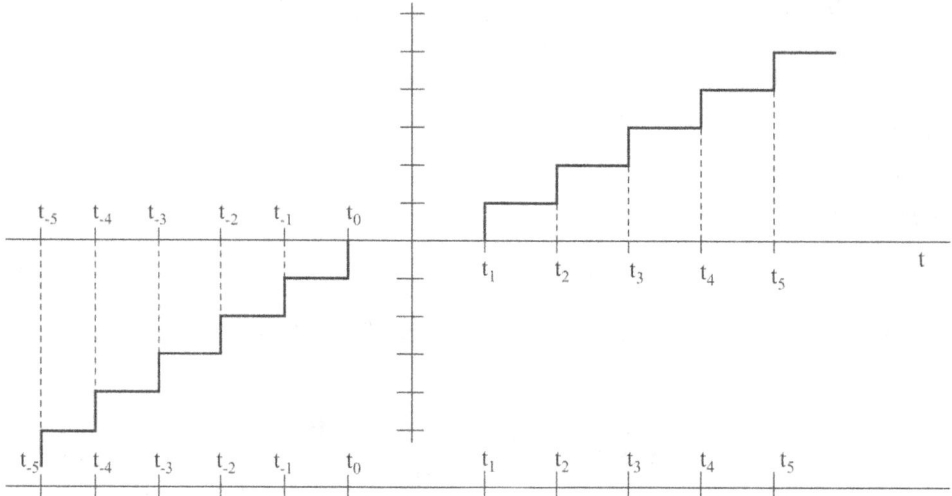

Fig. 6.5(a) and (b)

The number of event occurrences in any finite interval of time is described by Poisson distribution where the average rate of occurrences is denoted by λ

6.23 Probability Density Function

The probability of exactly k occurrence over a time interval $(0, t)$ is given by

$$P\,[X(t) = k] = \frac{(\lambda t)^k\, e^{-\lambda t}}{k!} \qquad k = 0, 1, 2,\ 3, \ldots \qquad \text{.....(6.57)}$$

By using $b = \lambda t$

The probability density of the number of occurrence is

$$f_X(x) = \sum_{k=0}^{\infty} \frac{(\lambda t)^k\, e^{-\lambda t}}{k!}\ \delta(x - k) \qquad \text{.....(6.58)}$$

The mean and variance of a poisson random variable are equal to λt. The second moment is given by

$$E\left[X^2(t)\right] = \sigma_X^2 + \left\{E[X(t)]\right\}^2 \qquad \text{.....(6.59)}$$

$$= \lambda t + (\lambda t)^2$$

The above facts are used to establish useful equations by evaluating the mean and second moment

$$E[X(t)] = \int_{-\infty}^{\infty} x f_X(x) dx = \int_{-\infty}^{\infty} x \sum_{k=0}^{\infty} \frac{(\lambda t)^k e^{-\lambda t}}{k!} \delta(x-k) dx$$

$$= \sum_{k=0}^{\infty} \frac{k(\lambda t)^k e^{-\lambda t}}{k!} = \lambda t \qquad \qquad .....(6.60)$$

$$E[X^2(t)] = \int_{-\infty}^{\infty} x^2 f_X(x) dx$$

$$= \sum_{k=0}^{\infty} \frac{k^2 (\lambda t)^k e^{-\lambda t}}{k!}$$

$$= \lambda t [1 + \lambda t] \qquad \qquad .....(6.61)$$

6.24 Joint Probability Density

The probability of k_1 event occurrences over $(0, t)$ is given by

$$P [X(t_1) = k_1] = \frac{(\lambda t_1)^{k_1} e^{-\lambda t_1}}{k_1!} \quad k_1 = 0, 1, 2, 3, \dots \qquad .....(6.62)$$

This is obtained by observing from equation (6.62).

The conditional probability of k_2 events occurred over $(0, t)$ is just the probability that $(k_2 - k_1)$ events occurred over (t_1, t_2) and is given by

$$P [X(t_2) = k_2 / X(t_1) = k_1] = \frac{\lambda(t_2 - t_1)^{k_2 - k_1} e^{-\lambda(t_2 - t_1)}}{(k_2 - k_1)!} \qquad .....(6.63)$$

for $\qquad \qquad k_2 \geq k_1$

The joint probability of k_2 occurrence at time t_2 and k_1 occurrence at time t_1 is the product of equations (6.6-6) and (6.6-7) and is given by

$$P(k_1, k_2) = P [X(t_2) = k_2 / X(t_1) = k_1] \ \ p [X(t_1) = k_1]$$

$$= \frac{(\lambda t_1)^{k_1} [\lambda(t_2 - t_1)]^{k_2 - k_1} e^{-\lambda t_2}}{k_1! (k_2 - k_1)!} \qquad \text{for} \quad k_2 \geq k_1 \qquad .....(6.64)$$

The joint density now becomes

$$f_X(x_1, x_2) = \sum_{k_1=0}^{\infty} \sum_{k_2=k_1} P(k_1, k_2) \ \delta(x_1 - k_1) \ \delta(x_2 - k_2) \qquad .....(6.65)$$

For the process random variables

$$X(t_1) = X_1 \ \ X(t_2) = X_2$$

Examples

***Example* 1:** Consider a random process $X(t) = A \cos(\omega t + \theta)$ where A and ω are constants and θ is random variable uniformly distributed over $(0, 2\pi)$. Determine whether $X(t)$ is a wide sense stationary or not.

Sol: The mean value is

$$E[X(t)] = \int_0^{2\pi} A \cos(\omega t + \theta) \frac{1}{2\pi} \, d\theta$$

$$= A \int_0^{2\pi} \frac{1}{2\pi} \cos(\omega t + \theta) \, d\theta$$

$$= 0$$

The auto correlation function is

$$R_{XX}(\tau) = E[X(t) \times X(t + \tau)]$$

$$= E[A \cos(\omega t + \theta) A \cos(\omega t + \omega\tau + \theta)]$$

$$= \frac{A^2}{2} E[\cos(\omega t) + \cos(2\omega t + \omega\tau + 2\theta)]$$

$$= \frac{A^2}{2} \cos(\omega t)$$

The second term $\dfrac{A^2}{2} E[\cos(2\omega t + \omega\tau + 2\theta)] = 0$

The auto correlation function $R_{x\,x}(\tau)$ depends only on τ and the mean value is a constant. So $X(t)$ is wide sense stationary random process.

***Example* 2:** A random process $X(t)$ is defined as $X(t) = A \sin(\omega t + \theta)$ where ω and θ are constants and A is a random variable determined whether $X(t)$ is a wide sense stationary process or not

Sol: The mean value $E[X(t)]$ is

$$E[X(t)] = E[A \sin(\omega t + \theta)]$$

$$= A E[\sin(\omega t + \theta)]$$

The auto correlation function $R_{XX}(\tau)$ is

$$R_{XX}(\tau) = E[X(t) \times X(t + \tau)]$$

$$= E[A \sin(\omega t + \theta) A \sin(\omega t + \omega\tau + \theta)]$$

$$= E[A^2] \ \sin(\omega t + \theta) \ \sin(\omega t + \omega \tau + \theta)$$

$$= E[A^2] \ [\cos(\omega_0 \tau) - \cos(2\omega t + \omega \tau + 2\theta)]$$

Both mean and auto correlation functions are dependent on time. Hence $X(t)$ is not a wide sense stationary process.

***Example* 3:** Let the random process $X(t)$ is defined by $X(t) = A\cos(\omega t) + B\sin(\omega t)$ where ω is a constant and A and B are uncorrelated zero mean random variable having different density functions but the same variable prove that $X(t)$ is a wide sense stationary.

Sol: If A and B are uncorrelated zero mean random variable then

$$E[A\,B] = E[A] = E[B] = 0$$

Given that Variance of $A = E[A^2] - [E(A)]^2$

Mean of $\quad E[X(t)] = E[A\cos(\omega t) + B\sin(\omega t)]$

$$= E[A]\cos \omega t + E[B]\sin \omega t$$

$$= 0$$

The auto correlation function $R_{XX}(\tau)$ is

$$R_{XX}(\tau) = E[X(t) \times X(t + \tau)]$$

$$= E\big[\{A\cos(\omega t) + B\sin \omega t\}\{A\cos(\omega t + \omega \tau) + B\sin(\omega t + \omega \tau)\}\big]$$

$$= E[A^2 \cos(\omega t)\cos(\omega t + \omega \tau) + AB\cos(\omega t)\sin(\omega t + \omega \tau)$$

$$+ AB\sin(\omega t)\cos(\omega t + \omega \tau) + B^2 \sin(\omega t)\sin(\omega t + \omega \tau)]$$

Since $\quad E[A^2] = E[B^2] = \sigma^2$ and $E[AB] = 0$

$$= \sigma^2 [\cos(\omega t)\cos(\omega t + \omega \tau) + \sin(\omega t)\sin(\omega t + \omega \tau)]$$

$$= \sigma^2 [\cos(\omega t - \omega t - \omega \tau)]$$

$$= \sigma^2 \ \cos \omega \tau$$

So the random process $X(t)$ is a wide sense stationary random process

***Example* 4:** Determine whether the constant process $X(t) - A$, where A is random variable with mean $\overline{A}$ and variance σ_A^2 is mean ergodic.

Sol: Given random process $X(t) = A$ where A is random variable

$$E[X(t)] = E[A] = \overline{A}$$

Variance of $X(t) = \sigma_A^2$

The process becomes mean ergodic if and only if

$$E[X(t)] = <X(t)> \quad \text{as } T \to \infty$$

$$E[X(t)] = \overline{A}$$

$$<X(t)> = \underset{T\to\infty}{Lt} \frac{1}{T} \int_{-T/2}^{T/2} X(t)dt$$

$$= \underset{T\to\infty}{Lt} \frac{1}{T} \int_{-T/2}^{T/2} A dt$$

$$= \underset{T\to\infty}{Lt} \frac{1}{T} AT$$

$$= A$$

Since $E[X(t)] = <X(t)> = A$

Since $X(t)$, the process becomes mean ergodic.

Example 5: If $X(t)$ is a stationary random process having a mean value $E[X(t)] = 2$ and autocorrelation function $R_{XX}(\tau) = 5 + 3e^{-|\tau|}$, find

1. The mean value

2. The variance of the random variable $Y = \int_0^3 X(t)\, dt$

Sol: Given that, for the random process $X(t)$

$$E[X(t)] = 2$$

$R_{XX}(\tau) = 5 + 3e^{-|\tau|}$ i.e., $X(t)$ is WSS random process $Y = \int_0^3 X(t)\, dt$

1. The mean value of $Y(t)$ is

$$E[Y(t)] = E\left[\int_0^3 X(t)\, dt\right]$$

$$= \int_0^3 E[X(t)]dt$$

$$= \int_0^3 2\,dt$$

$$= \left[2t\right]_0^3$$

$$= 2(3\text{-}0)$$

$$= 6$$

2. The variance of $Y(t)$ i.e. σ_Y^2

$$\sigma_Y^2 = E\left[Y^2\right] - \left\{E\left[Y\right]\right\}^2$$

$$E\left[Y^2\right] = E\left[\int_0^3 X(t)\,dt \int_0^3 X(t)\,dt\right]$$

$$= \int_0^3 \int_0^3 E\left[X^2(t)\right]dt\,dt$$

$$= \int_0^3 \int_0^3 R_{XX}(0)dt\,dt$$

$$= \int_0^3 \int_0^3 (5 + e^{-0})dt\,dt$$

$$= \int_0^3 \int_0^3 8\,dt\,dt$$

$$= 8 \int_0^3 \left[\int_0^3 dt\right]dt$$

$$= 8 \int_0^3 \left[t\right]_0^3 dt$$

$$= 24 \int_0^3 dt$$

$$= 24\left[t\right]_0^3$$

$$= 24 \times 3$$

$$= 72$$

Example 6: For a stationary ergodic process with the periodic components, the autocorrelation function is given as $R_{XX}(\tau) = 36 + \dfrac{4}{1+5\tau^2}$. find the mean and variance of the process $X(t)$

Sol: Given $R_{XX}(\tau) = 36 + \dfrac{4}{1+5\tau^2}$.

Since $X(t)$ is ergodic with no periodic components,

We have $\underset{|T|\to\infty}{Lt}\ R_{XX}(\tau) = \overline{X}^2 = [E(X)]^2$

$$\underset{|T|\to\infty}{Lt}\ R_{XX}(\tau) = \underset{|T|\to\infty}{Lt}\ 36 + \frac{4}{1+5T^2}$$

$$= 36 + 0$$

$$= 36$$

$$\overline{X}^2 = 36$$

$\overline{X}$ =Mean value = ± 6

We know that

$$R_{XX}(0) = E\left[X^2(t)\right]$$

$$= 36 + \frac{4}{1+0}$$

$$= 40$$

$$E\left[X^2(t)\right] = 40$$

For a *WSS* process, variance does not depend on time and is given by

$$\sigma_X^2 = R_{XX}(0) - \overline{X}^2$$

$$= 40 - 36$$

$$= 4$$

$$\sigma_X^2 = \text{variance} = 4$$

Example 7: Prove that $\left|R_{XX}(\tau)\right| \le R_{XX}(0)$

Sol: We know that $R_{XX}(t_1, t_2) = E\left[X(t_1).X(t_2)\right]$

If $t_2 - t_1 = \tau$ then

$R_{XX}(t_1, t_2) = R_{XX}(\tau)$ for WSS random process

If $t_1 = t_2$ then $R_{XX}(0) = E\left[X^2(t_1)\right] = E\left[X^2(t_2)\right]$

Let $Y(t) = \left[X(t_1) \pm X(t_2)\right]^2$ then

$E[Y(t)] = E\left[\{X(t_1) \pm X(t_2)\}^2\right] \geq 0$ non-negative

$E\left[X^2(t_1) + X^2(t_2) \pm 2R_{XX}(\tau)\right] \geq 0$

$R_{XX}(0) + R_{XX}(0) \pm 2R_{XX}(\tau) \geq 0$

$2R_{XX}(0) \geq \mp 2R_{XX}(\tau)$

This implies $R_{XX}(\tau) \leq R_{XX}(0)$

Hence proved.

Example 8: Show that random process $X(t) = A\cos(\omega t + \theta)$ is correlation ergodic where A and ω are constants and θ is uniformly distributed random variable in the interval $(-\pi, \pi)$

Sol: Given random process $X(t) = A\cos(\omega t + \theta)$

Condition for correlation ergodic process is

$$R_{XX}(\tau) = E[X(t_1).X(t_1 + \tau)] = <X(t_1) \times X(t_1 + \tau)>$$

We consider $<X(t_1) \times X(t_1 + \tau)>$

$$<X(t_1) \times X(t_1 + \tau)> = \underset{T \to \infty}{Lt} \frac{1}{T} \int_{-T/2}^{T/2} A\cos(\omega t + \theta) A\cos \omega\{(t + \tau) + \theta\} dt$$

$$= \underset{T \to \infty}{Lt} \frac{1}{T} \frac{A^2}{2} \int_{-T/2}^{T/2} \left[\cos(2\omega t + \omega\tau + 2\theta) + \cos(-\omega\tau)\right] dt$$

$$= \frac{A^2}{2} \underset{T \to \infty}{Lt} \frac{1}{T} \left[\frac{\sin(2\omega t + \omega\tau + 2\theta)}{2\omega} + t\cos(-\omega\tau)\right]_{-\frac{T}{2}}^{\frac{T}{2}}$$

$$= \frac{A^2}{2} \underset{T \to \infty}{Lt} \frac{1}{T} \left[\frac{\sin(\omega T + \omega\tau + 2\theta) - \sin(-\omega T + \omega\tau + 2\theta)}{2\omega T} + T\cos(\omega\tau)\right]$$

$$= \frac{A^2}{2} \left[0 - 0 + \cos(\omega\tau)\right]$$

$$= \frac{A^2}{2} \cos\omega\tau$$

$$E[X(t_1) \times X(t_1 + \tau)] = \int_{-\infty}^{\infty} X(t_1) X(t_1 + \tau) f(\theta) d\theta$$

Where $f(\theta)$ is probability density function and is defined as

$$f(\theta) = \begin{cases} \dfrac{1}{2\pi} & -\pi < \theta < \pi \\ 0 & elsewhere \end{cases}$$

$$E[X(t_1) \times X(t_1 + \tau)] = \int_{-\pi}^{\pi} A\cos(\omega t + \theta) A\cos\{(\omega t + \omega\tau) + \theta\} \frac{1}{2\pi} d\theta$$

$$= \frac{A^2}{4\pi} \int_{-\pi}^{\pi} \cos(2\omega t + \omega\tau + 2\theta) + \cos\{(-\omega\tau)\} d\theta$$

$$= \frac{A^2}{4\pi} \left[\frac{\sin(2\omega t + \omega\tau + 2\theta)}{2} + \cos(\omega\tau)\theta \right]_{-\pi}^{\pi}$$

$$= \frac{A^2}{4\pi} \left[\frac{1}{2}\{\sin(2\omega t + \omega\tau + 2\pi) - \sin(2\omega t + \omega\tau + 2\pi)\} + \cos(\omega\tau)\{2\pi\} \right]$$

$$= \frac{A^2}{4\pi} \cos(\omega\tau)\, 2\pi$$

$$= \frac{A^2}{2} \cos(\omega\tau)$$

We have $R_{XX}(\tau) = E[X(t_1) \times X(t_1 + \tau)] = \ <X(t_1) \times X(t_1 + \tau)>$

Hence the given random process is correlation ergodic.

Quiz Questions

1. If the future value of a sample function cannot be predicted based on its past values, the process is referred to a

   (a)   Deterministic process   (b)   non-deterministic process

   (c)   Independent process   (d)   statistical process

2. If sample of X(t) is a random variables then the cumulative distribution function $F_X(x_1,\ t_1)$ is

   (a) $p\{X(t_1)\}$ (b) $p\{X(t_1) \le 0\}$

   (c) $p\{X(t_1) \le x_1\}$ (d) $p\{X(t_1) \ge x_1\}$

3. The random process $X(t)$ and $Y(t)$ are said to be independent, if $f_{XY}(x_1, y_2; t_1, t_2)$ is

(a) $f_X(x_1, t_1)$ (b) $f_Y(y_1, t_2)$

(c) $f_X(x_1, t_1)f_Y(y_1, t_2)$ (d)   0

4. A random process is defined as $X(t) = \cos(\omega_0 t + \theta)$ where θ is a uniform random variable over $(-\pi, \pi)$. The second movement of the process is

(a)   0   (b) $\dfrac{1}{2}$ (c) $\dfrac{1}{4}$ (d)   1

5. For the random process $X(t) = A\cos(\omega t + \theta)$ where w is a constant and A is a uniform random variable over $(0,1)$ the mean square value is

(a) $\dfrac{1}{3}$ (b) $\dfrac{1}{3}\cos(\omega t)$ (c) $\dfrac{1}{3}\cos^2(\omega t)$ (d) $\dfrac{1}{9}$

6. In ergodic process ensemble and time averages are

(a)   opposite to each other   (b)   different

(c)   identical   (d)   none of these

7. The auto correlation function $R_X(\tau)$ satisfies which one of the following properties

(a) $R_X(\tau) = -R_X(-\tau)$ (b) $R_X(\tau) = R_X(-\tau)$

(c) $R_X(\tau) \geq R_X(0)$ (d) $R_X(\tau) \geq 1$

8. The auto correlation function $R_X(\tau)$ of the signal $X(t) = A\sin(\omega t)$ is given by

(a) $\dfrac{1}{2}A^2 \cos(\omega t)$ (b) $A^2 \cos(\omega t)$

(c) $A^2 \cos^2(\omega t)$ (d) $2A^2 \cos(\omega t)$

9. If the value of auto correlation function at t_0 is $R(t_0)$, then $R(0)$ is

(a) $\leq R(t_0)$ (b) $\geq R(t_0)$

(c) $= R(t_0)$ (d)   of any value of $R(t_0)$

10. The auto correlation function of a function $X(t)$ is defined as

    (a) $R(t) = \dfrac{1}{T} \displaystyle\int_0^T x(t - t_0) x(t + t_0) dt$ (b) $R(T) = \dfrac{1}{T} \displaystyle\int_0^T x(t) x(t + t_0) dt$

    (c) $R(t_0) = \dfrac{1}{T} \displaystyle\int_0^T x(t) x(t + t_0) dt$ (d) $R(t) = \dfrac{1}{T} \displaystyle\int_0^T x(t) x(t + t_0) dt$

11. The cross correlation function $\left| R_{XY}(\tau) \right|$ is bounded by

    (a) $R_{XY}(0)$ (b) $R_{XY}(0) R_{YX}(0)$

    (c) $\sqrt{R_{XX}(0) R_{YY}(0)}$ (d) $\dfrac{1}{2}\left[R_{XX}(0) R_{YY}(0) \right]$

12. The auto correlation function $R_{XX}(\tau)$ is given interms of power spectral density $S_{XY}(\omega)$ as

    (a) $R_{XX}(\tau) = \displaystyle\int_{-\infty}^{\infty} S_{XX}(\omega)\, e^{jw\tau} d\tau$ (b) $R_{XX}(\tau) = \dfrac{1}{2\pi} \displaystyle\int_{-\infty}^{\infty} S_{XX}(\omega)\, e^{jw\tau} d\tau$

    (c) $R_{XX}(\tau) = \displaystyle\sum_{-\infty}^{\infty} S_{XX}(\omega)\, e^{jw\tau} d\tau$ (d) $R_{XX}(\tau) = \dfrac{1}{2\pi} \displaystyle\int_{-\infty}^{\infty} S_{XX}(\omega)\, e^{-jw\tau} d\tau$

13. The random process $X(t)$ and $Y(t)$ are jointly wide sense stationary (WSS) if and only if

    (a)   Either $X(t)$ or $Y(t)$ is WSS
    (b)   Neither $X(t)$ nor $Y(t)$ is WSS
    (c)   Both $X(t)$ and $Y(t)$ are strict sense stationary
    (d)   Both $X(t)$ and $Y(t)$ are WSS

14. If statistical parameters do not change with the choice of time origin, the process is said to be

    (a)   Stationary   (b)   Non-stationary
    (c)   Ergodic   (d)   none of these

15. For ergodic process

    (a)   mean is necessarily zero
    (b)   mean square value is infinity
    (c)   all time averages are zero
    (d)   mean square value is independent of time

16. Let $X(t)$ is a random process which is wide sense stationary, then
   (a) $E[X(t)]$= constant
   (b) $E[X(t).X(t+T)]=R_{XX}(\tau)$
   (c) $E[X(t)]$=constant and $E[X(t).X(t+T)]=R_{XX}(\tau)$
   (d) $E[X^2(t)]=0$

17. Time average of quantity $X(t)$ is defined as $A(x(t))$
   (a) $\int\limits_{-T}^{T} X(t)\,dt$
   (b) $\dfrac{1}{2T}\int\limits_{-T}^{T} X(t)\,dt$
   (c) $\underset{T\to\infty}{Lt}\ \dfrac{1}{2T}\int\limits_{-T}^{T} X(t)\,dt$
   (d) $\underset{T\to\infty}{Lt}\ \int\limits_{-T}^{T} X(t)\,dt$

18. A stationary random process $X(t)$ is periodic with periodic $2T$. Its auto correlation function is
   (a) non-periodic
   (b) periodic with period T
   (c) periodic with period $2T$
   (d) periodic with period $\dfrac{T}{2}$

19. The auto covariance $C(t_1,t_2)$ of a process $X(t)$ is (assume that $E[X(t)]\le n(t)$)
   (a) $C(t_1,t_2)=R(t_1,t_2)$
   (b) $C(t_1,t_2)=R(t_1,t_2)-|n(t)|^2$
   (c) $C(t_1,t_2)=R(t_1,t_2)+|n(t)|^2$
   (d) none

20. Two processes $X(t)$ and $Y(t)$ are called mutually orthogonal if for every t_1 and t_2
   (a) $R_{XY}(t_1,t_2)=0$
   (b) $R_{XY}(t_1,t_2)>0$
   (c) $R_{XY}(t_1,t_2)<0$
   (d) $R_{XY}(t_1,t_2)=1$

21. Two processes $X(t)$ and $Y(t)$ are called on correlated if
   (a) $C_{XY}(t_1,t_2)=0$
   (b) $C_{XY}(t_1,t_2)>0$
   (c) $C_{XY}(t_1,t_2)<0$
   (d) $C_{XY}(t_1,t_2)=1$

22. The mean square value of a random process whose auto correlation function $\dfrac{n}{4Rc}\,e^{-|\tau|Rc}$ is
   (a) $\dfrac{n}{4}$
   (b) $\dfrac{n}{4Rc}$
   (c) $\dfrac{n\tau}{4Rc}$
   (d) $\dfrac{n|\tau|}{4Rc}$

23. A random process is a random variable that is a function of

    (a) time   (b) Temperature

    (c) Both (a) and (b)   (d) none

24. A random process $X(t)$ is said to be mean ergodic or ergodic in the mean sense, if its statistical average is ... to time average

    (a) equal   (b) Not equal

    (c) Both   (d) None

25. Correlation coefficient $R_{XX}(t_1,t_2) =$

    (a) $\dfrac{C_{XY}(t_1,t_2)}{\sqrt{C_{XX}(t_1,t_1)}\sqrt{C_{XX}(t_2,t_2)}}$ (b) $\dfrac{C_{XX}(t_1,t_2)}{\sqrt{C_{XY}(t_1,t_1)}\sqrt{C_{XY}(t_2,t_2)}}$

    (c) $\dfrac{C_{XX}(t_1,t_2)}{\sqrt{C_{XX}(t_1,t_1)}\sqrt{C_{XX}(t_2,t_2)}}$ (d) None

26. ... averges are computed by considering all sample functions

    (a) time   (b) Ensemble

    (c) Both   (d) None

27. If $R_{XX}(\tau \pm T) = R_{XX}(\tau)$ then it is

    (a) periodic   (b) Non-periodic

    (c) Both   (d) None

28. A random process is defined as $X(t) = \cos(w_0 t + \theta)$ where θ is a uniform random variable over $(-\pi,\pi)$. The second movement of the process is

    (a) 0   (b) $\dfrac{1}{2}$ (c) $\dfrac{1}{4}$ (d) 1

29. $R_{XY}(\tau) = \ldots$

    (a) $R_{YX}(\tau)$ (b) $R_{YX}(-\tau)$ (c) $R_{XX}(\tau)$ (d) None

Answers

1. (b)	2. (c)	3. (c)	4. (b)	5. (c)
6. (c)	7. (b)	8. (a)	9. (b)	10.
11. (c)	12. (b)	13. (d)	14. (a)	15. (d)
16. (c)	17. (c)	18. (c)	19. (b)	20. (a)
21. (a)	22. (b)	23. (a)	24. (a)	25. (c)
26. (b)	27. (a)	28. (b)	29. (b)	

Review Quiz Questions

1. If the future value of a sample function can be predicted based on its past values, the Process is referred to
   - (a) Deterministic process
   - (b) non-deterministic process
   - (c) statistical process
   - (d) none

2. If sample of X(t) is a random variables then the cumulative distribution function $F_X(x_1, t_1)$ is
   - (a) $p\{X(t_1)\}$
   - (b) $p\{X(t_1) \le 0\}$
   - (c) $p\{X(t_1) \le x_1\}$
   - (d) $p\{X(t_1) \ge x_1\}$

3. The random process X(t) and Y(t) are said to be dependent, if $f_{XY}(x_1, y_2; t_1, t_2)$ is not equal to
   - (a) $f_X(x_1, t_1)$
   - (b) $f_Y(y_1, t_2)$
   - (c) $f_X(x_1, t_1)f_Y(y_1, t_2)$
   - (d) none

4. A random process is defined as $X(t) = \cos(\omega_0 t + \theta)$ where θ is a uniform random variable over $(-\pi, \pi)$. The third movement of the process is
   - (a) 0
   - (b) $\dfrac{1}{2}$
   - (c) $\dfrac{1}{4}$
   - (d) 1

5. For the random process $X(t) = A\cos(\omega t + \theta)$ where w is a constant and A is a uniform random variable over $(-1,1)$ the mean square value is
   - (a) $\dfrac{1}{3}$
   - (b) $\dfrac{1}{3}\cos(\omega t)$
   - (c) $\dfrac{1}{3}\cos^2(\omega t)$
   - (d) $\dfrac{1}{9}$

6. $R_X(\tau) = R_X(-\tau)$ is one of the properties of auto correlation function. Is it true

    (a)   Yes   (b)   No   (c)   can't say   (d)   none

7. The auto correlation function $R_X(\tau)$ of the signal $X(t) = A\cos(\omega t)$ is given by

    (a) $\dfrac{1}{2}A^2\cos(\omega t)$ (b) $A^2\cos(\omega t)$

    (c) $A^2\cos^2(\omega t)$ (d)   none

8. If the value of auto correlation function at t_0 is $R(t_0)$, then $R(0)$ is

    (a) $\leq R(t_0)$ (b) $\geq R(t_0)$

    (c) $= R(t_0)$ (d)   of any value of $R(t_0)$

9. The cross correlation function $|R_{XY}(\tau)|$ is bounded by

    (a) $R_{XY}(0)$ (b) $R_{XY}(0)R_{YX}(0)$

    (c) $\sqrt{R_{XX}(0)R_{YY}(0)}$ (d) $\dfrac{1}{2}[R_{XX}(0)R_{YY}(0)]$

10. The power spectral density $S_{XY}(\omega)$ is given interms of auto correlation function $R_{XX}(\tau)$ as

    (a) $\displaystyle\int_{-\infty}^{\infty} S_{XX}(\omega)\,e^{j\omega\tau}\,d\tau = R_{XX}(\tau)$ (b) $\dfrac{1}{2\pi}\displaystyle\int_{-\infty}^{\infty} S_{XX}(\omega)\,e^{j\omega\tau}\,d\tau = R_{XX}(\tau)$

    (c) $\displaystyle\sum_{-\infty}^{\infty} S_{XX}(\omega)\,e^{j\omega\tau}\,d\tau = R_{XX}(\tau)$ (d) $\dfrac{1}{2\pi}\displaystyle\int_{-\infty}^{\infty} S_{XX}(\omega)\,e^{-j\omega\tau}\,d\tau = R_{XX}(\tau)$

11. For a stationary process . . . do not change with the choice of time origin, is

    (a)   statistical parameters   (b)   statistical parameters

    (c)   ergodic   (d)   none of these

12. If mean square value is independent of time, then it is said to be

    (a)   ergodic process zero   (b)   ergodic process is infinity

    (c)   all time averages are zero   (d)   ergodic process

13. Let $Y(t)$ is a random process which is wide sense stationary, then

    (a) $E[Y(t)] = $ constant

    (b) $E[Y(t). Y(t+T)] = R_{YY}(\tau)$

    (c) $E[Y(t)] = $ constant and $E[Y(t). Y(t+T)] = R_{YY}(\tau)$

    (d) $E[Y^2(t)] = 0$

14. Time average of quantity $Y(t)$ is defined as $A(y(t))$

    (a) $\displaystyle\int_{-T}^{T} Y(t)\, dt$ (b) $\dfrac{1}{2T} \displaystyle\int_{-T}^{T} Y(t)\, dt$

    (c) $\displaystyle\operatorname*{Lt}_{T\to\infty} \dfrac{1}{2T} \int_{-T}^{T} Y(t)\, dt$ (d) $\displaystyle\operatorname*{Lt}_{T\to\infty} \int_{-T}^{T} Y(t)\, dt$

15. A stationary random process X(t) is periodic with periodic T. Its auto correlation function is

    (a)   non-periodic   (b)   periodic with period T

    (c)   periodic with period 2T   (d)   none

16. The auto covariance $C(t_1, t_2)$ of a process $X(t)$ is (assume that $E[X(t)] \geq n(t)$)

    (a) $C(t_1, t_2) = R(t_1, t_2)$ (b) $C(t_1, t_2) = R(t_1, t_2) - |n(t)|^2$

    (c) $C(t_1, t_2) = R(t_1, t_2) + |n(t)|^2$ (d)   none

17. Two processes $X(t)$ and $Y(t)$ are said to be correlated if

    (a) $C_{XY}(t_1, t_2) = 0$ (b) $C_{XY}(t_1, t_2) > 0$

    (c) $C_{XY}(t_1, t_2) < 0$ (d) $C_{XY}(t_1, t_2) = 1$

18. The mean square value of a random process whose auto correlation function $\dfrac{n}{4Rc} e^{-|\tau|Rc}$ is

    (a) $\dfrac{n}{4}$ (b) $\dfrac{n}{4Rc}$ (c) $\dfrac{n\tau}{4Rc}$ (d) $\dfrac{n|\tau|}{4Rc}$

19. A random process Y(t) is said to be mean ergodic or ergodic in the mean sense, if its statistical average is . . . to time average

    (a) equal   (b) Not equal   (c) Both   (d) None

20. Correlation coefficient $\dfrac{C_{XX}(t_1,t_2)}{\sqrt{C_{XX}(t_1,t_1)}\sqrt{C_{XX}(t_2,t_2)}}$ is

    (a) $R_{YX}(t_1,t_2)$ (b) $R_{Xy}(t_1,t_2)$ (c) $R_{XX}(t_1,t_2)$ (d) None

21. . . . averges are computed by considering all sample fucntions

    (a) time   (b) Ensemble   (c) Both   (d) None

22. If $R_{XX}(\tau \pm T) = R_{XX}(\tau)$ then it is

    (a) periodic   (b) Non-periodic   (c) Both   (d) None

23. A random process is defined as $X(t) = \cos(\omega_0 t + \theta)$ where θ is a uniform random variable over $(0, \pi)$. The second movement of the process is

    (a) 0   (b) $\dfrac{1}{2}$ (c) $\dfrac{1}{4}$ (d) none

24. $R_{XY}(-\tau) = $ . . .

    (a) $R_{YX}(\tau)$ (b) $R_{YX}(\tau)$ (c) $R_{XX}(\tau)$ (d) None

Exercise Questions

1. A random process is defined by $X(t) = A$ where A is a continuous random variable uniformly distributed on (0,1). Determine whether it is wide sense stationary.

2. A random process is given by $X(t) = A\sin(\omega t) + B\cos(\omega t)$ + where A and B are random variables. Obtain the necessary condition on A and B for $X(t)$ to be wide sense stationary.

3. Random process $X(t)$ and $Y(t)$ are defined by $X(t) = A\cos(\omega_0 t + \theta)$, $Y(t) = B\sin(\omega_0 t + \theta)$ where A, B and ω_0 are constants while θ is a random variable uniformly on$(0, 2\pi)$.determine the cross correlation function $R_{XY}(t, t + \tau)$ and show that $X(t)$ and $Y(t)$ are wide sense stationary

4. The auto correlation function of a stationary random variable $X(t)$ is given by

    $R_{XX}(\tau) = 25 + \dfrac{4}{1 + 6T^2}$. Find the mean and variance of the system

    [mean 5, variance 4]

5. Assume that the ergodic random process $X(t)$ has an autocorrelation function
$$R_{XX}(\tau) = 18 + \frac{2}{6+T^2}[1 + 4\cos(12\tau)]$$ find (i) $|\overline{X}|$ (ii) dos the process have a periodic component (iii) what is the average power in $X(t)$

6. Determine if the constant process $X(t) = A$ where A is a random variable with mean $\overline{A}$ and variance σ_A^2 is mean ergodic

7. Define a random process by $X(t) = A\cos(\pi t)$ where A is Gaussian random variable with zero mean and variance

   I   Find the density functions of $X(0)$ and $X(1)$

   II   Is $X(t)$ stationary in any sense

8. A random process $X(t)$ is known to be wide sense stationary with $E[X^2(t)] = 11$. Give one or more reasons why each of the following expressions can not be autocorrelation function of the process

   (a) $R_{XX}(t, t+\tau) = \cos(8t)e^{-(t+\tau)^2}$

   (b) $R_{XX}(t, t+\tau) = -11e^{-|\tau|}$

   (c) $R_{XX}(t, t+\tau) = \dfrac{\sin(2\tau)}{1+\tau^2}$

9. Telephone calls are imitated through an exchange at the average rate of 75 per minute and are described by poisson process. Find the probability that more than 3 calls are imitated in any 5 second period.

10. Show that the auto correlation function of the poisson process is
$$R_{XX}(t_1, t_2) = \begin{cases} \lambda t_1[1 + \lambda t_2] & t_1 < t_2 \\ \lambda t_2[1 + \lambda t_1] & t_1 > t_2 \end{cases}$$

11. Determine the auto covariance function of the poisson process.

12. Find the auto correlation function for the white noise shown in figure

13. A random process is $X(t)$ has the auto correlation function
$$R_X(\tau) = 15e^{-2|\tau|} \qquad -\infty < \infty$$
The random process $Y(t)$ is $Y(t) = X(t) - 3$

   I   What is the auto correlation of $Y(t)$

   II   What are the total power, d.c power and a.c power of $Y(t)$

   III   What is the cross correlation $R_{XY}(\tau)$

14. A random process is described by $X(t) = A^2 \cos^2(\omega_c t + \theta)$ where a and w_c are constants and θ is random variable uniformly distributed between $\pm \pi$. Is $X(t)$ wide sense stationary? If not, then why? If so, then what are the mean and the auto correlation function for the random process.

15. Show that $R_{XX}(\tau)$ is an even function

16. Show that $R_{XY}(\tau) = R_{YX}(-\tau)$

17. If $X(t)$ and $Y(t)$ are two WSS process , show that $R_{XY}(\tau) \leq \sqrt{R_{XX}(0)R_{YY}(0)}$

18. For the given WSS random process with autocorrelation $R_{XY}(\tau) = e^{-a|\tau|}$. Determine the second moment of two random variable $X(5) - X(3)$

19. Prove that $|R_{XY}(\tau)| \leq \dfrac{1}{2}[R_{XX}(0) + R_{YY}(0)]$

20. Consider a random process having the auto correlation function of

$$R_X(\tau) = 10e^{-2|\tau|} - 5e^{-4|\tau|}$$

   I Find the mean and variance of this process

   II Is the process differentiable? Why

Review Exercise Questions

1. Explain stationary and ergodic random processes
2. Discuss in detail about stationary and independence
3. Define and explain the random process
4. Distinguish between the random variable and random process
5. Mention the classification of random processes with examples
6. Describe autocorrelation and how this is different from simple linear correlation
7. State the properties of autocorrelation
8. Explain autocorrelation and cross correlation of random processes
9. State the conditions of wide sense stationary random process
10. State and prove the properties of cross correlation of two stationary random processes $X(t)$ and $Y(t)$

11. Distinguish clearly between sample averages and time averages of random processes and hence discuss the stochastic continuity of the process

12. Define ergodic processes and hence discuss the important characteristics of ergodic process

13. Distinguish between stationary and non stationary random process

14. Define stochastic process and hence differentiate clearly between stationary, wide sense stationary and non stationary processes.

15. Discuss the property of ergodicity of a random process. Is speech stationary or non stationary process justify

16. Find auto correlation of a random binary telegraph wave

17. Distinguish between strict sense stationary process and wide sense stationary process

18. What is correlation? Distinguish between auto correlation, cross correlation and covariance

19. Prove that $R_{XX}(\tau)$ is an even function of τ

20. Explain the condition for a random process to become ergodic

21. Distinguish correlation from covariance

22. Explain mean ergodic process

23. Distinguish between mean ergodic process and correlation ergodic process

24. Write short notes on Gaussian random process with N pdf's

25. Explain Poisson random process

26. Discuss the joint Pdf of Poisson random process.

CHAPTER 7

Random Processes-
Spectral Characteristics

7.1 Introduction

So far wide sense stationary random processes are described using Auto correlation, cross correlation and covariance function. Both the domain and frequency analysis methods exist for analyzing linear systems and deterministic wave forms. The main purpose of this chapter is to introduce random processes in frequency domain. The spectral description of a deterministic wave form is obtained by Fourier transforming the wave form.

The Fourier transform $X(\omega)$ of deterministic signal $x(t)$ is given by

$$X(\omega) = \int\limits_{-\infty}^{\infty} x(t)e^{-j\omega t}d\omega \qquad .....(7.1)$$

The function $X(\omega)$, sometimes called "the spectrum of $X(t)$, has units of volts per Hertz when $X(t)$ is a voltage. $X(\omega)$ can be considered to be a "voltage density spectrum" of $X(t)$ and it represents both amplitudes and phases of frequencies present in $X(t)$.

The deterministic signal $X(t)$ can be recovered by means of the "Inverse Fourier Transform" of $X(\omega)$,

i.e., $$X(t) = \frac{1}{2\pi} \int\limits_{-\infty}^{\infty} X(\omega)e^{j\omega t}d\omega \qquad .....(7.2)$$

$X(\omega)$ may not exist for most of the sample functions of the random process. So we conclude that spectral characterization of random process utilizing a voltage density spectrum is not feasible because such a spectrum may not exist.

The power in the random process as a function of frequency instead of voltage will be described as explained in the next section.

7.2   The Power Density Spectrum

By its nature, a Fourier transform cannot be applied directly to wide sense stationary signals. The Fourier transform requires a function $X(t)$ to be absolutely integrable before its transform has meaning. A sample function from a wide sense stationary random process exists for all time. Therefore, the requirement

$$\int_{-\infty}^{\infty} |X(t)| \, dt < \infty \qquad\qquad .....(7.3)$$

Cannot be satisfied with an arbitrary function.

To overcome this difficulty, we devise a window of length 2T and only view $X(t)$ through that window. $X(t)$ with the window applied to it called $X_T(t)$.

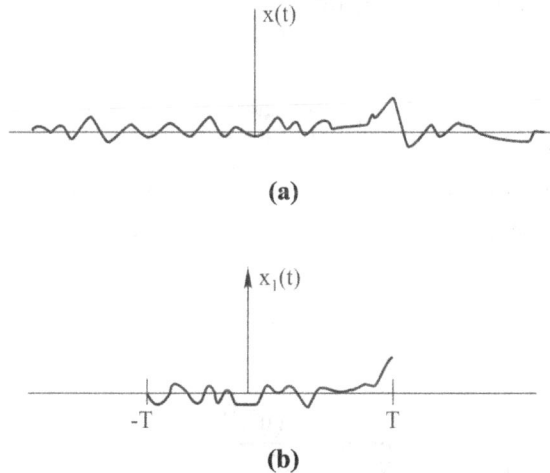

(a)

(b)

Fig. 7.1

A Sample function $X(t)$ from a random process $X(t)$ is illustrated in (a). In (b), a window of width $2T$ allows us to view $X(t)$ during $-T < t < T$. The windowed portion is called $X_T(t)$.

The window is illustrated in Fig. 7.1. The width of the window is arbitrary provided it is large enough we denote this by letting $T \to \infty$.

As long as T is finite, the Fourier transform can be applied to the window sample function. The resulting Fourier transform pair

$$X_T(\omega) = \int\limits_{-\infty}^{\infty} X_T(t)e^{-j\omega t}\, dt \qquad .....(7.4)$$

$$X_T(t) = \frac{1}{2\pi}\int\limits_{-\infty}^{\infty} X_T(\omega)e^{j\omega t}\, d\omega \qquad .....(7.5)$$

The energy contained in $X(t)$ in the interval $(-T, T)$ is defined as

$$E(T) = \int\limits_{-T}^{T} x_T^2(t)\, dt \qquad .....(7.6)$$

$$E(T) = \int\limits_{-T}^{T} x^2(t)\, dt \qquad .....(7.7)$$

By using parsevel's theorem, we can write E(T) as

$$E(T) = \int\limits_{-T}^{T} x^2(t)\, dt$$

$$= \frac{1}{2\pi}\int\limits_{-\infty}^{\infty} \left|X_T(w)\right|^2\, dw \qquad .....(7.8)$$

Since $x_T(t)$ is fourier transformable, its energy must also related to $X_T(\omega)$

The average power $P_{XX}(T)$ is obtained by dividing equation (7.8) by '2T'

$$P_{XX}(T) = \frac{1}{2T}\int\limits_{-T}^{T} x^2(t)\, dt$$

$$= \frac{1}{2\pi}\int\limits_{-\infty}^{\infty} \frac{\left|X_T(w)\right|^2}{2T}\, dw \qquad .....(7.9)$$

The 'Average Power' $P_{XX}(T)$ in the random process $X(t)$ is obtained by taking limit as $T\to\infty$. And the expected value

$$P_{XX}(T) = \underset{T\to\infty}{Lt}\ \frac{1}{2T}\int\limits_{-T}^{T} E\left[x^2(t)\right]\, dt$$

$$= \underset{T\to\infty}{Lt}\ \left\{\frac{1}{2\pi}\int\limits_{-\infty}^{\infty} \frac{E\left[\left|X_T(w)\right|\right]^2}{2T}\, dw\right\} \qquad .....(7.10)$$

From equation (7.10), two important facts are noticed.

1. Average power P_{XX} in a random process $X(t)$ is given by the time average of its second moment.

$$P_{XX}(T) = \underset{T \to \infty}{Lt} \; \frac{1}{2T} \int_{-T}^{T} E\left[x^2(t)\right] dt \; = A \left\{E\left[x^2(t)\right]\right\} \ldots(7.11)$$

Where A indicates time average. If the random process is at least wide sense stationary (*WSS*)

$$\left\{E\left[x^2(t)\right]\right\} = \overline{X^2}, \text{ a constant.}$$

$$P_{XX}(T) = \underset{T \to \infty}{Lt} \; \frac{1}{2T} \int_{-T}^{T} \overline{X^2} \, dt$$

$$= \overline{X^2} \; \underset{T \to \infty}{Lt} \; \frac{1}{2T} \int_{-T}^{T} dt$$

$$= \overline{X^2} \qquad\qquad \ldots(7.12)$$

2. $P_{XX}(T)$ can be obtained by a frequency domain integration.

Let we define $S_{XX}(\omega)$ as power density Spectrum (P.D.S) for the random process.

$$S_{XX}(\omega) = \underset{T \to \infty}{Lt} \; \frac{E\left[|X_T(\omega)|^2\right]}{2T} \qquad\qquad \ldots(7.13)$$

The power of the random process is given by

$$P_{XX}(T) = \frac{1}{2\pi} \int_{-\infty}^{\infty} S_{XX}(\omega) \, d\omega \qquad\qquad \ldots(7.14)$$

Two examples will illustrate the above concepts.

Examples

Example 1

Consider a random process $X(t) = A\cos(\omega t + \theta)$ where 'ω' is a real constant and θ is uniform random variable in $(0, \frac{\pi}{2})$. Find the average power P_{XX} in $X(t)$.

Sol:

Mean squared value is

$$P_{XX} = E\left[A^2 \cos^2(\omega t + \theta)\right]$$

$$= \underset{T \to \infty}{Lt} \ \frac{1}{2T} \int_{-T}^{T} E\left[X^2(t)\right] dt$$

$$E\left[X^2(t)\right] = E\left[A^2 \cos^2(\omega t + \theta)\right]$$

$$= A^2 E\left[\frac{1 + \cos 2(\omega t + \theta)}{2}\right]$$

$$= A^2 E\left[\frac{1}{2}\right] + A^2 E\left[\frac{\cos 2(\omega t + \theta)}{2}\right]$$

Since θ is a uniform random variable in $(0, \frac{\pi}{2})$

$$f(\theta) = \begin{cases} \dfrac{2}{\pi} & 0 < \theta < \dfrac{\pi}{2} \\ 0 & else\ where \end{cases}$$

Now consider $E\left[\dfrac{\cos 2(\omega t + \theta)}{2}\right]$

$$E\left[\frac{\cos 2(\omega t + \theta)}{2}\right] = \frac{1}{\pi} \int_{0}^{\frac{\pi}{2}} \cos 2(\omega t + \theta) d\theta$$

$$= \frac{1}{2\pi} \left[\sin 2(\omega t + \theta)\right]_{0}^{\frac{\pi}{2}}$$

$$= \frac{1}{2\pi} \left[\sin(2\omega t + \pi) - \sin 2\omega t\right]$$

$$= -\frac{\sin 2\omega t}{2\pi}$$

$$E\left[X^2(t)\right] = \frac{A^2}{2} - \frac{A^2 \sin 2\omega t}{2\pi}$$

$$P_{XX} = \underset{T \to \infty}{Lt} \frac{1}{2T} \int_{-T}^{T} \left[\frac{A^2}{2} - \frac{A^2 \sin 2\omega t}{2\pi} \right] dt$$

$$= \underset{T \to \infty}{Lt} \frac{A^2}{2T} \left\{ \frac{1}{2} \int_{-T}^{T} dt - \frac{1}{\pi} \int_{-T}^{T} \sin 2\omega t \, dt \right\}$$

$$= \underset{T \to \infty}{Lt} \frac{A^2}{2T} \left\{ \frac{1}{2} [t]_{-T}^{T} - \frac{1}{\pi} \left[\frac{\cos 2\omega t}{2\omega} \right]_{-T}^{T} \right\}$$

$$= \underset{T \to \infty}{Lt} \frac{A^2}{2T} \left\{ \frac{1}{2} [T+T] - \frac{1}{\pi} [0] \right\}$$

$$= \underset{T \to \infty}{Lt} \frac{A^2}{2T} \frac{2T}{2}$$

$$P_{XX} = \frac{A^2}{2}$$

7.3 Properties of the Power Density Spectrum

The power density spectrum has the following properties

(1) $S_{XX}(\omega) \geq 0$ .....(7.15)

The power density spectrum of a WSS process is always non-negative.

(2) $S_{XX}(-\omega) = S_{XX}(\omega)$ for $X(t)$ real   .....(7.16)

The power density spectrum of a real valued random process is an even function of frequency.

(3) $S_{XX}(\omega)$ is always real since $X_T(\omega)$ is real

where $X_T(\omega) = E[X_T(t)]$ .....(7.17)

(4) $\dfrac{1}{2\pi} \int_{-\infty}^{\infty} S_{XX}(\omega) \, d\omega = A \left\{ E[x^2(t)] \right\}$ .....(7.18)

(5) $S_{X'X'}(\omega) = \omega^2 S_{XX}(\omega)$ .....(7.19)

The power density spectrum of the derivative $X^1(t) = \dfrac{d}{dt}X(t)$ is ω^2 times the power spectrum of $X(t)$

$$(6)\ \frac{1}{2\pi}\int_{-\infty}^{\infty} S_{XX}(\omega)e^{j\omega t}d\omega = A\left[R_{XX}(t,\ t+\tau)\right] \qquad .....(7.20)$$

It states that the power density spectrum and the time average of auto correlation function form a Fourier transform pair.

If $X(t)$ is at least WSS,

$$A\left[R_{XX}(t,\ t+\tau)\right] = R_{XX}(\tau) \qquad\qquad .....(7.21)$$

$$S_{XX}(\omega) = \int_{-\infty}^{\infty} R_{XX}(\tau)e^{-j\omega\tau}d\tau$$

$$R_{XX}(\tau) = \frac{1}{2\pi}\int_{-\infty}^{\infty} S_{XX}(\omega)e^{j\omega\tau}d\omega \qquad\qquad .....(7.22)$$

For a wide sense stationary process.

7.4   Bandwidth of power density spectrum

Except for the fact that the area of $S_{XX}(\omega)$ is not necessarily unit, the $S_{XX}(\omega)$ has characteristics similar to a probability density function.   A new function is formed by dividing $S_{XX}(\omega)$ by its area with the property that unity area which is analogous to a density function.

$$f_X(\omega) = \frac{S_{XX}(\omega)}{\displaystyle\int_{-\infty}^{\infty} S_{XX}(\omega)d\omega} \qquad\qquad .....(7.23)$$

$f_X(\omega) \ge 0$ for all w and the total area under $f_X(\omega)$ is unity.

The normalized power spectrum behaves similar to a PDF and is a measure of its spread that we call rms bandwidth.

We denote rms bandwidth as ω_{rms} which has units grade is

$$\omega_{rms}^2 = \frac{\displaystyle\int_{-\infty}^{\infty} \omega^2 S_{XX}(\omega)d\omega}{\displaystyle\int_{-\infty}^{\infty} S_{XX}(\omega)d\omega} \qquad .....(7.24)$$

Equation (7.24) is valid only for low pass random process $X(t)$ that its spectral components are clustered near w = 0

For a band pass random process (it has a band pass form of power spectrum with spectral components cluster near some frequencies ω_o and $-\omega_o$, then rms bandwidth ω_{rms} is given by

$$\omega_{rms}^2 = \frac{4\displaystyle\int_0^{\infty} (\omega - \omega_0)^2 S_{XX}(\omega)d\omega}{\displaystyle\int_0^{\infty} S_{XX}(\omega)d\omega} \qquad .....(7.25)$$

Where mean frequency $\bar{\omega}_o$ is defined as

$$\bar{\omega}_o = \frac{\displaystyle\int_o^{\infty} \omega\, S_{XX}(\omega)\,d\omega}{\displaystyle\int_o^{\infty} S_{XX}(\omega)\,d\omega}$$

7.5 Relationship between Power Spectrum and Auto Correlation Function

It was shown that the inverse Fourier transform of the power density spectrum is the time average of the autocorrelation function in section 7.1

$$\frac{1}{2\pi}\int_{-\infty}^{\infty} S_{XX}(\omega)\, e^{j\omega t}\, d\omega = A\left[R_{XX}(t,\, t+\tau) \right] \qquad .....(7.26)$$

Proof:

We know that $\quad X_T(\omega) = \displaystyle\int_{-T}^{T} X_T(t)e^{-j\omega t}\, dt \quad$ and

$$S_{XX}(\omega) = \lim_{T \to \infty} E\left[\frac{|X_T(\omega)|^2}{2T}\right]$$

Using $X_T(\omega)$ in $S_{XX}(\omega)$, we get

$$S_{XX}(\omega) = \lim_{T \to \infty} E\left[\frac{1}{2T} \int_{-T}^{T} X(t_1)e^{j\omega t_1}dt_1 \int_{-T}^{T} X(t_2)e^{-j\omega t_2}dt_2\right]$$

$$= \lim_{T \to \infty} \frac{1}{2T} \int_{-T}^{T}\int_{-T}^{T} E\left[X(t_1)\times X(t_2)\right] e^{-j\omega(t_2-t_1)}dt_2 dt_1 \quad .....(7.27)$$

Let the expectation in the integrand of equation (7.7) be identified as $R_{XX}(t_1, t_2)$ Autocorrelation function of $X(t)$

$$E\left[X(t_1).X(t_2)\right] = R_{XX}(t_1, t_2) \quad -T < (t_1 \text{ and } t_2) < T \quad .....(7.28)$$

We simply the equation (7.22) and

$$S_{XX}(\omega) = \lim_{T \to \infty} \frac{1}{2T} \int_{-T}^{T}\int_{-T}^{T} R_{XX}(t_1, t_2) e^{-j\omega(t_2-t_1)}dt_2 dt_1 \quad .....(7.29)$$

Applying the inverse transform on both sides of equation (7.29), we get

$$\frac{1}{2\pi} \int_{-\infty}^{\infty} S_{XX}(\omega)e^{j\omega \tau}d\omega$$

$$= \frac{1}{2\pi} \int_{-\infty}^{\infty} \lim_{T \to \infty} \frac{1}{2T} \int_{-T}^{T}\int_{-T}^{T} R_{XX}(t, t_1)e^{-j\omega(t_1-t)}dt dt_1 e^{j\omega \tau}d\omega \quad .....(7.30)$$

$$= \lim_{T \to \infty} \frac{1}{2T} \int_{-T}^{T}\int_{-T}^{T} R_{XX}(t_1, t_2)\frac{1}{2\pi} \int_{-\infty}^{\infty} e^{-j\omega(\tau-t_1+t)}d\omega dt_1 dt \quad .....(7.31)$$

We note that the inner integral of (7.31) as the impulse of

$$2\pi \, \delta(\tau - t_1 + t) = 2\pi \, \delta(t_1 - t - \tau), \text{ so the integral}$$

$$\frac{1}{2\pi} \int_{-\infty}^{\infty} S_{XX}(\omega)e^{j\omega \tau}d\omega$$

$$= \lim_{T \to \infty} \frac{1}{2T} \int_{-T}^{T}\int_{-T}^{T} R_{XX}(t, t_1) \, \delta(t_1 - t - \tau)dt_1 dt \quad .....(7.32)$$

From the definition of impulse, we rewrite equation (7.32) as

$$\frac{1}{2\pi} \int_{-\infty}^{\infty} S_{XX}(\omega) e^{j\omega\tau} d\omega$$

$$= \underset{T \to \infty}{Lt} \frac{1}{2T} \int_{-T}^{T} R_{XX}(t, t+\tau) \, dt \qquad -T < t+\tau < T \qquad .....(7.33)$$

The RHS of equation (7.32) is the time average of the process autocorrelation function, (i.e)

$$\underset{T \to \infty}{Lt} \frac{1}{2T} \int_{-T}^{T} R_{XX}(t, t+\tau) \, dt = A\left[R_{XX}(t, t+\tau) \right] \qquad .....(7.34)$$

Use of equation (7.33) in (7.34) we get,

$$\frac{1}{2\pi} \int_{-\infty}^{\infty} S_{XX}(w) e^{jw\tau} dw = A\left[R_{XX}(t, t+\tau) \right] \qquad .....(7.35)$$

Taking the direct fourier form an both sides of equation (7.34) we get

$$S_{XY}(\omega) = \int_{-\infty}^{\infty} A\left[R_{XX}(t, t+\tau) \right] e^{-j\omega\tau} d\tau \qquad .....(7.36)$$

Which shows that $S_{XY}(\omega)$ and $A\left[R_{XX}(t, t+\tau) \right]$

form a Fourier transform pair for $X(t)$ is at least WSS random process, then $A\left[R_{XX}(t, t+\tau) \right] = R_{XX}(\tau)$ and we get

$$S_{XY}(\omega) = \int_{-\infty}^{\infty} R_{XX}(\tau) e^{-j\omega\tau} d\tau \qquad .....(7.37)$$

$$R_{XY}(\tau) = \frac{1}{2\pi} \int_{-\infty}^{\infty} S_{XX}(\omega) e^{j\omega\tau} d\omega \qquad .....(7.38)$$

Or $\qquad R_{XX}(\tau) \leftrightarrow S_{XX}(\omega) \qquad .....(7.39)$

The equation (7.36) and (7.38) are usually called Wiener-Khinchine relations which gives the relation between time domain description (correlation functions) of the processes and their description in the frequency domain (power spectrum).

Examples

***Example:* 1** The auto correlation function of random process $X(t)$ is

$R_{XX}(\tau) = \dfrac{A^2}{2}\cos(\omega_0\tau)$ where A and ω_0 are constants. Find the power density spectrum.

Sol:

We know that

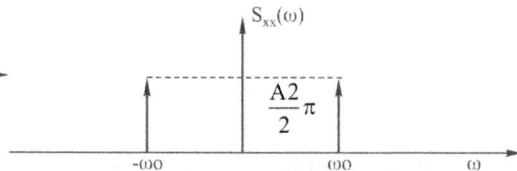

$F\left[R(\tau)\right] =$ power density spectrum

i.e., $\qquad \text{PDS} = F\left[\dfrac{A^2}{2}\cos(\omega_0\tau)\right]$

$$= F\left[\dfrac{A^2}{2}\dfrac{e^{j\omega_0\tau}+e^{-j\omega_0\tau}}{2}\right]$$

$$= \dfrac{A^2}{4}\left[Fe^{j\omega_0\tau}\right]+\dfrac{A^2}{4}\left[Fe^{-j\omega_0\tau}\right]$$

We know that

$$F[1] = 2\pi\ \delta(\omega)$$

$$1 \leftrightarrow 2\pi\ \delta(\omega)$$

$$1.e^{j\omega_0\tau} \leftrightarrow 2\pi\ \delta(\omega-\omega_0)$$

$$1.e^{-j\omega_0\tau} \leftrightarrow 2\pi\ \delta(\omega+\omega_0)$$

$$\text{PDS} = \dfrac{A^2}{4}\ 2\pi\ \left[\delta(\omega-\omega_0)+\delta(\omega+\omega_0)\right]$$

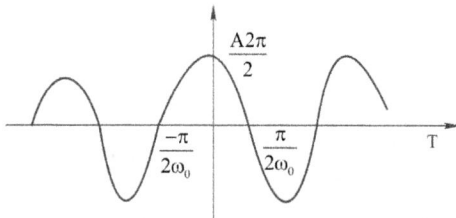

Fig. (a) Auto correlation function $\qquad$ Fig. (b) Power density spectrum

Fig.7.2

***Example:* 2** Consider an WSS random process $X(t)$ with auto correlation function

$$R_{XX}(\tau) = \begin{cases} A_0 \left[1 - \dfrac{|\tau|}{T} \right] & -T \le t \le T \\ 0 & elsewhere \end{cases}$$

where $T > 0$ and A_0 are constants

***Sol*:**

We know that $R_{XX}(\tau) \leftrightarrow S_{XX}(w)$

$$S_{XX}(\omega) = F\left[R_{XX}(\tau) \right] = A_0 T \frac{\sin^2(\omega T / 2)}{(\omega T / 2)^2}$$

The $R_{XX}(\tau)$ and $S_{XX}(\omega)$ are sketched below in Fig. (7.3)

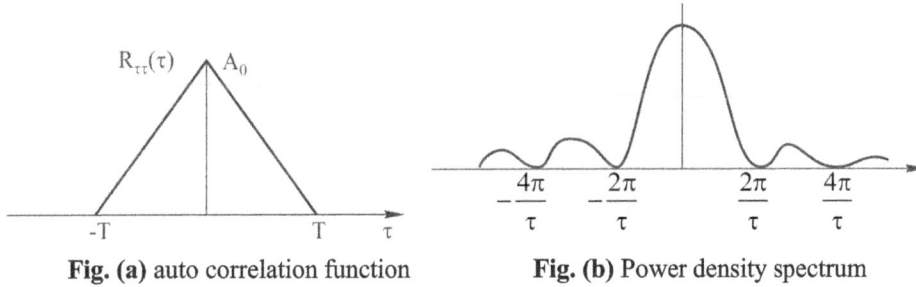

Fig. (a) auto correlation function

Fig. (b) Power density spectrum

Fig. 7.3

***Example:* 3**

Determine which of the following function can and can not be valid power density spectre. For those that are not, explain why

(i) $\dfrac{1}{\sqrt{1 - 3\omega^2}}$ (ii) $\dfrac{\omega^2}{\omega^6 + 3\omega^2 + 3}$

Sol:

(i) $S_{XX}(\omega) = \dfrac{1}{\sqrt{1 - 3\omega^2}}$

For all values of w, $S_{XX}(\omega) > 0$

$$S_{XX}(-\omega) = \frac{1}{\sqrt{1-3(-\omega)^2}} = \frac{1}{\sqrt{1-3\omega^2}} = S_{XX}(\omega)$$

Hence the given function is a valid PSD

(ii)

$$S_{XX}(-\omega) = \frac{(-\omega)^2}{(-\omega)^6 + 3(-\omega)^2 + 3}$$

$$= \frac{\omega^2}{\omega^6 + 3\omega^2 + 3} = S_{XX}(\omega)$$

$$S_{XX}(\omega) = S_{XX}(-\omega)$$

$$S_{XX} > 0$$

Hence the given function is a valid power density spectrum.

Example: 4

Find the auto correlation function of the process $X(t)$ where power density spectrum is

$$S_{XX}(\omega) = \begin{cases} 1+\omega^2 & for \quad |\omega| \le 1 \\ 0 & for \quad |\omega| > 1 \end{cases}$$

Sol:

$$R_{XX}(\tau) = F^{-1}\left[S_{XX}(\omega)\right]$$

$$= \frac{1}{2\pi} \int_{-\infty}^{\infty} S_{XX}(\omega) \, e^{j\omega\tau} \, d\omega$$

$$= \frac{1}{2\pi} \int_{-1}^{1} \left(1+\omega^2\right) e^{j\omega\tau} \, d\omega$$

$$= \frac{1}{2\pi} \left[\int_{-1}^{1} e^{j\omega\tau} \, d\omega + \int_{-1}^{1} \omega^2 e^{j\omega\tau} \, d\omega\right]$$

$$= \frac{1}{2\pi} \left[\left[\frac{e^{j\omega\tau}}{j\tau}\right]_{-1}^{1} + \omega^2 \left[\frac{e^{j\omega\tau}}{j\tau}\right]_{-1}^{1} + 2\omega \left[\frac{e^{j\omega\tau}}{\tau^2}\right]_{-1}^{1} - 2\left[\frac{e^{j\omega\tau}}{j\tau^3}\right]_{-1}^{1}\right]$$

$$= \frac{1}{2\pi} \left[\frac{e^{j\tau} - e^{-j\tau}}{j\tau} \right]$$

$$+ \frac{1}{2\pi} \left\{ \left[\frac{e^{j\tau} - e^{-j\tau}}{j\tau} \right] + \frac{2}{\tau^2} \left[e^{j\tau} + e^{-j\tau} \right] - \frac{2}{j\tau^3} \left[e^{j\tau} - e^{-j\tau} \right] \right\}$$

$$= \frac{1}{\pi} \frac{\sin \tau}{\tau} + \frac{1}{\pi} \frac{\sin \tau}{\tau} + \frac{2}{\pi} \frac{\cos \tau}{\tau^2} - \frac{2}{\pi} \frac{\sin \tau}{\tau^3}$$

$$R_{XX}(\tau) = \frac{1}{\pi} \left[\tau^2 \sin \tau + \tau \cos \tau - \sin \tau \right]$$

Example: 5

Find the PSD of a random process $X(t)$ if $E[X(t)] = 1$ and

$$R_{XX}(\tau) = 1 + e^{-\alpha|\tau|}$$

Sol:

$$S_{XX}(\omega) = \int_{-\infty}^{\infty} R_{XX}(\tau) e^{-j\omega\tau} d\tau$$

$$= \int_{-\infty}^{\infty} \left[1 + e^{-\alpha|\tau|} \right] e^{-j\omega\tau} d\tau$$

$$= \int_{-\infty}^{\infty} e^{-j\omega\tau} d\tau + \int_{-\infty}^{0} e^{\alpha\tau} e^{-j\omega\tau} d\tau + \int_{0}^{\infty} e^{-\alpha\tau} e^{-j\omega\tau} d\tau$$

$$= \delta(\omega) + \frac{1}{(\alpha - j\omega)} + \frac{1}{(\alpha + j\omega)}$$

$$= \delta(\omega) + \frac{2\alpha}{(\alpha^2 + \omega^2)}$$

Example: 6

The power density spectrum of a zero mean WSS process x(t) is given by

$$S_{XX}(\omega) = \begin{cases} 1 & for \quad |\omega| < \omega_0 \\ 0 & elsewhere \end{cases}$$

Find $R_{XX}(\tau)$ and show that x(t) and $x\left(t + \dfrac{\tau}{\omega_0}\right)$ are uncorrelated

Sol:

$$R_{XX}(\tau) = \frac{1}{2\pi} \int\limits_{-\infty}^{\infty} S_{XX}(\omega)\, e^{j\omega\tau}\, d\omega$$

$$= \frac{1}{2\pi} \int\limits_{-\omega_0}^{\omega_0} e^{j\omega\tau}\, d\omega$$

$$= \frac{1}{2\pi} \left[\frac{e^{j\omega\tau}}{j\tau} \right]_{-\omega_0}^{\omega_0}$$

$$= \frac{1}{\pi} \frac{\sin \omega_0 \tau}{\tau}$$

$$= \frac{\sin \omega_0 \tau}{\pi \tau}$$

Therefore $E\left[x(t + \dfrac{\pi}{\omega_0}) x(t) \right]$

$$= R_{XX}(\frac{\pi}{\omega_0})$$

$$R_{XX}(\frac{\pi}{\omega_0}) = \frac{\omega_0}{\pi} \sin \pi = 0$$

Since the mean of the process is zero

$$Cov\left[x(t + \frac{\pi}{\omega_0}) \times x(t) \right] = E\left[x(t + \frac{\pi}{\omega_0}) \times x(t) \right]$$

$x(t)$ and $x(t + \dfrac{\pi}{\omega_0})$ are uncorrelated

Example: 7

The power density spectrum of WSS process is given by

$$S_{XX}(\omega) = \begin{cases} \dfrac{b}{a}(a - |\omega|) & \text{for} \quad |\omega| \le a \\ 0 & |\omega| > a \end{cases}$$

Find auto correlation function of the process.

Sol:

$$R_{XX}(\tau) = \frac{1}{2\pi} \int_{-a}^{a} S_{XX}(\omega) e^{j\omega\tau} d\omega$$

$$= \frac{1}{2\pi} \int_{-a}^{a} \frac{b}{a} (a - |\omega|) e^{j\omega\tau} d\omega$$

$$= \frac{1}{\pi} \times \frac{b}{a} \int_{0}^{a} (a - \omega) \cos(\tau\omega) d\omega$$

$$= \frac{b}{\pi a} \left[(a - \omega) \frac{\sin(\tau\omega)}{\tau} - \frac{\cos(\tau\omega)}{\tau^2} \right]_{0}^{a}$$

$$= \frac{b}{\pi a\tau^2} \left[(1 - \cos a\tau) \right]$$

$$= \frac{ab}{2\pi} \left[\frac{\sin(a\tau/2)}{(a\tau/2)} \right]^2$$

7.6   Cross power density spectrum and its properties

The cross power density spectrum provides a measure of frequency interrelationship between two random processes.

Consider $x(t)$ and $y(t)$ be sample functions observed in random processes $X(t)$ and $Y(t)$. The truncated passion of these sample functions are in the interval $(-T, T)$

$$x_T(t) = \begin{cases} x(t) & -T < t < T \\ 0 & elsewhere \end{cases} \qquad \text{.....(7.40)}$$

$$y_T(t) = \begin{cases} y(t) & -T < t < T \\ 0 & elsewhere \end{cases} \qquad \text{.....(7.41)}$$

We know that $X_T(\omega)$ and $Y_T(\omega)$ are Fourier transforms of $x(t)$ and $y(t)$ respectively.

We define the cross power $P_{XY}(T)$ in the two process within the interval $(-T, T)$ is defined as

$$P_{XY}(T) = \frac{1}{2T} \int_{-T}^{T} x_T(t) y_T(t) dt = \frac{1}{2T} \int_{-T}^{T} x(t) y(t) dt \qquad \text{.....(7.42)}$$

Applying Parseval's theorem to equation (7.42), we get

$$P_{XY}(T) = \frac{1}{2T} \int_{-T}^{T} x(t)y(t)dt = \frac{1}{2\pi} \int_{-\infty}^{\infty} \frac{X_T^*(\omega)Y_T(\omega)}{2T} d\omega \quad ......(7.43)$$

$P_{XY}(T)$ is a random quantity since its value will vary depending on which sample functions are considered.

By taking the expected value in equation (7.43), we obtain the average cross power

$$P_{\overline{XY}}(T) = \underset{T \to \infty}{Lt} \frac{1}{2T} \int_{-T}^{T} R_{XY}(t,t) dt$$

$$= \frac{1}{2\pi} \int_{-\infty}^{\infty} \frac{E\left[X_T^*(\omega)Y_T(\omega)\right]}{2T} d\omega \quad .....(7.44)$$

Now the total average cross power P_{XY} is obtained by letting $T \to \infty$

$$P_{XY} = \underset{T \to \infty}{Lt} \frac{1}{2T} \int_{-T}^{T} R_{XY}(t,t) dt = \frac{1}{2\pi} \int_{-\infty}^{\infty} \underset{T \to \infty}{Lt} E\left[\frac{X_T^*(\omega)Y_T(\omega)}{2T}\right] d\omega \quad .....(7.45)$$

It is evident that the integrand involving w can be defined as a cross power density spectrum and is denoted by $S_{XY}(\omega)$

$$S_{XY}(\omega) = \underset{T \to \infty}{Lt} \frac{E\left[X_T^*(\omega)Y_T(\omega)\right]}{2T} \quad .....(7.46)$$

We obtain the cross power P_{XY} using the cross power density spectrum $S_{XY}(\omega)$

$$P_{XY} = \frac{1}{2\pi} \int_{-\infty}^{\infty} S_{XY}(\omega) \ d\omega \quad .....(7.47)$$

We can also find another cross power density spectrum by

$$S_{YX}(\omega) = \underset{T \to \infty}{Lt} \frac{E\left[Y_T^*(\omega)X_T(\omega)\right]}{2T} \quad .....(7.48)$$

Cross power is given by

$$P_{YX} = \frac{1}{2\pi} \int_{-\infty}^{\infty} S_{YX}(\omega) \ d\omega = P_{XY}^* \quad .....(7.49)$$

7.7   Properties of cross power density spectrum

Some properties of cross power spectrum are listed below

(1) $S_{XY}(\omega) = S_{YX}(-\omega) = S_{YX}^{*}(\omega)$ .....(7.50)

(2) $\text{Re}\left[S_{XY}(\omega)\right]$ and $\text{Re}\left[S_{YX}(\omega)\right]$ are even functions of ω .....(7.51)

(3) Imaginary parts of $S_{XY}(\omega)$ and $S_{YX}(\omega)$ are odd functions of ω .....(7.52)

(4) $S_{XY}(\omega) = 0$ and $S_{YX}(\omega) = 0$ if $X(t)$ and $Y(t)$ are orthogonal. .....(7.53)

(5) If $X(t)$ and $Y(t)$ are uncorrelated and have constant means $\overline{X}$ and $\overline{Y}$

$$S_{XY}(\omega) = S_{YX}(\omega) = 2\pi \, \overline{X} \, \overline{Y} \, \delta(\omega) \qquad .....(7.54)$$

(6) Cross power density spectrum and the time average of cross correlation functions are Fourier transform pair

$$S_{XY}(\omega) \leftrightarrow A\left[R_{XY}(t, \ t+\tau)\right] \qquad .....(7.55)$$

$$S_{YX}(\omega) \leftrightarrow A\left[R_{YX}(t, \ t+\tau)\right] \qquad .....(7.56)$$

For joint wide sense stationary processes, equation (7.55) and (7.56) reduce to useful forms as

$$S_{XY}(\omega) = \int_{-\infty}^{\infty} R_{XY}(\tau) \, e^{-j\omega\tau} d\tau \qquad .....(7.57)$$

$$S_{YX}(\omega) = \int_{-\infty}^{\infty} R_{YX}(\tau) \, e^{-j\omega\tau} d\tau \qquad .....(7.58)$$

$$R_{XY}(\tau) = \frac{1}{2\pi} \int_{-\infty}^{\infty} S_{XY}(\omega) e^{j\omega\tau} d\omega \qquad .....(7.59)$$

$$R_{YX}(\tau) = \frac{1}{2\pi} \int_{-\infty}^{\infty} S_{YX}(\omega) e^{j\omega\tau} d\omega \qquad .....(7.60)$$

Examples

Example: 1

The cross power spectrum is given by

$$S_{XX}(\omega) = \begin{cases} a + \dfrac{jb\omega}{W} & -W < \omega < W \\ 0 & elsewhere \end{cases}$$

Sol:

Where $w > 0$, 'a' and 'b' are real constants using the equation (7.59) to find cross correlation as

$$R_{XY}(\tau) = \frac{1}{2\pi}\int\limits_{-W}^{W}\left(a + \frac{jb\omega}{W}\right) e^{j\omega\tau} d\omega$$

$$= \frac{a}{2\pi}\int\limits_{-W}^{W} e^{j\omega\tau} dw + \frac{jb}{2\pi W}\int\limits_{-W}^{W} \omega\, e^{j\omega\tau} d\omega$$

The $R_{XY}(\tau)$ is simplified as

$$R_{XY}(\tau) = \frac{a}{2\pi}\left[\frac{e^{jw\tau}}{j\tau}\right]_{-W}^{W} + \frac{jb}{2\pi W}\left\{e^{j\omega\tau}\left[\frac{\omega}{j\tau} - \frac{1}{(j\tau)^2}\right]\right\}_{-W}^{W}$$

$$= \frac{1}{\pi W \tau^2}\left[(aW\tau - b)\sin(W\tau) + bW\tau\cos(W\tau)\right]$$

Example: **2**

Determine the cross correlation function corresponding to cross power density spectrum

$$S_{XX}(\omega) = \frac{8}{(\alpha + j\omega)^3} \text{ where } \alpha > 0 \text{ is a constant}$$

Solution:

$$S_{XX}(\omega) = \frac{8}{(\alpha + j\omega)^3} \text{ where } \alpha > a \text{ is a constant}$$

$$R_{XY}(\tau) \overset{F}{\leftrightarrow} S_{XY}(\omega)$$

$$R_{XY}(\tau) = \frac{1}{2\pi}\int\limits_{-a}^{a} S_{XY}(\omega)e^{j\omega\tau} d\omega$$

$$R_{XY}(\tau) = \frac{1}{2\pi}\int\limits_{-a}^{a}\frac{8}{(\alpha + j\omega)^3} e^{j\omega\tau} d\omega$$

We know that

$$u(t)t^2 e^{-\alpha t} \overset{F}{\leftrightarrow} \frac{2}{(\alpha + j\omega)^3}$$

$$\frac{8}{(\alpha + j\omega)^3} \quad \overset{F}{\leftrightarrow} \quad 4\, u(t)\, t^2 e^{-\alpha t}$$

$$R_{XY}(\tau) = 4\, u(\tau)\, \tau^2 e^{-\alpha \tau}$$

7.8 Relationship between cross power spectrum and cross correlation functions

We know that

$$S_{XY}(\omega) = \int_{-\infty}^{\infty} \left\{ \underset{T \to \infty}{Lt} \frac{1}{2T} \int_{-T}^{T} R_{XY}(t, t+\tau)\, dt \right\} e^{-j\omega\tau}\, d\tau \qquad \text{.....(7.61)}$$

The transforms of truncated processes are given by

$$X_T(\omega) = \int_{-T}^{T} X(t)\, e^{-j\omega t}\, dt \qquad \text{.....(7.62)}$$

$$Y_T(\omega) = \int_{-T}^{T} Y(t_1)\, e^{-j\omega t_1}\, dt_1 \qquad \text{.....(7.63)}$$

We use (7.4-2) and (7.4-3) to form

$$X_T^*(\omega)\, Y_T(\omega) = \int_{-T}^{T} X(t)\, e^{j\omega t}\, dt \int_{-T}^{T} Y(t_1)\, e^{-j\omega t_1}\, dt_1 \qquad \text{.....(7.64)}$$

The cross power density spectrum is given by

$$S_{XY}(\omega) = \underset{T \to \infty}{Lt} \frac{E\left[X_T^*(\omega) Y_T(\omega) \right]}{2T} \qquad \text{.....(7.65)}$$

On substituting the equation (7.64) in equation (7.65), we get

$$S_{XY}(\omega) = \underset{T \to \infty}{Lt} \frac{1}{2T} E\left[\int_{-T}^{T} X(t)\, e^{j\omega t}\, dt \int_{-T}^{T} Y(t_1)\, e^{-j\omega t_1}\, dt_1 \right]$$

$$= \underset{T \to \infty}{Lt} \frac{1}{2T} \left[\int_{-T}^{T}\int_{-T}^{T} R_{XY}(t, t_1)\, e^{-j\omega(t_1 - t)}\, dt\, dt_1 \right] \qquad \text{.....(7.67)}$$

We inverse Fourier transform both sides of equation (7.67), we get

$$\frac{1}{2\pi}\int_{-\infty}^{\infty}S_{XY}(\omega)e^{j\omega\tau}dw = \frac{1}{2\pi}\left[\int_{-\infty}^{\infty}\underset{T\to\infty}{Lt}\frac{1}{2T}\int_{-T}^{T}\int_{-T}^{T}R_{XY}\left(t,t_1\right)e^{-j\omega(t_1-t)}dt\,dt_1e^{j\omega\tau}d\omega\right] \quad .....(7.68)$$

$$= \underset{T\to\infty}{Lt}\frac{1}{2T}\left[\int_{-T}^{T}\int_{-T}^{T}R_{XY}\left(t,t_1\right)\frac{1}{2\pi}\int_{-\infty}^{\infty}e^{jw(\tau-t_1+t)}d\omega dtdt_1\right]$$

$$= \underset{T\to\infty}{Lt}\frac{1}{2T}\int_{-T}^{T}\int_{-T}^{T}R_{XY}\left(t,t_1\right)\delta(t_1-\tau-t)dt_1dt \quad\quad .....(7.69)$$

Use of the impulse definition i.e.,

$$\int_{-\infty}^{\infty}\delta(x)\phi(x)dx = \phi(0)$$

In equation .(7.69), we obtain the solution for integral over t_1 as

$$\frac{1}{2\pi}\int_{-\infty}^{\infty}S_{XY}(\omega)e^{jw\tau}d\omega = \underset{T\to\infty}{Lt}\frac{1}{2T}\int_{-T}^{T}R_{XY}\left(t,t+\tau\right)dt \quad\quad .....(7.70)$$

This is valid for $-T < t+\tau < T$

From the equation (7.70) it is evident that the cross power spectrum and time average cross correlation function form Fourier transform pair.

For jointly wide sense stationary processes, however, the cross correlation function $R_{XY}\left(\tau\right)$ can be found from $S_{XY}\left(\omega\right)$ since its time average is just $R_{XY}\left(\tau\right)$

Examples

Example: 1

Let the cross correlation function of two processes $X(t)$ and $Y(t)$ be

$$R_{XY}\left(t,t+\tau\right) = \frac{AB}{2}\left\{\sin(\omega_0\tau)+\cos\left[\omega_0(2t+\tau)\right]\right\}$$ where A, B and ω_0 are constants. Find the cross power spectrum.

Sol:

We find the cross power spectrum by use of equation (7.61).

First the time average is formed

$$\underset{T\to\infty}{Lt}\frac{1}{2T}\int_{-T}^{T}R_{XY}\left(t,t+\tau\right)dt = \frac{AB}{2}\sin(\omega_0\tau)+\frac{AB}{2}\underset{T\to\infty}{Lt}\frac{1}{2T}\int_{-T}^{T}\cos\left[\omega_0(2t+\tau)\right]dt$$

The integral is readily evaluated and is found to be zero.

$$S_{XY}\left(\omega\right) = f\left\{\frac{AB}{2}\sin(\omega_0\tau)\right\}$$

$$= \frac{-j\pi\,AB}{2}\left[\delta(\omega-\omega_0)-\delta(\omega+\omega_0)\right]$$

Example: 2. Determine which of the following functions are valid power density spectrums and why

a) $\dfrac{\cos 8\omega}{2+\omega^4}$ b) $e^{-(\omega-1)^2}$ c) $\dfrac{\omega^2}{\omega^6+3\omega^2+3}$

Solution

a)

Given $S_{xx}\left(\omega\right) = \dfrac{\cos 8\omega}{2+\omega^4}$

From the properties of PSD the given function $s_{xx}(w)$ is real and positive

$$S_{xx}(-\omega) = \frac{\cos 8(-\omega)}{2+\left(-\omega^4\right)}$$

$$= \frac{\cos 8\omega}{2+\omega^4}$$

$$= S_{xx}(\omega)$$

the function is even and hence the given function is valid PSD

b)

Given $S_{xx}(\omega) = e^{-(\omega-1)^2}$

From the properties of PSD

$$S_{xx}(-\omega) = e^{-(\omega-1)^2}$$

$$= e^{-(\omega+1)^2}$$

$$\neq S_{xx}(\omega)$$

Hence the given function is not valid PSD

c)

Given
$$Sxx(\omega) = \frac{\omega^2}{\omega^6 + 3\omega^2 + 3}$$

From the properties of PSD, the Given function is real and positive

$$S_{xx}(-\omega) = \frac{(-\omega)^2}{(-\omega)^6 + 3(-\omega)^2 + 3}$$

$$= \frac{\omega^2}{\omega^6 + 3\omega^2 + 3}$$

$$= S_{xx}(\omega)$$

Hence the given function is valid PSD

Example 3: The PSD of $X(f)$ is given by $S_{xx}(\omega) = 1+\omega^2$ for $|\omega| < 1$

$$= 0 \quad \text{otherwise}$$

Find out the auto correlation function

Solution:

The Auto Correlation Function $Rxx(\tau)$ is given by

$$R_{xx}(\tau) = \frac{1}{2\pi} \int_{-x}^{x} Sxx(\omega) e^{t\omega t} d\omega$$

$$= \frac{1}{2\pi} \int_{-1}^{1} (1+\omega)^2 e^{J\omega} d\omega$$

$$= \frac{1}{2\pi} \left[\int_{-1}^{1} e^{J\omega} d\omega + \int_{-1}^{1} \omega^2 e^{J\omega} d\omega \right]$$

We take the integral

$$\int_{-1}^{1} e^{j\omega\tau} d\omega = \left[\frac{e^{j\omega\tau}}{j\tau} \right]_{-1}^{1} = \frac{e^{j\tau} - e^{-j\tau}}{j\tau}$$

The second integral is evaluated as follows

$$\int_{-1}^{1} \omega^2 e^{j\omega\tau} d\omega = \left[\frac{e^{j\omega\tau}}{j\tau} \omega^2 \right]_{-1}^{1} - \int_{-1}^{1} \frac{e^{j\omega t}}{j\tau} (2\omega) d\omega$$

$$= \frac{e^{j\tau} - e^{-j\tau}}{j\tau} - \left[\frac{2e^{j\omega\tau}}{(j\tau)^2} \omega \right]_{-1}^{1} - \int_{-1}^{1} \frac{2e^{j\omega\tau}}{(j\tau)^2} d\omega$$

$$= \frac{e^{j\tau} - e^{-j\tau}}{j\tau} - 2\frac{e^{j\tau} + e^{-j\tau}}{-\tau^2} + \frac{2}{-\tau^2} \left[\frac{e^{j\omega\tau}}{j\tau} \right]_{-1}^{1}$$

$$= \frac{e^{j\tau} - e^{-j\tau}}{j\tau} - 2\frac{\left(e^{j\tau} + e^{-j\tau}\right)}{-\tau^2} + 2\frac{e^{j\tau} - e^{-j\tau}}{j\tau^3}$$

By substituting the results of above integrals, we get

$$R_{xx}(\tau) = \frac{1}{2\pi} \left[\frac{e^{j\tau} - e^{-j\tau}}{j\tau} + \frac{e^{j\tau} - e^{-j\tau}}{j\tau} + \frac{2}{\tau^2} \left(\frac{e^{j\tau} - e^{-j\tau}}{\tau^2} \right) - \frac{2}{j\tau^3} \left(e^{j\tau} - e^{j\tau} \right) \right]$$

$$= \frac{1}{2\pi} \left[\frac{2\sin\tau}{\tau} + \frac{2\sin\tau}{\tau} + \frac{4\cos\tau}{\tau^2} - \frac{4\sin\tau}{\tau^3} \right]$$

$$= \frac{1}{\pi} \left[\frac{2\sin\tau}{\tau} + \frac{2\cos\tau}{\tau^2} - \frac{2\sin\tau}{\tau^3} \right]$$

$$R_{xx}(\tau) = \frac{2}{\pi\tau^3} \left[\tau^2 \sin\tau + \tau\cos\tau - \sin\tau \right]$$

Example 4: Find out the PSD of *WSS* random process $X(t)$ whose auto correlation function is $R_{xx}(\tau) = a\, e^{-b|\tau|}$

Solution:

$$S_{xx}(\omega) = \int_{-\alpha}^{\alpha} Rxx^{(\tau)} e^{-j\omega\tau} d\tau$$

$$= \int_{-\alpha}^{\alpha} a\, e^{-b|\tau|} e^{-j\omega\tau} d\tau$$

$$S_{xx}(\omega) = a \int_{-\alpha}^{0} e^{b\tau} e^{-j\omega\tau} d\tau + a \int_{0}^{\alpha} e^{-b\tau} e^{-j\omega\tau} d\tau$$

$$= a \left[\int_{-\alpha}^{0} e^{(b-j\omega)\tau} d\tau + \int_{0}^{\alpha} e^{-(b+j\omega)\tau} d\tau \right]$$

$$= a \left[\frac{e^{(b-j\omega)\tau}}{(b-j\omega)} \right]_{-\infty}^{0} + a \left[\frac{e^{-(b+j\omega)\tau}}{-(b+j\omega)} \right]_{0}^{\infty}$$

$$= \frac{a}{b-j\omega}(1-0) + \frac{-a}{(b+j\omega)}(0-1)$$

$$= \frac{a}{b - j\omega} + \frac{a}{(b + j\omega)}$$

Therefore $S_{xx}(\omega) = \dfrac{2ab}{b^2 + \omega^2}$

Example 5: Find the PSD of a random process $Z(t) = X(t) + Y(t)$ where $X(t)$ and $Y(t)$ are zero mean, individual random process.

Solution

$$\begin{aligned}
R_{zz}(t, t+\tau) &= E\big[Z(t)\, Z(t+\tau)\big] \\
&= E\big[\{X(t) + Y(t)\}\{X(t+\tau) + Y(t+\tau)\}\big] \\
E X(t) X(t+\tau) &+ X(t)\, Y(t+\tau) + Y(t)X(t+\tau) + Y(t)Y(t+\tau) \\
&= R_{xx}(\tau) + Rxy(\tau) + Ryx(\tau) + Ryy(\tau)
\end{aligned}$$

Since $X(t)$ and $Y(t)$ are zero mean, independent random processes

i.e $E_{xy}(\tau) = Rxy(\tau) = 0$

$\therefore R_{ZZ}(\tau) = Rxx(\tau) + RYY(\tau)$

Taking the Fourier transform on both sides, we get

$$Szz(\tau) = Sxx(\tau) + Syy(\tau)$$

Example: 6. If the auto correlation function of a WSS process is $R(\tau) = k\,e^{-k|\tau|}$ show

that its special density is given by $S(\omega) = \dfrac{2}{1 + \left(\dfrac{\omega}{k}\right)^2}$

Solution:

$$S(\omega) = F[R(\tau)] = \int_{-\infty}^{\infty} ke^{-k|\tau|} e^{-j\omega\tau}\, d\tau$$

$$= k \int_{-\alpha}^{0} e^{k\tau}\, e^{-j\omega\tau}\, d\tau + k \int_{0}^{\alpha} e^{-k\tau}\, e^{-j\omega\tau}\, d\tau$$

$$= k \left\{ \int_{-\infty}^{0} e^{(k-j\omega)\tau}\, d\tau + \int_{0}^{\infty} e^{-(k+j\omega)\tau}\, d\tau \right\}$$

$$= k \left\{ \left[\frac{e^{(k-j\omega)\tau}}{(k-j\omega)} \right]_{-\infty}^{0} + \left[\frac{e^{-(k+j\omega)\tau}}{-(b+j\omega)} \right]_{0}^{\infty} \right\}$$

$$= \frac{k}{k - j\omega} + \frac{k}{k + j\omega}$$

$$= \frac{k\left[(k + j\omega) + k - j\omega\right]}{k^2 + \omega^2}$$

Therefore $\quad S(\omega) = \dfrac{2k^2}{\omega^2 + k^2} = \dfrac{2}{1 + \left(\dfrac{\omega}{k}\right)^2}$

***Example* 6:** Find the PSD of random process $X(t)$ is $E[X(t)] = 1$ and $R_{xx}(\tau) = 1 + e^{-\alpha|\tau|}$

Solution:

Since $\quad\quad S_{xy}(\omega) = F[R(\tau)]$

$$= \int_{-\alpha}^{\alpha}\left[1 + e^{-\alpha|\tau|}\right]e^{-j\omega t}\,d\tau$$

$$= \int_{-\alpha}^{\alpha} 1.e^{-j\omega t}\,d\tau + \int_{-\alpha}^{0} e^{\alpha\tau}\,e^{-j\omega\tau}\,d\tau + \int_{0}^{\alpha} e^{-\alpha\tau}\,e^{-j\omega\tau}\,d\tau$$

$$= 2\pi\,\delta(\omega) + \left[\frac{e^{(\alpha - j\omega)\tau}}{\alpha - j\omega}\right]_{-\infty}^{0} + \left[\frac{e^{-(\alpha + j\omega)\tau}}{-(\alpha + j\omega)}\right]_{0}^{\infty}$$

$$= 2\pi\,\delta(\omega) + \frac{1}{(\alpha - j\omega)} + \frac{\alpha}{(\alpha + j\omega)}$$

$$= 2\pi\,\delta(\omega) + \frac{2\alpha}{(\alpha^2 + \omega^2)}$$

Example: 8. The power spectral density of random processes is given by

$$S_{xx}(\omega) = \pi \quad |\omega| < 1 = 0 \quad \text{Elsewhere}$$

Find its auto correlation function

Solution:

$$R_{xx}(\tau) = \frac{1}{2\pi}\int_{-\sigma}^{\alpha} Sxx(\omega)\,e^{j\omega\tau}\,d\omega$$

$$= \frac{1}{2\pi}\int_{-1}^{1}\pi\,e^{j\omega\tau}\,d\omega$$

$$= \frac{1}{2} \left[\frac{e^{j\omega\tau}}{j\omega} \right]_{-1}^{1}$$

$$= \frac{1}{2} \frac{e^{j\tau} - e^{-j\tau}}{j\tau}$$

$$= \frac{1}{\tau} \sin\tau = \frac{\sin\tau}{\tau}$$

Therefore $R_{xx}(\tau) = \sin\tau$

Example: 9. Show that if $Y(t) = X(t+a) - X(t-a)$ then

i) $R_{yy}(\tau) = 2 R_{xx}(\tau) - R_{xx}(\tau+2a) - R_{xx}(\tau-2a)$

ii) $S_{yy}(\omega) = 4 S_{xx}(\omega) \; \text{Sin}^2 a\omega$

Solution:

$$R_{yy}(\tau) = E\{[x(t+a) - X(t-a)] [X(t+\tau+a) - X9t+\tau-a]\}$$

$$= E\{X(t+a)X(t+\tau+a)\} - E\{X(t+a)X(t+\tau-a)\}$$

$$- E\{X(t-a)X(t+\tau+a)\} + E\{X(t-a)X(t+\tau-a)\}$$

$$= R_{xx}(\tau) - E\{X(t+a)X[t+a+(\tau-2a)]\}$$

$$- E\{X(t-a)X[t-a+(\tau+2a)]\} + R_{xx}(\tau)$$

$$R_{yy}(\tau) = 2R_{xx}(\tau) - R_{xx}(\tau-2a) - R_{xx}(\tau+2a)$$

(ii) $$S_{yy}(\omega) = \int_{-\alpha}^{\alpha} R_{yy}(\tau) e^{-j\omega\tau} \, d\tau$$

$$= 2 \int_{-\alpha}^{\alpha} R_{xx} e^{-j\omega\tau} \, d\tau - \int_{-\alpha}^{\alpha} R_{xx}(\tau-2a) e^{-j\omega\tau} \, d\tau$$

$$- \int_{-\alpha}^{\alpha} R_{xx}(\tau+2a) e^{-j\omega\tau} \, d\tau$$

$$= 2 S_{xx}(w) - - \int_{-\alpha}^{\alpha} R_{xx}(\tau-2a) e^{-j\omega\tau} \, d\tau$$

$$- \int_{-\alpha}^{\alpha} R_{xx}(\tau+2a) e^{-j\omega\tau} \, d\tau$$

For the first integral let $(\tau - 2a) = \tau_1$

$$d\tau = d\tau_1$$

$$\tau = \tau_1 + 2a$$

For the second integral let $\tau + 2a = \tau_2$

$$d\tau = d\tau_2$$

$$\tau = \tau_2 - 2a$$

$$S_{yy}(\omega) = 2 S_{xx}(\omega) - \left[\int_{-\alpha}^{\alpha} R_{xx}(\tau_1) e^{-j\omega\tau_1} e^{j2\omega a} d\tau_1 \right]$$

$$- \left[-\int_{-\alpha}^{\alpha} R_{xx}(\tau_2) e^{-j\omega\tau_2} e^{j2\omega a} d\tau_2 \right]$$

$$S_{yy}(\omega) = 2 S_{xx}(\omega) - e^{j2\omega a} \left[\int_{-\alpha}^{\alpha} R_{xx}(\tau_1) e^{-j\omega\tau_1} d\tau_1 \right]$$

$$- e^{j2\omega a} \left[\int_{-\alpha}^{\alpha} R_{xx}(\tau_2) e^{-j\omega\tau_2} d\tau_2 \right]$$

$$= 2 S_{xx}(\omega) - e^{j2\omega a} S_{xx}(\omega) - e^{-j2\omega a} s_{xx}(\omega)$$

$$= S_{xx}(\omega) \left\{ 2 - \left[e^{j2\omega a} + e^{-j2\omega a} \right] \right\}$$

$$= S_{xx}(\omega) \left\{ 2 - 2\cos(2\omega a) \right\}$$

$$= 2 S_{xx}(\omega) \left\{ 1 - \cos(2\omega a) \right\}$$

$$= 2 S_{xx}(\omega) 2 \sin^2 a\omega$$

Therefore $S_{yy}(\omega) = 4 S_{xy}(\omega) \operatorname{Sin}^2 a\omega$

Example: 10. Find the cross spectral density of a $R_{xx}(\tau) = \dfrac{A^2}{2} \sin \omega_0 \tau$

Solution:

Given $R_{xx}(\tau) = \dfrac{A^2}{2} Sin \, \omega_0 \, \tau$

The cross power spectral density is

$$S_{xy}(\omega) = F\left[\frac{A^2}{2} \sin \omega_0 \tau \right]$$

$$= \frac{jA^2\pi}{2} \left[8(\omega + \omega_0) - 8(\omega - \omega_0) \right]$$

Example :11 Find out the Cross correlation function for the PSD $S_{xy}(\omega) = \dfrac{1}{25 + \omega^2}$

Solution:

Given   PSD $S_{xy}(\omega) = \dfrac{1}{25 + \omega^2}$

The cross correlation function

$$R_{xy}(\tau) = F^{-1}\left[S_{XX}(\omega)\right]$$

$$= F^{-1}\left[\frac{1}{25 + \omega^2}\right]$$

We know that from the Fourier transform $\dfrac{2a}{a^2 + \omega^2} \leftrightarrow e^{-a(t)}$, $a > 0$ Constant

$$\frac{1}{a^2 + \omega^2} \leftrightarrow \frac{1}{2a} e^{-a(t)}$$

$$\left[\frac{1}{25 + \omega^2}\right] \leftrightarrow \frac{1}{2 \times 5} e^{-5(t)}$$

$$R_{xy}(\tau) = \frac{1}{10} e^{-5(t)}$$

Review Quiz Questions

1. The power density spectrum is

   a)   square function   b)   Linear function

   c)   odd function   d)   Even function

2. The power density spectrum represents power per unit

   a)   wavelength   b)   bandwidth

   c)   Frequency   d)   none of the above

3. For a periodic function, the power spectral density and the auto correlation function are

   a)   one and the same thing   b)   Fourier transform pair

   c)   Laplace transform pair   d)   none of these

4. If S (ω) denotes the power spectral density spectrum, then the average power of a signal is given by

a) $\dfrac{1}{2\pi}\displaystyle\int_{-\infty}^{\infty} \delta(\omega)\, d\omega$

b) $\dfrac{1}{2\pi}\displaystyle\sum_{-\infty}^{\infty} \delta^2(\omega)\, d\omega$

c) $\dfrac{1}{\pi}\displaystyle\int_{-\infty}^{\infty} \delta(\omega)\, d\omega$

d) $\dfrac{1}{\pi}\displaystyle\int_{-\infty}^{\infty} \delta^2(\omega)\, d\omega$

5. The auto correlation function $R_{XX}(\tau)$ given in terms of P S D $S_{XX}(\omega)$ as

a) $R_{XX}(\tau) = \displaystyle\int_{-\infty}^{\infty} S_{XX}(\omega) e^{j\omega t}\, d\tau$

b) $R_{XX}(\tau) = \dfrac{1}{2\pi}\displaystyle\int_{-\infty}^{\infty} S_{XX}(\omega) e^{j\omega t}\, d\tau$

c) $R_{XX}(\tau) = \displaystyle\sum_{-\infty}^{\infty} S_{XX}(\omega) e^{-j\omega t}\, d\tau$

d) $R_{XX}(\tau) = \dfrac{1}{2\pi}\displaystyle\int_{-\infty}^{\infty} S_{XX}(\omega) e^{-j\omega t}\, d\omega$

6. The PSD of a random process whose auto correlation function is $a\, e^{-b|\tau|}$ is

a) $\dfrac{a}{a^2 + \omega^2}$

b) $\dfrac{2ab}{a^2 + \omega^2}$

c) $\dfrac{2ab}{b\left(a^2 + \omega^2\right)}$

d) $\dfrac{2ab}{a^2 + \omega^4}$

7. The auto correlation function of a random process power spectral density $S_{XX}(\omega)$ is

$\dfrac{4}{1 + \dfrac{\omega^2}{4}}$ is

a) $e^{-2|\tau|}$

b) $2e^{-2|\tau|}$

c) $3e^{-2|\tau|}$

d) $4e^{-2|\tau|}$

8. Time average of auto correlation function and the power spectral density form – pair

a) Fourier transform

b) Laplace transform

c) Z- transform

d) convolution

9. power spectral density of Wss is always

a) Negative

b) non negative

c) Finite

d) can be negative or positive

10. The average power of a periodic random process whose auto correlation function

$$R_{XX}(\tau) = Exp\left[-\frac{\tau^2}{2\sigma^2}\right] \text{ is}$$

a) 0   b) 1

c) 2   d) 3

11. For a WSS process, PSDat zero frequency gives

a) The area under the graph of Psd

b) the area under the graph of Auto correlation

c) mean of the process

d) Variance of the process

12. The mean square value of Wss process equals

a) area under the graph of Psd

b) area under the graph of ACF of process

c) Zero

d) mean of the process

13. $X(t) = A\cos(\omega_0 t + \theta)$ where A and w_0 are constants and θ is random variable uniformly distributed over $(0, \pi)$, then the average power of $X(t)$ is

a) 0   b) $\dfrac{A^2}{2}$

c) $\dfrac{A^2}{4}$ d) $\dfrac{A^2}{8}$

14.   The real part and imaginary part of $S_{YY}(\omega)$ $\ldots$ and $\ldots$ function of w respectively.

a) odd, odd   b)   odd, even

c) even, odd   d)   even, even

15. If X(t) and Y(t) are orthogonal then

a) $S_{XY}(\omega) = 0$ b) $S_{XY}(\omega) = 1$

c) $S_{XY}(\omega) > 1$ d) $S_{XY}(\omega) < 1$

16. The cross spectral density of tow random processes $X(t)$ and $Y(t)$ is $S_{XY}(\omega) = a + \dfrac{jb\omega}{k}$, the area enclosed by their autocorrelation fucntion is

   a) $\dfrac{a}{k}$

   b) ak

   c) a

   d) $a+k$

17. The means of two independent WSS processes $X(t)$ and $Y(t)$ are 2 and 3 respectively, their cross spectral density is

   a) $6\delta(\omega)$

   b) $12\pi\delta(\omega)$

   c) $2\pi\delta(\omega)$

   d) None

18. $S_{XY}(\omega) = 0$ if $X(t)$ and $Y(t)$ are

   a) Orthogonal

   b) Parallel

   c) vertical

   d) none

19. If $R_{XX}(\tau) \xleftrightarrow{FT} S_{XX}(\omega)$, the relation is called as

   a) Parseval's

   b) density fucntion

   c) Wiener-Khinchins

   d) none

20. The curve $\dfrac{S_{XX}(\omega)}{A}$ always encloses ... one

   a) Double

   b) Unit

   c) Tripple

   d) none

Answers

   1. (d)   2. (b)   3. (b)   4. (c)   5. (b)   6. (b)

   7. (d)   8. (a)   9. (b)   10. (b)   11. (b)   12. (a)

   13. (b)   14. (c)   15. (a)   16. (a)   17. (b)   18. (a)

   19. (c)   20. (b)

Review Quiz Questions

1. The PSD of a random process whose auto correlation function is $3\,e^{-2|\tau|}$ is

   a) $\dfrac{3}{9+\omega^2}$

   b) $\dfrac{12}{9+\omega^2}$

   c) $\dfrac{12}{2\left(9+\omega^2\right)}$

   d)   none

2. The auto correlation function of a random process power spectral density $S_{XX}(\omega)$ is
   $\dfrac{2}{1+\dfrac{\omega^2}{2}}$ is

   a) $e^{-2|\tau|}$

   b) $2e^{-2|\tau|}$

   c) $3e^{-2|\tau|}$

   d)   none

3. Time average of auto correlation function and the power spectral density form – pair

   a) Fourier transform

   b)   Laplace transform

   c) Z- transform

   d)   convolution

4. The average power of a periodic random process whose auto correlation function
   $R_{XX}(\tau) = Exp\left[-\dfrac{\tau^2}{4\sigma^2}\right]$ is

   a) 0

   b)   1

   c) 2

   d)   none

5. The mean square value of Wss process equals

   a) area under the graph of Psd

   b) area under the graph of ACF of process

   c) Zero

   d) mean of the process

6. $X(t) = A\sin(\omega_0 t + \theta)$ where A and w_0 are constants and θ is random variable uniformly distributed over $(0, \pi)$, then the average power of $X(t)$ is

   a) 0

   b) $\dfrac{A^2}{2}$

   c) $\dfrac{A^2}{4}$

   d) none

7. The real part and imaginary part of $R_{YY}(\tau)$ ... and ... function of w respectively.

   a) odd, odd

   b) odd, even

   c) even, odd

   d) none

8. If X(t) and Y(t) are orthogonal then

   a) $S_{XY}(\omega) = 0$

   b) $S_{XY}(\omega) = 1$

   c) $S_{XY}(\omega) > 1$

   d) $S_{XY}(\omega) < 1$

9. The cross spectral density of tow random processes $X(t)$ and $Y(t)$ is $S_{XY}(\omega) = 2 + \dfrac{jb\omega}{k}$, the area enclosed by their autocorrelation function is

   a) $\dfrac{2}{k}$

   b) $2k$

   c) 2

   d) none

10. The means of two independent WSS processes $X(t)$ and $Y(t)$ are 4 and 9 respectively, their cross spectral density is

    a) $6\delta(\omega)$

    b) $12\pi\delta(\omega)$

    c) $2\pi\delta(\omega)$

    d) None

11. If $R_{XX}(\tau) \xleftrightarrow{FT} S_{XX}(\omega)$, the relation is called as

    a) Parseval's

    b) density function

    c) Wiener-Khinchins

    d) none

Exercise Questions

1. Find the power spectral density (PSD) for a processes for which $Rxx\,(\tau) = 1$ for all τ

2. A random processes has a PSD function given by $S(f) = \dfrac{1}{\left(1+\left(\dfrac{f}{B}\right)^2\right)^3}$

   (i) Find the absolute Bandwidth
   (ii) Find the 3db band width
   (iii) Find the Rms band Width

3. A random process $X(f)$ has PSD given by $Sxx\,(\omega) = \begin{cases} 4-\dfrac{\omega^2}{9} & |\omega| \le 6 \\ 0 & Otherwise \end{cases}$

   Determine (a) average power (b) the Auto correlation function of the processes

4. A random process $Z\,(t)$ has the auto correlation function given by

$$R_{zz}\,(\tau) = \begin{cases} 1+\dfrac{\tau}{\tau_0} & -\tau_0 \le \tau \le 0 \\ 1-\dfrac{\tau}{\tau_0}, & 0 \le \tau \le \tau_0 \\ 0 & otherwise \end{cases}$$

   Where τ_0 is a constant calculate the PSD of the Process

5. Give reasons why the function given below can not be the power spectral density of a *WSS* random Process

   a) $S_{xx}\,(\omega) = \dfrac{\sin\omega}{\omega}$ b) $S_{xx}\,(\omega) = \dfrac{\cos\omega}{\omega}$

   c) $S_{xx}\,(\omega) = \dfrac{8}{\omega^2+16}$ d) $S_{xx}\,(\omega) = \dfrac{8\omega}{1+3\omega^2+4\omega4}$

6. A band limited white noise has the power spectral density defend by

$$S_{xx}\,(\omega) = \begin{cases} 0.01 & 400\pi \le |\omega| \le 500\pi \\ 0 & Otherwise \end{cases}$$

   Find the mean structure value of the process

7. A wise sense stationary random process $N(t)$ has the autocorrelation function

    $R_{NN} = A\ e^{-4|\tau|}$ where A is a constant determine the power spectral density

8. Two jointly stationary random Processes $X(t)$ and $Y(t)$ have the crocs correlation function given by Rxy $(\tau) = 2\ e^{-\tau}$, $\tau \geq 0$ Determine the following

    (a) The cross power spectral density $S_{xy}(\omega)$

    (b) The cross power spectral density $S_{yx}(\omega)$

9. Two Jointly stationary random process $X(t)$ and $Y(t)$ have the cross power spectral density given by $S_{xy}(\omega)\ \dfrac{1}{-\omega^2 + j4\omega + 4}$ find the corresponding cross correlation function

10. A zero mean *WSS* random process $X(t)$, $-\alpha < t < 0$ has the following *PSD*

    $S_{xx}(\omega) = \dfrac{2}{1+\omega^2} - \alpha < \omega < \alpha$ the random processes $Y(t)$ is defined by $Y(t) = \sum_{k=0}^{2} X(t+k)$

    (a) Find the mean of $Y(t)$

    (b) Find the variance of $Y(t)$

11. If the auto correlation function of a *WSS* process is $R(\tau) = k.e^{-k|\tau|}$. Show that its spectral density is given by $S(\omega) = \dfrac{2}{1+(\omega/t)^2}$.

12. Assume that an ergodic random process $X(t)$ has an autocorrelation function $R_{XX}(\tau)$

    $= 18 + \dfrac{2}{6+\tau^2}[1+4\cos(12\tau)]$. Find the average power in $X(t)$.

13. For a random process $R(\tau) = a.e^{-b|\tau|}$. Show that the *PSD* is given by

    $$S_{XX}(\omega) = \dfrac{2ab}{b^2 + \omega^2}.$$

14. Determine which of the following function can and can not be valid power density spectrums. For those that are not, explain why

    (i) $\exp\left[-(\omega-1)^2\right]$

    (ii) $\dfrac{\omega^2}{\omega^4 +1} - 8\,(\omega)$

(iii) $\dfrac{\cos(3\omega)}{1+\omega^2}$

(iv) $\dfrac{1}{\left(1+\omega^2\right)^2}$

(v) $\dfrac{1}{\sqrt{1-3\omega^2}}$

(vi) $\dfrac{|\omega|}{1+2\omega+\omega^2}$

15. Find the rms bandwidth of power spectrum

$$S_{XX}(\omega) = \begin{cases} p\ \cos(\dfrac{\pi\omega}{2W}) & |\omega| \le W \\ 0 & elsewhere \end{cases}$$

Where $W > 0$ and $p > 0$ constants

16. The auto correlation function of a random process $X(t)$ is

$$R_{XX}(\tau) = 3 + 2\exp(-4\tau^2)$$

(i) Find the power spectrum of $X(t)$

(ii) What is the average power in $X(t)$

(iii) What fraction of the power lies in the frequency band $\dfrac{-1}{\sqrt{2}} \le w \le \dfrac{1}{\sqrt{2}}$

17. Determine the cross power density spectrum corresponding to the cross correlation function

$$R_{XX}(\tau) = u(-\tau)\dfrac{e^{b\tau}}{a+b} + u(\tau)\dfrac{e^{-b\tau}}{a^2-b^2}\left[a+b-2be^{-(a-b)\tau}\right]$$

Where $a > 0$ and $b > 0$ are constants.

18. Find the cross correlation function corresponding to the cross power spectrum

$$S_{XY}(\omega) = \dfrac{6}{\left(9+\omega^2\right)\left(3+j\omega\right)^2}$$

19. If $X(t)$ and $Y(t)$ are real random processes, determine which of the following function can be valid. For those that are not -----. State at least one reason why

    (a) $R_{XX}(\tau) = \exp(-|\tau|)$

    (b) $|R_{XX}(\tau)| \le j\sqrt{R_{XX}(0)\, R_{YY}(0)}$

    (c) $R_{XX}(\tau) = 2\sin(3\tau)$

    (d) $S_{XX}(\omega) = \dfrac{6}{6+\tau\,\omega^3}$

    (e) $S_{XX}(\omega) = \dfrac{4\exp(-3|\tau|)}{1+\omega^2}$

    (f) $S_{XY}(\omega) = (3+j\omega)^2$

    (g) $S_{XY}(\omega) = 188\,(\omega)$

20. Define two random processes by

    $$X(t) = A\cos(\omega_0 t + \theta)$$

    $$Y(t) = W(t)\cos(\omega_0 t + \theta)$$

    Where 'A' and ω_0 are real positive constants, θ is a random variable independent of $\omega(t)$ and $W(t)$ is a random process with a constant mean value $\overline{\omega}$.

    Show that $S_{XY}(\omega) = \dfrac{a\overline{\omega}\pi}{2}\left[\delta(\omega-\omega_0)+\delta(\omega+\omega_0)\right]$ regardless of the form of the probability density function of θ.

21. Consider the random processes of problem (10),

    (i)   Show that the cross correlation function is given by

    $$R_{XY}(t,t+\tau) =$$

    $$\dfrac{a\overline{\omega}}{2}\left\{\cos(\omega_0\tau) + E\left[\cos(2\theta)\right]\cos(2\omega_0 t + \omega_0\tau) - E\left[\sin(2\theta)\right]\sin(2\omega_0 t + \omega_0\tau)\right\}$$

    Where the expectation is w.r.t θ only

    (ii)   Find the time average of $R_{XY}(t,t+\tau)$ and determine the cross power density spectrum $S_{XY}(\omega)$.

Review Exercise Questions

1.   What is meant by Spectral Analysis

2.   Define the power spectral density function of a stationary process.

3.   State and prove the properties of the PSD of a stationary process

4.   Define the cross power spectral density and state any two its properties.

5.   What is the use of Wiener-Kchinchine theorem? Explain

6.   Show that the rms bandwidth of a low pass random process $X(t)$ as given by

$$R_{XX}(\tau) = 3 + 2e^{-4\tau^2} \quad \text{can also be obtained from}$$

$$w_{rms}^2 = \frac{-1}{R_{XX}(\tau)|_{\tau=0}} \left[\frac{d^2 R_{XX}(\tau)}{d\tau^2} \right]_{\tau=0}$$

Where $R_{XX}(\tau)$ is the auto correlation function of $X(t)$

7.   Find PSD of the random binary transmission process whose auto correlation function

$$R(\tau) = \begin{cases} 1 - \dfrac{|\tau|}{\tau} & |\tau| \le T \\ 0 & elsewhere \end{cases}$$

8.   Find the Auto correlation function of a stationary process, whose PSD is given by

$$S(\omega) = \begin{cases} \omega^2 & |\omega| \ge 1 \\ 0 & elsewhere \end{cases}$$

9.   Determine which of the following functions can and can not be valid power density spectrum. For those that are not, explain why

(a) $\dfrac{\cos(5\omega)}{1+\omega^2}$ (b) $\dfrac{|\omega|}{1+2\omega+\omega^2}$

10.   Prove that the power Spectrum and Auto correlation of random processes form Fourier transform pair

11.   What is the relationship between cross power spectrum and cross correlation function? Prove the relationship.

12.   Find the cross correlation function corresponding to the cross power spectrum

$$S_{XY}(\omega) = \frac{6}{\left(9+\omega^2\right)\left(3+j\omega\right)^2}$$

13.   Determine the rms bandwidth of the power spectrums given by

(i) $S_{XY}(\omega) = \begin{cases} p & |\omega| \leq W \\ 0 & |\omega| > W \end{cases}$ (ii) $S_{XY}(\omega) = \begin{cases} p\left[1 - \left|\dfrac{\omega}{W}\right| \right] & |\omega| \leq W \\ 0 & |\omega| > W \end{cases}$

14.   If the power spectrum of band pass white noise is defined as shown in Fig. 7.3

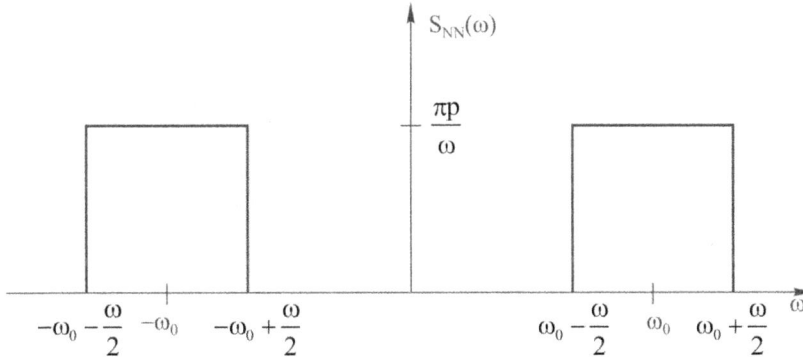

Fig. 7.3

15.   A wide sense stationary noise process $N(t)$ has an autocorrelation function $R_{NN}(\tau) = e^{-3|\tau|}$ where p is a constant. Find its power spectrum.

16.   Determine the cross correlation function corresponding to the cross power density spectrum

$$S_{XY}(\omega) = \frac{8}{(\alpha + j\omega)^3} \text{ where } \alpha > 0 \text{ is a constant.}$$

17.   The cross correlation of jointly WSS stationary processes $X(t)$ and $Y(t)$ is assumed to   be $R_{XX}(\tau) = B\, u(\tau)\, e^{-\omega\, \tau}$ Where $B > 0$ and $W > 0$ are constants. Find $R_{YX}(\tau)$, $S_{XY}(\omega)$ and $S_{YX}(\omega)$

18.   Find the cross power spectrum for the two processes $X(t)$ and $Y(t)$ for which the cross correlation function is

$$R_{XY}(t, t+\tau) = \frac{AB}{2}\left\{ \sin(\omega_0\tau) + \cos\left[(2t + \tau)\omega_0 \right] \right\}$$

Where A, B and ω_0 are constants.

19. The band limited white noise is defined by having a non zero and constant power spectrum over a finite frequency band as shown in Fig. 7.4. Determine its auto correlation function

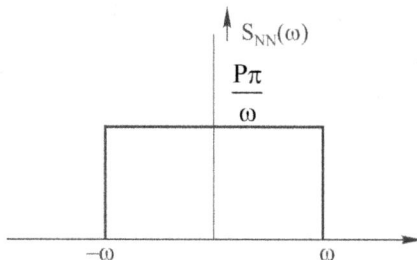

Fig. 7.4

CHAPTER 8

Linear Systems
with Random Inputs

8.1 Introduction

The objective of this chapter is to introduce the basic concepts of Linear systems and to explain how these random signals interact with linear systems. We also describe the methods of determining the response of a Linear system when the random signals are applied as input.

8.2 Linear System Fundamentals

Consider a linear system which provides response $y(t)$ to be different from the input signal $x(t)$. The input $x(t)$ and output $y(t)$ are related by

$$y(t) = L\{x(t)\} \qquad \qquad \text{......(8.1)}$$

Where L is operator representing the action of linear system on the input $x(t)$.

A linear system is said to linear system if its response to a sum of inputs

$x_n(t)$, n = 1, 2, . . . k is equal to the sum of responses taken individually.

If $x_n(t)$ causes a response $y_n(t)$, n = 1, 2, . . . N, then we write the output $y(t)$ as

$$y(t) = L\left[\sum_{n=1}^{k} \alpha_n x_n(t) \right]$$

$$= \sum_{n=1}^{k} \alpha_n L\left[x_n(t) \right]$$

513

$$= \sum_{n=1}^{k} \alpha_n y_n(t) \qquad \ldots(8.2)$$

Which is valid only when α_n are arbitrary constant and N is infinite. We write x(t) as

$$x(t) = \int_{-\infty}^{\infty} x(\tau)\delta(t-\tau)d\tau \qquad \ldots(8.3)$$

from the definition and properties of impulse function. By substituting the equation (8.3) in equation (8.1), and observing that the Operator L functions on the time function, we get

$$y(t) = L\{ x(t)\} = L\left[\int_{-\infty}^{\infty} x(\tau)\delta(t-\tau)d\tau \right]$$

$$= \int_{-\infty}^{\infty} x(\tau)L\big[\delta(t-\tau)\big]d\tau \qquad \ldots(8.4)$$

We define $h(t,\tau)$ as impulse response of linear system

i.e., $\qquad L\big[\delta(t-\tau)\big] = h(t,\tau) \qquad \ldots(8.5)$

use of equation (8.5) in (8.4), we write output $y(t)$ as

$$y(t) = \int_{-\infty}^{\infty} x(\tau)h(t,\tau)d\tau \qquad \ldots(8.6)$$

which shows that the response of a general system is completely determined by the impulse response.

8.3 Linear Time Invariant System

A linear system is said to time invariant if the impulse response $h(t,\tau)$ does not depend on the time that the impulse is applied. Thus if an impulse $\delta(t)$ occurring at $t = 0$, causes the response $h(t)$, then impulse $\delta(t-\tau)$ occurring at $t = \tau$, must cause the response $h(t-\tau)$ if the system is time invariant.

$$h(t,\tau) = h(t-\tau) \qquad \ldots(8.7)$$

For a linear time invariant system, the equation (8.6) can be written as

$$y(t) = \int_{-\infty}^{\infty} x(\tau)h(t-\tau)d\tau \qquad \ldots(8.8)$$

which is known as the convolution integral of $x(t)$ and $h(t)$.

In short form, we write equation (8.8) as

$$y(t) = x(t) \times h(t) \qquad \qquad .....(8.9)$$

Alternatively, equation (8.8) can be written as

$$y(t) = \int_{-\infty}^{\infty} h(t)x(t-\tau)d\tau \qquad \qquad .....(8.10)$$

8.4 Time Invariant system Transfer Function

From the equation (8.8) and (8.10), it is evident that the linear time invariant system is completely characterized by its impulse response which is a temporal characterization.

We may derive an equivalent characterization of Linear time invariant system in frequency domain by the use Fourier transformation of $y(t)$ is

$$Y(w) = \int_{-\infty}^{\infty} y(t)e^{-j\omega t}dt$$

$$= \int_{-\infty}^{\infty} \left[\int_{-\infty}^{\infty} x(\tau)h(t-\tau)d\tau \right] e^{-j\omega t}dt$$

$$= \int_{-\infty}^{\infty} x(\tau) \left[\int_{-\infty}^{\infty} h(t-\tau)e^{-j\omega(t-\tau)}dt \right] e^{-j\omega\tau}d\tau$$

$$= \int_{-\infty}^{\infty} x(\tau)H(\omega)e^{-j\omega\tau}d\tau$$

$$Y(\omega) = X(\omega)\,y(\omega) \qquad \qquad .....(8.11)$$

Where $X(\omega)$, $y(\omega)$ and $H(\omega)$ are respective Fourier transforms of $x(t)$, $y(t)$ and $h(t)$.

The function $H(\omega)$ is called transfer function of the system. It is noted from equation (8.11) that the Fourier transform of response of any Linear time invariant system is equal to the production of transform of input signal and the transform of the network impulse response.

If the input is an exponential signal

i.e., $\qquad \qquad x(t) = e^{j\omega t} \qquad \qquad .....(8.12)$

it can be shown that

$$H(\omega) = \frac{L\left[e^{j\omega t}\right]}{e^{j\omega t}} = \frac{y(t)}{x(t)} \qquad \qquad .....(8.13)$$

Where $\qquad \qquad y(t) = L\left[e^{j\omega t}\right] \qquad \qquad .....(8.14)$

Examples

***Example:* 1.** Find the transfer function $H(\omega)$ for the network shown in Fig. 8.1.

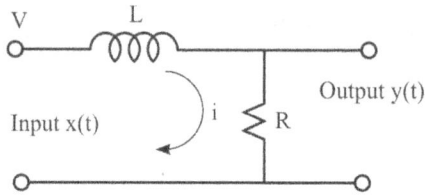

Fig. 8.1

Sol:

We write $x(t)$ as

$$x(t) = L\frac{di}{dt} + y(t)$$

But $y(t) = iR$ so

$$\frac{dy(t)}{dt} = R\frac{di}{dt} \qquad \Rightarrow \qquad \frac{di}{dt} = \frac{1}{R}\frac{dy(t)}{dt}$$

$$x(t) = \frac{L}{R}\frac{dy(t)}{dt} + y(t)$$

With $x(t) = exp\ (j\omega t)$, then

$$y(t) = H(\omega)x(t)$$

$$\frac{dy(t)}{dt} = H(\omega)j\ w\ x(t)$$

$$x(t) = \frac{L}{R}H(\omega)j\ w\ x(t) + H(w)x(t)$$

Finally we write $H(\omega)$ as

$$H(\omega) = \frac{1}{1+(j\omega L/R)}$$

Example 2: As a second example, we prove (8.13) by direct use equation (8.10)

Sol:

   If $x(t) = exp\ (j\omega t)$, we get output as

$$y(t) = \int_{-\infty}^{\infty} h(\tau)e^{j\omega(t-\tau)}d\tau$$

$$= x(t) \int_{-\infty}^{\infty} h(\tau)e^{-j\omega\backslash\tau}d\tau$$

But the integral is $H(\omega)$, the Fourier transform of h(t), so

$y(t) = H(\omega)x(t)$ which gives (8.13)

8.5 Causal and Stable System

A linear time invariant system is said to be causal if does not respond prior to the application of input signal.

Mathematically, $y(t) = 0$ for $t < t_0$ if $x(t) = 0$ for $t < t_0$, where t_0 is any real constant. From equation (8.10), this condition requires that

$$h(t) = 0 \ \text{ for } t < 0 \qquad\qquad\qquad .....(8.15)$$

A physical realizable system should satisfy the condition given by equation (8.15)

A linear time invariant system is said to be stable if its response to nay bounded input is bounded

i.e., if $|x(t)| < M$, then $|y(t)| < M \, I$ where M is some constant and I is also constant

which is independent of the input $I = \int_{-\infty}^{\infty} |h(t)|dt < \infty$

8.6 Random Signal Response of Linear Systems

We describe the temporal characteristics such as mean value MSEC mean squared value of the response, its auto correlation function and cross correlation function of a stable, linear, time invariant system as shown in Fig. 8.2 when the input is an sample function

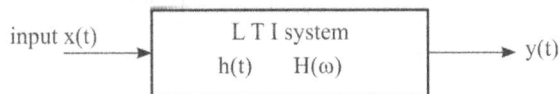

input x(t) → | L T I system h(t)   H(ω) | → y(t)

Fig. 8.2 Linear time invariant system

$x(t)$ of a random process $X(t)$, we assume that the system is impulse response $h(t)$ is a real function.

8.7 System Response

When $x(t)$ is a random signal, the *LTI* system response $y(t)$ is given by convolution integral

$$y(t) = \int_{-\infty}^{\infty} x(\tau) \, h(t - \tau) \, d\tau \qquad \qquad .....(8.16)$$

Or
$$y(t) = \int_{-\infty}^{\infty} h(\tau) \, x(t - \tau) \, d\tau \qquad \qquad .....(8.17)$$

We may interpret equation (8.17) as an operation on sample function $x(t)$ of the random process $X(t)$ that produces a sample function of a new process $Y(t)$. We define the new random process $Y(t)$ in terms of the process $X(t)$ from equation (8.17)

$$Y(t) = \int_{-\infty}^{\infty} h(\tau) \, X(t - \tau) \, d\tau \qquad \qquad .....(8.18)$$

We may view the system as accepting the random process $X(t)$ as its input and responding with the new process $Y(t)$ according to equation (8.18).

Mean of the output

If $X(t)$ is a random signal, the system response $Y(t)$ is given by

$$Y(t) = \int_{-\infty}^{\infty} h(\tau) \, X(t - \tau) \, d\tau$$

Taking the Expected operation on both sides, we get

$$E\big[Y(t)\big] = E\left[\int_{-\infty}^{\infty} h(\tau) \, X(t - \tau) \, d\tau \right]$$

$$= \int_{-\infty}^{\infty} h(\tau) \, E\big[x(t - \tau)\big] \, d\tau$$

$$= \overline{X} \int_{-\infty}^{\infty} h(\tau) \, d\tau$$

$$= \overline{Y} \quad \text{(constant)} \qquad \qquad .....(8.19)$$

It is noted from equation (8.19) that the mean value of $Y(t)$ equal to the mean value of $X(t)$ times the area under the impulse response if $X(t)$ is wide sense stationary.

8.8 Mean Square Value of the Output

We find the mean squared value of $Y(t)$ as

$$E\left[Y^2(t)\right] = E\left[\int_{-\infty}^{\infty} h(\tau_1)\, X(t-\tau_1)\, d\tau_1 \int_{-\infty}^{\infty} h(\tau_2)\, X(t-\tau_2)\, d\tau_2\right]$$

$$= \int_{-\infty}^{\infty}\int_{-\infty}^{\infty} E\left[X(t-\tau_1)X(t-\tau_2)\right] h(\tau_1)\, h(\tau_2)\; d\tau_1\, d\tau_2 \qquad .....(8.20)$$

If the input is wide sense stationary, then

$$E\left[X(t-\tau_1)X(t-\tau_2)\right] = R_{XX}(\tau_1-\tau_2) \qquad\qquad .....(8.21)$$

Substitution of equation (8.21) in equation (8.20), we get

$$\overline{Y}^2 = E\left[Y^2(t)\right] = \int_{-\infty}^{\infty}\int_{-\infty}^{\infty} R_{XX}(\tau_1-\tau_2)\, h(\tau_1)\, h(\tau_2)\; d\tau_1\, d\tau_2 \qquad ..... (8.22)$$

This equation (8.22) gives the power in $Y(t)$ but it is tedious to calculate this power in most of the cases.

Example: 1

Calculate the power for a system having white noise as input.

$$R_{XX}(\tau_1-\tau_2) = \left(\frac{N_0}{2}\right)\delta(\tau_1-\tau_2) \text{ where } N_0 \text{ is a positive real constant.}$$

Sol:

From equation (8.22), we get $\overline{Y}^2$ as

$$\overline{Y}^2 = \int_{-\infty}^{\infty}\int_{-\infty}^{\infty}\left(\frac{N_0}{2}\right)\delta(\tau_1-\tau_2)h(\tau_1)\, h(\tau_2)\; d\tau_1\, d\tau_2$$

$$= \frac{N_0}{2}\int_{-\infty}^{\infty} h^2(\tau_2)\, d\tau_2$$

Output power becomes proportional to the area under the square of $h(t)$ in this case.

8.9   Auto Correlation Function of Response

Let $x(t)$ be wide sense stationary random process the Autocorrelation function of output $y(t)$ is

$$R_{YY}(t, t+\tau) = E\left[y(t)y(t+\tau)\right]$$

$$= E\left[\int\limits_{-\infty}^{\infty} h(\tau_1) X(t-\tau_1)\, d\tau_1 \int\limits_{-\infty}^{\infty} h(\tau_2) X(t+\tau-\tau_2)\, d\tau_2\right]$$

$$= \int\limits_{-\infty}^{\infty}\int\limits_{-\infty}^{\infty} E\left[X(t-\tau_1)X(t+\tau-\tau_2)\right] h(\tau_1)\, h(\tau_2)\, d\tau_1\, d\tau_2 \qquad \dots\dots(8.23)$$

Which is simplified as

$$R_{YY}(\tau) = \int\limits_{-\infty}^{\infty} R_{XX}(\tau+\tau_1-\tau_2)\, h(\tau_1)\, h(\tau_2)\, d\tau_1\, d\tau_2 \qquad \dots\dots(8.24)$$

Because $X(t)$ is assured to wide sense stationary. Two facts result from equation (8.24)

1. $Y(t)$ is wide sense stationary if $x(t)$ is WSS because $R_{YY}(\tau)$ does not depend (WSS) on time t and $E\left[Y(t)\right]$ is constant.

2. It is evident from equation (8.24) that $R_{YY}(\tau)$ is two fold convolution of the input autocorrelation function with network's impulse response.

$$R_{YY}(\tau) = R_{XX}(\tau) * h(-\tau) * h(\tau) \qquad \dots\dots(8.25)$$

8.10   Cross Correlation Function of Input and Output

The cross correlation function of $X(t)$ and $Y(t)$

$$R_{XY}(t, t+\tau) = E\left[X(t)Y(t+\tau)\right]$$

$$= E\left[X(t) \int\limits_{-\infty}^{\infty} h(\tau) X(t+\tau-\tau_1)\, d\tau_1\right]$$

$$= \int\limits_{-\infty}^{\infty} E\left[X(t)X(t+\tau-\tau_1)\right] h(\tau)\, d\tau_1 \qquad \dots\dots(8.26)$$

If $X(t)$ is wide sense stationary, equation (8.26) simplifies to

$$R_{XY}(\tau) = \int\limits_{-\infty}^{\infty} R_{XX}(\tau - \tau_1) \, h(\tau_1) \, d\tau_1 \qquad \qquad .....(8.27)$$

Which is the convolution $R_{XY}(\tau) = R_{XX}(\tau) * h(\tau)$ .....(8.28)

Similarly it can be shown that

$$R_{YX}(\tau) = \int\limits_{-\infty}^{\infty} R_{XX}(\tau - \tau_1) \, h(-\tau) \, d\tau_1 \qquad \qquad .....(8.29)$$

Or

$$R_{YX}(\tau) = R_{XX}(\tau) * h(-\tau) \qquad \qquad .....(8.30)$$

From equations (8.27) and (8.29) it is evident that the cross correlation functions depend on 'τ' and not on absolute time 't'. As a consequence of this fact, $X(t)$ and $Y(t)$ are jointly WSS if $X(t)$ is *WSS*. Because we are already shown $Y(t)$ to be *WSS*.

By substituting equation (8.27) in to equation (8.24), Auto correlation function and cross correlation function are seen to be related by

$$R_{YY}(\tau) = \int\limits_{-\infty}^{\infty} R_{XY}(\tau + \tau_1) \, h(\tau_1) \, d\tau_1 \qquad \qquad .....(8.31)$$

Or

$$R_{YY}(\tau) = R_{XY}(\tau) * h(-\tau) \qquad \qquad .....(8.32)$$

A similar substitute of equation (8.29) in to (8.24),

we get

$$R_{YY}(\tau) = \int\limits_{-\infty}^{\infty} R_{YX}(\tau - \tau_2) \, h(\tau_2) \, d\tau_2 \qquad \qquad .....(8.33)$$

Or

$$R_{YY}(\tau) = R_{YX}(\tau) * h(\tau) \qquad \qquad .....(8.34)$$

Example: 2

Consider example 1, we find the cross correlation function $R_{XY}(\tau)$ and $R_{YX}(\tau)$

Sol:

From equation (8.27), we get $R_{XY}(\tau)$ as

$$R_{XY}(\tau) = \int_{-\infty}^{\infty} \left(\frac{N_0}{2}\right) \delta(\tau - \tau_1) h(\tau_1) \, d\tau_1$$

$$= \frac{N_0}{2} h(\tau)$$

From equation (8.29), we get $R_{XY}(\tau)$ as

$$R_{YX}(\tau) = \int_{-\infty}^{\infty} \left(\frac{N_0}{2}\right) \delta(\tau - \tau_1) h(-\tau_1) \, d\tau_1$$

$$= \frac{N_0}{2} h(-\tau)$$

$$= R_{XY}(-\tau)$$

8.11   Spectral Characteristics of System Response

In previous section, the temporal characteristics of Linear Systems are described in terms of auto correlation and cross correlation. The spectral characteristics of *LTI* systems are derived in this section using the Fourier transforms. Indeed, this approach is conceptually valid however the evaluation of integrals may be difficult from the practical point of view.

8.12   Power Density Spectrum of Response

We prove that the power density spectrum $S_{YY}(w)$ of the response of a *LTI* system having a transfer function H(w) is given by

$$S_{YY}(\omega) = S_{XX}(\omega) \, |H(\omega)|^2 \qquad\qquad .....(8.35)$$

Where $S_{XX}(\omega)$ is the power spectrum of the input process $X(t)$

Proof:

The Fourier transform of output correlation function is given by

$$S_{YY}(\omega) = \int_{-\infty}^{\infty} R_{YY}(\tau) \, e^{-j\omega\tau} d\tau \qquad\qquad .....(8.36)$$

Substituting the equation (8.24) in equation (8.36), we get

$$S_{YY}(\omega) = \int\limits_{-\infty}^{\infty} h(\tau_1) \int\limits_{-\infty}^{\infty} h(\tau_2) \int\limits_{-\infty}^{\infty} R_{XY}(\tau + \tau_1 - \tau_2)\, e^{-j\omega\tau}\, d\tau\, d\tau_2 d\tau_1 \qquad .....(8.37)$$

Let $\tau = \tau + \tau_1 - \tau_2,\ d\tau = d\tau$

$$S_{YY}(\omega) = \int\limits_{-\infty}^{\infty} h(\tau_1) e^{j\omega\tau_1} d\tau_1 \int\limits_{-\infty}^{\infty} h(\tau_2) e^{-j\omega\tau_2} d\tau_2 \int\limits_{-\infty}^{\infty} R_{XY}(\tau)\, e^{-j\omega\tau}\, d\tau \qquad .....(8.38)$$

Those integrals are recognized as $H^*(\omega)$, $H(\omega)$ and $S_{XX}(\omega)$ respectively. Hence

$$S_{XY}(\omega) = H^*(\omega)\, H(\omega)$$

$$= S_{XX}(\omega)\, |H(\omega)|^2 \qquad ..... (8.39)$$

And hence equation (8.35) is proved.

The average power, denoted P_{YY} is obtained by using equation (8.39)

$$P_{YY} = \frac{1}{2\pi} \int\limits_{-\infty}^{\infty} S_{XX}(\omega)\, |H(\omega)|^2\, d\omega \qquad .....(8.40)$$

Example: 3

Find the power spectrum and average power of the response of the network shown in example. When the $X(t)$ is white noise.

Sol:

The power spectrum of input process $X(t)$ is given by

$$S_{XX}(\omega) = \frac{N_0}{2}$$

Here the transfer function of the network is given by

$$H(\omega) = \left[1 + \left(\frac{j\omega L}{R}\right)\right]^{-1}$$

$$|H(\omega)|^2 = \frac{1}{1 + \left(\dfrac{\omega L}{R}\right)^2}$$

The power Spectrum of the response of network is given by

$$S_{YY}(\omega) = S_{XX}(\omega) \; |H(\omega)|^2$$

$$= \frac{\dfrac{N_0}{2}}{1 + \left(\dfrac{\omega L}{R}\right)^2}$$

The average power in $Y(t)$ is given by

$$P_{YY} = \frac{1}{2\pi} \int\limits_{-\infty}^{\infty} S_{YY}(\omega) d\omega$$

$$= \frac{N_0}{4\pi} \int\limits_{-\infty}^{\infty} \frac{d\omega}{1 + \left(\dfrac{\omega L}{R}\right)^2}$$

$$= \frac{N_0 R}{4L} \qquad\qquad \text{after the use of indefinite integral}$$

As a check on the calculation of P_{YY}, we note that

$$h(t) = \left(\frac{R}{L}\right) \mu(t) \, e^{-Rt/L} \;\leftrightarrow\; H(\omega) = \frac{1}{1 + \left(\dfrac{j\omega L}{R}\right)}$$

for this network and using the result of example (8.16), we obtain

$$P_{YY} = \overline{Y}^2 = \left(\frac{N_0}{2}\right) \int\limits_{-\infty}^{\infty} \left(\frac{R}{L}\right)^2 e^{-2Rt/L} \; dt$$

$$= \frac{N_0 R}{4L} \qquad \text{the two powers are in agreement.}$$

8.13   Cross Power Density Spectrums of Input and Output

It is easily shown that the Fourier transforms of the cross correlation functions of (8.27) and (8.29) may be

$$S_{XY}(\omega) = S_{XX}(\omega) \; H(\omega) \qquad\qquad .....(8.41)$$

$$S_{YX}(\omega) = S_{XX}(\omega) \; H(-\omega) \qquad\qquad .....(8.42)$$

8.14   Band Pass, Band Limited and Narrow Band Processes

A random process $N(t)$ will be called 'band pass' if its Psd $S_{NN}(w)$ has its significant component confined in a bandwidth W (rad/s) that does not include $w = 0$

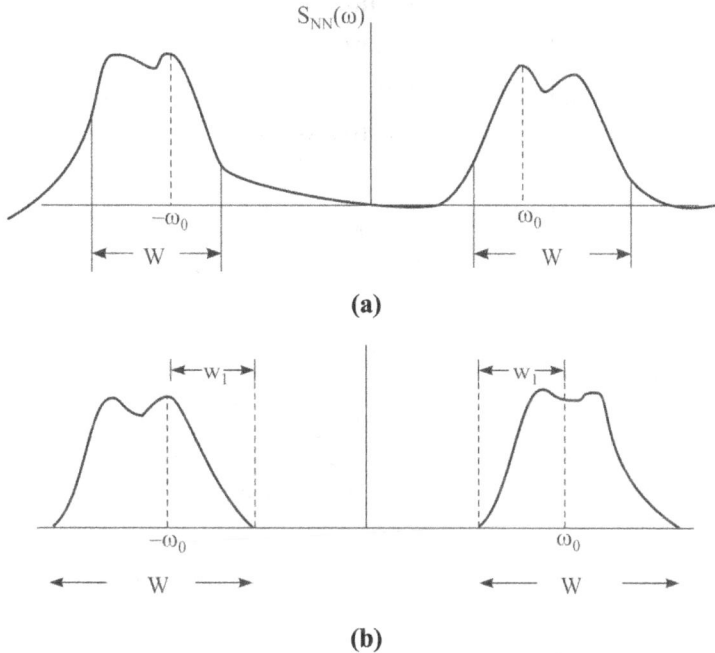

(a)

(b)

Fig. 8.3

Fig. 8.3 power density spectrum (a) for a band pass random process and (b) for a band limited band pass process

8.15 Band Limited Processes

If the power spectrum of a band pass random process is zero outside some frequency band of width W (rad/s) that does not include $w = 0$, then the process is called band limited.

The power spectrum of Band pass and Band limited processes are shown in Fig. 8.3.

8.16 Narrow Band Processes

Narrow band process is said to be narrow band if band width is small i.e., $\omega << \omega_o$ is chosen frequency near band center or near where the power spectrum is at its maximum. The power spectrum of narrow band process is shown in Fig. 8.4 (a).

A typical sample function might look like an arbitrary waveform as shown in Fig. 8.4(b)

We present a narrow band random process $N(t)$ in terms of envelope and phase components. Specifically we may write

$$N(t) = A(t) \; \cos\left[\omega_0 t + \phi(t)\right] \qquad \text{..... (8.43)}$$

Where A(t) is a random process representing the slowly varying amplitude and $\phi(t)$ is a process representing the slowly varying phase.

$A(t)$ and $\phi(t)$ have Rayleigh and Uniform (over 2M) first order probability density functions for Gaussian noise $N(t)$ and these processes are not statistically independent.

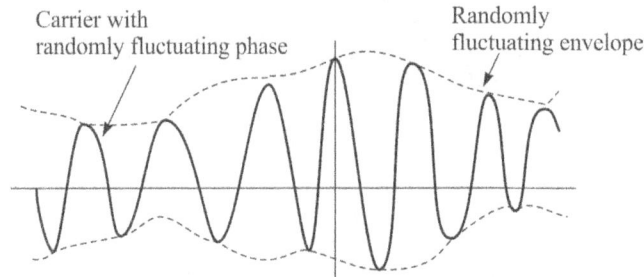

Fig. 8.4 (a) a power spectrum of narrow band random process $N(t)$

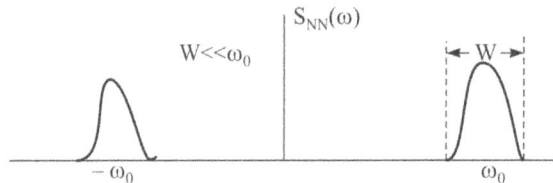

Fig. 8.4 (b) Typical ensemble member $n(t)$

Equation (8.43) is a general form for $N(t)$ and it is convenient to write the $N(t)$ in terms of in phase component $X(t)$ and quadrature component $Y(t)$ of the process $N(t)$

$$N(t) = X(t) \; \cos\left(\omega_0 t\right) - Y(t) \; \sin\left(\omega_0 t\right) \qquad \text{.....(8.44)}$$

Given the random process $N(t)$, we may extract the inphase component $X(t)$ and the quadrature component $Y(t)$ except for scaling factor using the averagement shown in Fig. 8.5.

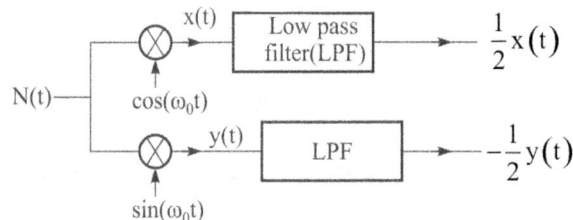

Fig. 8.5 Extraction of inphase and quadrature component of a narrow band process

The random processes A(t) and $\phi(t)$ are related to inphase component $X(t)$ and quadrature component $Y(t)$ of the process $N(t)$ as

$$A(t) = \sqrt{X^2(t) + Y^2(t)} \qquad\qquad .....(8.45)$$

$$\phi(t) = \tan^{-1}\left[\frac{Y(t)}{X(t)}\right] \qquad\qquad .....(8.46)$$

8.17 Properties of Band Limited Process

The expressions (8.43) and (8.44) are more in general and they can be applied to any band limited random process.

Consider any band limited wide sense stationary real random process $N(t)$ with zero mean and a power density spectrum that satisfies

$$S_{NN}(\omega) \begin{cases} \neq 0 & if\ 0 < \omega_0 - \omega_1 < \ |\omega| \ < \ \omega_0 - \omega_1 + \omega \\ = 0 & else\ where \end{cases} \qquad .....(8.47)$$

Where W_1 and W_2 are real positive constants.

The random processes $N(t)$ represented by right side of equation (8.44) where $X(t)$ and $Y(t)$ have the following properties.

1. $X(t)$ and $Y(t)$ are jointly wide sense stationary $\qquad\qquad .....(8.48)$
2. $X(t)$ and $Y(t)$ have zero means, i.e., $E\{X(t)\} = E\{Y(t)\} = 0 \qquad .....(8.49)$
3. $X(t)$ and $Y(t)$ have same powers $E\{X^2(t)\} = E\{Y^2(t)\} = E\{N^2(t)\} \qquad .....(8.50)$
4. $X(t)$ and $Y(t)$ have same Auto correlation and power spectrums

$$R_{YY}(\tau) = R_{XX}(\tau) \qquad\qquad .....(8.51)$$

$$S_{YY}(\omega) = S_{XX}(\omega) \qquad\qquad .....(8.52)$$

Where

$$R_{XX}(\tau) = \frac{1}{\pi}\int_0^\infty S_{NN}(\omega)\cos\big[(\omega - \omega_0)\tau\big]d\omega \qquad .....(8.53)$$

$$S_{XX}(\omega) = L_P\ \big[S_{NN}(\omega - \omega_0) + S_{NN}(\omega + \omega_0)\big] \qquad .....(8.54)$$

5. Random variables defined for the process $X(t)$ and $Y(t)$ are orthogonal

$$R_{XY}(0) = E\ \{\ X(t)Y(t)\ \} = 0\ , \quad R_{YX}(0) = 0 \qquad .....(8.55)$$

Where
$$R_{XY}(\tau) = \frac{1}{\pi} \int_0^\infty S_{NN}(\omega) \sin\big[(\omega - \omega_0)\tau\big] d\omega \qquad \text{.....(8.56)}$$

6) The cross correlation and cross power spectrum of random process are given by

$$R_{YX}(\tau) = -R_{XY}(\tau) \qquad R_{XY}(\tau) = -R_{XY}(-\tau) \qquad \text{.....(8.57)}$$

$$S_{XY}(\omega) = JL_P \big[S_{NN}(\omega - \omega_0) - S_{NN}(\omega + \omega_0)\big] \qquad \text{.....(8.58)}$$

If $N(t)$ has a power spectrum with components having even symmetry about $w = \pm w_0$, then $X(t)$ and $Y(t)$ will be orthogonal. As a consequence the cross power Spectrums of $X(t)$ and $Y(t)$ are zero.

8.18 Modeling of Noise Source

So far we discussed the finding the response of a linear system when a random signs was applied ad input. It is assumed that system is considered with out any noise source i.e free of any internally generate noise. In real world, this assumption is not justified because all systems generate noise internally. We begin the nearest section by developing models for noise source.

Noise Source

The various source of random noise are classified ad

  (i) External noise

  (ii) Internal noise

External Noise

This type of Noise generated outside the linear system which on predictable in nature and out of human control.

Atmospheric Noise

This type of noise (i) not a regular phenomenon and is caused by lightning discharges in thunder stores and other natural electric disturbance in atmosphere . This noise is also referred as static noise which becomes less sever at frequencies above 30 mz

Man made Noise

This noise source such as automobile ignitron electric motors and leakage from high voltage lines ete will come under man made noise.

Internal Noise

Internal Noise is created by the active and passive components present in the circuit itself.

Thermal noise (or) Resistor Noise

The thermal noise or Johnson noise is the random noise which is generated in a resistor on the resistive component of a complex impedance due to rapid and random motion of the molecules, atoms and electrons as a result of thermal energy.

The analysis of thermal noise is based on the kinetic theory which shows that the temperature of a particle   is a way of expressing its kinetic energy. This the noise power generated in a reason is proponranal to its absolute temperature and the bandwidth over which noise is misaimed

The expression for maximum noise power output of a reason is given as

$$P_n \propto T.B$$
$$P_n = KTB \tag{8.59}$$

where K is Batsmen constant

$$= 1.38 \times 10^{-23} \text{ joule / deg } .K$$

T is   absolute temperature

B is bandwidth of interest in H_z

8.19 Models of Noisy Resistor

The voltage / current models or equivalent circuit or a noisy resistor are shown as in Fig. 8.6.

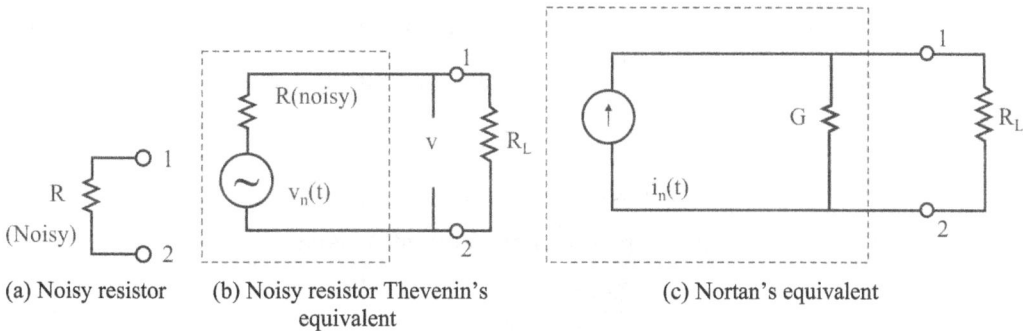

(a) Noisy resistor   (b) Noisy resistor Thevenin's equivalent   (c) Nortan's equivalent

Fig. 8.6

According to maximum power transfer theorem for maximum transfer of power from noise voltage source V_n to load resistor RL, we must have $RL=R$.

Then maximum noise power so transferred will be given

$$P_n = \frac{v^2}{RL}$$

But $\qquad R_L = R,$ so $P_n = \dfrac{v^2}{R}$

Applying voltage deviser method in Fig. 8.5 (b), we get

$$V = \frac{V_n}{2} \text{ so that } P_n = \frac{V^2}{R}$$

$$P_n = (V_n/2)^2 / R$$

$$= \frac{v_n^2}{4R}$$

$$Vn^2 = 4R\, Pn \qquad\qquad .....(8.60)$$

But we know that $P_n = KTB$

Putrefy was value of P_n in (5.2), we get

$$Vn^2 = 4R\,(KTB) = 4RKTB$$

$$Vn = Vn(t) = \sqrt{4KTBR} \qquad\qquad .....(8.61)$$

From equation 8.61, it is evident that of rms of noise voltage associated with a resistor is proportional to the absolute temperature T of the resistor, value R or the resistor and the bandwidth B over which noise is being measured. It may be noted the noise voltage is independent of the frequency at which it is measured.

Nortors there may be used to fine an equivalent current generator as known in figure 8.5.3(K). Using conductance $a = \dfrac{1}{R}$, the rms non current in for the current model of a noisy resistor will be expressed as

$$J_n = \sqrt{4GKTB} \qquad\qquad .....(8.62)$$

The power resistor spectrum of random noise current i) given by

$$S_i(\omega) = \frac{2KTG}{1 + (\frac{\omega}{d})^2} \qquad\qquad ..... (8.63)$$

α = the average number of conversions per second of an electron

G = Conductance of the resistor (μ)

T = the ambient temperature (O_K)

W = Radian frequency in rad /sec

The power density spectrum of current using themes noise is shown Fig. 8.7 as a function of $\left(\dfrac{\omega}{\alpha}\right)$

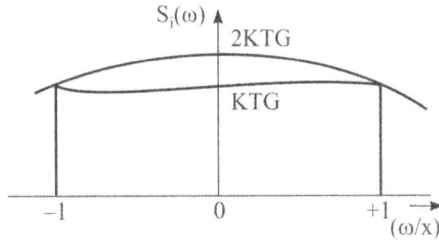

Fig. 8.7

Example

An amplifier operation over the frequency range from 18 to 20 mhz has a 10 kilo ohm input so this amplifier

$$Vn = \sqrt{4KTSFR}$$

$$= \sqrt{4x1.38x10^{-23}(27+273)x(20-18)10^6 x10^4}$$

$$= 18.2 \ \mu V$$

8.20 Arbitrary Noise Sources, Effective Noise Temperature

8.21 Series combination of resistor at same temperature

Here we assume that two resistor R_1 and R_2 both at the same temperature 7, connected in series as shown in Fig. 8.8.

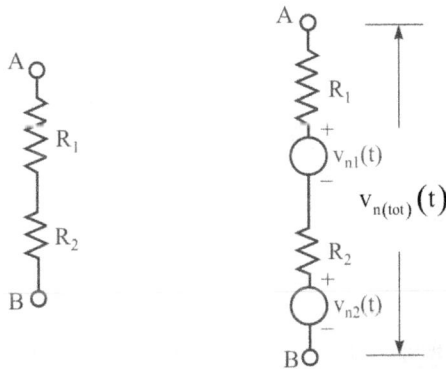

(a) series combination of noisy resistor　　(b) Thevenin's equivalent

Fig. 8.8

Fig 8.8 noise in resistor connected in series

$$\mu_{n_1} = \sqrt{4KTSFR_1} \qquad \qquad .....(8.64)$$

$$\mu_{n_2} \sqrt{4KTSFR_2} \qquad \qquad .....(8.65)$$

The sum of two such rms voltage in series i) given by the square root of the sum of their square, so that we get the total noise voltage a known the series combination of R_1 and R_2

$$Vn_{(tot)}(t) = \sqrt{V_{n_1}^{~2} + V_{n_2}^{~2}}$$

$$= \sqrt{4KTSFR_{tot}} \qquad \qquad .....(8.66)$$

Where $R_{tot} = R_1 + R_2$

Similarily the rms short current i) given by

$$In_{tot}~(t) = \sqrt{4KTBG} \qquad \qquad .....(8.66)$$

Where $$G = \frac{1}{R_{tot}} = \frac{1}{R_1 + R_2}$$

8.22   Series Combination of Resistor at Different Temperature

Let us consider two resistors R_1 and R_2 are connected in series and they are at different temperature i.e., T_1 and T_2. The noisy resistor R_1 and R_2 are replaced by their Theremins equivalent models as shown in Fig. 8.9.

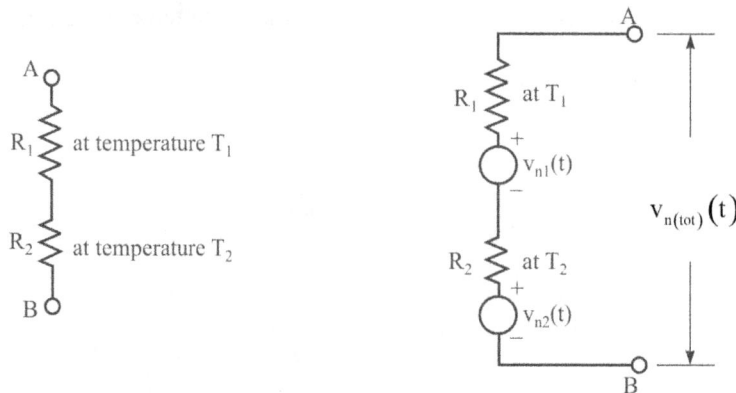

(a) Noisy resistors at different temperature   (b) Thevenin's model of Fig (a) R_1 R_2 are noiseless resistors

Fig. 8.9

The mean square noise voltage is given by

$$V_{2_{tot}}^2(t) = V_{n_1}^2(t) \quad + \quad V_{n_2}^2(t)$$

$$= 4\,kT_1BR_1 + 4\,kT_2\,B\,R_2$$

$$= 4kB\,(T_1R_1 \;+\; T_2R_{2)} \qquad\qquad .....(8.67)$$

We know that $\qquad V_{n\,tot}^2\;(t) \quad = \quad 4kB\,TS\,R_{tot} \qquad\qquad .....(8.68)$

Where T_s is effective noise temperature of the some by substrata ex (8.68) in Ex (8.6.5), we get $4kBTSR_{tot} = 4kB\,(T_1R_1 + T_2R_2)$

The effective temperature of noise some (i) given by $T_s = \dfrac{(T1R1 + T2R2)}{R_{tot}} \qquad .....(8.69)$

Where $\qquad R_{ot} = R_1 \;+\; R_2$

$$T_s = \frac{(T1R1 + T2R2)}{(R_1 + R_2)} \qquad\qquad .....(8.70)$$

The effective noise temperature not only dependence on individual noise temperature but also on resistors. Hence it shown that the effective noise temperature of a some is not necessary equal to its physical temperature

8.23 Network with Reactive Elements

Let us consider a network consisting of a combination of resistors, capacitors and inductors as shown in Fig. 8.10.

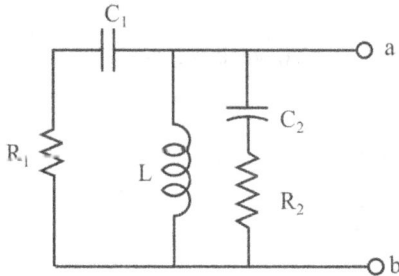

(a) Network consisting a combination of resistors, capacitors   (b) Thevenin's model of Fig. (a)

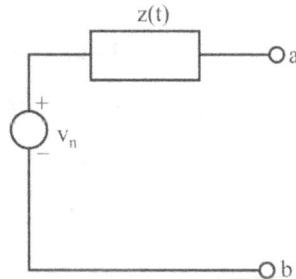

Fig. 8.10

8.24 Combination of Resistors, Capacitors and Inductors

The network in Fig. 8.10 (a) i) replaced by its Thevenin's equivalent circuit with equivalent impedance z(f) and noise voltage $V_n(t)$ as shown in Fig. 8.10(b).

$Z(f) = R(f) + J X(t)$ where $R(f)$ is resistive real component of the impedance and $X(t)$ is the reaction (imaginary) Component.

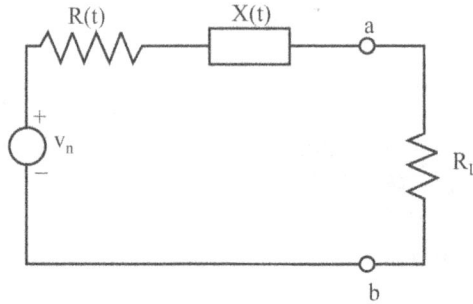

Fig. 8.10(c) Equivalent circuit of Fig. (b)

Let us consider a load resistor RL the terminals (a,b) as shown in Fig. 8.10(a). now we replace the noisy resistor RL by its Thevenin's equivalent as shown in Fig. 8.10 (d).

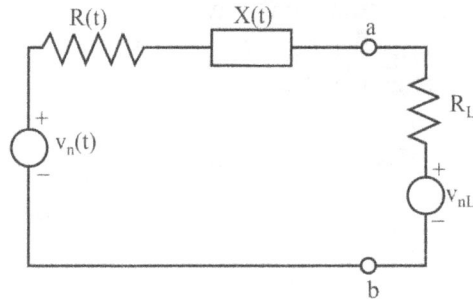

Fig. 8.10(d) R_L in Fig. (d) is replaced by its Thevenin's model

The effective impedance Zeff is given by

$$Zeff(t) = R(f) + JX(t) + R_L \qquad .....(8.71)$$

$$|Zeff(t)| = \sqrt{(R(t) + R_L)^2 + x^2(t)} \qquad .....(8.72)$$

The rms current flowing through the circuit due to $V_{n_2}(t)$ is $In(t)$ and is given by

$$In(t) = \frac{v_n(t)}{|zeff|} = \frac{V_n}{\sqrt{(r(t) + R_L)^2 + x^2(t)}} \qquad .....(8.73)$$

Similarly the rms current flowing through the circuit due to $V_{n_2}(t)$ i) given by

$$i_{n_L}(t) = \frac{V_{n_L}(t)}{|zeff|} = \frac{V_{nL}}{\sqrt{(R(t)+R_L)^2 + x^2(t)}} \quad .....(8.74)$$

From the condition that there be no net transfer of power we have power delivered by

$$Z(t) \text{ to } R_L = i_n{}^2 R_L = \frac{V_n{}^2 x R_L}{(R(t)+R_L)^2 + x^2(t)} \quad .....(8.75)$$

Power delivered by R_L to $X(t)$

$$= i_{n_L}{}^2 R(t) = \frac{V_{nL}{}^2 x R_L}{(R(t)+R_L)^2 \times x^2(t)} \quad ..... (8.76)$$

Equating the equations (8.75) and (8.76), we get

$$\frac{V_n{}^2 R_L}{(R(t)+R_L)^2 + x^2 x^2(t)} = \frac{V_{nL}{}^2 \ R(t)}{(R(t)+R_L)^2 + x^2(t)}$$

$$V_n{}^2 \ R_L = V_{nL}{}^2 \ R(t) \quad .....(8.77)$$

But rms value of noise voltage V_{nL} of pure resistors R_L is given by

$$V_{nL}(t) = 4kTB \ R_L \quad .....(8.78)$$

Substituting equation (8.78) in (8.77), we get

$$V_n{}^2(t) R_L = 4kTB \ R_L \ R(t)$$

$$V_n{}^2(t) = 4kTB \ R(t)$$

$$V_n(t) = \sqrt{4kTB \ R(t)} \quad .....(8.79)$$

The rms of noise voltage $V_n(t)$ of the reactive network is equal to rms value of those developed by its resistive part only.

8.25   Noisy two port Network (Amplifier)

Amplifier is used to amplify the input signal but it also amplifies the noise generated by the signal source. In addition to this, additional noise is generated in the amplifies. All these noises are added to the amplified signal that is delivered to the load. Hence the noise in the output of the amplifies is contributed by the signal as well as by the amplifies.

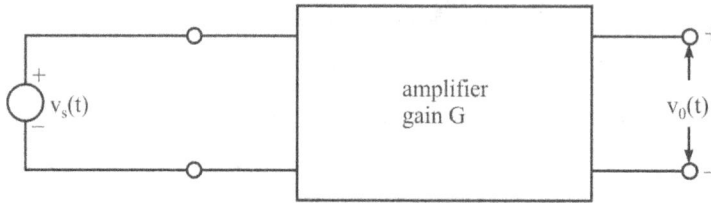

Fig. 8.11 single stage Amplifies

In Fig. 8.11,

$$V_i(t) = V_{si}(t) \pm V_{n\,i}(t) \qquad \qquad \text{.....(8.80)}$$

where $V_{si}(t)$ is the signal input and $V_{n\,i}$ is the noise input.

Now

$$V_o(t) = G[V_i(t) \pm V_{na}(t)]$$

$$= G[V_{si}(t) \pm V_{ni}(t)] + V_{na}(t)$$

$$= G[V_{si}(t)] + V_{no}(t) \qquad \qquad \text{.....(8.81)}$$

Where $V_{no}(t) =$ amplifiers output noise voltage

$$= G[V_{ni}(t)] + V_{na}(t)$$

$V_{na}(t) =$ noise voltage generated in the amplifies

If we neglect the noise generated in the amplifies, we have

$$V_{no}(t) = G[V_{ni}(t)]$$

$$V_o(t) = = G[V_{si}(t) + V_{ni}(t)] \qquad \qquad \text{.....(8.82)}$$

8.26 Available Noise Power and Available Noise Power Spectral density

Available power of a source is defined as the maximum power that can be drawn from the source.

Consider a linear noise free two pork network having a source impedance Z with a load impendence Z across the output terminals as shown in Fig. 8.12.

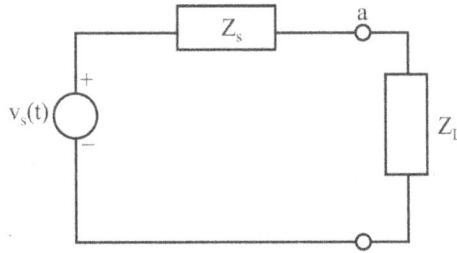

Fig. 8.12

Let $Z_s = R_s + J \times S$

$$Z_L = R_L + J \times L \qquad \qquad .....(8.83)$$

Maximum power transfer theorem states that the maximum power is drawn from source when the load impedance is complex conjugate of some impedance

i.e., $Z_L = Z_s^*$

$$R_L + J \times L = \left(R_s + J \times S \right)^* = \left(R_s - J \times S \right)$$

i.e., $R_L = R_j, \ X_L = X_j$ $\qquad \qquad .....(8.84)$

Under maximum power transfer conditions, no reactive part exists and total resistance is

$$R_j + R_L = 2 R_j$$

$i(t) =$ current flowing the circuit

$$= \frac{V_s(t)}{2R_j} \qquad \qquad .....(8.85)$$

And power across the load is P_0,

$$P_0 = i^2(t) R_L$$

$$= \left(\frac{V_s(t)}{R_j + R_L} \right)^2 R_L$$

$$= \left(\frac{V_s(t)}{2R_j} \right)^2 R_L$$

$$= \frac{V_s^2(t)}{4R_j^2} R_L$$

$$P_0 \approx \frac{V_s^2(t)}{4R_j} \qquad \because R_j = R_L \qquad\qquad .....(8.86)$$

From each equation (8.86), it is noted that the available power depends only on resistive component of source impedance.

Incremental Modeling of Noisy Networks

8.27 Available Power Gain

Consider linear noise free two port network heavy an input impedance Z_i when the output port is open circuited.

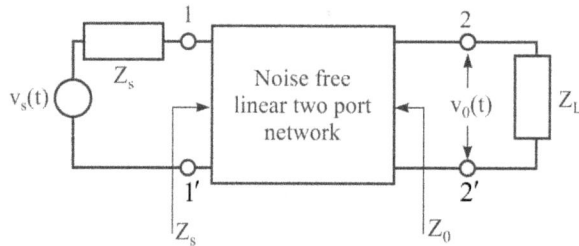

Fig. 8.13 Two port network with source na load convected

Z_0 is the output impedance by looking back in to the output port when source with Z_s is convercted

Available power at the input port of the network is

$$P_{ai} = \frac{\overline{V_s^2(t)}}{4R_j} \qquad\qquad .....(8.87)$$

Similarly, available power at the output due to the source is

$$P_{a0} = \frac{\overline{V_s^2(t)}}{4R_0} \qquad\qquad .....(8.88)$$

Available power gain G_a is defined as

$$G_a = \frac{Available\ output\quad power\,(P_{a0})}{Available\ input\quad power\,(P_{ai})}$$

$$= \frac{\overline{V_0^2(t)}}{\overline{V_j^2(t)}} \times \frac{R_s}{R_0} \qquad\qquad .....(8.89)$$

Available power gain for a two port networks can be also defined as

$$G_a = \frac{Available\ output\quad power\ spectral\ density}{Available\ input\quad power\ spectral\ density}$$

$$= \frac{S_0(\omega)}{S_i(\omega)} \quad\quad\quad\quad .....(8.90)$$

Available power Gain is useful in determining the output power. E.g., if the available source power spectral density at the input port is $S_i(w)$ and gain of the network is $G_a(w)$, the available output power spectral density is given by the product of inpt power spectral density and Gain

i.e. $\quad\quad S_0(\omega) = S_i(\omega) \times G_a(\omega) \quad\quad\quad\quad .....(8.91)$

Available power is

$$P_{a0} = \int_{-a}^{a} S_0(\omega)d\omega = \int_{-\infty}^{\infty} S_i(\omega)G_a(\omega)d\omega \quad\quad .....(8.92)$$

If the source is a thermal noise generation, then

$$S_N(\omega) = \frac{d(P_n)}{d\omega} = \frac{d(KTB)}{d\omega} = KT \quad\quad\quad .....(8.93)$$

The two sided available noise power spectral density is given by

$$S_i(\omega) = \frac{KT}{2} \quad\quad\quad\quad .....(8.94)$$

And available output noise power

$$P_{a0} = \frac{1}{2\pi} \frac{KT}{2} \int_{-\infty}^{\infty} G_a(\omega)d\omega \quad\quad\quad .....(8.95)$$

8.28 Equivalent Networks, Effective Input Noise Temperature

The noise temperature is referred at the input of a two port network which accounts, for the intervals noise produced by the network and thereafter the network is considered to be noise free.

Consider a linear two port network as shown in the Fig. 8.14.

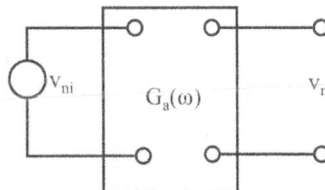

Fig. 8.14

The input is white noise source U_{ni} with the available power density $S_{ni}{}^{/}(\omega) = \dfrac{KT_s}{2}$,

let $G_a(\omega)$ be the available power gain of the network. If the network is noise free, then available output power density is given by

$$S_{no}{}^{/}(\omega) = G_a(\omega)\,\frac{KT_s}{2} \qquad\qquad .....(8.96)$$

The $S_{no}(\omega)$ at the output of a noisy network, is higher than $S_{no}{}^{/}(\omega)$ of a noisy free network. The increase in noise power density may be accounted interms of noise temperature at the input and the network may be considered noise free.

The input noise temperature of the network which accounts for the interval none generated by the network is known as effective input noise temperature denoted by T_c , which depends on the source as well as on network.

The output power density $S_{no}(\omega)$ is given by

$$S_{no}(\omega) = G_a(\omega)\,\frac{KT_s}{2} + G_a(\omega)\,\frac{KT_e}{2}$$

$$= (T_s + T_e)\frac{K}{2}G_a(\omega) \qquad\qquad .....(8.97)$$

Where T_s = noise temperature of the source

T_e = noise temperature of the network generated noise

For noise less system $T_e = 0$ then

$$S_{no}(\omega) = = G_a(\omega)\,\frac{KT_s}{2} \qquad\qquad .....(8.98)$$

8.29   Noise Figure

Noise figure gives the amount of noise internally generated by the system and is defined as the ratio of power density of the available noise at the output of the network to the power density at the output only due to the input noise source. Which gives the measure of the system performance of the noise

$$F = \frac{S_{no}(\omega)}{S_{no}{}^{/}(\omega)} = \frac{S_{no}{}^{/}(\omega) + S_{no}{}^{//}(\omega)}{S_{no}{}^{/}(\omega)}$$

$$= 1 + \frac{S_{no}{}^{//}(\omega)}{S_{no}{}^{/}(\omega)} \qquad\qquad .....(8.99)$$

For a noiseless system $S_{n0}''(\omega) = 0$

The $F = 1$ .....(8.100)

If $F > 1$ then, the system is noisy.

Noise figure interms of Available power Gain

$$F = \frac{S_{n0}(\omega)}{S_{n0}'(\omega)} \qquad .....(8.101)$$

$$S_{n0}(\omega) = G_a(\omega)\, S_{n0}'(\omega) \qquad .....(\,8.102)$$

$$= G_a(\omega)\, \frac{KT_s}{2}$$

Therefore $\qquad F = \dfrac{S_{n0}(\omega)}{G_a(\omega)\, \dfrac{KT_s}{2}}$

$$S_{n0}(\omega) = G_a(\omega)F\, \frac{KT_s}{2} \qquad .....(\,8.103)$$

Noise figure in terms of effective input temperature

$$S_{n0}(\omega) = G_a(\omega) \times S_{ni}(\omega) \qquad .....(\,8.104)$$

$$S_{ni}(\omega) = S_{ni}'(\omega) + S_{ni}''(\omega) \qquad .....(\,8.105)$$

$$S_{ni}(\omega) = G_a(\omega)\, \frac{KT_s}{2} + G_a(\omega)\, \frac{KT_e}{2}$$

$$= G_a(\omega)K\left(\frac{T_s + T_e}{2}\right) \qquad .....(\,8.106)$$

Comparing equation (8.12.4) and (8.12.7), we get

$$G_a(\omega)F\, \frac{KT_s}{2} = G_a(\omega)K\left(\frac{T_s + T_e}{2}\right)$$

$$F = \left(\frac{T_s + T_e}{T_s}\right) \qquad .....(8.107)$$

$$= 1 + \frac{T_e}{T_s}$$

This implies $T_e = (F-1)T_s$ .....(8.108)

F *in* as $= 10 \, log_{10} \, F$

For noiseless system, noise figure is $F = UdB$

8.30   Noise Figure in Terms of Signal to Noise Ratio

We know that signal power spectral density at the output

$$S_{n0}(\omega) = G_a(\omega) \times S_{si}(\omega)$$

$$S_{n0}{}'(\omega) = G_a(\omega) \times S_{ni}(\omega)$$

From equation (8.101), we get

$$S_{n0}(\omega) = F \times S_{n0}{}'(\omega)$$

$$= F \times G_a(\omega) \, S_{ni}(\omega)$$

$$= \frac{S_{n0}(\omega)}{S_{si}(\omega)} F \, S_{ni}(\omega) \qquad \qquad .....(8.109)$$

From equation (8.109), we get noise figure as

$$F = \frac{S_{si}(\omega) \Big/ S_{ni}(\omega)}{S_{\delta0}(\omega) \Big/ S_{si}(\omega)} \qquad \qquad .....(8.110)$$

Noise figure is defined as signal to noise ratio of input to signal to noise ratio of the output

If $\left(S/N \right) i = \left(S/N \right)$ 0 then F = 1 for ideal system

$\left(S/N \right)$ is always greater than $\left(S/N \right)$ 0 then F is always greater than 1.

8.31   Modeling of Practical Noisy Networks

In the previous sections, we have defined the noise figure which is constant. But in practical cases, the noise figure depends on the frequency. So the average noise figure of the system should be considered.

8.32 Average Noise Figure

The average noise figure F is defined as the ratio of the total output available noise power Nao from the network divided by the total output available noise power Naos due to the source alone. Thus

$$\overline{F} = \frac{Na_0}{Na_0 s} \qquad \qquad .....(8.111)$$

$Na_0 s$ is found by integration of

$$Na_0 s = = \frac{k}{2\pi} \int_0^\infty T_s G_a(\omega) d\omega \qquad \qquad .....(8.112)$$

From equation (8.111), $Na_0 = \overline{F} \; Na_0 s$

$$Na_0 s = = \frac{k}{2\pi} \int_0^\infty \overline{F} \; T_s G_a(\omega) d\omega \qquad \qquad .....(8.113)$$

We write average noise figure $\overline{F}$ as

$$\overline{F} = \frac{\int_0^\infty \overline{F} \, T_s G_a d\omega}{\int_0^\infty T_s G_a d\omega} \qquad \qquad .....(8.114)$$

In many cases, the source temperature is approximately constant. The average noise figure $\overline{F}$ then becomes

$$\overline{F} = \frac{\int_0^\infty \overline{F} \, G_a d\omega}{\int_0^\infty G_a d\omega} \qquad \qquad .....(8.11.5)$$

8.33 Output Noise Power and System Noise Power

The output noise power is equal to the sum of the noise power available at the output due ot external noise and noise power of the system

$$V_0 = G_a(\omega) N_i + N_{Syst} \qquad \qquad .....(8.116)$$

$$F = \frac{S_i/N_i}{S_0/N_0} = \frac{N_0}{G_a(\omega)N_i}$$

$$V_0 = G_a(\omega)\,N_i F \qquad\qquad .....(8.117)$$

Equating (8.116) and (8.117) we get

$$G_a(\omega)\,N_i F = G_a(\omega)\,N_i + N_{Syst}$$

Noise power of the system is

$$N_{sys} = G_a(\omega)\,N_i(F-1) \qquad\qquad .....(8.118)$$

8.34 Noise is Cascade Amplifiers

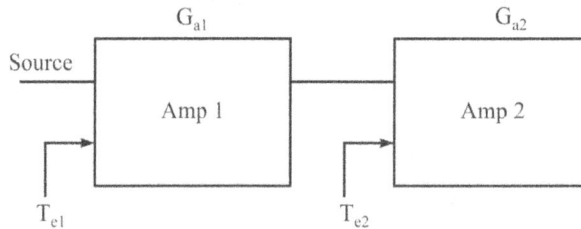

Fig. 8.15

Consider two amplifiers are cascaded as shown in Fig. 8.14 where

$Ga1$ - Power gain of amplifier 1

$Ga2$ - power gain of amplifier 2

$Te1$ - Equivalent noise input temperature of amplifier 1

$Te2$ - Equivalent noise input temperature of amplifier 2

$F1$ - noise figure of amplifier 1

$F2$ - noise figure of amplifier 2

The overall gain $G_a = G_{a1} \times G_{a2}$ $\qquad\qquad .....(8.119)$

The output noise power is given by

$$N_0 = N_{01} + N_{02} + N_{03} \qquad\qquad .....(8.120)$$

N_{01} = output noise power due to input noise power N_i

$$= G_{a1}\,G_{a2}\,N_i$$

$$= G_a\,N_i \qquad\qquad .....(8.121)$$

N_{02} = output Noise power due to noise power generated internally by first amplifier

From equation (8.118),

$$N_{syst} = G_{a1} \ N_i(F_1 - 1)G_{a2}$$

$$\therefore \qquad N_{02} = G_{a1} \ N_i(F_1 - 1) \qquad\qquad .....(8.122)$$

N_{03} = output Noise power due to noise power generated by second amplifiers

$$= G_{a2} \ N_i(F_2 - 1) \qquad\qquad .....(8.123)$$

Overall Noise figure is

$$\therefore \qquad N_{02} = G_a \ N_i F \qquad\qquad .....(8.124)$$

Now

$$N_0 = N_{01} + N_{02} + N_{03}$$

$$G_a \ N_i F = G_a \ N_i + G_{a1} \ N_i(F_1 - 1) + G_{a2} \ N_i(F_2 - 1)$$

$$F = 1 + (F_1 - 1) + (F_2 - 1)\frac{G_{a2}}{G_{a1}G_{a2}} \qquad\qquad .....(8.125)$$

$$F = F_1 + (F_2 - 1)\frac{1}{G_{a1}} \qquad\qquad .....(8.126)$$

For N amplifiers, we can write F as

$$F = F_1 + \frac{F_2 - 1}{G_{a1}} + \frac{F_3 - 1}{G_{a1}G_{a2}} + \frac{F_4 - 1}{G_{a1}G_{a2}G_{a3}} + \cdots + \frac{F_n - 1}{G_{a1}G_{a2}\cdots G_{an-1}}$$

This equation is called Friss formula. It shows that the contribution to overall noise figure is mainly by the First stage.

8.35 Noise Figure in Terms of Noise Temperatures

We know that $\qquad\qquad F = 1 + \frac{T_e}{T} \qquad\qquad .....(8.127)$

$$F_1 = 1 + \frac{T_{e1}}{T}$$

$$F_2 = 1 + \frac{T_{e2}}{T}$$

The overall Noise figure is $F = F_1 + (F_2 - 1)\dfrac{1}{G_{a1}}$

$$1 + \frac{T_e}{T} = 1 + \frac{T_{e1}}{T} + 1 + \frac{T_{e2}}{Ga_1 T} \qquad \text{.....(8.128)}$$

$$T_e = T_{e1} + \frac{T_{e2}}{Ga_1} \qquad \text{.....(8.129)}$$

For N amplifiers

$$T_e = T_{e1} + \frac{T_{e2}}{Ga_1} + \frac{T_{e3}}{Ga_1 Ga_2} + \cdots + + \frac{T_{en}}{Ga_1 Ga_1 \cdots Ga_{n-1}} \qquad \text{.....(8.130)}$$

Additional Examples

Example: **1** An amplifier operating over the frequency require of 450 to 460 khz has a 230 kg input resistor. what is the rms voltage at the input to this amplifier if the ambient l\temperature is 17^0 c

Solution

Bandwidth $= (460 - 455 \ ) KHz \ = 10 \ KHz \ = 104$ given

$$R = 20 \ kr$$

$$T = 17^0 c = (273 + 17 \ ) k = 290^0 \ k$$

Rms noise voltage $Vn = \sqrt{4KTBR}$

$$= Vn = \sqrt{4 \times 1.38 \times 10^{-23} \times 290 \times 10^4 \times 23 \times 10^4}$$

$$= 6.06 \ \mu \ v$$

Example: **2.** A receiver connected to an antenna whose resistance is $60r$ has an equivalent nose resistance of $30r$. Calendar the noise figure in QB and its equivalent noise resistance.

Solution

Given equivalent noise resistance of the receives $R_{eq.} = 30 \ r$

Antenna resistance $R_a = 60 \ r$

Noise figure F_a is given by

$$F_a = 1 + \frac{Req}{Ra}$$

$$= 1 + \frac{30}{60} = 1.5$$

F_a in dB $= 10 \ log_{10} \ 1.5$

$$= 1.75 \ dB$$

Equivalent noise temperature D

$$T_e = T_o (F_a - 1)$$
$$= 290 (1.5 - 1)$$
$$= 145K$$

Example: 3 The noise present at the output of a two part network is 1μ w. The noise figure of the network D 0.8 dB and its gain is 10^{10}. Calculate (i) Available noise power contributed by two port (ii) Available output noise power

Solution

Noise due to two port is

$$N_{tp} = g_a (F - 1) N_i$$

We have $10 \log F = 0.8$; $F = 1. 202$

$$N_{tp} = 10^{10} (1.202 - 1) \times 1 \times 10^{-6}$$
$$= 2.02 \text{ kilo wats}$$

(ii) We have noise figure

$$F = \frac{1}{g_a} \frac{N_0}{N_i}$$

Available Noise figure

$$N_0 = g_a F N_i$$
$$N_0 = 10^{10} \times 1.202 \times 10^{-6}$$
$$= 12.02 \text{ kilo wats}$$

Example: 4 An antenna has noise temp tan $t = 6^0$ k. Which is connected to a receiver that is having noise temperature $T_e = 100^0$ k. The midband available gain of the receiver is 10^{10} and it can be represented by a parallel *RLC* filter having $3dB$ band width of 10 *MHz*. Find the noise bandwidth and available output noise power

Solution

For Parallel RLC filter, we have the transfer function $H(f)$ as

$$H(f) = \frac{1}{\dfrac{1}{R} + j\omega c + \dfrac{1}{j\omega L}}$$

$$|H(f)|^2 = \left| \frac{1}{\dfrac{1}{R} + J\omega C + \dfrac{1}{J\omega l}} \right|$$

$$= \left| \frac{R}{1 + J\omega RC + \dfrac{R(\pi c)}{J(\omega L)2\pi c}} \right|^2$$

We have 3 *dB* bandwidth of *RLC* filter

$$B = \frac{1}{2\pi RC} \text{ and center frequency to } \frac{1}{2\pi\sqrt{LC}}$$

$$|H(f)|^2 = \left| \frac{R}{1 + \dfrac{Jf}{B} - \dfrac{Jf_0^{\;2}}{}} \right|^2$$

$$= \left| \frac{R}{1 + \dfrac{jf}{B}\left[1 - \left(\dfrac{f_o}{f}\right)^2\right]} \right|_2$$

$$= \frac{R^2 B^2 f^2}{f^4 - (2f_0^{\;2} - B^2)\;f_0^{\;2} + fo^4} \qquad .....(1)$$

Noise band width $=$ $\quad B = \dfrac{1}{|H(f_o)|_2} \displaystyle\int_0^\infty |H(f)|^2\, df$

Substituting $|H(f)|^2$ expression in the above integral, we get

$$B = \frac{R^2 B^2 f_o^{\;2}}{f_o^{\;4} - (2f_o^{\;2} - B^2)f_o^{\;2} + f_o^{\;4}}$$

$$= \frac{R^2 B^2 f_o^{\;2}}{f_o^{\;2} B^2} = R^2$$

$$\therefore \qquad B_N = \frac{1}{R^2}\int_0^\infty \frac{B^2 R^2 f^2}{f^4 - \left(2f_o^2 - B^2\right)f^2 + f_o^4}dt$$

$$= \frac{1}{R^2}\left(\pi R^2 B\right) = \pi B.$$

$$= \pi B.$$

Noise band width of the receiver

$$B_N = \pi B.$$

$$B_N = \pi \times 10 \times 10^6$$

$$= 31.42 \text{ MHz}$$

Available output noise power Pa_0 is

$$Pa_0 = Ga_0 KT_e B_N$$

$$T_e = T_{ant} + T_{eq}$$

$$= 100 + 6$$

$$= 106$$

$$Pa_0 = 10^{10} \times 1.38 \times 10^{-23} \times 106 \times 31.42 \times 10^6$$

$$= 1.38 \times 106 \times 31.42 \times 10^{-7}$$

$$= 459.6 \, \mu \text{ watts}$$

***Example:* 5** If two resistors R_1=600 r and R_2= 300 r at temperatures T_1=40^{0}c and T_e=100^{0}c are connected in parallel. Find the noise temperature of the combination.

Solution

If two resistors are connected in parallel, then equivalent noise temperature is given by

$$T_n = \frac{T_1 R_2 + T_2 R_1}{R_1 + R_2}$$

$$T_1 = 40^0 c = 40 + 270 = 310^0 K$$

$$T_2 = 100^0 c = 100 + 270 = 370^0 K$$

$$T_n = \frac{310 \times 300 + 370 \times 600}{900}$$

$$= \frac{315000}{900} = 350^0 K$$

$$= 80^0 c$$

Example 6: The first stage of a two stage amplifier has a voltage fain of 15, a 500 *r* input resistor, a 1200 *r* equivalent noise resistance and a 25 *kr* output resistor. For the second stage, these values are 30, 75 *kr*, 13 *kr* and 1 *Mega ohm* respectively. Calculate the input noise resistance of this two stage amplifier.

Solution

$$R_1 = 500 + 1200 = 1700r$$

$$R_2 = \frac{25 \times 75}{25 + 75} + 13$$

$$= 18.75 + 13$$

$$= 31.75$$

$$R_3 = 1Mr$$

$$R_{eq} = R_1 + \frac{R_2}{G_1^2} + \frac{R_3}{G_1^2 G_2^2}$$

$$R_{eq} = 1700 + \frac{31750}{15^2} + \frac{1,000,111}{15^2 \times 30^2}$$

$$= 1700 + 141.1 + 4.94$$

$$= 1846.04r$$

Example 7: Find the rms value of the noise voltage at $27^0 c$ developed across the capacitor terminal of the circuit in figure below.

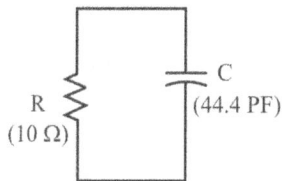

Fig.(a)

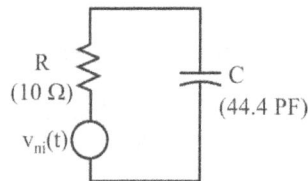

Fig.(b)

Solution

The only noise source is the circuit is resistors *R* and hence it can be replaced by an input noise voltage source $V_{ni}(t)$ and noiseless resistor R as shown in figure (*b*).

The power spectrum density $S_{ni}(\omega)$ is given by

$$S_{ni}(\omega) = 2KTR$$

The transfer function $H(w)$ is given by

$$H(\omega) = \frac{\dfrac{1}{j\omega c}}{R + \dfrac{1}{j\omega c}}$$

$$= \frac{1}{1 + j\omega c}$$

$$= \frac{1 - j\omega cR}{1 + \omega^2 R^2 c^2}$$

And the magnitude $|H(w)|$ is given by

$$|H(\omega)| = \frac{\sqrt{1 + \omega^2 c^2 R^2}}{1 + \omega^2 c^2 R^2}$$

$$= \frac{1}{\sqrt{1 + \omega^2 c^2 R^2}}$$

The power density spectrum of the output noise voltage $V_{n0}(t)$ is given by

$$S_{n0}(\omega) = |H(\omega)|^2 \, S_{ni}(w)$$

$$= \frac{1}{1 + \omega^2 c^2 R^2} \, 2KTR$$

$$= \frac{2KTR}{1 + \omega^2 c^2 R^2}$$

The average power of mean square value of the output noise voltage $V_{n0}(t)$

$$P_0 = \overline{V_{n0}^2}$$

$$= \frac{2}{\pi} \int\limits_{-\infty}^{\infty} S_{n0}(\omega) d\omega$$

$$= \frac{2}{\pi} \int\limits_{-\infty}^{\infty} \frac{2KTR}{1 + \omega^2 c^2 R^2} d\omega$$

$$= \frac{2KTR}{\pi} \int\limits_{0}^{\infty} \frac{1}{1 + \omega^2 c^2 R^2} d\omega$$

$$\overline{V_{n0}}^2 = \frac{2KTR}{\pi RC}\left[\tan^{-1}\omega RC\right]_0^\infty$$

$$= \frac{2KTR}{\pi RC}\left(\frac{\pi}{2}\right)$$

$$= \frac{KT}{C}$$

$$V_{rms} = \sqrt{\overline{V_{n0}}^2} = \sqrt{\frac{KT}{C}}$$

$$K = 1.38\times10^{-23}\,J/^0k; \quad T = 273 + 27 = 300^0k; \quad c = 44.4Pj$$

$$V_{rms} = \sqrt{\overline{V_{n0}}^2} = \sqrt{\frac{KT}{C}}$$

$$= \frac{1.38\times10^{-23}\times300}{44.4\times10^{-12}}$$

$$= 9.66\mu v$$

Example 8: The noise output of a resistor is amplified by a noiseless amplifier having a gain of 50 and a bandwidth of 20KHz. A resistor is convected to the output of the amplifier reads 1*m volt rms*.

(i) The bandwidth of the amplifier is reduced to 5 *KHz* its gain remaining constant. What does the meter read now?

(ii) If the resistor is operated at 90^0c, what is its resistance?

Solution

(i) Since the bandwidth is reduced to $1/4^{th}$ of its value, therefore the noise voltage will be

$$V_n \ \alpha \ \sqrt{B}$$

Noise voltage at 5 *KHz* will become half its value at 20 *KHz* bandwidth

$$V_n \ = \ 0.5m \ volt$$

(ii) Now $V_n = \sqrt{2KTBR}$

$$V_{n0} = \frac{10^{-3}}{50} \ volt,$$

$$B = 20 \ KHz, \qquad T = 98c \ = 90 + 273 = 363^0c$$

$$V_n = \sqrt{4KTBR}$$

$$= \sqrt{4 \times 1.38 \times 10^{-23} \times 363 \times 20 \times 10^3 \times R}$$

$$\frac{10^{-3}}{50} = \sqrt{4 \times 1.38 \times 10^{-23} \times 363 \times 20 \times 10^3 \times R}$$

After simplification, we get

$$R = 998.12 kr$$

Example 9: In *TV* receiver, the antenna is often mounted on a tall mask and a long lossy cable is used to connect the antenna and receiver. To over cover the effect of noisy cable, a preamplifier is mounted on the antenna. The parameters of the direct stages are Preamplifier gain = 20 *dB*; Preamplifier noise figure = 6 *dB*; Lossy cable noisy figure = 3*dB* Cable loss = –20 *dB*; Receiver front end gain = 60 *dB*; Receiver noise figure = 16 *dB*. Determine the overall noise figure of the system

Solution:

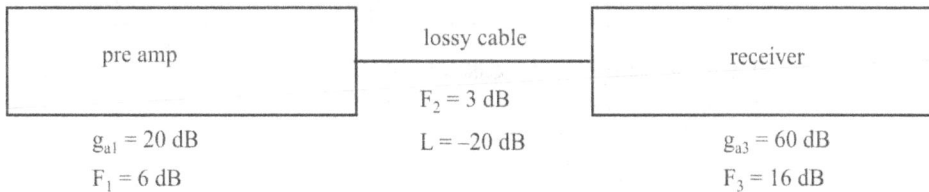

pre amp	lossy cable	receiver
	$F_2 = 3$ dB	
$g_{a1} = 20$ dB	$L = -20$ dB	$g_{a3} = 60$ dB
$F_1 = 6$ dB		$F_3 = 16$ dB

From the given data,

$$ga_1 = 20dB \Rightarrow ga_1 = 10^2$$

$$F_1 = 6dB \Rightarrow F_1 = 10^{0.6} = 3.98$$

$$F_2 = 3dB \Rightarrow F_2 = 10^{0.3} = 1.99$$

$$L = -20dB \Rightarrow L = 10^{-2}$$

$$ga_2 = \frac{1}{L} = 10^2$$

$$ga_3 = 60dB \Rightarrow ga_3 = 10^6$$

$$F_3 = 16dB \Rightarrow F_3 = 10^{1.3} = 39.8$$

Total noise figure of TV receivers is

$$F = F_1 + \frac{F_2 - 1}{ga_1} + \frac{F_3 - 1}{ga_1 ga_2}$$

$$= 3.98 + \frac{1.99 - 1}{10^2} + \frac{39.8 - 1}{10^2 \times 100}$$

$$= 3.98 + 0.0099 + 0.0038$$

$$= 3.9937$$

$$= 6 dB$$

Example 10: An amplifier has three stages fro which $T_{e1} = 150^{\circ} k$ (first stage), $T_{e2} = 300^0 k$ and $T_{e3} = 500^{\circ} k$ (output stage). Available gain of the first stage is 10 and overall input effective noise temperature is 190 a) What is the overall power gain of second stage (b) What is cascade's noise figure c) What is the cascade noise figure when used with a source of noise temperature $T_s = 50^{\circ} k$

Solution

$Te_1 = 150^0 k$ $g_{a1} = 10$	$Te_2 = 300^0 k$	$Te_3 = 500^0 k$

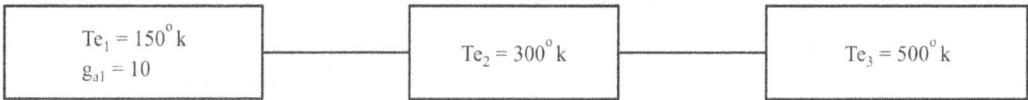

(a)   Effective noise temperature of the cascade is

$$T_e = T_{e1} + \frac{T_{e2}}{ga_1} + \frac{T_{e3}}{ga_1 ga_2}$$

$$\Rightarrow \qquad 190 = 150 + \frac{300}{10} + \frac{500}{10 ga_2}$$

$$\Rightarrow \qquad 190 = 180 + \frac{50}{ga_2}$$

$$\Rightarrow \qquad \frac{50}{ga_2} = 10$$

$$\Rightarrow \qquad ga_2 = 5$$

(b)   Overall noise figure

$$F = 1 + \frac{T_e}{T_0}$$

$$= 1 + \frac{190}{290} \quad = 1.655$$

(c) $T = T_s + T_e = 50 + 190 = 240^0 k$

Noise figure

$$F = 1 + \frac{T}{T_0}$$

$$F = 1 + \frac{240}{290}$$

$$= 1.827$$

Example 11: A random process $X(t)$ whose mean square value is 2 and autocorrelation function $R_{XX}(\tau) = 4.e^{-2|\tau|}$ is applied to a system whose transfer function is $\frac{1}{2+j\omega}$. Find out the mean value, auto correlation, power density spectrum and average power of o/p signal $Y(t)$

Sol:

Given that $R_{XX}(\tau) = 4.e^{-2|\tau|}$, $\overline{X} = 2$ and $H(w) = \frac{1}{2+j\omega}$

We know that $S_{XX}(w) = F[R_{XX}(\tau)]$

$$= F\left[4.e^{-2|\tau|}\right]$$

$$S_{XX}(\omega) = 4 \times \frac{4}{4+\omega^2}$$

$$= \frac{16}{4+\omega^2}$$

The mean value of $Y(t)$ is

$$E[Y(t)] = E[X]\, H(U)$$

$$\overline{Y} = \overline{X} H(\omega)$$

$$= 2 \times \frac{1}{2} = 1$$

The Psd of $Y(t) = S_{XX}(\omega)$

$$= |H(\omega)|^2\, S_{XX}(\omega)$$

$$= \frac{1}{4+\omega^2} \times \frac{16}{4+\omega^2}$$

$$= \frac{16}{\left(4+\omega^2\right)^2}$$

The ACF of $Y(t)$ is

$$R_{YY}(\tau) = F^{-1}[S_{YY}(\omega)]$$

$$F^{-1} = \left[\frac{16}{\left(4 + \omega^2\right)^2} \right]$$

$$= \frac{16}{4} \tau \, e^{-2|\tau|}$$

$$= 4\tau \, e^{-2|\tau|}$$

$$R_{XX}(0) = 4$$

***Example* 12:** A random process $X(t)$ is the input to a linear system whose impulse response is $h(t) = 2 \, e^{-t} \quad t \geq 0$. If the auto correlation function of the process is $R_{XX}(\tau) = e^{-2|\tau|}$. Find the Psd of the output process $Y(t)$.

Sol:

The Psd of input process is

$$R_{XX}(\tau) = 2.e^{-2|\tau|}$$

$$S_{XX}(\omega) = \int_{-\infty}^{\infty} R_{XX}(\tau) e^{-j\omega\tau} d\tau$$

$$= \int_{-\infty}^{\infty} e^{-2|\tau|} e^{-j\omega\tau} d\tau$$

$$= \int_{-\infty}^{0} e^{2\tau} e^{-j\omega\tau} d\tau + \int_{0}^{\infty} e^{-2\tau} e^{-j\omega\tau} d\tau$$

$$= \frac{1}{2 - j\omega} + \frac{1}{2 + j\omega}$$

$$= \frac{4}{4 + \omega^2}$$

The transfer function of linear system is given by

$$H(\omega) = \int_{-\infty}^{\infty} h(t) e^{-j\omega t} dt$$

$$= \int_0^\infty 2e^{-t}e^{-j\omega y}\,dt$$

$$= 2\int_0^\infty e^{-(1+j\omega)t}\,dt$$

$$= \frac{2}{1+j\omega}$$

The Psd of output process is given by

$$S_{YY}(\omega) = H^*(\omega)H(\omega)S_{XX}(\omega)$$

$$= \left|H^*(\omega)\right|^2 S_{XX}(\omega)$$

$$= \left|\frac{2}{1+j\omega}\right|^2 \frac{4}{4+\omega^2}$$

$$= \frac{16}{\left(\omega^2+1\right)\left(\omega^2+4\right)}$$

***Example* 13:** A random process $n(t)$ has a power spectral density $G(t) = \dfrac{\eta}{2}$ for $-\infty < t < \infty$. The Random process is passed through a low pass filter which has a transfer function

$$H(f) = \begin{cases} 2 & for \ -f_n \le f \le f_n \\ 0 & Otherwise \end{cases}$$

Sol:

Given $X(t)$ is a random process with input

$$\text{Psd} \qquad G(t) = \frac{\eta}{2} \qquad -\infty < t < \infty$$

$$\text{O/P Psd} == \left|H(f)\right|^2 \left((i/p)\ Psd\right)$$

Given that $\qquad \left|H(f)\right| = 2$

this implies $\qquad \left|H(f)\right|^2 = 4$.

$$\text{O/P Psd} = 4 \times \frac{\eta}{2}$$

$$= 2\eta$$

Example 14: Consider a linear system shown as follows

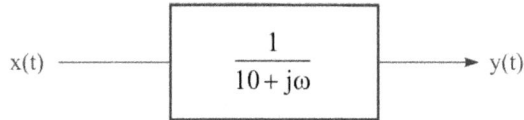

x(t) ────────→ | $\dfrac{1}{10 + j\omega}$ | ────────→ y(t)

$X(t)$ is the input and $R_{XX}(\tau) = 5\,\delta(\tau)$. Find the Psd, correlation function and mean square value of the output $Y(t)$.

Sol:

Given

$$R_{XX}(\tau) = 5\,\delta(\tau)$$

$$S_{XX}(\omega) = F\big[R_{XX}(\tau)\big]$$

$$= F\big[5\delta(\tau)\big]$$

$$= 5$$

$$H(j\omega) = \frac{1}{10 + j\omega} |H(j\omega)|^2$$

$$= \frac{1}{100 + \omega^2}$$

$$\text{Output Psd} = S_{YY}(\omega)$$

$$= |H(\omega)|^2 \, S_{XX}(\omega)$$

$$S_{YY}(\omega) = \frac{5}{100 + \omega^2}$$

(ii)   we know that $R_{YY}(\tau) = 5\,\delta(\tau)$

$$= F^{-1}\big[S_{YY}(\omega)\big]$$

$$= F^{-1}\left[\frac{5}{10^2 + \omega^2}\right]$$

$$R_{YY}(\tau) = \frac{5}{20} F^{-1}\left[\frac{5\times2\times10}{10^2+\omega^2}\right]$$

$$R_{YY}(\tau) = \frac{5}{20} e^{-10|\tau|}$$

(iii)   Mean square value of y(t) = $E[y^2(t)]$

$$= R_{YY}(0)$$

$$= \frac{5}{20}$$

Example: 15. Develop an expression for the power spectral density for the noise voltage equation for the network shown

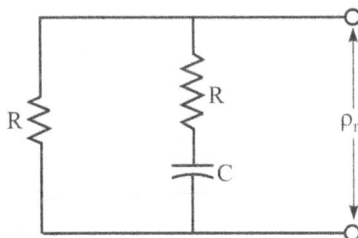

Sol:

$$\text{Admittance} \quad Y = 1 + \cfrac{1}{1+\cfrac{1}{j\omega c}}$$

$$= 1 + \frac{j\omega c}{1+j\omega c}$$

$$= \frac{1+j\omega c+j\omega c}{1+j\omega c}$$

$$= \frac{1+2j\omega c}{1+j\omega c}$$

$$Z = \frac{1+j\omega c}{1+2j\omega c}$$

$$= \frac{(1+j\omega c)\,(1-2j\omega c)}{(1+2j\omega c)\,(1-2j\omega c)}$$

$$= \frac{1 - 2j\omega c + j\omega c + 2\omega^2 c^2}{\left(1 + 4\omega^2 c^2\right)}$$

$$Z = \frac{1 + 2\omega^2 c^2}{1 + 4\omega^2 c^2} - \frac{-j\omega c}{1 + 4\omega^2 c^2}$$

$$Z = R + jY$$

$$R = \text{Re}(Z) = \frac{1 + 2\omega^2 c^2}{1 + 4\omega^2 c^2}$$

Psd of noise voltage is given by

$$S_n(\omega) = 2KTR$$

$$= \frac{2KT\left(1 + 2\omega^2 c^2\right)}{1 + 4\omega^2 c^2}$$

Quiz Questions

1. Thermal noise power is

    a) proportional to B

    b) proportional to $\sqrt{B}$

    c) proportional to $1/B^2$

    d) proportional to B^2

2. If the value of a resistor is doubled, the resistor noise power generated

    a) Gets halved

    b) Gets doubled

    c) Gets quadrupled

    d) Remains unchanged

3. Johnson noise is due to random

    a) matron of atoms and molecules

    b) matron of free dectrons

    c) vibration of atoms their positron inside the conductring medians

    d) None of the above

4. Thermal noise is independent of

    a) band width

    b) temperature

    c) center frequency

    d) BoHzman's constant

5. Parallel combination of a resistance R and a capacity C develops a noise voltage varies

    a) proportional to R

    b) Inversely proportional to C

    c) inversely proportional to square root of C d) proportional to RC

6. A noise voltage some has a resistance of 10 *ohms* its power density spectrum is 0.24×10^{-5}. The corresponding available power density is

    a) 2.6×10^{-5} b) 0.025   c) 26×10^{-5} d) 6×10^{-8}

7. In a certain system, the signal power is 13 *dBm* and noise power is $-1dBm$. The *SNR* will be

    a) 14 *dB*   b) $-13\ dB$

    c) 12 *dBm*   d) 12*dB*

8. The co-variance function of a band limited while noise is
    a)   A dirac delta function
    b)   an exponentially decreasing function
    c)   a *SMC* function
    d) SMC^2 function

9. An amplifier having noise figure of 20 *dB* and available power gain of 15 *dB* is followed by a miner   circuit having noise figure of 9 *dB*. The oversell noise figure as refereed to input in *dB* is

    a)   11.7   b)   10.44   c)   20.02   d)   0.63

10. The noise figure in an amplifier with a gain of 5000, input of 5 $\mu\ v$, output note of 200 *nv* is given

    a)   6   b)   4   c)   8   d)   10

11. In an certain receiving systems, the noise power available from the aerial within a 10 *kHZ* band is equal to 10^{-17} *w*. the aerial noise temp is

    a) $62.5^0\ k$ b) $72.5^0\ k$ c) $27.5^0\ k$ d) $75.2^0\ k$

12. The noise figure of an amplifier i0 3 *db*. Its noise temperature will be about

    a)   145 *k*   b)   290 *k*   c)   580 *k*   d)   870 *k*

13. If each stage had a gain of 10 *dB* and noise figure 10 *dB*, then the overall noise figure of a two stage cascade amplifier will be

    a)   10   b)   1.09   c)   1.0   d)   10.9

14. The noise figure of a receiver is 1.6, its equivalent noise temperature is

    a) 464 *k*   b)   174 *k*   c)   108.75 *k*   d)   181.25

15. The power special density of while noise is
    a)   varies as square root of frequency   b)   Varies as square of frequency
    c)   varies as square of frequency   d)   constant with frequency

16.   A system has a receiver noise resistance of 50 Ω. It is connected to an antennae with an output resistance of 50 Ω. The noise figure of the system is

a)   1   b)   2   c) 50   d) 101

17.   Which one of the following typer of noise gains importance at high frequency

a)   shot noise   b)   random noise

c)   impulse noise   d)   transit noise

18.   White noise with two sided psd n/2 is Passed through an RC low pass network with time constant $\Im = RC$ and there after through an ideal amplifier with voltage gain 10. The expression for mean square value of output noise is

a) B.   C.25 n   d) $\dfrac{RC}{n}$

19.   For an $LT\,I$ system, the impulse response and transfer function from a . . . pair

a)   convolution   b) Fourier   c) Laplace   d) Fourier & Laplace

20.   If $S_{xx}(\omega)$ is the power spectrum of the input process and $\left|H(\omega)\right|^2$ is the transfer function of the system, then the average power Pxx is

a) $\displaystyle\int_{-\alpha}^{\alpha} S_{xx}(\omega)\left|H(\omega)\right|^2 d\omega$

b) $\dfrac{1}{2\pi}\displaystyle\int_{-\alpha}^{\alpha} S_{xx}(\omega)\left|H(\omega)\right|^2 d\omega$

c) $\displaystyle\int_{-\alpha}^{\alpha} \dfrac{S_{xx}(\omega)}{\left|H(\omega)\right|^2} d\omega$

d) $\dfrac{1}{2\pi}\displaystyle\int_{-\alpha}^{\alpha} S_{xx}(\omega)/\left|H(\omega)\right|^2 d\omega$

21.   If the power density spectrum $S_{NN}(\omega)$ of a random process has its significant component, clustered in a bandwidth B_N r/s that does not include $w = 0$, then the it is called . . . process

a)   band limited   b)   band pass

c)   narrow band   d)   statronary

22.   Let $n(t)$ be narrow band representation of noise where

$n(t) = n_c(t) \cos w_c t - n_s(t) \sin w_c t$ and let p_1, p_2, p_3 are the powers of the $n(t)$, $n_c(t)$ and $n_s(t)$ respectively, then p_1 is related by ....

a) $p_1 = 2p_2 = 3p_3$

b) $p_1 = \dfrac{p_2}{2} = \dfrac{p_3}{3}$

c) $p_1 = p_2 = p_3$

d) $3p_1 = 2p_2 = p_3$

23.   A Gaussian filter has the transfer function

$$H(t) = e^{-k^2 f^2} \quad \text{for } (f) < \alpha$$
$$= 0 \qquad \text{elsewhere}$$

Its 3 dB band width is

a) $\dfrac{0.58}{k}$ b) $(0,58) k$ c) 0.58   d) $k^2 (0.58)$

24.   The noise bandwidth of Practical system whose transfer function is $H(\omega)$, B_N is

a) $\dfrac{1}{\left|H(\omega_o)^2\right|} \int_o^\alpha |H(\omega)|^2 \, d\omega$ b) $|H(\omega)|^2 \int_o^\alpha |H(\omega)|^2 \, d\omega \, d\omega$

c) $\dfrac{1}{\left|H(\omega_o)^2\right|} \int_o^\alpha |H(\omega)|^2 \, d\omega$ d) $|H(\omega)|^2 \int_o^\alpha |H(\omega)|^2 \, d\omega$

25.   If the input power spectral density of a system is $S_{xx}(\omega) = N_0$ and its output Psd is

$$S_{xx}(\omega) = N_0 \left[\dfrac{\omega^2 + 9}{\omega^2 + 16}\right] \quad \text{then the transfer function of the system is}$$

a) $\dfrac{\omega - j3}{\omega - j4}$ b) $\dfrac{\omega + j3}{\omega + j4}$

c) $\dfrac{\omega + j4}{\omega + j3}$ d) $\dfrac{\omega - j9}{\omega - j(3)}$

Answers

1. (a)	2. (d)	3. (b)	4. (c)	5. (c)
6. (d)	7.	8. (a)	9.	10. (c)
11. (b)	12. (b)	13. (d)	14. (b)	15. (d)
16. (b)	17. (d)	18. ..	19.	20.
21. (b)	22. (c)	23. (a)	24. (a)	25. (b)

Review Quiz Questions

Fill up the blanks:

1.   A system is said to be linear if satisfies .... Principle
2.   If $S(t)$ is the input, the impulse response of the system is ....

3. A general linear system is said to be if its input–output characterstics do not change with time

4. The input response of the system can be obtained by taking *IFT* of .... function

5. A *LTI* system is said to be .... if it does not respond to the application of input signal

6. The power density spectrum $S_{yy}(\omega)$ of the response of a *LTI* having a transfer function $H(\omega)$ is given by $S_{xx}(\omega)$..................

7. A band limited random process is said to be narrow band if ....

8. For a noisless system, Te is ....

9. In an ideal network, Te = 0 so F = ... for any real network

10. If $R_{xx}(\tau) = 38(\tau)$ and $H(\omega) = \dfrac{1}{6 + j\omega}$ the Mean square value of $Y(t)$ is $\dfrac{1}{4}$

....

Answers

1. Homogeneity

2. $h(t, \tau) = 8(t, \tau)$

3. time variant

4. transfer

5. Causal

6. $|H(\omega)|^2$

7. $W \ll w_0$

8. zero

9. ..

10. 0

Exercise Questions

1. A *WSS* process $X(t)$ with $R_{xx}(\tau) = A\,e^{-a|\tau|}$ where '*A*' and a real positive constant 0 applied to the *I/P* of an *LTI* system with $h(t) = e^{-bt}\,U(t)$ where b is a real positive constant. Find the *PSD* of the *O/P* of the system.

2. A random process $n(t)$ has a *PSD* $G(t) = 10^{-4}$ for $-\alpha \leq f \leq \alpha$. Which is passes through an *LPF* whose transfer function is

$$H(t) = \begin{array}{ll} 100 & -fn \leq f \leq fn \\ 0 & otherwise \end{array}$$

Find the *PSD* of the wave form at the *O/P* of the filter.

3. A random processes a *PSD* of $S_{xx}(t) = 1/(1+f^2)$ is input to a filter. The filter is to be designed so that the output a whitening filter

   a) Find the transfer function of the whitening for this input process. Be score that the filter is causal.

   b) Sketch a cricket that well realize this transfer function

4. Consider a linear system having transfer function $H(\omega) = \dfrac{1}{G + jc}$ $X(t)$ is the input and $Y(t)$ is the output of the system. The auto correlation of $X(t)$ is $R_{xx}(\tau) = 3.8$ (τ). Find the *PSD*, auto correlation function and mean square value of the output $Y(t)$.

5. Find the rms noise voltage across a *iMF* capacitor over the ensure frequency band when the capacitors is shafted by a $1k\Omega$ resistro maintained at $300^0 k$

6. Complete the input thermal noise voltage of a *TV* receiver, whose equivalent noise resistance 200Ω and the source resistance 300Ω. The receiver bandwidth is 6 *MHZ* and the temperature & $27^0 c$

7. In a certain reciving system the noise power available from the aerial within a 10 *KHZ* band is 10^{-17} w. evaluate the aerial noise temperature.

8. Calculate the rms noise voltage generated in a bandwise of 10 *KHZ*, by a resistor of $1000\ \Omega$ maintained at M also find the available noise power over this bandwise determine the corresponding values when resistor value is increased to 10K Ω.

9. An ampliafier width $G_a = 40$ *dB* and $B_{01} = 20$ *KHX* is found to have $T_o = 10$ K. find T_e and noise figure.

10. The noise present at the input to a two port is 1μ the noise figure F is 0.5 *dB*. The receives gain $G_a = 10^{10}$ Calculate.

    a) The available noise power Contributed by the two – port and

    b) The output available power.

A WSS process $X(t)$ with $R_{xx}(\tau) = A\ e^{-\alpha|\tau|}$ where 'A' and a real positive constant 0 applied to the I/P of an *LTI* system with $h(t) = e^{-\alpha|\tau|} u(t)$ where b is a real positive constant. Find the *PSD* of the *O/P* of the system.

11. A random process $n(t)$ has a PSD $G(f) = 10^{-4}$ for $-\alpha \leq f \leq \alpha$. Which is passes through an *LPF* whose transfer function is

$$H(t) = \begin{array}{ll} 100 & -fn \leq f \leq fn \\ 0 & otherwise \end{array}$$

Find the *PSD* of the wave form at the *O/P* of the filter.

12. A random processes a PSD of $S_{xx}(t) = 1/(1+f^2)$ is input to a filter. The filter is to be designed so that the output process is (constant *PSO*).   This is called a whitening filter

    a) Find the transfer function of the whitening for this input process. Be score that the filter is causal.

    b) Sketch a circuit that well realize this transfer function

13. Consider a linear system having transfer function $H(w) = \dfrac{1}{2+J\omega}$ $X(t)$ is the input and $Y(t)$ is the output of the system. The auto correlation of $X(t)$ is $R_{xx}(\tau) = e^{-2}|\tau|$. Find the *PSD*, auto correlation function and mean square value of the output $Y(t)$.

14. Find the rms noise voltage across a *iMF* capacitor over the ensure frequency band when the capacitors is shafted by a $1k\Omega$ resistor maintained at 300^0K

15. Complete the input thermal noise voltage of a *TV* receiver, whose equivalent noise resistance 200Ω and the source resistance 300Ω. The receiver bandwidth is 6 *MHZ* and the temperature & 27^0c

16. In a certain receiving system the noise power available from the aerial within a 10KHZ band is 10^{-17} w. evaluate the aerial noise temperature.

17. Calculate the rms noise voltage generated in a bandwidth of 10 *KHZ*, by a resistor of 1000 Ω maintained at M also find the available noise power over this bandwidth determine the corresponding values when resistor value is increased to 10K Ω.

18. An amplifier width $G_a = 40$ *dB* and $B_{0l} = 20$ *KHX* is found to have $T_e = 10$ K. find T_e and noise figure.

19. The noise present at the input to a two port is 1μ the noise figure F is 0.5 *dB*. The receives gain $G_a = 10^{10}$ Calculate.

    a) The available noise power Contributed by the two – port and

    b) The output available power.

Review Exercise Questions

1. A random process $n(t)$ has a *PSD* $G(t) = 10^{-3}$ for $--\alpha \le f \le \alpha$. Which is passes through an *LPF* whose transfer function is

    $$H(t) = \begin{array}{ll} 100 & -fn \le f \le fn \\ 0 & otherwise \end{array}$$

    Find the *PSD* of the wave form at the *O/P* of the filter.

2. A random processes a *PSD* of $S_{xx}(\tau) = 1/(1-f^2)$ is input to a filter. The filter is to be designed so that the output a whitening filter.

    Find the transfer function of the whitening for this input process. Be score that the filter is causal.

3. Consider a linear system having transfer function $H(w) = \dfrac{1}{4 + J\omega}$ $X(t)$ is the input and $Y(t)$ is the output of the system. The auto correlation of $X(t)$ is $R_{xx}(\tau) = 4 + e^{-3|\tau|}$. Find the mean.

4. Find the rms noise voltage across a *iMF* capacitor over the ensure frequency band when the capacitors is shafted by a $1k\Omega$ resistor maintained at $250^0 k$

5. Complete the input thermal noise voltage of a *TV* receiver, whose equivalent noise resistance 100Ω and the source resistance 150Ω. The receiver bandwidth is 3 *MHZ* and the temperature & $52^0 c$

6. In a certain receiving system the noise power available from the aerial within a 12 *KHZ* band is 10^{-12} w. evaluate the aerial noise temperature.

7. Calculate the rms noise voltage generated in a bandwise of 20 *KHZ*, by a resistor of 2000 Ω maintained at *M* also find the available noise power over this bandwise determine the corresponding values when resistor value is increased to 20K Ω.

8. An ampliafier width $G_a = 80$ *dB* and $B_{0l} = 40$ *KHX* is found to have $T_e = 20$ *K*. find T_e and noise figure.

9. The noise present at the input to a two port is 2μ the noise figure F is 0.4 *dB*. The receives gain $G_a = 10^{11}$ Calculate.

    a) The available noise power Contributed by the two – port and

    b) The output available power.

10. A WSS process $X(t)$ with $R_{xx}(\tau) = k\ e^{-\alpha|\tau|}$ where k and a real positive constant 0 applied to the I/P of an *LTI* system with $h(t) = e^{-bt}\ u(t)$ where b is a real positive constant. Find the *PSD* of the *O/P* of the system.

    a) The available noise power Contributed by the two – port and

    b) The output available power.

Statistical Tables

Table 1 Standard Normal Distribution Function

$$F(z) = \frac{1}{\sqrt{2\pi}} \int_{-\infty}^{z} e^{-t^2/2} dt$$

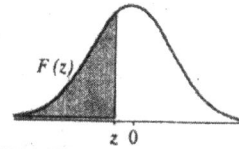

$F(z)$

z	0.00	0.01	0.02	0.03	0.04	0.05	0.06	0.07	0.08	0.09
-5.0	0.0000003									
-4.0	0.00003									
-3.5	0.0002									
-3.4	0.0003	0.0003	0.0003	0.0003	0.0003	0.0003	0.0003	0.0003	0.0003	0.0002
-3.3	0.0005	0.0005	0.0005	0.0004	0.0004	0.0004	0.0004	0.0004	0.0004	0.0003
-3.2	0.0007	0.0007	0.0006	0.0006	0.0006	0.0006	0.0006	0.0005	0.0005	0.0005
-3.1	0.0010	0.0009	0.0009	0.0009	0.0008	0.0008	0.0008	0.0008	0.0007	0.0007
-3.0	0.0013	0.0013	0.0013	0.0012	0.0012	0.0011	0.0011	0.0011	0.0010	0.0010
-2.9	0.0019	0.0018	0.0018	0.0017	0.0016	0.0016	0.0015	0.0015	0.0014	0.0014
-2.8	0.0026	0.0025	0.0024	0.0023	0.0023	0.0022	0.0021	0.0021	0.0020	0.0019
-2.7	0.0035	0.0034	0.0033	0.0032	0.0031	0.0030	0.0029	0.0028	0.0027	0.0026
-2.6	0.0047	0.0045	0.0044	0.0043	0.0041	0.0040	0.0039	0.0038	0.0037	0.0036
-2.5	0.0062	0.0060	0.0059	0.0057	0.0055	0.0054	0.0052	0.0051	0.0049	0.0048
-2.4	0.0082	0.0080	0.0078	0.0075	0.0073	0.0071	0.0069	0.0068	0.0066	0.0064
-2.3	0.0107	0.0104	0.0102	0.0099	0.0096	0.0094	0.0091	0.0089	0.0087	0.0084
-2.2	0.0139	0.0136	0.0132	0.0129	0.0125	0.0122	0.0119	0.0116	0.0113	0.0110
-2.1	0.0179	0.0174	0.0170	0.0166	0.0162	0.0158	0.0154	0.0150	0.0146	0.0143
-2.0	0.0228	0.0222	0.0217	0.0212	0.0207	0.0202	0.0197	0.0192	0.0188	0.0183
-1.9	0.0287	0.0281	0.0274	0.0268	0.0262	0.0256	0.0250	0.0244	0.0239	0.0233
-1.8	0.0359	0.0351	0.0344	0.0336	0.0329	0.0322	0.0314	0.0307	0.0301	0.0294
-1.7	0.4446	0.0436	0.0427	0.0418	0.0409	0.0401	0.0392	0.0384	0.0375	0.0367
-1.6	0.0548	0.0537	0.0526	0.0516	0.0505	0.0495	0.0485	0.0475	0.0465	0.0455
-1.5	0.0668	0.0655	0.0643	0.0630	0.0618	0.0606	0.0594	0.0582	0.0571	0.0559
-1.4	0.0808	0.0793	0.0778	0.0764	0.0749	0.0735	0.0721	0.0708	0.0694	0.0681
-1.3	0.0968	0.0951	0.0934	0.0918	0.0901	0.0885	0.0869	0.0853	0.0838	0.0823
-1.2	0.1151	0.1131	0.1112	0.1093	0.1075	0.1056	0.1038	0.1020	0.1003	0.0985
-1.1	0.1357	0.1335	0.1314	0.1292	0.1271	0.1251	0.1230	0.1210	0.1190	0.1170
-1.0	0.1587	0.1562	0.1539	0.1515	0.1492	0.1469	0.1446	0.1423	0.1401	0.1379
-0.9	0.1841	0.1814	0.1788	0.1762	0.1736	0.1711	0.1685	0.1660	0.1635	0.1611
-0.8	0.2119	0.2090	0.2061	0.2033	0.2005	0.1977	0.1949	0.1922	0.1894	0.1867
-0.7	0.2420	0.2389	0.2358	0.2327	0.2296	0.2266	0.2236	0.2206	0.2177	0.2148
-0.6	0.2743	0.2709	0.2676	0.2643	0.2611	0.2578	0.2546	0.2514	0.2483	0.2451
-0.5	0.3085	0.3050	0.3015	0.2981	0.2946	0.2912	0.2877	0.2843	0.2810	0.2776
-0.4	0.3446	0.3409	0.3372	0.3336	0.3300	0.3264	0.3228	0.3192	0.3156	0.3121
-0.3	0.3821	0.3783	0.3745	0.3707	0.3669	0.3632	0.3594	0.3557	0.3520	0.3483
-0.2	0.4207	0.4168	0.4129	0.4090	0.4052	0.4013	0.3974	0.3936	0.3897	0.3859
-0.1	0.4602	0.4562	0.4522	0.4483	0.4443	0.4404	0.4364	0.4325	0.4286	0.4247
-0.0	0.5000	0.4960	0.4920	0.4880	0.4840	0.4801	0.4761	0.4721	0.4681	0.4641

Table 2 Standard Normal Distribution Function

$$F(z) = \frac{1}{\sqrt{2\pi}} \int_{-\infty}^{z} e^{-t^2/2}\, dt$$

z	0.00	0.01	0.02	0.03	0.04	0.05	0.06	0.07	0.08	0.09
0.0	0.5000	0.5040	0.5080	0.5120	0.5160	0.5199	0.5239	0.5279	0.5319	0.5359
0.1	0.5398	0.5438	0.5478	0.5517	0.5557	0.5596	0.5636	0.5675	0.5714	0.5753
0.2	0.5793	0.5832	0.5871	0.5910	0.5948	0.5987	0.6026	0.6064	0.6103	0.6141
0.3	0.6179	0.6217	0.6255	0.6293	0.6331	0.6368	0.6406	0.6443	0.6480	0.6517
0.4	0.6554	0.6591	0.6628	0.6664	0.6700	0.6736	0.6772	0.6808	0.6844	0.6879
0.5	0.6915	0.6950	0.6985	0.7019	0.7054	0.7088	0.7123	0.7157	0.7190	0.7224
0.6	0.7257	0.7291	0.7324	0.7357	0.7389	0.7422	0.7454	0.7486	0.7517	0.7549
0.7	0.7580	0.7611	0.7642	0.7673	0.7704	0.7734	0.7764	0.7794	0.7823	0.7852
0.8	0.7881	0.7910	0.7939	0.7967	0.7995	0.8023	0.8051	0.8078	0.8106	0.8133
0.9	0.8159	0.8186	0.8212	0.8238	0.8264	0.8289	0.8315	0.8340	0.8365	0.8389
1.0	0.8413	0.8438	0.8461	0.8485	0.8508	0.8531	0.8554	0.8577	0.8599	0.8621
1.1	0.8643	0.8665	0.8686	0.8708	0.8729	0.8749	0.8770	0.8790	0.8810	0.8830
1.2	0.8849	0.8869	0.8888	0.8907	0.8925	0.8944	0.8962	0.8980	0.8997	0.9015
1.3	0.9032	0.9049	0.9066	0.9082	0.9099	0.9115	0.9131	0.9147	0.9162	0.9177
1.4	0.9192	0.9207	0.9222	0.9236	0.9251	0.9265	0.9279	0.9292	0.9306	0.9319
1.5	0.9332	0.9345	0.9357	0.9370	0.9382	0.9394	0.9406	0.9418	0.9429	0.9441
1.6	0.9452	0.9463	0.9474	0.9484	0.9495	0.9505	0.9515	0.9525	0.9535	0.9545
1.7	0.9554	0.9564	0.9573	0.9582	0.9591	0.9599	0.9608	0.9616	0.9625	0.9633
1.8	0.9641	0.9649	0.9656	0.9664	0.9671	0.9678	0.9686	0.9693	0.9699	0.9706
1.9	0.9713	0.9719	0.9726	0.9732	0.9738	0.9744	0.9750	0.9756	0.9761	0.9767
2.0	0.9772	0.9778	0.9783	0.9788	0.9793	0.9798	0.9803	0.9808	0.9812	0.9817
2.1	0.9821	0.9826	0.9830	0.9834	0.9838	0.9842	0.9846	0.9850	0.9854	0.9857
2.2	0.9861	0.9864	0.9868	0.9871	0.9875	0.9878	0.9881	0.9884	0.9887	0.9890
2.3	0.9893	0.9896	0.9898	0.9901	0.9904	0.9906	0.9909	0.9911	0.9913	0.9916
2.4	0.9918	0.9920	0.9922	0.9925	0.9927	0.9929	0.9931	0.9932	0.9934	0.9936
2.5	0.9938	0.9940	0.9941	0.9943	0.9945	0.9946	0.9948	0.9949	0.9951	0.9952
2.6	0.9953	0.9955	0.9956	0.9957	0.9959	0.9960	0.9961	0.9962	0.9963	0.9964
2.7	0.9965	0.9966	0.9967	0.9968	0.9969	0.9970	0.9971	0.9972	0.9973	0.9974
2.8	0.9974	0.9975	0.9976	0.9977	0.9977	0.9978	0.9979	0.9979	0.9980	0.9981
2.9	0.9981	0.9982	0.9982	0.9983	0.9984	0.9984	0.9985	0.9985	0.9986	0.9986
3.0	0.9987	0.9987	0.9987	0.9988	0.9988	0.9989	0.9989	0.9989	0.9990	0.9990
3.1	0.9990	0.9991	0.9991	0.9991	0.9992	0.9992	0.9992	0.9992	0.9993	0.9993
3.2	0.9993	0.9993	0.9994	0.9994	0.9994	0.9994	0.9994	0.9995	0.9995	0.9995
3.3	0.9995	0.9995	0.9995	0.9996	0.9996	0.9996	0.9996	0.9996	0.9996	0.9997
3.4	0.9997	0.9997	0.9997	0.9997	0.9997	0.9997	0.9997	0.9997	0.9997	0.9998
3.5	0.9998									
4.0	0.99997									
5.0	0.9999997									

Table 3 Values of t_α

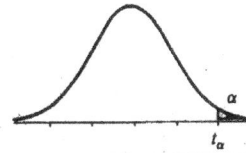

v	$\alpha=0.10$	$\alpha=0.05$	$\alpha=0.025$	$\alpha=0.01$	$\alpha=0.00833$	$\alpha=0.00625$	$\alpha=0.005$	v
1	3.078	6.314	12.706	31.821	38.204	50.923	63.657	1
2	1.886	2.920	4.303	6.965	7.650	8.860	9.925	2
3	1.638	2.353	3.182	4.541	4.857	5.392	5.841	3
4	1.533	2.132	2.776	3.747	3.961	4.315	4.604	4
5	1.476	2.015	2.571	3.365	3.534	3.810	4.032	5
6	1.440	1.943	2.447	3.143	3.288	3.521	3.707	6
7	1.415	1.895	2.365	2.998	3.128	3.335	3.499	7
8	1.397	1.860	2.306	2.896	3.016	3.206	3.355	8
9	1.383	1.833	2.262	2.821	2.934	3.111	3.250	9
10	1.372	1.812	2.228	2.764	2.870	3.038	3.169	10
11	1.363	1.796	2.201	2.718	2.820	2.981	3.106	11
12	1.356	1.782	2.179	2.681	2.780	2.934	3.055	12
13	1.350	1.771	2.160	2.650	2.746	2.896	3.012	13
14	1.345	1.761	2.145	2.624	2.718	2.864	2.977	14
15	1.341	1.753	2.131	2.602	2.694	2.837	2.947	15
16	1.337	1.746	2.120	2.583	2.673	2.813	2.921	16
17	1.333	1.740	2.110	2.567	2.655	2.793	2.898	17
18	1.330	1.734	2.101	2.552	2.639	2.775	2.878	18
19	1.328	1.729	2.093	2.539	2.625	2.759	2.861	19
20	1.325	1.725	2.086	2.528	2.613	2.744	2.845	20
21	1.323	1.721	2.080	2.518	2.602	2.732	2.831	21
22	1.321	1.717	2.074	2.508	2.591	2.720	2.819	22
23	1.319	1.714	2.069	2.500	2.582	2.710	2.807	23
24	1.318	1.711	2.064	2.492	2.574	2.700	2.797	24
25	1.316	1.708	2.060	2.485	2.566	2.692	2.787	25
26	1.315	1.706	2.056	2.479	2.559	2.684	2.779	26
27	1.314	1.703	2.052	2.473	2.553	2.676	2.771	27
28	1.313	1.701	2.048	2.467	2.547	2.669	2.763	28
29	1.311	1.699	2.045	2.462	2.541	2.663	2.756	29
inf.	1.282	1.645	1.960	2.326	2.394	2.498	2.576	inf.

Table 4 Values of X_α^2

v	$\alpha=0.995$	$\alpha=0.99$	$\alpha=0.975$	$\alpha=0.95$	$\alpha=0.05$	$\alpha=0.025$	$\alpha=0.01$	$\alpha=0.005$	v
1	0.0000393	0.000157	0.000982	0.00393	3.841	5.024	6.635	7.879	1
2	0.0100	0.0201	0.0506	0.103	5.991	7.378	9.210	10.597	2
3	0.0717	0.115	0.216	0.352	7.815	9.348	11.345	12.838	3
4	0.207	0.297	0.484	0.711	9.488	11.143	13.277	14.860	4
5	0.412	0.554	0.831	1.145	11.070	12.832	15.056	16.750	5
6	0.676	0.872	1.237	1.635	12.592	14.449	16.812	18.548	6
7	0.989	1.239	1.690	2.167	14.067	16.013	18.475	20.278	7
8	1.344	1.646	2.180	2.733	15.507	17.535	20.090	21.955	8
9	1.735	2.088	2.700	3.325	16.919	19.023	21.666	23.589	9
10	2.156	2.558	3.247	3.940	18.307	20.483	23.209	25.188	10
11	2.603	3.053	3.816	4.575	19.675	21.920	24.725	26.757	11
12	3.074	3.571	4.404	5.226	21.026	23.337	26.217	28.300	12
13	3.565	4.107	5.009	5.892	22.362	24.736	27.688	29.819	13
14	4.075	4.660	5.629	6.571	23.685	26.119	29.141	31.319	14
15	4.601	5.229	6.262	7.261	24.996	27.488	30.578	32.801	15
16	5.142	5.812	6.908	7.962	26.296	28.845	32.000	34.267	16
17	5.697	6.408	7.564	8.672	27.587	30.191	33.409	35.718	17
18	6.265	7.015	8.231	9.390	28.869	31.526	34.805	37.156	18
19	6.844	7.633	8.907	10.117	30.144	32.852	36.191	38.582	19
20	7.434	8.260	9.591	10.851	31.410	34.170	37.566	39.997	20
21	8.034	8.897	10.283	11.591	32.671	35.479	38.932	41.401	21
22	8.643	9.542	10.982	12.338	33.924	36.781	40.289	42.796	22
23	9.260	10.196	11.689	13.091	35.172	38.076	41.638	44.181	23
24	9.886	10.856	12.401	13.484	36.415	39.364	42.980	45.558	24
25	10.520	11.524	13.120	14.611	37.652	40.646	44.314	46.928	25
26	11.160	12.198	13.844	15.379	38.885	41.923	45.642	48.290	26
27	11.808	12.879	14.573	16.151	40.113	43.194	46.963	49.645	27
28	12.461	13.565	15.308	16.928	41.337	44.461	48.278	50.993	28
29	13.121	14.256	16.047	17.708	42.557	45.772	49.588	52.336	29
30	13.787	14.953	16.791	18.493	43.773	46.979	50.892	53.672	30
40	20.706	22.164	24.433	26.509	55.758	59.342	63.691	66.766	40
50	27.991	29.707	32.357	34.764	67.505	71.420	76.154	79.490	50
60	35.535	37.485	40.482	43.118	79.082	83.298	88.379	91.952	60
70	43.275	45.442	48.758	51.739	90.531	95.023	100.425	104.215	70
80	51.172	53.540	57.153	60.391	101.879	106.629	112.329	116.321	80
90	59.196	61.754	65.646	69.126	113.145	118.136	124.116	128.299	90
100	67.328	70.065	74.222	77.929	124.342	129.561	135.807	140.169	100

Table 5(a) Values of $F_{0.05}$

v_1 = Degrees of freedom for numerator

v_2 = Degrees of freedom for denominator	1	2	3	4	5	6	7	8	9	10	12	15	20	25	30	40	60	120	∞
1	161	200	216	225	230	234	237	239	241	242	244	246	248	249	250	251	252	253	254
2	18.51	19.00	19.16	19.25	19.30	19.33	19.35	19.37	19.38	19.40	19.41	19.43	19.45	19.46	19.46	19.47	19.48	19.49	19.50
3	10.13	9.55	9.28	9.12	9.01	8.94	8.89	8.85	8.81	8.79	8.74	8.70	8.66	8.63	8.62	8.59	8.57	8.55	8.53
4	7.71	6.94	6.59	6.39	6.26	6.16	6.09	6.04	6.00	5.96	5.91	5.86	5.80	5.77	5.75	5.72	5.69	5.66	5.63
5	6.61	5.79	5.41	5.19	5.05	4.95	4.88	4.82	4.77	4.74	4.68	4.62	4.56	4.52	4.50	4.46	4.43	4.40	4.37
6	5.99	5.14	4.76	4.53	4.39	4.28	4.21	4.15	4.10	4.06	4.00	3.94	3.87	3.83	3.81	3.77	3.74	3.70	3.67
7	5.59	4.74	4.35	4.12	3.97	3.87	3.79	3.73	3.68	3.64	3.57	3.51	3.44	3.40	3.38	3.34	3.30	3.27	3.23
8	5.32	4.46	4.07	3.84	3.69	3.58	3.50	3.44	3.39	3.35	3.28	3.22	3.15	3.11	3.08	3.04	3.01	2.97	2.93
9	5.12	4.26	3.86	3.63	3.48	3.37	3.29	3.23	3.18	3.14	3.07	3.01	2.94	2.89	2.86	2.83	2.79	2.75	2.71
10	4.96	4.10	3.71	3.48	3.33	3.22	3.14	3.07	3.02	2.98	2.91	2.85	2.77	2.73	2.70	2.66	2.62	2.58	2.54
11	4.84	3.98	3.59	3.36	3.20	3.09	3.01	2.95	2.90	2.85	2.79	2.72	2.65	2.60	2.57	2.53	2.49	2.45	2.40
12	4.75	3.89	3.49	3.26	3.11	3.00	2.91	2.85	2.80	2.75	2.69	2.62	2.54	2.50	2.47	2.43	2.38	2.34	2.30
13	4.67	3.81	3.41	3.18	3.03	2.92	2.83	2.77	2.71	2.67	2.60	2.53	2.46	2.41	2.38	2.34	2.30	2.25	2.21
14	4.60	3.74	3.34	3.11	2.96	2.85	2.76	2.70	2.65	2.60	2.53	2.46	2.39	2.34	2.31	2.27	2.22	2.18	2.13
15	4.54	3.68	3.29	3.06	2.90	2.79	2.71	2.64	2.59	2.54	2.48	2.40	2.33	2.28	2.25	2.20	2.16	2.11	2.07
16	4.49	3.63	3.24	3.01	2.85	2.74	2.66	2.59	2.54	2.49	2.42	2.35	2.28	2.23	2.19	2.15	2.11	2.06	2.01
17	4.45	3.59	3.20	2.96	2.81	2.70	2.61	2.55	2.49	2.45	2.38	2.31	2.23	2.18	2.15	2.10	2.06	2.01	1.96
18	4.41	3.55	3.16	2.93	2.77	2.66	2.58	2.51	2.46	2.41	2.34	2.27	2.19	2.14	2.11	2.06	2.02	1.97	1.92
19	4.38	3.52	3.13	2.90	2.74	2.63	2.54	2.48	2.42	2.38	2.31	2.23	2.16	2.11	2.07	2.03	1.98	1.93	1.88
20	4.35	3.49	3.10	2.87	2.71	2.60	2.51	2.45	2.39	2.35	2.28	2.20	2.12	2.07	2.04	1.99	1.95	1.90	1.84
21	4.32	3.47	3.07	2.84	2.68	2.57	2.49	2.42	2.37	2.32	2.25	2.18	2.10	2.05	2.01	1.96	1.92	1.87	1.81
22	4.30	3.44	3.05	2.82	2.66	2.55	2.46	2.40	2.34	2.30	2.23	2.15	2.07	2.052	1.98	1.94	1.89	1.84	1.78
23	4.28	3.42	3.03	2.80	2.64	2.53	2.44	2.37	2.32	2.27	2.20	2.13	2.05	2.00	1.96	1.91	1.86	1.81	1.76
24	4.26	3.40	3.01	2.78	2.62	2.51	2.42	2.36	2.30	2.25	2.18	2.11	2.03	1.97	1.94	1.89	1.84	1.79	1.73
25	4.24	3.39	2.99	2.76	2.60	2.49	2.40	2.34	2.28	2.24	2.16	2.09	2.01	1.96	1.92	1.87	1.82	1.77	1.71
30	4.17	3.32	2.92	2.69	2.53	2.42	2.33	2.27	2.21	2.16	2.09	2.01	1.93	1.88	1.84	1.79	1.74	1.68	1.62
40	4.08	3.23	2.84	2.61	2.45	2.34	2.25	2.18	2.12	2.08	2.00	1.92	1.84	1.78	1.74	1.69	1.64	1.58	1.51
60	4.00	3.15	2.76	2.53	2.37	2.25	2.17	2.10	2.04	1.99	1.92	1.84	1.75	1.69	1.65	1.59	1.53	1.47	1.39
120	3.92	3.07	2.68	2.45	2.29	2.18	2.09	2.02	1.96	1.91	1.83	1.75	1.66	1.60	1.55	1.50	1.43	1.35	1.25
∞	3.84	3.00	2.60	2.37	2.21	2.10	2.01	1.94	1.88	1.83	1.75	1.67	1.57	1.51	1.46	1.39	1.32	1.22	1.00

Table 5(b) Values of $F_{0.01}$

v_1 = Degrees of freedom for numerator

v_2 = Degrees of freedom for denominator	1	2	3	4	5	6	7	8	9	10	12	15	20	25	30	40	60	120	∞
1	4,052	5,000	5,403	5,625	5,764	5,859	5,928	5,982	6,023	6,056	6,106	6,157	6,209	6,240	6,261	6,287	6,313	6,339	6,366
2	98.50	99.00	99.17	99.25	99.30	99.33	99.36	99.37	99.39	99.40	99.42	99.43	99.45	99.46	99.57	99.47	99.48	99.49	99.50
3	34.12	30.82	29.46	28.71	28.24	27.91	27.67	27.49	27.35	27.23	27.05	26.87	26.69	26.58	26.50	26.41	26.32	26.22	26.13
4	21.20	18.00	16.69	15.98	15.52	15.21	14.98	14.80	14.66	14.55	14.37	14.20	14.02	13.91	13.84	13.75	13.65	13.56	13.46
5	16.26	13.27	12.06	11.39	10.97	10.67	10.46	10.29	10.16	10.05	9.89	9.72	9.55	9.45	9.38	9.29	9.20	9.11	9.02
6	13.75	10.92	9.78	9.15	8.75	8.47	8.26	8.10	7.98	7.87	7.72	7.56	7.40	7.30	7.23	7.14	7.06	6.97	6.88
7	12.25	9.55	8.45	7.85	7.46	7.19	6.99	6.84	6.72	6.62	6.47	6.31	6.16	6.06	5.99	5.91	5.82	5.74	5.65
8	11.26	8.65	7.59	7.01	6.63	6.37	6.18	6.03	5.91	5.81	5.67	5.52	5.36	5.26	5.20	5.12	5.03	4.95	4.86
9	10.56	8.02	6.99	6.42	6.06	5.80	5.61	5.47	5.35	5.26	5.11	4.96	4.81	4.71	4.65	4.57	4.48	4.40	4.31
10	10.04	7.56	6.55	5.99	5.64	5.39	5.20	5.06	4.94	4.85	4.71	4.56	4.41	4.31	4.25	4.17	4.08	4.00	3.91
11	9.65	7.21	6.22	5.67	5.32	5.07	4.89	4.74	4.63	4.54	4.40	4.25	4.10	4.01	3.94	3.86	3.78	3.69	3.60
12	9.33	6.93	5.95	5.41	5.06	4.82	4.64	4.50	4.39	4.30	4.16	4.01	3.86	3.76	3.70	3.62	3.54	3.45	3.36
13	9.07	6.70	5.74	5.21	4.86	4.62	4.44	4.30	4.19	4.10	3.96	3.82	3.66	3.57	3.51	3.43	3.34	3.25	3.17
14	8.86	6.51	5.56	5.04	4.69	4.46	4.28	4.14	4.03	3.94	3.80	3.66	3.51	3.41	3.35	3.27	3.18	3.09	3.00
15	8.68	6.36	5.42	4.89	4.56	4.32	4.14	4.00	3.89	3.80	3.67	3.52	3.37	3.28	3.21	3.13	3.05	2.96	2.87
16	8.53	6.23	5.29	4.77	4.44	4.20	4.03	3.89	3.78	3.69	3.55	3.41	3.26	3.16	3.10	3.02	2.93	2.84	2.75
17	8.40	6.11	5.18	4.67	4.34	4.10	3.93	3.79	3.68	3.59	3.46	3.31	3.16	3.07	3.00	2.92	2.83	2.75	2.65
18	8.29	6.01	5.09	4.58	4.25	4.01	3.84	3.71	3.60	3.51	3.37	3.23	3.08	2.98	2.92	2.84	2.75	2.66	2.57
19	8.18	5.93	5.01	4.50	4.17	3.94	3.77	3.63	3.52	3.43	3.30	3.15	3.00	2.91	2.84	2.76	2.67	2.58	2.49
20	8.10	5.85	4.94	4.43	4.10	3.87	3.70	3.56	3.46	3.37	3.23	3.09	2.94	2.84	2.78	2.69	2.61	2.52	2.42
21	8.02	5.78	4.87	4.37	4.04	3.81	3.64	3.51	3.40	3.31	3.17	3.03	2.88	2.79	2.72	2.64	2.55	2.46	2.36
22	7.95	5.72	4.82	4.31	3.99	3.76	3.59	3.45	3.35	3.26	3.12	2.98	2.83	2.73	2.67	2.58	2.50	2.40	2.31
23	7.88	5.66	4.76	4.26	3.94	3.71	3.54	3.41	3.30	3.21	3.07	2.93	2.78	2.69	2.62	2.54	2.45	2.35	2.26
24	7.82	5.61	4.72	4.22	3.90	3.67	3.50	3.36	3.26	3.17	3.03	2.89	2.74	2.64	2.58	2.49	2.40	2.31	2.21
25	7.77	5.57	4.68	4.18	3.85	3.63	3.46	3.32	3.22	3.13	2.99	2.85	2.70	2.60	2.54	2.45	2.36	2.27	2.17
30	7.56	5.39	4.51	4.02	3.70	3.47	3.30	3.17	3.07	2.98	2.84	2.70	2.55	2.45	2.39	2.30	2.21	2.11	2.01
40	7.31	5.18	4.31	3.83	3.51	3.29	3.12	2.99	2.89	2.80	2.66	2.52	2.37	2.27	2.20	2.11	2.02	1.92	1.80
60	7.08	4.98	4.13	3.65	3.34	3.12	2.95	2.82	2.72	2.63	2.50	2.35	2.20	2.10	2.03	1.94	1.84	1.73	1.60
120	6.85	4.79	3.95	3.48	3.17	2.96	2.79	2.66	2.56	2.47	2.34	2.19	2.03	1.93	1.86	1.76	1.66	1.53	1.38
∞	6.63	4.61	3.78	3.32	3.02	2.80	2.64	2.51	2.41	2.32	2.18	2.04	1.88	1.77	1.70	1.59	1.47	1.32	1.00

Bibliography

1. D. Kannan, An introduction to Stochastic processes, North Holland, New York, 1979.

2. Papoulis A., Probability, Random variables and stochastic processes, 3rd Edition, Mc Graw Hill Inc., 1991.

3. Peebles Jr. Probability Random variables and Random signal principles, second Edition, Mc Graw Hill International Edition, 1987.

4. L. Blank, Statistical procedures for Engineering, Management and Science, McGraw – Hill Book Company, 1980.

5. S.K. Campbell, Applied business statistics – Text, Problems and cases, Harper & Row, Publishers, New York, 1987.

6. K.L. Chung, Elementary Probability Theory with Stochastic processes, Narosa publishing house, 1997.

7. J.L. Devore, Probability and Statistics for Engineering and the Sciences, Brooks/Cole Publishing Company, California, 1982.

8. Emanuel poucen, Stochastic Processes, Holden – day, 1967.

9. S.N. Ethier and T.G. Kurtz, Markov Processes – Characterization and Convergence, John Wiley & Sons, 1986.

10. S.C. Gupta and V.K. Kapoor, Fundamentals of Mathematical Statistics, Sultan Chand & Sons, 1980.

11. John B. Thomas, Introduction to probability, Springer – Verlag, 1986.

12. R.I. Levin and D.S. Rubin, Statistics for Management, Seventh Edition, P.H.I., 1997.

13. Montgomery, Introduction & Statistical Quality Control.

14. Richard A. Johnson, Probability & Statistics for Engineers, 5th Edition, P.H.I., 1994.

15. A.E. Scheerer, Probability on Discrete Sample spaces with applications, International Textbook Company, Pennsylvania, 1969.

16. Vijayakumar & Sreenivasan, Stochatic Processes and Applications, Narosa Publishing House, New Delhi, 1999.

17. Haykin, Digital Communications, John Wiley & Sons, 1988.

18. Jeruchim M.C., Balaban P. and K. Sam Shanmugam, Simulation of Communication Systems, Plenum Press, New York, 1992.

19. Lathi B.P., Modern Digital and analog communication systems, Press Books private Ltd., Barods, India, 1993.

20. Martin S. Randen, Analog and Digital Communication Systems, Orentice Hall of India Ltd., New Delhi, 1994.

21. Michael K. Ochi, Applied probability and stochastic processes, John Wiley and Sons, 1990.

22. O Flynn. M., Probabilities, Random Variables and Random Processes, Harper and Row Publishers, New York, 1982.

23. Paul Mayer, Introductory probability and statistical Applications, Oxford and IBH publishing company, 1970.

24. Proakis J.K., Digital Communications, Mc Graw Hill, New York, 1988.

25. Sam K. Shanmugam, Digital and Analog communication systems, John Wiley and Sons, 1994.

26. Taub and Schilling, principles of communication systems, Tata Mc Graw Hill Company Ltd., New Delhi, 1995.

27. An introduction to probability theory and its applications feller, vol. 3rd edition, John Wiley & sons.

28. K. Murugesan, P. Gurusawmy – probability, statistics & random process - Anuradha Publications.

29. Ronald E. Walpole Probability and statistics for Engineers & Scientists – Pearson Education.

30. Miller & Freund's "Probability and Statistics for Engineers- Pearson Education.